T0191778

ASTRONOMY AND
ASTROPHYSICS LIBRARY

Springer-Verlag Berlin Heidelberg GmbH

Physics and Astronomy

ONLINE LIBRARY

springeronline.com

Michael Stix

The Sun

An Introduction

Second Edition

With 225 Figures, Including 16 Color Figures,
and 27 Tables

 Springer

Dr. Michael Stix
Kiepenheuer-Institut für Sonnenphysik
Schöneckstrasse 6
79104 Freiburg, Germany

Cover picture: Sunspot and solar photosphere, observed in the G band at 430 nm with the German Vacuum Tower Telescope, Tenerife. An adaptive optics system of the National Solar Observatory, Tucson, was used, and the image was further processed by a speckle technique at the Universität Göttingen. Courtesy O. von der Lühe

---

Cataloging-in-Publication Data applied for

Bibliographic information published by Die Deutsche Bibliothek. Die Deutsche Bibliothek lists this publication in the Deutsche Nationalbibliografie; detailed bibliographic data is available in the Internet at <http://dnb.ddb.de>.

Second Edition 2002. Corrected Second Printing 2004

ISSN 0941-7834

ISBN 978-3-642-62477-3   ISBN 978-3-642-56042-2 (eBook)
DOI 10.1007/978-3-642-56042-2

springeronline.com

© Springer-Verlag Berlin Heidelberg 2002
Originally published by Springer-Verlag Berlin Heidelberg New York in 2002
Softconer erprint of the hardcover 2nd edition 2002

Typesetting: Frank Herweg, Leutershausen
Cover design: *design & production* GmbH, Heidelberg

Printed on acid-free paper   SPIN 10973264   55/3141/ba - 5 4 3 2 1 0

To Jakob

Preface to the Second Edition

Since this introduction to the physics of the Sun first appeared in 1989, significant progress has been made in solar research. New insights grew especially from the results of space missions, above all SOHO, the Solar and Heliospheric Observatory, which in 1996 reached its orbit around the Lagrangian point L_1 between the Earth and the Sun. Refined theoretical models have been advanced as well, but clearly, in my view, the lead was on the observational side.

For the present edition I have retained the subdivision into nine chapters, and much of the original text. However, all nine chapters have been revised with many adjustments, and with a number of major additions. Chapter 1, for example, now includes the 11-year variation of the solar luminosity, Chap. 2 presents an improved standard solar model and a discussion of the neutrino experiments of the past 13 years. In Chap. 3 the sections on image reconstruction and adaptive optics, on narrow-band filters, and on polarimetry have been enlarged; the table of chemical element abundances in Chap. 4 has been revised. The later chapters take advantage of the advances made in several directions, in particular helioseismology, numerical simulation, and observation from space. Thus, we now know more about the internal rotation of the Sun, about the hydrodynamics of the convection zone, and about the source region of the solar wind, to name only a few prominent examples.

What has been said in the first edition about the intention of this book, about the audience to which it is addressed, the style, equations, units, etc., remains valid for this edition. The preface of 1989 is therefore reprinted here. As for the first edition, I have selected the subjects according to my personal interest, and again I apologize for omitting topics that might be considered very important by other colleagues working in the field. On the other hand I think that a general introductory monograph on the Sun could never cover the wealth of results published in the journals. Also, good books are on the market that go into more detail in one or other direction, such as *The Sun from Space* by K. R. Lang or *Solar and Stellar Magnetic Activity* by C. J. Schrijver and C. Zwaan. The bibliographical notes at the end of each chapter have been complemented, but by no means in an exhaustive manner. Powerful programs for literature searches are available in the Internet.

Colleagues and students often inquired for solutions to the problems that are posed in the text. In many cases I could not comply, because a collection of written-up solutions did not exist. In fact, while preparing such a collection, I realized that some of the problems had to be formulated more carefully. On my home page – http://www.kis.uni-freiburg.de/~stix/ – a set of solutions can be found; comments and suggestions concerning these solutions are most welcome.

I wish to thank Prof. Oskar von der Lühe and the colleagues and students at the Kiepenheuer-Institut, as well as many colleagues at other institutions, for illuminating discussions, for reading sections of the text, and especially for help with the figures. Springer-Verlag provided the typographic style files and useful advice, Peter Caligari and Reiner Hammer helped whenever I had a computer problem, Ilsa David took care of the figure scanning, and Markus Roth converted the first edition into TEX, so that I could proceed. Thanks to all of them.

Freiburg, April 2002   *Michael Stix*

Preface to the First Edition

"The Sun will not break the rules", says Heraclitus of Ephesus; if it does, "the Erinyes, the aids of justice, will take vengeance". And Giorgos Seferis, in his 1963 Nobel speech at Stockholm, wondered whether a modern scientist could take advantage of this message.

I think he can. In fact, have not most of us always regarded astronomy in general, and solar physics in particular, as quite "normal" physics? As a field where the same rules apply that are valid on Earth? Enigmatic spectral lines have found their explanation without recourse to such mysterious new elements as "nebulium" or "coronium"; likewise the present puzzles, above all the "solar neutrino problem", will certainly find their solutions within the rules of general physics. Of course, we must keep in mind that we may not yet know all those rules with sufficient precision. But in this case the message is equally clear: the Sun will help us to comprehend the rules.

The present book aims at illustrating the application of the rules of physics to a star like the Sun. When the publisher approached the Kiepenheuer-Institut in search of an author, we had some debate about the best way of achieving this aim. I felt that many good and up-to-date review articles on many branches of solar physics were available, and that quite a number of excellent monographs on special topics also existed. In addition, a large number of beautifully illustrated non-specialist volumes about the Sun have appeared in recent years. Thus, when it became clear that I was to be the writer, I decided to try a coherent yet more technical text on the whole Sun, which I believed had not been available for some time.

I have written this introduction for students of physics at an intermediate level. For example, I assume that the reader knows the basic laws of thermodynamics and of hydrodynamics, and Maxwell's equations. Even the equation of radiative transfer is not rigorously deduced. Nevertheless, I have tried to familiarize the reader with all equations on the grounds of their physical meaning. Thus the audience may include scientists from many diverse branches, astronomers who perhaps have not yet specialized in solar astronomy, and in fact everybody who is interested in the Sun and is not afraid of formulas; yet I believe that the mathematics used is always quite elementary.

In a field as wide as solar astronomy a choice of subjects must be made, and, of course, I have selected my own favorite subjects. I hope that the specialists whose preferred ideas I have not treated will forgive me. Likewise, I ask all my colleagues and friends for forgiveness if they have not even been quoted. In general I have tried not to replace an argument by a reference. But this way of writing takes much space. As a compromise, therefore, a section headed "bibliographical notes" is appended to each chapter, where the classic textbooks as well as some more specialized contributions are at least mentioned. Often it was not clear (to me) where a particular idea was advanced first. In this case an "e.g." may precede the quotation.

Each chapter contains a number of "problems". The purpose of these problems is, of course, to include more material in the text than would have been possible otherwise. In addition, I could thereby avoid the often deceiving phrase "as can easily be seen". In fact, some of the problems are very easy. Others may require the reader to consult the literature; however, studying the existing literature may be worthwhile in itself.

Generally, I adhere to the international system of units. Nevertheless, I think that the gauss and the ångstrøm are perfect units, in particular, when a magnetic field strength or the width of a spectral line are such that these units make the numbers involved easy to handle. The important point is that physical equations should be written in a form which does not depend on the choice of units. I believe that I have observed this rule throughout the text, and hope that, in case I have still confused the reader, the list of symbols at the end of the book will help (a number of rather "local" variables, however, is not included in this list).

I wish to express my deep gratitude to Prof. E.-H. Schröter, as well as to all my colleagues at the Kiepenheuer-Institut, for granting to me the privilege to write while they had to work. In addition, I wish to thank them for the numerous discussions we had, for the many hints and written notes I obtained, and for the time and labour spent by those who read parts of the manuscript. In preparing the original photographs and drawings I was helped greatly both by the staff here in Freiburg and by many friends at other institutions.

My family remained extremely patient and helpful during the course of my writing. Most credit goes to Christian, without whose skillful and diligent typing the manuscript would never have been completed.

Freiburg, January 1989   *Michael Stix*

Table of Contents

1. Characteristics of the Sun

We all have been fortunate enough at times to see and to contemplate the starry sky on a clear night. We have seen bright stars and faint stars. And of course we have been told that stars can be bright either because they are close to us or because their *luminosity* is really large.

Perhaps not everyone has realized that the stars differ not only in brightness, but also in *color*. The stellar colors are clearly discernable with the naked eye: some are bluish, others are reddish. If you have not seen this, watch for it next time. As with an ordinary lamp, the color varies with temperature. Hence, when we observe the color of a star, what we see is in fact a measure of the *temperature* at the surface of that star.

When an astronomer studies a certain star for the first time, he determines the luminosity and the surface temperature, if this is possible in some way: these are the coordinates of the *Hertzsprung–Russell diagram*. Having placed a star in this diagram he will in most cases be able to tell its radius, its method of energy generation and energy transport, the periods and growth rates of its pulsations, etc. Even the average rotation of stars, their capability to conserve

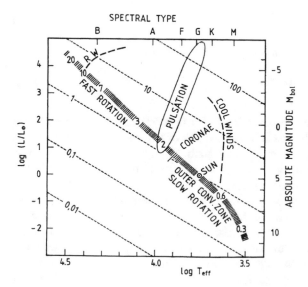

Fig. 1.1. Hertzsprung–Russell diagram. Numbers in the shaded main-sequence band are stellar masses in units of the solar mass. *Dotted lines* are curves of constant radius, labelled $r/r_\odot$; RW radiatively driven winds. Spectral types B, A, etc., correspond to main-sequence stars

or generate magnetism, or the occurrence of hot coronae seem to vary in a systematic way across the Hertzsprung–Russell diagram, as illustrated in Fig. 1.1.

From a theoretical point of view, the star's mass, its age, and its chemical composition are considered as more fundamental, and may in fact be used in order to *predict* the luminosity and the surface temperature.

Information about basic quantities such as the solar mass, the luminosity, or the surface temperature will be collected in this first chapter, although much detail must be postponed until later. I shall begin with the Sun's distance from Earth, which by itself is of course incidental; but its knowledge is nevertheless necessary for the determination of many fundamental intrinsic properties of the Sun.

1.1 Distance

Until recently, distances in the solar system were measured by triangulation. The nearer the object is whose distance we want to know, the better this method works. Therefore, in place of the Sun, planets or asteroids were used, notably Venus (at the times when it passed in front of the solar disc) and Eros, which occasionally approaches the Earth as close as 0.15 times the mean solar distance. In principle, a single measurement of a linear distance during such an occasion is sufficient to derive *all* distances between the planets and the Sun. This is because, by precise determination of angular coordinates at various precisely defined epochs, the *apparent* orbits and the periods, T_i, of planets and asteroids may be known. In addition, by Kepler's third law,

$$\frac{a^3}{T^2} = \frac{Gm_\odot}{4\pi^2}(1 + m/m_\odot) ,$$
(1.1)

the *ratios* of their semi-major axes a_i can be derived:

$$\left(\frac{a_1}{a_2}\right)^3 = \left(\frac{T_1}{T_2}\right)^2 \frac{1 + m_1/m_\odot}{1 + m_2/m_\odot} .$$
(1.2)

We see from this equation that it is also necessary to know the masses m_1 and m_2 of the two orbiting bodies in units of the solar mass, $m_\odot$. These masses are obtained from the mutual perturbations of planetary orbits, and since $m_i/m_\odot \ll 1$, these ratios need only be known with lesser accuracy (still, 5 or 6 decimal places are usually given).

The triangulation method always deals with very small angular differences and therefore has limited accuracy. More accurate results have been obtained since 1961 by the measurement of *radar echos*. Again the Sun itself is not directly used. Because the solar corona is rather inhomogeneous and variable, it is far from being a well-defined reflector of radio waves. Hence, the travel times of echos from planetary surfaces are measured. These travel times are then, again with the help of Kepler's law, converted to the *light time for unit distance*, τ_A, i.e., to the travel time for the astronomical unit (to be defined presently). The result is

$$\tau_A = 499.004782 \pm 0.000006 \text{ s} . \tag{1.3}$$

The time τ_A can be determined with such a small standard error that in 1976 the International Astronomical Union adopted it as one of the primary quantities in the system of astronomical constants. The unit distance or *astronomical unit A* in this system is *dynamically* defined in such a way that the semi-major axis a (in astronomical units), and the sidereal period T (in days), of an elliptical orbit of a body of mass m (in units of $m_\odot$) obey the relation $4\pi^2 a^3/T^2 = k^2(1+m)$, where k, the *Gaussian gravitational constant*, by definition has the value 0.01720209895. As the velocity of light, also by definition (since 1983), is $c = 299792458$ m/s, we obtain from (1.3) the astronomical unit

$$A = 149597870 \pm 2 \text{ km} . \tag{1.4}$$

The error essentially arises from the inhomogeneity of the radar reflecting layer and from the uncertain distance of this layer to the center of gravity of the orbiting body.

When the sidereal year and the mass of the Earth–Moon system are substituted into the defining equation for the astronomical unit, one finds $a = 1.000000036$. Thus, strictly speaking, the semi-major axis of the Earth's orbit, or the Earth's mean distance from the Sun, exceeds the astronomical unit by a few parts in 10^8. However, for all purposes of solar physics the difference is of no concern. In most cases it is quite sufficient to use the rounded value 1.496×10^{11} m for the mean solar distance.

The variation of the solar distance is between 1.471×10^{11} m (at perihelion, in January) and 1.521×10^{11} m (at aphelion, in July). At these distances, the angle of $1''$ – often used by astronomers as a length unit for the solar atmospheric fine structure – corresponds to 710 and 734 km, respectively, at the center of the Sun's disc.

Problem 1.1. In which sense of the word "mean" is the semi-major axis of an elliptical orbit the mean distance?

1.2 Mass

Once distances in the solar system are known, we may use Kepler's law once more in order to determine the Sun's mass, $m_\odot$. The high precision of time and distance measurements yields, however, an equally high precision only for the *product*, $Gm_\odot$, of the solar mass with the constant of gravitation: $Gm_\odot = (132712438 \pm 5) \times 10^{12}$ m^{3}s^{-2}. Laboratory measurements of G give

$$G = (6.67259 \pm 0.00085) \times 10^{-11} \text{ m}^3\text{kg}^{-1}\text{s}^{-2} \; ;$$

the mass of the Sun is therefore

$$m_\odot = (1.9889 \pm 0.0003) \times 10^{30} \text{ kg} \; . \tag{1.5}$$

This is the present value of $m_\odot$. The Sun's luminosity $L_\odot$, i.e., the total energy radiated into space, corresponds to a mass loss $\dot{m}_\odot = L_\odot/c^2$, or about 4×10^9 kg/s (based on the value of $L_\odot$ given below). As we shall see in Chap. 9, the solar wind carries another 10^9 kg/s away. Thus, during the Sun's life of about 1.5×10^{17} s (cf. Chap. 2) the total mass loss has been less than 10^{27} kg; this is of the same order as the uncertainty of today's $m_\odot$, and is usually neglected. However, mass loss may have been (or will be) significant in very early or very late stages of the Sun's evolution when stronger winds did (or will) blow than presently. Of course, during the period of star formation, the *accretion* of mass was an important, and at times dominant, process.

1.3 Radius

In order to determine the radius $r_\odot$ of the Sun, we must use the solar distance and a measurement of the angular diameter of the visible disc. For the photoelectric method we define this diameter as the angular distance between the inflection points of the intensity profiles at two opposite limbs. Recent results from ground-based and balloon-borne instruments (Sofia et al. 1994, Neckel 1995, Laclare et al. 1996, Wittmann 1997, Brown and Christensen-Dalsgaard 1998) cluster remarkably close around the value 959.63″ of the angular semidiameter at mean solar distance, which was already obtained from (visual) heliometer measurements during the 19th century (Auwers 1891), and which is used until today in the yearly *Astronomical Almanac*. Visual drift-scan measurements yield values that are larger by $\approx 0.5''$ because the human eye sees the limb at an intensity which is lower than the intensity at the inflection point (Wittmann 1997); however, this seems not to be the case with visual methods where two images are brought into contact, such as the old heliometer measurements or the visual results of Laclare et al. (1996).

The statistical error of some of the new results is as small as $0.01''$; however, the differences between the diverse experiments indicate a systematic uncertainty of at least $0.1''$. Hence, we may take the mentioned canonical value, but a conservative error estimate is appropriate. With (1.4), we thus have

$$r_\odot = (6.960 \pm 0.001) \times 10^8 \text{ m} \tag{1.6}$$

as the distance of the inflection point from the solar center.

The inflection point of the limb intensity profile corresponds to an optical depth $\tau \approx 0.004$ (Wittmann 1974b). This can be seen when a solar model atmosphere, such as described in Chap. 4, is used to calculate the total emergent intensity along various tangential lines of sight (cf. also Ulrich and Rhodes 1983). On the other hand, the boundary of an interior solar model is usually placed at $\tau = 2/3$, where the temperature equals the effective temperature defined below. The difference between the two levels amounts to slightly over 300 km, cf. Table 4.1. Therefore, in a precise solar model the boundary condition should be applied at

$$r_{\tau=2/3} = 6.957 \times 10^8 \, \mathrm{m} \ .$$

The comparison of theoretical and observed frequencies of the solar f-mode oscillation confirms this value of the solar radius at $\tau = 2/3$ (Sect. 5.3.7).

Measurements in narrow spectral bands show that the solar radius depends on wavelength: in a continuum band at $\lambda = 700$ nm, the Sun appears larger than at $\lambda = 400$ nm by $\approx 0.1''$ (e.g., Neckel 1995); still larger differences are observed for spectral bands including absorption lines. The value given in (1.6) corresponds to wavelengths of about 500 nm to 600 nm.

Calculations of the Sun's main-sequence evolution, cf. Chap. 2, show that presently the change of the solar radius is $\dot{r}_\odot \approx 2.4$ cm/year. This is a long-term change, i.e., an average over 10^8 years or more. Superimposed on this extremely slow change there could be faster variations, e.g., connected to the 11-year activity cycle, but the evidence for such fast variations remains controversial.

Knowing its mass and radius we calculate the Sun's mean density:

$$\bar{\rho} = 1.408 \ \mathrm{g/cm}^3. \tag{1.7}$$

In addition, using the value of $Gm_\odot$ given above, we find the gravitational acceleration at the solar surface, $g_\odot = Gm_\odot/r_\odot{}^2$, or

$$g_\odot = 274 \ \mathrm{m/s}^2 \ . \tag{1.8}$$

The gravitational acceleration has a substantial influence on the structure of a stellar atmosphere.

1.4 Luminosity

The solar luminosity $L_\odot$ or – more precisely – the photo-luminosity is defined as the total energy output (per unit time) in the form of electromagnetic radiation, or photons. Under the assumption of spherical symmetry, it is related to the total flux S at mean distance of the Earth from the Sun, also called the total *irradiance at mean distance* or, for short, the *solar constant*, by

$$S = L_\odot/4\pi A^2 \ . \tag{1.9}$$

The quantities $L_\odot$ and S are integrals over the electromagnetic spectrum. As shown in Fig. 1.2, the Earth's atmosphere strongly attenuates (in the infrared) or even completely blocks (in the ultraviolet below $\lambda = 300$ nm) the solar radiation. Accurate measurements of S are therefore possible only from above the atmosphere. The instrument used, the pyrheliometer (Sect. 3.6.1), is essentially a black cavity which totally absorbs the incident radiation. Results have been obtained from balloon, rocket, and satellite experiments. The Solar Maximum Mission, a solar satellite that orbited the Earth from 1980 to 1989, measured a mean value of $1367\,\mathrm{W/m^2}$. As Fig. 1.3 illustrates, the Earth Radiation Budget experiment on the Nimbus 7 satellite (HF) measured a value of S that is larger by $\approx 3\mathrm{W/m^2}$, whereas the Earth Radiation Budget Satellite (ERBS) determined a value smaller by $\approx 2\mathrm{W/m^2}$. We may use these differences for a conservative estimate of the systematic error and write

$$S = 1367 \pm 3 \ \mathrm{W/m^2} \ . \tag{1.10}$$

Using this result and (1.4), we find the solar luminosity

$$L_\odot = (3.844 \pm 0.010) \times 10^{26} \ \mathrm{W} \ . \tag{1.11}$$

A luminosity such as (1.11) is quite normal for a main-sequence star of spectral type G2 like the Sun.

The apparent brightness of stars is usually given in the logarithmic scale of magnitudes m. Stars visible with the naked eye have *apparent* magnitudes $m = 0$ to $m = 6$, and the scale is such that $\Delta m = 5$ exactly means a factor $1/100$. The *absolute* magnitude M of a star is equal to the apparent

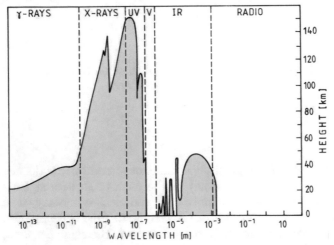

Fig. 1.2. Absorption in the Earth's atmosphere. The edge of the shaded area marks the height where the radiation is reduced to $1/2$ of its original strength. UV ultraviolet; V visible; IR infrared

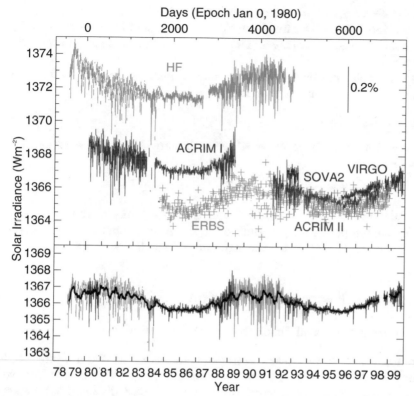

Fig. 1.3. Total solar irradiance, measured by space experiments between 1978 and 1999 (*upper panel*). The *lower panel* shows a composite constructed by adjusting different measurements within the overlapping periods, after correction for degradation and operational effects as described in Fröhlich and Lean (1998). Courtesy C. Fröhlich

magnitude that the star would have at a distance of 10 parsec, i.e., ten times the distance from which the astronomical unit subtends an angle of $1''$. The absolute (bolometric, i.e., integrated over the whole spectrum) magnitude of the Sun is $M_\odot = 4.74$.

The name "solar constant" for S is misleading. We shall see in Chap. 2 that the theory of stellar evolution tells us that the luminosity has increased from about 72% of today's $L_\odot$ to the present value over the Sun's life of about 4.6×10^9 years. Moreover, the same instruments on the Solar Maximum Mission and other satellites that measured the *absolute* irradiance quoted above, recorded *relative* variations of up to 0.2% over several days, and variations of order 10^{-6} over time intervals of minutes. We must discuss the latter in the context of the solar oscillations in Chap. 5, and we shall see in Chap. 8 that the former provide a clue to the energy balance in sunspots. On an intermediate time scale, Fig. 1.3 clearly demonstrates a variation with the 11-year activity cycle of the Sun, as first reported by Willson and Hudson (1988). This variation is *in phase* with the cycle: the maxima of the irradiance and of the number of sunspots coincide. The amplitude of the variation is $\approx 0.09\%$, peak-to-peak.

Since the solar radius is known, we may, as another measure of the luminosity, define the effective temperature T_{eff} by

$$L_\odot = 4\pi r_\odot^2 \sigma T_{\text{eff}}^4 \; , \tag{1.12}$$

where $\sigma = 5.67051 \times 10^{-8}$ W/m^{2}K^4 is the Stefan–Boltzmann constant. For the Sun we have

$$T_{\text{eff}} = 5778 \pm 3 \text{ K} \; . \tag{1.13}$$

The uncertainty is predominantly due to the uncertainty in the luminosity. The effective temperature is, in addition to the gravitational acceleration, the second essential parameter for the structure of a stellar atmosphere.

1.5 Spectral Energy Distribution

1.5.1 Energy Flux and Intensity

Let us now consider the solar electromagnetic *spectrum*. We are interested here in the *absolute* solar emission in various wavelength regions; but we shall be content with low spectral resolution. We must distinguish between two quantities: the first is the *energy flux* $F(\lambda)$ at the solar surface, i.e., the energy emitted per unit area, time, and wavelength interval. Under the assumption of spherical symmetry the energy flux is related to the *spectral irradiance* $S(\lambda)$ through

$$r_\odot^2 F(\lambda) = A^2 S(\lambda) \; . \tag{1.14}$$

The second quantity is the *intensity* $I(\theta, \lambda)$, i.e., the energy emitted per unit area, time, wavelength interval, and solid angle. The intensity depends on the angular distance θ from the direction perpendicular to the solar surface. The integral of $I(\theta, \lambda) \cos \theta$ over all outward directions, i.e., those with $\cos \theta > 0$, yields the energy flux $F(\lambda)$.

Problem 1.2. Show that the energy flux $F(\lambda)$ is π times the intensity averaged over the visible solar disc.

With $\mu = \cos \theta$ the intensity averaged over the visible disc is

$$\bar{I}(\lambda) = 2 \int_0^1 I(\mu, \lambda) \mu \, d\mu \; , \qquad \text{or} \tag{1.15}$$

$$\bar{I}(\lambda) = 2I(1,\lambda) \int\limits_{0}^{1} \frac{I(\mu,\lambda)}{I(1,\lambda)} \mu d\mu \ . \tag{1.16}$$

Hence, in order to obtain $F(\lambda)$ or, which is the same, $\pi\bar{I}(\lambda)$, one possibility is to measure $\bar{I}(\lambda)$ directly. Diffuse light, i.e., radiation emerging from *all* parts of the solar disc, must be used in such an experiment. The second possibility is to employ an imaging system, and to measure both the *central intensity*, $I(1,\lambda)$, and the *limb-darkening function*, $I(\mu,\lambda)/I(1,\lambda)$; for the latter, a *relative* measurement is sufficient.

Instead of the wavelength λ, often the frequency ν is used as the independent argument. It is common to write I_λ or I_ν, with the argument as a subscript. For a distribution such as the intensity this subscript at the same time indicates the appropriate interval: a transformation of the intensity integral (over wavelength or frequency) yields $I_\lambda d\lambda = -I_\nu d\nu$; since $\lambda = c/\nu$ and $d\lambda = -c\nu^{-2}d\nu$, we also have $\lambda I_\lambda = \nu I_\nu$.

1.5.2 The Visible

The visible and near-infrared parts of the spectrum contain most of the solar energy. For these two spectral regions both types of experiments have been carried out, mostly from high mountain sites such as the Jungfraujoch Observatory in Switzerland, or from aircraft. The required correction for atmospheric extinction is done by the classical *Langley* method, i.e., by measurements at various elevations of the Sun (or various amounts of air mass), and subsequent extrapolation to zero air mass.

Problem 1.3. How does the atmospheric extinction depend on the zenith angle?

Figure 1.4 shows the visible spectrum. On the red side of the maximum the mean intensity $\bar{I}(\lambda)$ approximately follows a black-body spectrum, with $T = T_{\text{eff}} = 5778$ K. However, even at the low resolution used in this figure, some of the stronger absorption lines, such as $H\alpha$ at $\lambda = 656.3$ nm, are clearly discernible. On the blue side of the maximum, absorption lines dominate the spectrum.

From 1961 onwards, D. Labs and H. Neckel measured the mean intensity by means of the limb-darkening function and the absolute intensity $I(1,\lambda)$ at disc center. The latter, which is also shown in Fig. 1.4 (upper curve), can be used to calibrate a solar spectrum with higher resolution. The Fourier transform spectrometer (Sect. 3.3.2) has a sensitivity which varies in a very smooth fashion over a wide spectral range and therefore yields intensity spectra that are particularly suited for such calibration. In their experiments, Labs and Neckel selected bands of 2 nm width by means of a double monochromator (Sect. 3.4.6) and compared the integrated intensity of each of these bands to

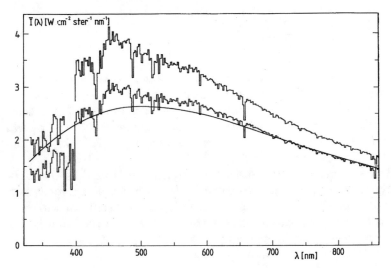

Fig. 1.4. Central intensity $I(1, \lambda)$ – *upper curve* – and mean intensity $\bar{I}(\lambda)$ – *lower curve* –, averaged over 2-nm bands. Data from Neckel and Labs (1984). The *smooth curve* is a black-body spectrum for $T = 5778$ K, the effective temperature of the Sun

the corresponding intensity of a standard lamp. The lamp itself was calibrated in the laboratory by means of the radiation of a black body of given temperature. The measurements show a scatter of about 0.2%, and systematic errors are estimated to be less than 0.5%.

1.5.3 The Infrared

Next to the visible spectrum the infrared, which is shown in Fig. 1.5, makes the second largest contribution to the solar energy output: about 44% of the electromagnetic radiation is emitted at wavelengths longer than 0.8 μm. The spectrum is approximately thermal, although high-resolution spectra show a large number of lines. The infrared spectrum is rather well represented by the Rayleigh–Jeans (i.e., long-wavelength) approximation of Planck's law,

$$B_\lambda = \frac{2hc^2}{\lambda^5(e^{hc/\lambda kT} - 1)} \ , \qquad \text{i.e.,} \quad B_\lambda \simeq \frac{2ckT}{\lambda^4} \ , \tag{1.17}$$

so that

$$S(\lambda) \simeq 2\pi ckT\lambda^{-4}(r_\odot/A)^2 \ . \tag{1.18}$$

In particular, in the far infrared above 10 μm, the double logarithmic spectrum of Fig. 1.5 deviates very little from a straight line. The brightness temperature, T_B, defined by $I_\lambda = B_\lambda(T_B)$ [where I_λ is the observed absolute intensity at wavelength λ, and $B_\lambda(T)$ is the Kirchhoff–Planck function],

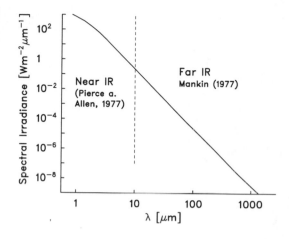

Fig. 1.5. Spectral irradiance at 1 astronomical unit for the infrared solar radiation

changes from $\approx 5100\,\mathrm{K}$ at $10\,\mu\mathrm{m}$ to $\approx 4440\,\mathrm{K}$ at $100\,\mu\mathrm{m}$, and to $\approx 6000\,\mathrm{K}$ at $1\,\mathrm{mm}$.

The main difficulty with measurements in the infrared is absorption by water vapor in the Earth's atmosphere. Still, the spectral irradiance has been determined with an uncertainty of 1% or less.

Problem 1.4. Calculate the spectral irradiance for the wavelength and brightness temperature values given in the text. Estimate the error arising from the Rayleigh–Jeans approximation of Planck's law.

Problem 1.5. Use the Rayleigh–Jeans approximation for an estimate of the entire solar irradiance above any given wavelength. Show that about 0.06% of the Sun's energy is emitted in the far infrared, at $\lambda > 10\,\mu\mathrm{m}$.

1.5.4 The Radio Spectrum

Figure 1.6 illustrates the solar radio spectrum, which begins in the microwave region at $\lambda = 1$ mm. As is common in radio astronomy, the energy flux is given per frequency interval rather than per wavelength interval. The thermal *quiet-Sun* spectrum, which smoothly continues from the infrared into the radio region, therefore has a slope of -2 in the double logarithmic representation. This slope is, however, not constant everywhere: between $\lambda = 1\,\mathrm{cm}$ and $\lambda = 1\,\mathrm{m}$ there is a transition of the brightness temperature from about $10^4\,\mathrm{K}$ to about $10^6\,\mathrm{K}$!

Superimposed on the quiet solar radio emission is a spectrum of great variability. The *s-component* (for slowly varying) is correlated to the solar 11-year activity cycle. Indeed, the radio flux at 10.7 cm is used as an index of the solar activity. The spectrum of the s-component is approximately thermal, and its flux normally is 1 or 2 orders of magnitude below the quiet-Sun flux. On the other hand, there are rapid *bursts* of radio emission, on time scales of

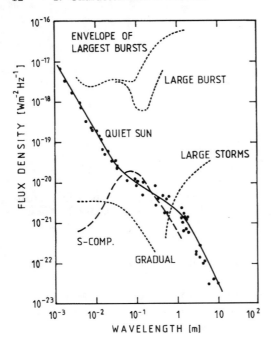

Fig. 1.6. Solar radio emission. *Dots* and *solid curve*: quiet Sun; *dashed curve*: slowly varying component; *dotted curves*: typical rapid events. Composite after Shimabukuro (1977)

seconds to days. During such events the flux may exceed the quiet-Sun level by several orders of magnitude, with substantial deviations from the thermal spectrum. The frequency of occurrence of radio bursts is again correlated to the 11-year cycle.

The solar radio emission was first detected by G. C. Southworth and by J. S. Hey around 1942. Its absolute flux is now generally measured with an uncertainty of about 10%. Since the Earth's atmosphere has a broad radio window (Fig. 1.2), ground-based antennas can be used.

1.5.5 The Ultraviolet

The solar ultraviolet spectral irradiance is shown in Fig. 1.7. As in the blue and violet parts of the visible spectrum, absorption lines are the dominant feature down to about 210 nm, although only a few of the strongest lines, such as the Mg I line at $\lambda = 285.2$ nm or the Mg II h and k lines at $\lambda = 280.3$ nm and $\lambda = 279.6$ nm, can be identified at the low spectral resolution of Fig. 1.7. The sharp decrease near $\lambda = 210$ nm and the continuum which follows towards shorter wavelengths are due to the ionization of Al I. The whole band between 200 nm and 150 nm is approximately represented by a brightness temperature of 4700 K.

Below 150 nm, *emission* lines dominate the spectrum. Most prominent is the Lyman α line of hydrogen, a line of about 0.1 nm width centered at

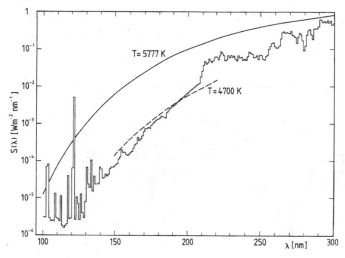

Fig. 1.7. Solar spectral irradiance in the ultraviolet, averaged over 1-nm bands. The *solid* and *dashed curves* are black-body spectra. Data from Heath and Thekaekara (1977)

121.57 nm; the average irradiance in this line alone is 6 mW/m^2, and this is as much as in the whole spectrum below 150 nm besides Lyman α!

Towards shorter wavelengths, the ultraviolet irradiance is increasingly *variable*. For example, on a time scale comparable to the Sun's synodic period of rotation (27 days), changes of up to 25% are observed at 120 nm. These changes are partly true temporal variations and partly a manifestation of the non-uniform distribution of the sources in the solar atmosphere; these sources pass across the visible hemisphere as the Sun rotates. Still larger in amplitude – up to a factor 2 – is a variation which correlates with the 11-year activity cycle.

Absorption in the Earth's atmosphere, mainly by O_2, makes it necessary to use rockets or satellites for observations in the ultraviolet. A major additional problem for absolute measurements are the standards. Recent results (Thuillier et al. 1997) still have an uncertainty of 2.2% at 350 nm, and of 4% at 200 nm. For the total luminosity of the Sun the errors – and also the mentioned variations – are of minor importance: the bands from 300 nm to 330 nm, 210 nm to 300 nm, and 150 nm to 210 nm contribute only 1.5%, 1%, and 0.01%, respectively.

1.5.6 Extreme Ultraviolet and X-rays

Below $\lambda = 120$ nm we have the extreme ultraviolet, or EUV. It is highly variable, and characterized by a large number of emission lines which stem from neutral atoms and from ions of widely differing ionization levels, up to, e.g., Fe XVI. Parts of the solar atmosphere comprising a wide temperature range, from about $8000\,\mathrm{K}$ to $4 \times 10^6\,\mathrm{K}$, can thus be studied. We shall see in Chap. 9 that EUV emission lines are a major source of information for the transition zone from the chromosphere to the corona.

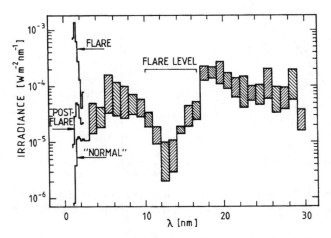

Fig. 1.8. Solar spectral irradiance between 1 nm and 30 nm. *Shaded intervals* represent the range of results from various rocket experiments between 1965 and 1972. Adapted from Manson (1977)

The corona is also a source of soft x-rays (0.1 nm to 10 nm). Figure 1.8 illustrates the variability of the solar spectral irradiance at these short wavelengths. In addition to the rocket experiments there has been continuous monitoring from space: between solar maximum and minimum the SOLRAD satellites found a variation by a factor of 200 in the 8–20 Å band, and by a factor of 20 in the 44–60 Å band.

The largest variations, often several orders of magnitude, occur during solar *flares*. Hard x-rays, below $\lambda = 1\,\text{Å}$, and even γ-rays are emitted in these events, which we shall briefly discuss in Chap. 9.

1.5.7 Color Indices

For comparison with other stars it is often convenient to express the spectral energy distribution in the rough form of *color indices*. These are defined as ratios of broad-band integrals over wavelength, viz.

$$\mathrm{U-B} = -2.5\left(\log \int_0^\infty S(\lambda)E_{\mathrm{U}}(\lambda)d\lambda - \log \int_0^\infty S(\lambda)E_{\mathrm{B}}(\lambda)d\lambda\right) + C_{\mathrm{UB}}\,, \quad (1.19)$$

$$\mathrm{B-V} = -2.5\left(\log \int_0^\infty S(\lambda)E_{\mathrm{B}}(\lambda)d\lambda - \log \int_0^\infty S(\lambda)E_{\mathrm{V}}(\lambda)d\lambda\right) + C_{\mathrm{BV}}\,. \quad (1.20)$$

The functions E_U, E_B, and E_V are defined by filter windows in the ultraviolet (and partly violet), blue, and visible. Their width is of the order of 100 nm and their central wavelengths are 365 nm, 440 nm, and 548 nm, respectively. The constants C_{UB} and C_{BV} are chosen in such a way that for stars of spectral type A0 both $U - B$ and $B - V$ are zero. In this system the Sun has the color indices $U - B = 0.195 \pm 0.005$ and $B - V = 0.650 \pm 0.005$ (Neckel 1986).

1.6 Bibliographical Notes

The 1986 adjustment of the fundamental physical constants is reviewed by Cohen and Taylor (1987). The IAU (1976) system of astronomical constants is described by Duncombe et al. (1977). Astronomical data are also published yearly in *The Astronomical Almanac*; many solar data are compiled in *Landolt-Börnstein* (1981), in the *Astrophysical Formulae* (Lang 1999), and in *Allen's Astrophysical Quantities* (Cox 2000).

Most of the material in this chapter concerning the luminosity and spectral energy distribution has been taken from White (1977). Ground-based irradiance measurements are most comprehensively described by Neckel and Labs (1984); Avresen et al. (1969) present the results of measurements from aircraft. Willson (1984) gives an account of space experiments, in particular including results obtained during the Solar Maximum Mission. Reviews on the irradiance in various spectral regions have appeared in the *Journal of Geophysical Research*, Vol. 92, No. D1 (1987). New conference proceedings on the subject include *Solar Physics*, Vol. 177 (1998) and *Space Science Review*, Vol. 94 (2000). Reviews and discussions are also presented by Pap (1997) and Pap et al. (1998). Foukal (1990) devotes a full chapter of his book on the Sun to the subject of solar variability, while *The Sun in Time* (Sonnet et al. 1991) provides results as well as theoretical treatments. The Sun among the stars, especially with respect to the color indices, is discussed by Hardorp (1982) and Neckel (1986).

2. Internal Structure

Except for the neutrinos, there is no direct observational means to study the Sun's interior. Theoretical considerations must therefore be made. These consist partly of rigorous physical laws, such as the conservation of energy or mass, and partly of plausible assumptions, concerning, e.g., the chemical composition, or the effect of magnetic and centrifugal forces. Sometimes the physics is known in principle, but it is very complicated to work out all the details, and uncertainties remain; examples are the opacity of the solar gas, or the equation of state.

The theory of stellar structure successfully explains the Hertzsprung–Russell diagram of star clusters and other observed basic facts. It must therefore be essentially correct. However, more and better observations exist for the Sun than for any other star. Theoretically constructed solar models can therefore be tested more critically than models for other stars. The neutrinos generated in the Sun's central region as a by-product of nuclear energy generation provide such a possibility; another, to be treated in Chap. 5, consists in the measurement and theoretical calculation of the oscillation eigenfrequencies.

2.1 Construction of a Model

2.1.1 The Evolutionary Sequence

The aim of this chapter is to give a description of the internal constitution of the *present* Sun. It will soon be clear that this requires knowledge of the Sun's history of evolution. The reason is that the Sun, like other main-sequence stars, supplies its energy needs by conversion of hydrogen to helium, and we cannot infer from observations of the solar surface alone how much helium has already been synthesized in the interior. An evolutionary sequence of models, each distinguished from its predecessor by a slightly increased helium to hydrogen ratio, must therefore be constructed. As the thermonuclear fusion proceeds at the fastest rate in the center, a gradient in helium content will develop. In thermal equilibrium the total radiation which is emitted during the time elapsed between two models will correspond to the total amount of helium generated in this time span.

The thermonuclear fusion process leads to a gradual increase of the mean molecular weight, μ. There must be a corresponding increase of temperature and (or) density in the deep interior so that the thermal pressure, $\rho \mathcal{R} T/\mu$, is kept strong enough to support the weight of the star. As the nuclear reaction rates depend on temperature and density, the energy generation, and therefore the Sun's luminosity, also increases. The general procedure then is to begin with a chemically homogeneous star of solar mass. For the initial helium content (fractional, by weight), Y_0, one could simply take the "primordial" value which was established shortly after the big bang, say $Y_0 = 0.25$; the age of the Sun would then be determined as the period of time which the model sequence needs to reach the present solar luminosity. However, as we shall see in the following section, the age of the solar system is rather well known from the study of meteorites. Moreover, it turns out that the luminosity of all models in the sequence depends in a rather sensitive way on the initial helium content. For these reasons one normally varies Y_0 until the luminosity of the most recent model, i.e. that with the given age of 4.6×10^9 years (see below), equals the 3.844×10^{26} W that are measured today. That this is possible with Y_0 values close to the above-mentioned $Y_0 = 0.25$ is a reassuring result of these entirely theoretical considerations.

But not only the luminosity must be correct. The model of the present Sun must also have the proper *radius* of 6.96×10^8 m. In general this requires the adjustment of a second parameter. Now the turbulent convection which dominates the energy transport in the outer layers of the Sun (and which manifests itself in the *granulation* visible at the surface, cf. Fig. 3.5) is commonly described in terms of the *mixing-length theory* which was developed by G. I. Taylor, W. Schmidt, and L. Prandtl from about 1915. We shall treat this formalism in Chap. 6 and we find that the ratio

$$\alpha = l/H_{\mathrm{P}} \tag{2.1}$$

of the mixing length, l, to the pressure scale height, $H_{\mathrm{P}} = -1/(d\ln P/dr)$, is a convenient free parameter in the solar model. Biermann (1932) and others applied the mixing-length concept to the problem of stellar convection; later it was proposed that α should be of order 1, i.e., that a parcel of convecting gas should travel a distance of the order of the local pressure scale height before it dissolves. Indeed, the value of α required to adjust the present Sun's radius lies between 1 and 2 for most models constructed so far.

In the actual calculations the luminosity and radius of the Sun must of course *simultaneously* be adjusted by a proper choice of Y_0 and α. But the luminosity is more sensitive to variations of Y_0, and the radius responds more strongly to changes of α. In the expansion (the index $\odot$ means "present Sun")

$$\ln L = \ln L_\odot + a(Y_0 - Y_{0\odot}) + b(\alpha - \alpha_\odot)$$

$$\ln r = \ln r_\odot + c(Y_0 - Y_{0\odot}) + d(\alpha - \alpha_\odot) \, , \tag{2.2}$$

the coefficients a to d can be determined from calculations of evolutionary sequences with various Y_0 and α. The model described in this chapter yields:

$$a \equiv \frac{\partial \ln L}{\partial Y_0} = 8.6 \qquad b \equiv \frac{\partial \ln L}{\partial \alpha} = 0.04$$

$$c \equiv \frac{\partial \ln r}{\partial Y_0} = 2.1 \qquad d \equiv \frac{\partial \ln r}{\partial \alpha} = -0.13 \; . \tag{2.3}$$

For the fit of the present Sun the original helium content is $Y_{0\odot} = 0.276$, and the ratio of mixing length to pressure scale height is $\alpha_\odot = 1.81$. These results are obtained with the stellar evolution code described by Kippenhahn et al. (1967), and the same code has been used to calculate some of the details which will be presented in the subsequent sections. Other authors find slightly different values of $Y_{0\odot}$ and $\alpha_\odot$, and of the coefficients (2.3) for their standard models.

2.1.2 The Standard Model

The *standard model* of the Sun can be defined as the model which is based on the most plausible assumptions and on the best available physical input. Above all, it is *spherically symmetric*, i.e., all physical quantities depend only on the distance, r, from the center. This means that the internal rotation is sufficiently slow (comparable to that observed at the surface), and that the internal magnetic field is sufficiently weak (in the solar interior a field of sunspot strength would be weak in this sense), that the forces resulting from these two agents have negligible influence.

In the standard model the abundances of the elements other than hydrogen and helium are usually combined into a number Z, which is their fractional abundance by weight, so that

$$X + Y + Z = 1 \; , \tag{2.4}$$

where X and Y, the fractional abundances of hydrogen and helium, vary with depth in the star and with time, because hydrogen is fused into helium, and because – to a small amount – the heavier helium diffuses towards the solar center, cf. Sect. 2.3.3. Such gravitational settling also leads to a slight variation of Z. For the calculations of the present text only helium settling is taken into account, while Z is kept constant. I have chosen $Z = 0.02$, which closely represents the heavy element abundances observed on the solar surface (cf. Chap. 4), and which is typical for a population I star. But (2.4) is slightly modified to $X + Y + Y_3 + Z = 1$, where Y_3, the mass fraction of ^{3}He, also evolves as a function of depth and time, starting from the constant $Y_{30} = 4 \times 10^{-5}$. The correct value of Y_{30} is not precisely known, but Rood et al. (1984) and Balser et al. (1999) found values similar to the one used here when they studied the ^{3}He$^+$ hyperfine line at 8.7 GHz in galactic H II regions.

The detailed abundances of the heavy elements are not considered explicitly, with two exceptions: The opacity tables (Sect. 2.3.7) – these are based on abundances determined spectroscopically at the solar surface or inferred from elsewhere in the solar system (Table 4.2) –, and the CNO energy generation rate which depends on the initial abundances of ^{12}C and ^{14}N (cf. Sect. 2.3.6) in the solar core.

The relatively large content of heavy elements comes from the fact that the Sun was formed as a second-generation star out of matter already processed in stars of earlier generations. Some of the latter, the population II stars, are still observed today, in particular in the very old globular clusters. The heavy element content of these stars is generally smaller than the Sun's by an order of magnitude, or even more.

In the standard model, convection and mixing will occur only if the Schwarzschild criterion for the stability of stratification, to be discussed in Chap. 6, is violated. If this is the case, the model will employ the mixing length formalism already mentioned. Convection and radiation in unstable layers, and radiation alone in stably stratified layers, will be the only mechanisms of energy transport (a small amount of heat conduction will be included into the radiative part).

The standard model is not rigorously defined. A number of modifications have been presented in the literature, e.g. slightly different values of Z, or a more detailed treatment of the heavy elements and their ionization than is given here, or various versions of the equation of state (see below). More drastic modifications are considered as non-standard; some of them will be treated at the end of this chapter.

2.2 Age and Pre-Main-Sequence Evolution

Meteorites are the oldest bodies in the solar system studied so far in the laboratory. Their age is determined from the decay of radioactive isotopes, such as ^{87}Rb, which has a half-life of 4.8×10^{10} years. The abundances of both the parent and the daughter (^{87}Sr) isotopes are measured relative to the (constant) abundance of ^{86}Sr, and samples from many different meteorites, in particular chondrites, yield essentially the same age: $(4.55 \pm 0.05) \times 10^9$ years. The decay of uranium and thorium into lead isotopes confirms this result. According to recent determinations the error might be even 10 times smaller (Wasserburg 1995). Hence the age of meteorites, or, more precisely, the time elapsed since the condensation and chemical differentiation of these bodies began, is rather well known.

Problem 2.1. Show how two samples with different Rb/Sr abundance ratios can be used to determine their (common) age.

The problem is, then, to relate this epoch to the start of the Sun's main-sequence life, i.e., to the ignition of hydrogen burning in the center of the homogeneous initial star. To this end one should know the sequence and duration of various processes during the formation of the Sun and the solar system.

First, an interstellar cloud of, say, 10^4 solar masses is triggered into its gravitational collapse. That is, some external agency must have generated an initial compression of the cloud so that self-gravitation could overcome the internal gas pressure. The external agency could have been the shock wave of the galactic gas in a spiral arm: the fact that most young stars are found in or near spiral arms supports this hypothesis. It could also have been the shock wave generated by a nearby supernova explosion. This alternative would explain the large abundance of the ^{26}Mg isotope found in connection with ^{27}Al in the Allende meteorite of 1969: ^{26}Mg is a decay product of ^{26}Al, which has a half-life of only 7×10^5 years and should have disappeared if it was synthesized long before the formation of the solar system. If, on the other hand, the ^{26}Al came from a nearby supernova explosion then it would naturally yield the anomaly of ^{26}Mg.

The collapse begins when the initial compression is sufficiently strong for gravitation to win against the internal pressure. This is the case if the *Jeans criterion* is satisfied, i.e., if

$$\frac{Gm_{\rm c}}{r} > \frac{\mathcal{R}T}{\mu} , \tag{2.5}$$

where $\mathcal{R}$ is the gas constant, and μ, T, $m_{\rm c}$, r are the cloud's mean molecular weight, temperature, mass, and radius.

Problem 2.2. Transform (2.5) into a condition for the cloud mass, at given interstellar density and temperature. Show that about 10^3 solar masses is the minimum required for instability, at typical interstellar conditions: $T = 50\,{\rm K}$ and $\rho = 10^{-20}\,{\rm kg/m^3}$. Show that, as the collapse goes on, the conditions become *more favorable* for further collapse.

Problem 2.3. Calculate the free-fall time $t_{\rm ff}$, i.e., the time which a spherically symmetric cloud of initial density ρ_0 and negligible internal pressure needs for complete collapse.

For an initial density of $10^{-20}\,{\rm kg/m^3}$ the spherically symmetric collapse would take about $(\rho_0 G)^{-1/2} \approx 3 \times 10^7$ years (Problem 2.3). But rotation and the magnetic field cause complications: The interstellar gas is in irregular motion and therefore has large angular momentum per unit mass, typically $10^{18}\,{\rm m^2/s}$. In comparison, the present solar system's angular momentum per unit mass is only about $10^{16}\,{\rm m^2/s}$. Most of the angular momentum therefore must have been removed. In addition, the interstellar gas is permeated by a magnetic field of 10^{-6} to 10^{-5} G. If the flux crossing the material from which the Sun was formed would had been conserved during a spherically symmetric collapse, the field would have been amplified proportionally to $\rho^{2/3}$, to a value of 10^{10} G or more. That is, virtually *all* the magnetic flux must have been lost as well.

Problem 2.4. Use the solar model of Table 2.4 to calculate the Sun's moment of inertia, and the solar spin angular momentum. Compare the result with the orbital angular momenta of the planets.

Now it happens that the magnetic field provides a very effective lever arm for the torque required to remove angular momentum – as we shall see in Chap. 9 in the context of the solar wind. A plausible hypothesis, therefore, is that during a first phase of the collapse magnetic braking takes place. Later, when the density has become sufficiently large (and the temperature is still sufficiently low) so that neutral particles can form by recombination of the originally ionized material, the cloud is no longer coupled to the magnetic lines of force. The field may then escape back into the interstellar space.

A further complication is that, of course, the cloud must *fragment* in order to form the 10^3 to 10^4 stars of a typical galactic cluster. Spontaneous fragmentation may occur because, as the density increases, the Jeans criterion is satisfied for *smaller* masses – provided there is no significant increase of the temperature, cf. Problem 2.2. The presence of *molecules* prevents such temperature rise, because molecules effectively radiate the energy gained during the collapse back into space. Hence subsystems of the original cloud can begin their own collapse. The Sun, with its planets, must have formed out of such a subsystem.

The details of the successive removal of angular momentum and magnetic flux, and of the process of fragmentation, are not known. Probably there was an intermediate state on a stellar *accretion disc*, with two jets emanating perpendicular to the disc in the two directions along the axis. Observations in the infrared have revealed such discs around several young stars, e.g., Wega, β Pictoris, or HL Tauri, and jets also have been observed. As far as theoretical models and numerical simulations exist, they confirm that the duration of this whole early history is still essentially the free-fall time, $t_{\rm ff}$.

For a single fragment such as the pre-solar nebula the density is typically 10^{-17} to 10^{-15} kg/m^3, and the free-fall time is correspondingly shorter. Detailed numerical simulations exist for this phase (e.g., Bodenheimer 1983). While the bulk of the fragment collapses in free fall, the central part becomes optically thick, heats up, and thus is able to set itself into hydrostatic equilibrium: a *protostar* is formed.

It takes less than 10^6 years for the accretion of virtually the whole envelope on to the hydrostatic core. At the end of this phase, a low-density, slowly contracting cool star (that is: a gaseous sphere in *hydrostatic equilibrium*) is born. Its effective temperature is ≈ 3000 K, its radius is about four times the present solar radius, and its luminosity is several times the present luminosity. In the Hertzsprung–Russell diagram it evolves with decreasing luminosity along the *Hayashi line*, the location of fully convective stars. The central temperature is still below 10^6 K, much too low for the major nuclear reactions to take place at a significant rate. Instead, the source of the luminosity is the release of gravitational energy during the slow contraction.

According to the virial theorem (cf. Problem 2.10) the slowly contracting star gains an amount of internal energy which is equivalent to the energy radiated into space. Hence it becomes hotter, until finally hydrogen burning sets in (Sect. 2.3.6). The Sun has then arrived at the *zero-age main sequence*, i.e., at the main sequence of *chemically homogeneous* stars.

The duration of the slow contraction phase is the ratio of supply to consumption. This time scale is also called the *Kelvin–Helmholtz time*,

$$t_{KH} = \frac{Gm^2}{rL} \,, \tag{2.6}$$

and is about 3×10^7 years for the Sun. An example of an evolutionary track in the Hertzsprung–Russell diagram is shown in Fig. 2.1. It begins with the formation of the core (prior to this event the object is optically thin and an effective temperature cannot be defined) and covers the protostar collapse and the final hydrostatic approach to the main sequence.

As far as the Sun's age is concerned, the conclusion is as follows: The initial cloud collapse, and the removal of angular momentum and magnetic flux, took about 3×10^7 years. After a rapid collapse a time span of similar length was spent in the final hydrostatic contraction. Sometime during this whole evolution (or shortly after its end) the material from which meteorites

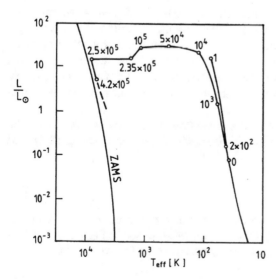

Fig. 2.1. Hertzsprung–Russell diagram showing the proto-stellar evolution of the Sun towards the zero-age main sequence (ZAMS). The numbers on the track give the time (in years) since the formation of the hydrostatic core. After Winkler and Newman (1980)

are formed condensed. Thus, the uncertainty in the dating of the origin of meteorites with respect to the Sun's pre-main-sequence evolution is larger than the uncertainty in the age of the meteorites itself. For the solar age we may therefore take the result of Wasserburg (1995), but with a more conservative error estimate:

$$t_\odot = (4.57 \pm 0.05) \times 10^9 \text{ years} . \tag{2.7}$$

This is the time span which the Sun has spent on the main sequence, converting hydrogen to helium. We shall see that the Sun's radius and luminosity both slowly increased during this main-sequence evolution, until finally the present $r_\odot$ and $L_\odot$ were reached.

2.3 Model Ingredients

We now discuss the laws and equations which govern the solar structure and evolution.

2.3.1 Conservation Laws

Conservation of *mass* is achieved in a most natural way by using the mass m interior to a sphere of radius r as the independent variable. Hence, instead of the familiar $\partial m/\partial r = 4\pi \rho r^2$ one usually uses

$$\frac{\partial r}{\partial m} = \frac{1}{4\pi \rho r^2} , \tag{2.8}$$

where the partial derivative indicates that in general r depends also on time.

The choice of m as the independent variable has an additional advantage: We know that the boundary conditions to our differential equations must be placed at $m = m_\odot$. Since we neglect mass loss, $m_\odot$ is known and constant. In contrast, the radius is generally unknown (except for the *present* Sun), and the application of boundary conditions would therefore be less straightforward.

For most of the present text, the conservation of *momentum* takes the simple form of the *hydrostatic equilibrium*, $\partial P/\partial r = -\rho g$. Using (2.8) and $g = -Gm/r^2$ we obtain

$$\frac{\partial P}{\partial m} = -\frac{Gm}{4\pi r^4} . \tag{2.9}$$

Here P is the total pressure required to balance the weight of the star. Its main contribution is the gas pressure but, as we shall see in Sect. 2.3.4, minor corrections must be taken into account at various depths in the Sun.

Problem 2.5. Use (2.9) to find a lower bound for the pressure at the center of the Sun. Compare the result with Table 2.4.

The hydrostatic equilibrium must be replaced by the more general *hydrodynamic* balance if processes such as the protostar collapse discussed in the preceding section are to be described. The appropriate equation is then

$$\frac{\partial^2 r}{\partial t^2} = -4\pi r^2 \frac{\partial P}{\partial m} - \frac{Gm}{r^2} \; . \tag{2.10}$$

In the present chapter we shall use the hydrostatic equilibrium (2.9) only. However, we shall return to (2.10) in Chap. 5, where we treat the solar oscillations.

Problem 2.6. Use (2.10) to show that a violation of (2.9) leads to a significant change of the solar radius in the *dynamical* time scale

$$t_{\mathrm{D}} \simeq (G\bar{\rho})^{-1/2} \approx 1 \text{ hour} \; , \tag{2.11}$$

where $\bar{\rho}$ is the mean density.

Problem 2.7. Calculate the Sun's gravitational energy, i.e., the total work required to disperse the solar matter over distances $r \gg r_{\odot}$. Use Table 2.4 for the function $r(m)$.

We now turn to the *energy* balance. Let $L(m)$ be the luminosity generated inside the sphere of mass m. We then define the energy generation ε per unit mass through

$$\frac{\partial L}{\partial m} = \varepsilon - T \frac{\partial S}{\partial t} \; , \tag{2.12}$$

where T is the temperature and S is the specific entropy (entropy per unit mass). The second term on the right ensures that, in addition to explicit sources, heating or cooling can affect the luminosity. As far as the main-sequence evolution is concerned, the Sun is very close to thermal equilibrium. The change in specific entropy is then very small in comparison to ε. The function $\varepsilon(\rho, T)$ itself represents the nuclear energy sources and will be considered in Sect. 2.3.6.

2.3.2 Energy Transport

At any depth in the Sun, the energy flux, F, is defined as the luminosity per unit area. Since we consider energy transport by radiation (F_{R}) and by convection (F_{C}), we may write

$$F = F_{\mathrm{R}} + F_{\mathrm{C}} = L/4\pi r^2 \; . \tag{2.13}$$

Let us first consider radiative energy transport. In the interior of stars the mean free path of a photon is very small, so small in fact that the places where the photon is emitted and absorbed have nearly the same temperature. That is, the conditions of local thermodynamic equilibrium (cf. Chap. 4) are very closely satisfied. Consequently, in the equation of radiative transfer we may replace the source function by the Kirchhoff–Planck function

$$B_\nu(T) = \frac{2h}{c^2} \frac{\nu^3}{e^{h\nu/kT} - 1} \,, \tag{2.14}$$

and obtain

$$\cos\theta \frac{dI_\nu}{dr} = -\kappa_\nu \rho (I_\nu - B_\nu) \,. \tag{2.15}$$

Here θ is the angle to the local vertical direction, and I_ν and κ_ν are the intensity and the absorption coefficient (cross section per unit mass) at frequency ν. Of course, we must not use $I_\nu = B_\nu$, because a completely isotropic radiation field would permit no net transport. But the expansion

$$I_\nu = B_\nu - \frac{\cos\theta}{\kappa_\nu \rho} \frac{dB_\nu}{dr} \tag{2.16}$$

satisfies (2.15) up to a term proportional to $\cos^2\theta$, which does not contribute to the flux. The energy flux is

$$F = \int_0^\infty F_\nu \, d\nu \,, \tag{2.17}$$

where F_ν is an integral over solid angle Ω:

$$F_\nu = \int \cos\theta I_\nu \, d\Omega = -\frac{4\pi}{3\kappa_\nu \rho} \frac{dB_\nu}{dr} = -\frac{4\pi}{3\kappa_\nu \rho} \frac{dB_\nu}{dT} \frac{dT}{dr} \,. \tag{2.18}$$

We combine (2.17), (2.18), and (2.13), set $F_C = 0$, and solve for the temperature gradient:

$$\frac{\partial T}{\partial m} = -\frac{3\kappa L}{256\pi^2 \sigma r^4 T^3} \,. \tag{2.19}$$

Here we have again used (2.8) in order to replace dr by dm, and have introduced the *opacity*, or *Rosseland mean absorption coefficient*, κ, defined through

$$\frac{1}{\kappa} = \frac{\int_0^\infty \frac{1}{\kappa_\nu} \frac{dB_\nu}{dT} \, d\nu}{\int_0^\infty \frac{dB_\nu}{dT} \, d\nu} \,. \tag{2.20}$$

The opacity is a *harmonic* mean, i.e., the average of $1/\kappa_\nu$. That is, more energy is transported at frequencies where the matter is more transparent. It is also a mean weighted with dB_ν/dT; this means that more energy is transported at frequencies where the radiation field is more temperature-dependent.

So far we have considered only energy transport by radiation. In the absence of convection the only other means of energy transport is heat conduction by electrons, which plays a small part in the solar interior. Since κ in (2.19) is defined essentially as an inverse conduction coefficient, we may take electron conduction into account by calculating once more the harmonic mean:

$$\frac{1}{\kappa} = \frac{1}{\kappa_{\mathrm{R}}} + \frac{1}{\kappa_{\mathrm{e}}} \,,$$

where κ_{R} is the radiative opacity considered above and κ_{e} is an "electron conduction opacity", formally defined in an equation of type (2.19).

Equation (2.19) yields the correct temperature gradient only in stably stratified layers. When the Schwarzschild criterion (Sect. 6.1) for stability is violated, energy is transported both by radiation and convection. The mixing-length theory of stellar convection, which has already been mentioned, then leads to a new expression for the temperature gradient – this expression will be deduced in Chap. 6. Here we write symbolically

$$\frac{\partial T}{\partial m} = \left(\frac{\partial T}{\partial m} \right)_{\mathrm{C}} \,, \tag{2.21}$$

which replaces (2.19) in a convection zone.

As we shall see convection is a very efficient means of energy transport. As a consequence, the temperature gradient in stellar convection zones is very close to the adiabatic gradient.

The energy transport equation is the last in the system of four differential equations describing the interior constitution of stars: Equations (2.8), (2.9) or (2.10), (2.12), and (2.19) or (2.21) define the gradients of r, P, L, and T. These equations must be supplemented by "constitutive" relations for the remaining four variables, ρ, ε, S, and κ. This will be done in the subsequent sections, but first we shall consider the diffusive transport of particles in the Sun.

2.3.3 Element Diffusion in the Interior

In the solar interior there is a large central region, of radius $\approx 0.7 r_\odot$, where energy is transported merely by radiation. Hence energy transport causes no mixing of matter in this region. The chemical composition would remain the same except for nuclear transmutations. However, driven mainly by the gravitational force, heavier particles tend to diffuse towards the solar center, while lighter ones tend to diffuse in the opposite direction, away from the center. This *gravitational settling* is a very slow process: its characteristic time exceeds the solar age by a factor 10^4, hence it makes only a small difference for the interior chemical composition of the present Sun. Nevertheless the process has been taken into account in many recent model calculations. This is appropriate in view of the high precision required for some helioseismological applications, or for the prediction of the solar neutrino flux.

Diffusion in the Sun has been treated already by Aller and Chapman (1960); Noerdlinger (1977) first showed that it lowers significantly the helium abundance at the solar surface, and raises the central helium abundance. In the present text I shall follow the treatment of Bahcall and Loeb (1990) and of Thoul et al. (1994).

In plasma physics a problem such as the present one is often treated in terms of a multi-component fluid, where each component is subject to a set of conservation equations that include the coupling to the other components. As an illustrative example we consider a three-fluid model consisting of hydrogen, helium, and electrons. Since the model is applied to the solar interior, where $T > 2 \times 10^6$ K, the assumption of complete ionization is in order.

The conservation of the hydrogen and helium particles, with densities n_{H} and n_{He}, is expressed – in spherical symmetry – by

$$\frac{\partial n_{\mathrm{H}}}{\partial t} = -\frac{1}{r^2}\frac{\partial}{\partial r}\left(r^2 n_{\mathrm{H}} v_{\mathrm{H}}\right) , \qquad \frac{\partial n_{\mathrm{He}}}{\partial t} = -\frac{1}{r^2}\frac{\partial}{\partial r}\left(r^2 n_{\mathrm{He}} v_{\mathrm{He}}\right) , \qquad (2.22)$$

where v_{H} and v_{He} are the diffusion velocities and where, on the right-hand sides, the particle density changes due to nuclear reactions must be added in a calculation of the Sun's evolution.

The coupling between the three fluid components appears in the momentum conservation equations. The hydrogen and helium fluids are coupled together by a friction force $f_{\mathrm{H,He}}$ mediated by *Coulomb collisions*, that is particle encounters with interactions due to their electric charge. Compared to the momentum transfer during such ion-ion interactions, the momentum transfer in ion-electron collisions is smaller by a factor $(m_{\mathrm{e}}/m_{\mathrm{i}})^{1/2}$, hence the latter type of friction is neglected. On the other hand there *is* coupling between ions and electrons, mediated by the electric field. With P_{H}, etc., denoting the partial pressures, the three momentum conservation equations are

$$-\frac{\partial P_{\mathrm{H}}}{\partial r} + n_{\mathrm{H}}(eE - m_{\mathrm{H}}g) - f_{\mathrm{H,He}} = 0 , \tag{2.23}$$

$$-\frac{\partial P_{\mathrm{He}}}{\partial r} + n_{\mathrm{He}}(2eE - m_{\mathrm{He}}g) + f_{\mathrm{H,He}} = 0 , \tag{2.24}$$

$$-\frac{\partial P_{\mathrm{e}}}{\partial r} - n_{\mathrm{e}}eE = 0 . \tag{2.25}$$

Since diffusion is a very slow process, we have omitted the acceleration terms. In addition, because of the small electron mass, the gravitational force has been omitted in the equation for the electrons. We recognize that the collisional friction between H and He has effects that are equal in magnitude, but opposite in sign, on these two fluid components.

Equations (2.23) to (2.25) must be used to determine the diffusion velocities that are required for the evaluation of (2.22). This is possible because the friction force $f_{H,He}$ is proportional to the diffusion velocity of the two species relative to each other, i.e., to $v_H - v_{He}$. In order to evaluate $f_{H,He}$ one must consider the Coulomb collisions in detail. The result is

$$f_{H,He} = \frac{mn_H n_{He} \ln \Lambda}{3} \left(\frac{e^2}{\pi kT \varepsilon_0}\right)^2 \left(\frac{2\pi kT}{m}\right)^{1/2} (v_H - v_{He}) , \qquad (2.26)$$

where $m = m_H m_{He}/(m_H + m_{He})$ is the reduced mass, and $\ln \Lambda$ is the *Coulomb logarithm* which takes care of the partial shielding of the ions by an electron cloud, cf. the following subsection. The Coulomb logarithm corresponds to the quantity $\ln \Lambda$ defined by Spitzer (1962) for the case of ion-electron collisions, which we shall need in Chaps. 8 and 9. In the solar interior the Coulomb logarithm for collisions of hydrogen with helium varies between 2.2 and 2.7 (Thoul et al. 1994, Fig. 10). Expression (2.26) is a plausible result as the two factors with brackets essentially are the cross section for Coulomb collisions and the thermal velocity, respectively.

In addition to the difference of the two diffusion velocities, their *ratio* is known since the total mass flow due to diffusion is zero:

$$m_H n_H v_H + n_{He} m_{He} v_{He} = 0 . \qquad (2.27)$$

We eliminate the electric field by (2.25), and express the diverse partial pressures in terms of the total pressure P and the hydrogen mass fraction X (which is easy in the case of the present three-fluid model with full ionization), and so obtain the diffusion velocities as linear functions of $\partial P/\partial r$ and $\partial X/\partial r$.

Thermal Diffusion. In addition to the gradients of pressure and species concentration, a *temperature gradient* constitutes a driving force for diffusion. This is because diffusion is governed by collisions, and particles coming from the direction of larger T have a slightly larger thermal velocity than particles coming from the opposite direction, of smaller T. It follows that the lighter species, here hydrogen, is driven in the direction of decreasing temperature, relative to the heavier species, helium. But there are not only consequences for the momentum transfer during the collisions, but also for the transfer of energy. Hence for each particle species an equation of energy conservation must be considered, in addition to Eqs. (2.22) to (2.25) above. For a multicomponent fluid Thoul et al. (1994) have solved this whole set of conservation equations numerically. For the solar application they also give approximate expressions of the resulting diffusion velocities of hydrogen, helium, oxygen, and iron; for hydrogen the result is

$$v_H = \frac{r_\odot^2 \rho_0}{\tau_0 \, \rho} \left(\frac{T}{T_0}\right)^{5/2} \left(A_P \frac{\partial \ln P}{\partial r} + A_T \frac{\partial \ln T}{\partial r} + A_H \frac{\partial \ln C_H}{\partial r}\right) , \qquad (2.28)$$

where $C_H = n_H/n_e = 2X/(1+X)$ and, with $\rho_0 = 100\,\text{g/cm}^3$, $T_0 = 10^7\,\text{K}$, and $\tau_0 = 6 \times 10^{13}$ years,

$$A_P = -2.09 + 3.15X - 1.07X^2 \ ,$$

$$A_T = -2.18 + 3.12X - 0.96X^2 \ , \tag{2.29}$$

$$A_H = -1.51 + 1.85X - 0.85X^2 \ .$$

The slowness of the diffusion process is manifest in the large value of the characteristic time τ_0.

Equations (2.28) and (2.29) have been used in the solar model of this text (Table 2.4). As compared to the same model without diffusion, the helium mass fraction at the surface decreases from $Y_s = 0.278$ to $Y_s = 0.245$, whereas at the center there is an increase from $Y_c = 0.631$ to $Y_c = 0.642$. Since the outer convection zone is always well-mixed, the surface value of Y is the same as in the outer part of the radiative core. Here it is too cool for nuclear fusion, hence there is only the small variation of X and Y that is due to diffusion; the calculation shows that ca. 2/3 of the helium settling is due to the pressure gradient, while ca. 1/3 is thermal diffusion, due to the temperature gradient. On the other hand, a concentration gradient of H and He has built up over the solar age in the central region. There, the pressure and temperature gradients together would cause about 1.6 times the actual diffusive helium increase, but the concentration gradient acts in the opposite direction and so compensates for the extra amount.

2.3.4 The Equation of State

For the solution of our differential equations we must express the density ρ in terms of P an T. It is more common, however, to consider the pressure P as a function of T and ρ, which is known as "the equation of state". We shall do the same here, too, but we must recall at the end that the relationship will be solved for ρ.

In a rigorous treatment the equation of state is obtained from a minimization of the free energy F via

$$P = -\left(\frac{\partial F}{\partial V}\right)_{T,n} . \tag{2.30}$$

In addition, the minimization of F yields the entropy $S = -(\partial F/\partial T)_{V,n}$, and the ionization equilibria through the derivatives with respect to the particle densities n (here n stands for the whole set of densities). We shall nevertheless use a more elementary approach here, sometimes under simplifying assumptions, in order to illustrate the physical input.

The pressure arises from transfer of momentum by particles and photons. Consequently, we divide the total pressure P into the gas pressure P_G and radiation pressure P_R. The latter depends only on temperature:

$$P_{\mathrm{R}} = \frac{a}{3} T^4 \; , \tag{2.31}$$

where $a = 4\sigma/c$. Except in the atmosphere, we have $P_{\mathrm{R}} \ll P_{\mathrm{G}}$ everywhere in the Sun. Usually a parameter β is defined through

$$P = P_{\mathrm{G}} + P_{\mathrm{R}} \; , \quad P_{\mathrm{G}} = \beta P \; , \quad P_{\mathrm{R}} = (1 - \beta)P \; ; \tag{2.32}$$

in actually calculated solar models we find that $1 - \beta$ is smaller than about 10^{-3} everywhere below the surface.

Perfect Gas Pressure. Thus, the pressure in the Sun's interior is practically identical to the *gas pressure*, which (to a reasonable approximation, as we shall see) can be described by the perfect gas law:

$$P_{\mathrm{PG}} = \frac{\rho \mathcal{R} T}{\mu} \; ; \tag{2.33}$$

here $\mathcal{R}$ is the gas constant, and μ is the mean molecular weight. In principle μ can be determined from the abundances of the constituent chemical elements and their respective degrees of ionization. However, I shall restrict the present discussion to the ionization of hydrogen and helium, while all other constituents are treated as if they were completely ionized; in view of the small value of the heavy element abundance Z the error will be small.

Let us call μ_0 the mean molecular weight in the case where hydrogen and helium are completely neutral. If we make the further approximation that the total number of particles contributed by a completely ionized atom equals half its atomic weight, then

$$\mu_0 = 1/(X + Y/4 + Z/2) \; . \tag{2.34}$$

Ionization of hydrogen and helium adds further particles; therefore the mean molecular weight is reduced:

$$\mu = \mu_0/(1 + E) \; , \tag{2.35}$$

where, by definition, E is the number of electrons set free by the ionization of H and He divided by the number of *all other* particles. Let η_{H}, η_{He}, and η_{He^+} be the degrees of ionization, i.e., the numbers of ionized particles divided by the *total* number of particles of the respective species. Then

$$E = \mu_0[\eta_{\mathrm{H}} X + (\eta_{\mathrm{He}} + 2\eta_{\mathrm{He}^+})Y/4] \; . \tag{2.36}$$

Problem 2.8. Confirm (2.34) to (2.36). Consider explicitly any one of the heavy elements, and modify (2.34) and (2.36) accordingly.

The degrees of ionization are determined by three Saha equations. For the densities n_{H^0}, n_{H^+}, n_{He^0}, n_{He^+}, $n_{\mathrm{He}^{++}}$ of neutral and ionized hydrogen and helium particles, these equations are

$$\frac{n_{H^+}}{n_{H^0}} = \frac{2(2\pi m_e)^{3/2}(kT)^{5/2}}{u_H h^3 P_e} \, e^{-\chi_H/kT} \; ,$$

$$\frac{n_{He^+}}{n_{He^0}} = \frac{2u_{He^+}(2\pi m_e)^{3/2}(kT)^{5/2}}{u_{He} h^3 P_e} \, e^{-\chi_{He}/kT} \; , \tag{2.37}$$

$$\frac{n_{He^{++}}}{n_{He^+}} = \frac{2(2\pi m_e)^{3/2}(kT)^{5/2}}{u_{He^+} h^3 P_e} \, e^{-\chi_{He^+}/kT} \; ,$$

where m_e, k, and h are the electron mass, the Boltzmann constant, and Planck's constant; P_e is the electron partial pressure, and the χ's are the ionization energies of H, He, and He^+. The partition functions, u_H, u_{He}, and u_{He^+}, of the particles which possess bound electrons are sums of the form

$$u_i = \sum_j g_{ij} \exp(-E_{ij}/kT) \; , \tag{2.38}$$

where g_{ij} is the statistical weight of the jth state, and E_{ij} is the energy of that state, relative to the ground state. For an isolated atom or ion the sum (2.38) has an infinite number of terms and diverges, but most of the terms have energies close to the ionization energy and are therefore cut off in a dense plasma where the particles perturb each other and so, in effect, lower the ionization potential. In fact, a common approximation is to replace the partition functions by the weights of the respective ground states, i.e., 2, 1, and 2 for H, He, and He^+.

It must also be mentioned that, as the electron pressure increases towards the stellar interior, ionization would finally *decrease* inwards according to (2.37). But the mutual interaction of the particles leads to additional ionization ("pressure ionization") which, in a crude approximation, is simulated by simply assuming complete ionization at all depths below the level where complete ionization is reached (or where ionization begins to decrease) according to (2.37).

Problem 2.9. Express the degrees of ionization, η_H ..., in terms of the particle densities, n_{H^0} ..., and calculate the electron pressure.

Problem 2.10. The virial theorem. Suppose the Sun consists of a perfect monatomic gas in hydrostatic equilibrium. Calculate the internal energy, and compare it to the gravitational energy (Problem 2.7). Find a lower bound for the mean (mass-weighted) temperature in the Sun. Compare the result with the temperature of the solar model (Table 2.4).

The Electrostatic Correction. In a perfect gas the particles interact only during collisions. However, *charged* particles also interact by means of their electrostatic force. The importance of this effect can be estimated by the ratio of the particles' mean electrostatic energy, $E_{ES} \equiv e^2/4\pi\varepsilon_0\bar{r}$ ($\bar{r}$ is the mean distance between the particles), and the average thermal energy, $3kT/2$. In

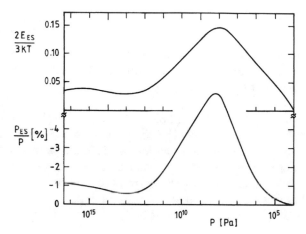

Fig. 2.2. Ratio of electrostatic to thermal energy (*upper curve*), and electrostatic pressure correction (*lower curve*), in a model of the Sun's interior

the Sun this ratio is of order 0.1 or smaller (Fig. 2.2). That is, the electrostatic interaction is not negligible, but small enough for the *Debye–Hückel* treatment to be valid. [The Debye–Hückel treatment is outlined in physics texts such as Landau and Lifschitz (1966 a, p. 246). The presentation here, essentially following Clayton (1968), will be short.]

In the neighborhood of an ion, the density n_Z of any other species with charge eZ (we set $Z = -1$ for electrons) deviates from the mean density $\bar{n}_Z$ in accord with a Boltzmann distribution:

$$n_Z = \bar{n}_Z \exp\left(-\frac{eZV}{kT}\right) . \tag{2.39}$$

The potential V in turn is determined via the Poisson equation from the combined densities of *all* charged particles. The result is

$$V = \frac{eZ}{4\pi\varepsilon_0 r} \exp(-r/r_D) , \tag{2.40}$$

where

$$r_D = \left(\frac{\varepsilon_0 kT}{e^2 \sum Z^2 \bar{n}_Z}\right)^{1/2} \tag{2.41}$$

is the *Debye radius*, and the sum is over all charged species. The density dependence of r_D follows from $\bar{n}_Z \propto \rho$.

We may now calculate the total *additional* energy density U_{ES}, due to the potential correction V_{ES} of V [expand the potential (2.40)!], by

$$U_{ES} = \frac{1}{2} \sum eZ\bar{n}_Z V_{ES} = -\frac{e^3 (\sum Z^2 \bar{n}_Z)^{3/2}}{8\pi\varepsilon_0 (\varepsilon_0 kT)^{1/2}} . \tag{2.42}$$

The additional energy density is *negative*. In correspondence, the pressure correction

$$P_{ES} = \frac{1}{3} U_{ES} \tag{2.43}$$

is also negative: due to the electrostatic interaction of the charged particles the plasma becomes "softer", i.e., at a given temperature and density the pressure is smaller than it would be for a perfect gas.

Problem 2.11. Expand the exponential in (2.39), and solve the Poisson equation to confirm (2.40) and (2.41). Use the condition of mean charge neutrality, the abundances X and Y, and the ionization degrees of hydrogen and helium to derive the explicit dependence of r_D and U_{ES} on density. Use the integrability condition for the entropy to confirm (2.43). Show that the condition of small electrostatic energy (compared to kT) is equivalent to a large Debye radius (compared to $\bar{r}$).

In the example shown in Fig. 2.2 the electrostatic pressure correction is about 1% or less of the total pressure in most of the solar interior, but reaches -5.9% at the depth where $T \approx 5 \times 10^4$ K and $P \approx 10^8$ Pa. Nevertheless, the total pressure in a solar model is practically the same whether or not the electrostatic interaction is included, because it is essentially determined by the hydrostatic equilibrium. Therefore, the effect of the electrostatic pressure correction is an *increased* perfect gas contribution. At the level where the effect is largest, the 5.9% change is accomplished by changes in temperature, density, and mean molecular weight of 1.7%, 2.6%, and -1.6%, respectively.

Partial Electron Degeneracy. The closer electrons are packed in physical space, the larger momenta they must have in order to comply with Pauli's principle that no more than two electrons must fill a cell of volume h^3 in the six-dimensional phase space. At the large density of the stellar interior the electron distribution may therefore deviate from the Maxwellian distribution. In general, states between p and $p+dp$ are occupied according to Fermi–Dirac statistics; introducing the degeneracy parameter ψ we may write

$$n_e(p)dp = \frac{8\pi p^2 dp}{h^3(e^{u/kT - \psi} + 1)}, \tag{2.44}$$

where

$$u = m_e c^2 \left(\sqrt{1 + (p/m_e c)^2} - 1 \right)$$

is the kinetic particle energy. When ψ has a large negative value we recover the Maxwellian distribution; when ψ is large and positive all states up to the Fermi limit p_F are occupied, and we have complete electron degeneracy. However, the parameter ψ is not known a priori; it must be determined together with the density ρ as a function of pressure P and temperature T.

Partial electron degeneracy occurs deep in the Sun's interior, where the density is high. We have already seen that in this region we may also expect complete ionization. For any X, Y, Z mixture, then, the electron density is

$\rho N_A (1 + X)/2$, where N_A is the Avogadro number, and this electron density must be equal to the integral over (2.44). Hence

$$\rho N_A (1 + X)/2 = \int_0^\infty n_e(p)\, dp \tag{2.45}$$

is a first equation for the unknowns ρ and ψ.

A second equation is obtained from the electron pressure P_e. On the one hand, P_e is the total momentum transfer, and thus can be calculated as an integral over $p n_e(p) v(p) dp/3$. On the other hand, $P_G \equiv \beta P = P_I + P_e$, where P_I is the ion contribution to the gas pressure. Because of their larger mass, the ions have larger momenta and therefore do not degenerate, at least not under solar conditions. Their pressure has perfect gas form: $P_I = \rho R T(X + Y/4 + Z/A_Z)$; here A_Z is the mean atomic weight of the elements represented by Z. Using $v = du/dp$ we thus obtain a second equation for ρ and ψ:

$$\beta P - \rho R T(X + Y/4 + Z/A_Z) = \frac{1}{3} \int_0^\infty p n_e(p)(du/dp)\, dp \ . \tag{2.46}$$

Of course, $\beta = 1 - aT^4/3P$ is also a known function of P and T. Equations (2.45) and (2.46) must be solved by iteration.

For a computed solar model Fig. 2.3 shows the degeneracy parameter as a function of depth; also shown is the pressure correction or, more precisely, the difference (in % of the total pressure) between the electron pressure and the pressure which the electrons would have if they were not degenerated. The correction becomes significant at $P \approx 10^{14}$ Pa, corresponding to $r \approx r_\odot/2$, and reaches about 1.7% at the center. As in the case of the electrostatic correction, the total pressure is of course again determined by the hydrostatic equilibrium. Only the relative importance of the contributions from ions and electrons changes when partial electron degeneracy is included into the model.

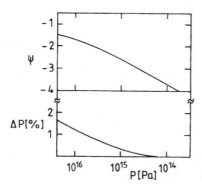

Fig. 2.3. Electron degeneracy in the solar core. *Upper curve*: degeneracy parameter; *lower curve*: pressure correction

We have seen that the equation of state in a solar model is an implicit equation for the density, and we have considered two corrections to the perfect gas law in some detail. But there are further corrections, all arising from particle interactions other than collisions. The already mentioned partition functions are an example. More and more details of two-particle and multi-particle interactions have been included, and the results are available in the form of tables $\rho(P,T)$. Such tables are commonly computed together with the opacity tables (Sect. 2.3.7), for which many detailed atomic processes must also be considered. The MHD and OPAL tables are most frequently used (Hummer and Mihalas 1988, Mihalas et al. 1988; Iglesias and Rogers 1996).

2.3.5 The Entropy

We shall now consider the second "equation of state", which determines the entropy. Only *changes* of the entropy are of interest, either in time [cf. (2.12)], or in depth – as we shall see when we treat the theory of stellar convection in Chap. 6. Hence the differential form common in thermodynamics is adequate, but in place of the extensive variables U and V (internal energy and volume) the intensive variables P and T are used. We write

$$dS = c_P \left(\frac{dT}{T} - \nabla_a \frac{dP}{P} \right) , \quad \text{where} \tag{2.47}$$

$$c_P = T \left(\frac{\partial S}{\partial T} \right)_P \quad \text{and} \tag{2.48}$$

$$\nabla_a = \left(\frac{\partial \ln T}{\partial \ln P} \right)_S \tag{2.49}$$

are the specific heat at constant pressure and the *adiabatic temperature gradient* (more precisely: the double-logarithmic isentropic temperature gradient). In order to obtain the entropy we must determine c_P and ∇_a as functions of P and T. We have already, in the preceding section, considered the function $\rho(P,T)$ and thus can derive

$$\delta = - \left(\frac{\partial \ln \rho}{\partial \ln T} \right)_P . \tag{2.50}$$

Let us again neglect the radiation pressure and consider only the perfect gas term in the equation of state; then

$$\delta = 1 - \left(\frac{\partial \ln \mu}{\partial \ln T} \right)_P , \tag{2.51}$$

where $\mu(P,T)$ is given by the ionization equilibrium (2.37). Using (2.50) we may express the adiabatic gradient in terms of the specific heat (cf. Problem 2.12):

$$\nabla_a = \frac{P\delta}{T\rho c_P} \, . \tag{2.52}$$

In order to find c_P itself, we use the definition (2.48) and the basic relation $T dS = dU + P dV$, where U is the specific internal energy and $V = 1/\rho$ is the specific volume (both per unit mass, in contrast to (2.42), where U_{ES} was an energy per unit volume). Then

$$c_P = \left(\frac{\partial U}{\partial T}\right)_P + \frac{P\delta}{\rho T} \, , \tag{2.53}$$

where again (2.50) has been used.

It remains to determine the specific internal energy U as a function of P and T. Its main contributions are the kinetic energy of free particles and the energy of ionization. If only hydrogen and helium are considered, then

$$U = \frac{3\mathcal{R}T}{2\mu} + \frac{1}{\rho}[n_{H^+}\chi_H + n_{He^+}\chi_{He} + n_{He^{++}}(\chi_{He} + \chi_{He^+})] \, . \tag{2.54}$$

The ionization energies, $\chi_H \ldots$, and the densities, $n_{H^+} \ldots$, of the various ions have been defined in the preceding section. The temperature dependence of the ion densities is defined by means of the three Saha equations (2.37).

Figure 2.4 shows c_P and ∇_a for a calculated solar model. Deviations from the perfect gas expressions $c_P = 5\mathcal{R}/2\mu$ and $\nabla_a = 2/5$ mainly occur in the layer where H and He are partially ionized, because there the degrees of ionization, which enter (2.54), depend on temperature. The main excursion is caused by the ionization of H and the first ionization of He, while the small hump near $P = 10^9$ Pa is due to the second ionization of He.

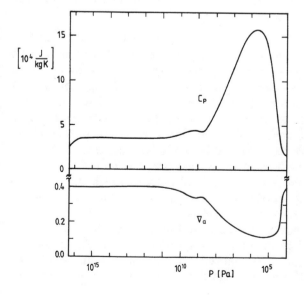

Fig. 2.4. Specific heat c_P and adiabatic temperature gradient ∇_a in the solar interior

Problem 2.12. Use the integrability condition for the enthalpy in order to prove (2.52).

Problem 2.13. Calculate the corrections to c_P and ∇_a which arise from the electrostatic interaction in the Debye–Hückel treatment.

2.3.6 Nuclear Energy Sources

In this section we shall treat the energy generation ε defined in (2.12). Since we already know the density ρ as a function of P and T, we may determine $\varepsilon(\rho, T)$; as will be clear presently, this is more natural than $\varepsilon(P, T)$ would be.

When it became clear that the Earth (and by implication the Sun) was much older than the Kelvin–Helmholtz time, energy sources other than gravitation or cooling had to be studied. Atkinson and Houtermans (1929) were among the first to consider nuclear energy, in particular the energy set free by the conversion of hydrogen to helium. Of all nuclear energy sources this "hydrogen burning" is the only lasting possibility for a star like the Sun. First of all, hydrogen is by far the most abundant element. Second, the mass defect of the helium nucleus – compared to the four constituent nucleons – is much larger than the mass defect of all other nuclei compared to their building blocks; therefore the energy release, 6.683 MeV per nucleon, is the largest available among all nuclear reactions. Third, the hydrogen nucleus has the smallest possible electric charge; the electrostatic barrier which must be overcome (or tunneled through) before a nuclear reaction can occur is therefore not as prohibitive as for heavier nuclei. Even at the high temperature of 1.5×10^7 K in the Sun's center this is a very important point: the thermal particle energy kT is only ≈ 1.3 keV, while the height of the electrostatic barrier generally is of order 1 MeV!

Hydrogen-Burning Reactions. The most important reactions in the Sun are those of the *proton-proton chains*, ppI, ppII, and ppIII. Let us use the notation $X(a, b)Y$, common in nuclear physics, where X is a target, a an incident particle, b an emitted particle (two particles may also appear in place of a or b), and Y is the residual nucleus. Thus, the reactants are before the comma, the resulting particles are after the comma. The hydrogen isotopes are designated by ^{1}H or p and ^{2}H or d, helium isotopes are ^{4}He or α and ^{3}He, and the other particles have their usual abbreviations. A superscript * means a nucleus in an excited state. The energy release per reaction is $Q = Q' + Q_\nu$, where Q' is the part delivered to the thermal bath in which the reactions occur, and Q_ν is the energy carried away by a neutrino. The pp chains are then described by the scheme shown in Table 2.1.

The ppII chain occurs as an alternative to the third reaction of ppI, and ppIII occurs as an alternative to the second reaction of ppII. The branching ratios depend on the reaction *rates*, and therefore on temperature. According to model calculations for the present Sun the global ratios are 85.2 : 14.8 for

Table 2.1. Nuclear reactions of the pp chains. Energy values according to Bahcall and Ulrich (1988) and Caughlan and Fowler (1988)

	Reaction	Q'[MeV]	Q_ν[MeV]	Rate symbol
ppI	$p(p,e^+\nu)d$	1.177	0.265	λ_{pp}
	$d(p,\gamma)^3He$	5.494		λ_{pd}
	$^3He(^3He,2p)\alpha$	12.860		λ_{33}
ppII	$^3He(\alpha,\gamma)^7Be$	1.586		λ_{34}
	$^7Be(e^-,\nu\gamma)^7Li$	0.049	0.815	λ_{e7}
	$^7Li(p,\alpha)\alpha$	17.346		λ'_{17}
ppIII	$^7Be(p,\gamma)^8B$	0.137		λ_{17}
	$^8B(,e^+\nu)^8Be^*$	8.367	6.711	λ_8
	$^8Be^*(,\alpha)\alpha$	2.995		λ'_8

the I : (II+III) branching, and 14.8 : 0.019 for the II : III branching (Problem 2.19).

The total energy release per produced α particle is 26.732 MeV in each case – note that the first two reactions must be counted twice for each α particle synthesized in ppI. The neutrino energies Q_ν (and hence also the corresponding Q' values) are averages. The averages are over continua for the p+p reaction and for the β-decay of ^8B, and over two lines at 384 keV and 862 keV for the electron capture of ^7Be. The neutrino spectrum will be considered again in Sect. 2.4.2 below.

Besides the pp chains, the *CNO cycle* is of some importance for the solar energy generation. In fact, when the reactions of the CNO cycle were first investigated (von Weizsäcker 1938; Bethe 1939), available solar models had central temperatures as high as 1.9×10^7 K, which made the energy generation of these reactions dominant. In contrast, present calculations show that this is true only for more massive main-sequence stars, while for the Sun the

Table 2.2. Nuclear reactions of the CNO cycle. Energy values according to Bahcall and Ulrich (1988) and Caughlan and Fowler (1988)

Reaction	Q'[MeV]	Q_ν[MeV]	Rate symbol
$^{12}C(p,\gamma)^{13}N$	1.944		λ_{p12}
$^{13}N(,e^+\nu)^{13}C$	1.513	0.707	λ_{13}
$^{13}C(p,\gamma)^{14}N$	7.551		λ_{p13}
$^{14}N(p,\gamma)^{15}O$	7.297		λ_{p14}
$^{15}O(,e^+\nu)^{15}N$	1.757	0.997	λ_{15}
$^{15}N(p,\alpha)^{12}C$	4.966		λ_{p15}

pp reactions, studied in detail by Bethe and Critchfield (1938), yield about 98.8% of the total energy.

The main characteristic of the CNO cycle is that isotopes of C, N, and O participate in the reactions, and are transmuted into each other, but are themselves not produced from lighter nuclei: they thus function like catalysts. The CNO cycle operates according to the scheme shown in Table 2.2.

The total energy release per produced α particle is again 26.732 MeV. The neutrino energies are averages over the continuous spectra that result from the β decay of ^{13}N and ^{15}O.

As an alternative to the last reaction of Table 2.2, ^{15}N may undergo a further (p,γ) reaction and thus initiate a secondary cycle. However, the ratio for this branch is less than 10^{-3} and is negligible for the Sun which, in any case, generates only about 1.2% of its energy needs by the whole CNO cycle. Since the most abundant isotope of oxygen, ^{16}O, is not involved in the fundamental loop, the cycle in this form is often called the "CN cycle".

Equations of Evolution. Nuclear reactions will occur in proportion to the densities of the particles involved. If two reactants are identical then for this species the density change will be proportional to the *square* of the density. We denote the number densities of the particles reacting in the pp chains and (or) in the CNO cycle by n_p, n_d, n_3, n_4, $n_{7\mathrm{Li}}$, $n_{7\mathrm{Be}}$, $n_{8\mathrm{B}}$, $n_{8\mathrm{Be}}$, and by $n_{12\mathrm{C}}$, $n_{13\mathrm{C}}$, $n_{13\mathrm{N}}$, $n_{14\mathrm{N}}$, $n_{15\mathrm{N}}$, $n_{15\mathrm{O}}$. The constants of proportionality, i.e., the reaction rates, will be denoted by the symbol λ with appropriate subscripts, as listed in Tables 2.1 and 2.2. This notation, essentially following Clayton (1968), indicates the reactants but is still short enough for a concise representation. The 15 reactions, then, give rise to the following density changes:

$$\dot{n}_p = -\lambda_{pp}n_p^2 - \lambda_{pd}n_p n_d + \lambda_{33}n_3^2$$
$$-\lambda'_{17}n_p n_{7\mathrm{Li}} - \lambda_{17}n_p n_{7\mathrm{Be}} - \lambda_{p12}n_p n_{12\mathrm{C}} \qquad (2.55)$$
$$-\lambda_{p13}n_p n_{13\mathrm{C}} - \lambda_{p14}n_p n_{14\mathrm{N}} - \lambda_{p15}n_p n_{15\mathrm{N}} \ ,$$

$$\dot{n}_d = \lambda_{pp}n_p^2/2 - \lambda_{pd}n_p n_d \ , \qquad (2.56)$$

$$\dot{n}_3 = \lambda_{pd}n_p n_d - \lambda_{33}n_3^2 - \lambda_{34}n_3 n_4 \ , \qquad (2.57)$$

$$\dot{n}_4 = \lambda_{33}n_3^2/2 - \lambda_{34}n_3 n_4 + 2\lambda'_{17}n_p n_{7\mathrm{Li}} + 2\lambda'_8 n_{8\mathrm{Be}} + \lambda_{p15}n_p n_{15\mathrm{N}} \ , \qquad (2.58)$$

$$\dot{n}_{7\mathrm{Be}} = \lambda_{34}n_3 n_4 - \lambda_{e7}n_e n_{7\mathrm{Be}} - \lambda_{17}n_p n_{7\mathrm{Be}} \ , \qquad (2.59)$$

$$\dot{n}_{7\mathrm{Li}} = \lambda_{e7}n_e n_{7\mathrm{Be}} - \lambda'_{17}n_p n_{7\mathrm{Li}} \ , \qquad (2.60)$$

$$\dot{n}_{8\mathrm{B}} = \lambda_{17}n_p n_{7\mathrm{Be}} - \lambda_8 n_{8\mathrm{B}} \ , \qquad (2.61)$$

$$\dot{n}_{8\mathrm{Be}} = \lambda_8 n_{8\mathrm{B}} - \lambda'_8 n_{8\mathrm{Be}} \ , \qquad (2.62)$$

$$\dot{n}_{12\mathrm{C}} = \lambda_{p15}n_p n_{15\mathrm{N}} - \lambda_{p12}n_p n_{12\mathrm{C}} \ , \qquad (2.63)$$

$$\dot{n}_{13\mathrm{N}} = \lambda_{\mathrm{p}12} n_{\mathrm{p}} n_{12\mathrm{C}} - \lambda_{13} n_{13\mathrm{N}} \ , \tag{2.64}$$

$$\dot{n}_{13\mathrm{C}} = \lambda_{13} n_{13\mathrm{N}} - \lambda_{\mathrm{p}13} n_{\mathrm{p}} n_{13\mathrm{C}} \ , \tag{2.65}$$

$$\dot{n}_{14\mathrm{N}} = \lambda_{\mathrm{p}13} n_{\mathrm{p}} n_{13\mathrm{C}} - \lambda_{\mathrm{p}14} n_{\mathrm{p}} n_{14\mathrm{N}} \ , \tag{2.66}$$

$$\dot{n}_{15\mathrm{O}} = \lambda_{\mathrm{p}14} n_{\mathrm{p}} n_{14\mathrm{N}} - \lambda_{15} n_{15\mathrm{O}} \ , \tag{2.67}$$

$$\dot{n}_{15\mathrm{N}} = \lambda_{15} n_{15\mathrm{O}} - \lambda_{\mathrm{p}15} n_{\mathrm{p}} n_{15\mathrm{N}} \ . \tag{2.68}$$

The factors $1/2$ and 2 appear in these equations whenever two particles of the same species are consumed or produced in a single reaction. The changes of ^{7}Be and ^{7}Li are in part due to electron capture of ^{7}Be and hence depend on the electron density n_e. Since the nuclear reactions occur in the hottest, central part of the Sun, we may use the expression for complete ionization, $n_\mathrm{e} = \rho N_\mathrm{A}(1 + X)/2$ (cf. Sect. 2.3.4).

The sum of the right-hand sides of (2.63) to (2.68) is zero. Of course this expresses the conservation of the total number of CNO nuclei.

Products of densities and the λ coefficients in (2.55) to (2.68) may be understood as inverse lifetimes. For example, $(\lambda_{\mathrm{pd}} n_\mathrm{d})^{-1}$ is the lifetime of protons against reactions with deuterons, $(\lambda_{\mathrm{pd}} n_\mathrm{p})^{-1}$ is the lifetime of deuterons against reactions with protons, and $(\lambda_{\mathrm{p}12} n_{12\mathrm{C}})^{-1}$ is the lifetime of protons against reactions with ^{12}C nuclei. In the case of β decay, i.e., λ_8, λ_{13} and λ_{15}, and in the case of the ^{8}Be* decay into two α particles, i.e., λ_8', the respective λ's are themselves inverse lifetimes.

Extreme differences between the various lifetimes are common. The β decays of ^{8}B, ^{13}C, and ^{15}O take only 0.8 s, 598 s, and 122 s, respectively. On the other hand, under the conditions in the Sun's center, the lifetime of ^{14}N against reactions with protons is about 10^9 years, and the lifetime of protons against protons, i.e., $(\lambda_{\mathrm{pp}} n_\mathrm{p})^{-1}$, is of order 10^{10} years. The proton-proton reaction is so slow because it involves the weak interaction of a β^+ decay of a proton into a neutron. Due to the temperature and density variations within the Sun, and the strong temperature dependence of the reaction rates (see below), wide differences occur even for single reactions at various depths in the Sun. An important example is ^{3}He which, at the Sun's center, is destroyed by reactions with itself and with ^{4}He in less than 10^6 years, but has a lifetime of order 10^9 years at $T \approx 7 \times 10^6$ K, about one-quarter of the solar radius away from the center.

We may utilize the large differences in the lifetimes to considerably simplify the reaction equations: deuterium sets itself into equilibrium in a matter of seconds; that is, the two terms on the right of (2.56) essentially cancel each other (each of them being large in comparison to the left-hand side!) and we may effectively write $\dot{n}_\mathrm{d} = 0$. Equally, the short lifetimes of ^{7}Be and ^{7}Li (of order 1 year) lead to rapid equilibrium of these nuclei, and to $\dot{n}_{7\mathrm{Be}} = 0$ and $\dot{n}_{7\mathrm{Li}} = 0$. Finally, the rapid decay of ^{8}B and of ^{8}Be* ensures $\dot{n}_{8\mathrm{B}} = 0$ and $\dot{n}_{8\mathrm{Be}} = 0$. Of course we must not confuse the neglect of all these time derivatives with an exact stationary state. The densities of d, ^{7}Be, etc., *do* change, but very slowly so as they continuously adjust their equilibrium densities to the species that vary less rapidly. We must also keep in mind that – except for the simple decay reactions – the rates depend on density and temperature, and hence on the solar model itself. After the calculation of a model it is therefore necessary in principle to check whether the approximations made were in fact permitted.

As a further simplification we shall now *assume* equilibrium also for the entire CNO cycle, although further below we shall discuss a model calculated without that assumption. The equilibrium of the CNO cycle is certainly not yet established at some distance from the Sun's center: at $T = 10^7$ K, for example, the lifetime of ^{14}N against reactions with protons is of order 10^{12} years. Nevertheless, the consequences for the solar energy generation are small because the CNO cycle contributes only a small fraction, and most of this comes from regions very close to the center. The assumption of CNO in equilibrium means $\dot{n}_{^{12}C} = \dot{n}_{^{13}N} = \ldots = \dot{n}_{^{15}N} = 0$.

Most of the unknown densities can now be eliminated, and we are left with the evolution in time of n_p, n_3, and n_4. The result (Problem 2.14) is:

$$\dot{n}_p = -\frac{3}{2}\lambda_{pp}n_p^2 + \lambda_{33}n_3^2 - \lambda_{34}n_3n_4 - 4\lambda_{p14}n_pn_{^{14}N} , \tag{2.69}$$

$$\dot{n}_3 = \frac{1}{2}\lambda_{pp}n_p^2 - \lambda_{33}n_3^2 - \lambda_{34}n_3n_4 , \tag{2.70}$$

$$\dot{n}_4 = \frac{1}{2}\lambda_{33}n_3^2 + \lambda_{34}n_3n_4 + \lambda_{p14}n_pn_{^{14}N} . \tag{2.71}$$

During nuclear reactions the *baryon number* is conserved. We may here neglect the small differences in total mass between the reactants and the reaction products (of course this must not be done when the energy release is calculated!). That is, we may consider the changes $\dot{n}_p$, $\dot{n}_3$, $\dot{n}_4$ as derivatives at constant mass density ρ. This is consistent with the general assumption (cf. Chap. 1) that the Sun's mass remains constant in spite of the mass loss due to hydrogen burning. With the substitution $n_p = \rho X/m_p$, $n_3 = \rho Y_3/m_3$, etc., we then obtain equations for the mass fractions X, Y_3, Y of H, ^{3}He, and ^{4}He:

$$\dot{X} = \frac{A_p}{N_A}(-3r_{pp} + 2r_{33} - r_{34} - 4r_{p14}) , \tag{2.72}$$

$$\dot{Y}_3 = \frac{A_3}{N_A}(r_{pp} - 2r_{33} - r_{34}) , \tag{2.73}$$

$$\dot{Y} = \frac{A_4}{N_A}(r_{33} + r_{34} + r_{p14}) , \tag{2.74}$$

where, for the reaction of species i and k,

$$r_{ik} = \frac{n_in_k}{\rho(1 + \delta_{ik})}\lambda_{ik} = \frac{\rho N_A^2 X_iX_k}{(1 + \delta_{ik})A_iA_k}\lambda_{ik} \tag{2.75}$$

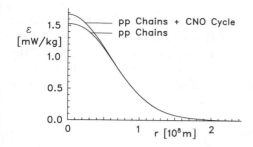

Fig. 2.5. Nuclear energy generation in the Sun

is now the reaction rate per *unit mass*; the $X_i = X$, Y_3, ... are the mass fractions, and the A_i are the atomic weights. The time derivatives in (2.72) to (2.74) are for fixed mass shells, i.e. Lagrangian derivatives (cf. Problem 2.15). Notice that for the CNO cycle in equilibrium $n_{14N} \gg n_{12C}$, so that $X_{14} \simeq X_C + X_N$, where X_C and X_N are the initial mass fractions of carbon and nitrogen.

Using the rates (2.75), we may now write down the energy generation per unit mass:

$$\varepsilon = \sum Q'_{ik} r_{ik} , \tag{2.76}$$

where the sum is over all the reactions. Due to the assumed equilibria of deuterium, ^{7}Be and ^{7}Li, etc., several of the r_{ik} coincide, and their energies Q'_{ik} may therefore be combined into a single term.

Figure 2.5 shows ε as a function of r in the Sun. Almost all of the energy is generated within 25% of the radius (i.e., 1.5% of the volume). Even more concentrated towards the center is the small contribution of $\approx 1.2\%$ made by the CNO cycle.

Problem 2.14. Confirm (2.69) to (2.71), and determine, as functions of n_p, n_3, n_4 and of the reaction rates, the number densities of all other reactants. Find the relationships between the rates r_{ik}; note that for simple decay reactions the definition of the rate r differs from (2.75).

Problem 2.15. Show that the time derivatives in equations (2.72) to (2.74) are Lagrangian derivatives, and that these equations remain correct when expansion or contraction of the star is allowed. The three equations slightly violate the condition $\dot{X} + \dot{Y}_3 + \dot{Y} = 0$. Often this is repaired by first calculating $\dot{Y}_3$ and $\dot{Y}$ according to (2.73) and (2.74), and then using $\dot{X} = -\dot{Y}_3 - \dot{Y}$, or by simply using the integer values 1, 3, and 4 for A_p, A_3, and A_4. Take the accurate mass balance of the nuclear reactions into account and find correct expressions for $\dot{X}$, $\dot{Y}_3$, and $\dot{Y}$.

Problem 2.16. The CNO cycle in the Sun has not yet reached full equilibrium. Use the simplifications suggested by the fast β decay of ^{13}N and ^{15}O and the reasonably short lifetime of ^{15}N, and write down the equations of chemical evolution and energy generation.

Reaction Rates. The rate at which a nuclear reaction takes place depends
on the relative velocity $v = |\boldsymbol{v}_i - \boldsymbol{v}_k|$ of the reactants, on their distribution
in velocity space, and on the cross section $\sigma(v)$, i.e., the probability that the
reaction occurs at a given particle flux density. If $dn_i(\boldsymbol{v}_i)$ and $dn_k(\boldsymbol{v}_k)$ are
the distributions of the two reactants with velocities $\boldsymbol{v}_i$ and $\boldsymbol{v}_k$, then the rate
per unit mass is

$$r_{ik} = \frac{1}{\rho(1 + \delta_{ik})} \iint v\sigma(v)\, dn_i dn_k \; , \tag{2.77}$$

where the integrals are over the two velocity spaces. In the stellar interior
the conditions of local thermodynamic equilibrium are satisfied to a very
high accuracy; the velocity distributions are therefore Maxwellian. The cross
section depends only on the *relative* velocity $\boldsymbol{v}$, but not on the velocity $\boldsymbol{V}$
of the center of gravity. We may therefore transform (2.77) into an integral
over the $\boldsymbol{v}$ and $\boldsymbol{V}$ spaces, and integrate over $\boldsymbol{V}$. Introducing the energy $E =
mv^2/2$, where $m = m_i m_k/(m_i + m_k)$ is the reduced mass, substituting the
Maxwellian distributions, and using (2.75) we find

$$\lambda_{ik} = \frac{(8/m\pi)^{1/2}}{(kT)^{3/2}} \int\limits_0^\infty \exp(-E/kT) E\sigma(E)\, dE \; . \tag{2.78}$$

The remaining integral can be evaluated most easily when the target
nucleus has a *resonance* at a particular energy $E_{\rm R}$, i.e., when $\sigma(E)$ is large
in a close neighborhood of $E_{\rm R}$. We may then treat σ as if it was a δ function
and simply set $E = E_{\rm R}$ in the other factors. For such reaction rates, the
dependence on temperature is of the characteristic form

$$\lambda_{ik} \propto T^{-3/2} \exp(-E_{\rm R}/kT) \; . \tag{2.79}$$

For the Sun resonant reaction rates are of minor importance, although reso-
nances do exist for the p reactions of ^{7}Li, ^{7}Be, ^{12}C, ^{13}C, ^{14}N and ^{15}N. Even at
the temperature of the solar center the exponential factor in (2.79) is so small
that these contributions are negligible; hence they are omitted in Table 2.3.

For non-resonant reactions the cross section $\sigma(E)$ is determined by a
combination of quantum-mechanical calculation and laboratory experiment.
First, one finds

$$\sigma(E) = \frac{1}{E} S(E) \exp\left(-\sqrt{\frac{m}{2E}} \frac{Z_i Z_k e^2 \pi}{\varepsilon_0 h}\right) \; . \tag{2.80}$$

It is plausible that the cross section is proportional to the square of the particle wavelength, $\lambda_{\rm p}$, which in quantum mechanics is related to the energy by
$\lambda_{\rm p}^2 = h^2/2mE$. This explains the $1/E$ factor in (2.80). The exponential factor
is a consequence of the electrostatic barrier between the two reactants with
charges $Z_i e$ and $Z_k e$. The wave function of a particle tunneling through such

a barrier is exponential rather than oscillatory. The probability of penetration through the barrier is therefore exponentially decreased.

For non-resonant reactions the cross-section factor $S(E)$ is a slowly varying function of E. On the other hand the two exponentials in (2.78) and (2.80) – the Maxwell–Boltzmann factor and the penetration factor – together form a sharply peaked function of the energy, the *Gamow peak*, named after G. Gamow who considered the tunnel effect in nuclear reactions for the first time in 1928. Introducing

$$b = \sqrt{m/2} Z_i Z_k e^2 \pi / \varepsilon_0 h$$

we find the maximum at

$$E_{\max} = (bkT/2)^{2/3} \; ;$$

its height is $\exp(-3E_{\max}/kT)$.

Problem 2.17. Approximate the Gamow peak by a Gaussian, and show that its full width at $1/e$ of the maximum is $4(E_{\max}kT/3)^{1/2}$. Calculate $E_{\max}$ and the width of the Gamow peak for the rates which appear in (2.69) to (2.71) at $T = 1.55 \times 10^7$ K, the temperature of the Sun's center.

When the Gamow peak is approximated by a Gaussian, and the constant $S(E_{\max})$ is used for the cross-section factor, the integral over E in (2.78) can be carried out. The result is

$$\lambda_{ik} = 4S(E_{\max}) \frac{(2/3m)^{1/2}(b/2)^{1/3}}{(kT)^{2/3}} \exp\left(-\frac{3(b/2)^{2/3}}{(kT)^{1/3}}\right) , \qquad (2.81)$$

and shows the characteristic temperature dependence of non-resonant reactions: $T^{-2/3}$ in front and $-T^{-1/3}$ in the exponent.

The rates listed in Table 2.3 have additional T-dependent factors of order 1. These corrections arise from the not exactly Gaussian shape of the Gamow peak.

The cross-section factor $S(E)$ itself is determined in laboratory experiments. The difficulty is that only at large energies of, say, several hundred keV, are the rates sufficiently large for a measurement of some precision. An extrapolation down to $E_{\max}$ (which is of order 20 keV, cf. Problem 2.17) is therefore necessary. As an example, the cross-section factor for the reaction ^{3}He$(\alpha,\gamma)^7$Be is shown in Fig. 2.6.

Table 2.3 gives a list of the rates that are most important for the Sun. They are all of the non-resonant form (2.81), with the exception of λ_{e7}, the rate for the electron capture of ^{7}Be. This reaction is of course not impeded by an electrostatic barrier. Only ionization of ^{7}Be lengthens its lifetime against electron capture. For $T < 10^6$ K, the rate $\lambda_{e7} = 1.51 \times 10^{-7}/n_e$ must therefore be used, in correspondence to the half-life of 53.3 days measured in the laboratory for not completely ionized ^{7}Be.

Table 2.3. Nuclear reaction rates for the Sun, $N_A\lambda$, in reactions per second and per (mole/cm^3), after Caughlan and Fowler (1988). T_9 is the temperature in units of 10^9 K

---

^{1}H(p,e$^+\nu$)d $\quad N_A\lambda_{\rm pp} = 4.01 \times 10^{-15}\, T_9^{-2/3} \exp(-3.380\, T_9^{-1/3})$
$$\times (1 + 0.123\, T_9^{1/3} + 1.09\, T_9^{2/3} + 0.938\, T_9)$$

^{3}He(^{3}He,2p)α $\quad N_A\lambda_{33} = 6.04 \times 10^{10}\, T_9^{-2/3} \exp(-12.276\, T_9^{-1/3}) \times (1 + 0.034\, T_9^{1/3}$
$$-0.522\, T_9^{2/3} - 0.124\, T_9 + 0.353\, T_9^{4/3} + 0.213\, T_9^{5/3})$$

^{3}He(α,γ)^{7}Be $\quad N_A\lambda_{34} = 5.61 \times 10^6 a^{-5/6}\, T_9^{-2/3} \exp(-12.826 a^{1/3} T_9^{-1/3}),$
$$\text{where } a = 1 + 0.0495\, T_9$$

^{7}Be(e$^-,\nu\gamma$)^{7}Li $\quad N_A\lambda_{\rm e7} = 1.34 \times 10^{-10}\, T_9^{-1/2}[1 - 0.537\, T_9^{1/3}$
$$+3.86\, T_9^{2/3} + 0.0027\, T_9^{-1} \exp(2.515 \times 10^{-3}/T_9)]^*$$

^{7}Li(p,α)α $\quad N_A\lambda'_{17} = 1.096 \times 10^9\, T_9^{-2/3} \exp(-8.472\, T_9^{-1/3})$
$$-4.830 \times 10^8 a^{-5/6}\, T_9^{-2/3} \exp(-8.472 a^{1/3} T_9^{-1/3}),$$
$$\text{where } a = 1 + 0.759\, T_9$$

^{7}Be(p,γ)^{8}B $\quad N_A\lambda_{17} = 3.11 \times 10^5\, T_9^{-2/3} \exp(-10.262\, T_9^{-1/3})$

^{14}N(p,γ)^{15}O $\quad N_A\lambda_{\rm p14} = 4.90 \times 10^7\, T_9^{-2/3} \exp(-15.228\, T_9^{-1/3}) \times (1 + 0.027\, T_9^{1/3}$
$$-0.778\, T_9^{2/3} - 0.149\, T_9 + 0.261\, T_9^{4/3} + 0.127\, T_9^{5/3})$$

---

* $\lambda_{\rm e7}$ must not exceed $1.51 \times 10^{-7}/n_{\rm e}$ (see text)

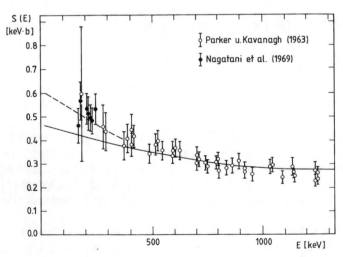

Fig. 2.6. Cross-section factor $S(E)$ for the reaction ^{3}He(α,γ)^{7}Be, in keV$\times$barn (1 barn $= 10^{-28}$ m^2)

Weak Electron Screening. The electrostatic barrier is slightly lowered by the formation of an electron "cloud" around the positively charged nuclei. The resulting *electron screening* leads to an increase of the reaction rates by a factor

$$f = 1 + \frac{Z_i Z_k e^2}{4\pi\varepsilon_0 r_D kT} \,, \tag{2.82}$$

where $Z_i e$ and $Z_k e$ are the charges of the two reactants and r_D is the Debye radius, defined in (2.41). The result (2.82), again valid only in the framework of the Debye–Hückel treatment, has been applied in the code which was used to calculate the solar model of the present chapter.

2.3.7 The Opacity

As the last ingredient required for the solution of the equations of stellar structure, (2.8), (2.9), (2.12), and (2.19), we must now determine the opacity κ as a function of variables which are already known. As in the case of ε, we choose ρ and T as independent variables for κ.

The opacity is an average over frequency ν; the dependence on ν must therefore be determined first. This is done by a calculation of the probability that a photon of energy $h\nu$ is absorbed or scattered by individual atoms, ions, or electrons. A time-dependent quantum-mechanical perturbation method is generally used, and the results are given in form of cross sections for the diverse processes. The present section is a short summary of such results.

Bound–Bound Absorption. Photons with a discrete frequency ν can be absorbed when an atom or ion undergoes a transition between two states of energy difference $h\nu$. In analogy to classical oscillators the cross section per atom is written in the form

$$\sigma_{bb}(\nu) = \frac{e^2}{4\varepsilon_0 m_e c} f \phi(\nu) \,, \tag{2.83}$$

where m_e is the electron mass; f is called the *oscillator strength* and contains the calculated transition probability; $\phi(\nu)$ is a normalized shape function of the absorption, or the *line profile*. The two most important profiles are the Gaussian Doppler broadening that arises from the Maxwellian velocity distribution of the absorbing particles,

$$\phi_D(\Delta\nu) = \frac{1}{\sqrt{\pi}\Delta\nu_D} \exp[-(\Delta\nu/\Delta\nu_D)^2] \,, \tag{2.84}$$

and the Lorentz profile due to collision broadening,

$$\phi_C(\Delta\nu) = \frac{\gamma}{(2\pi\Delta\nu)^2 + \gamma^2/4} \,. \tag{2.85}$$

Here $\Delta\nu = \nu - \nu_0$ is the frequency distance from the line center. The widths of the profiles are determined by the constant of collisional damping γ (or twice the effective collision frequency), and by the Doppler width

$$\Delta\nu_D = (2\mathcal{R}T/A)^{1/2}\nu_0/c .$$

The Doppler profile dominates in the core of the line, while the slowly decaying damping profile is characteristic for the wings. In general, both Doppler and collision broadening act together, and the two functions must be folded. We shall treat the resulting line profile in more detail in the context of line formation in the solar atmosphere (Chap. 4). For the opacity in the Sun's interior, line broadening is especially important because κ is a harmonic mean where not the lines themselves but the *windows* between the lines contribute. For good transparency windows must be kept clean.

Bound–Free Absorption. Bound-free absorption, or photoionization, can occur when the photon energy exceeds the ionization energy χ of an atom or ion. For hydrogen, or for hydrogen-like atoms and ions, $\chi = m_e e^4 Z^2/8(\varepsilon_0 h n)^2$, where Z is an effective charge number and n is the principal quantum number. The cross section per atom in state n is

$$\sigma_{\mathrm{bf}}(\nu) = \frac{m_e e^{10} Z^4 g_{\mathrm{bf}}(n,\nu)}{48\sqrt{3}\pi\varepsilon_0^5 ch^6 n^5 \nu^3} , \tag{2.86}$$

where the Gaunt factor g_{bf}, which only weakly depends on n and ν, results from the quantum-mechanical calculation of the ionization probability. For each atom (or ion) and state considered, (2.86) is valid for $\nu > \chi/h$, while $\sigma_{\mathrm{bf}} = 0$ for smaller ν. The characteristic ionization edges and adjacent ν^{-3} continua can be seen in the example shown in Fig. 2.7.

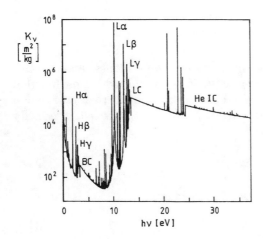

Fig. 2.7. Frequency-dependent absorption coefficient for a mixture with $X = 0.75$, $Y = 0.2321$, and $Z = 0.0179$, at $kT = 1.25\,\mathrm{eV}$ and $\rho = 1.88 \times 10^{-4}\,\mathrm{kg/m^3}$. BC, LC, and He IC are the Balmer, Lyman, and He I continua, showing the typical ν^{-3}-dependence. Adapted from Huebner (1986)

Free–Free Absorption. A *free* electron can also absorb a photon, but a *third particle* must be present: otherwise energy and momentum could not simultaneously be conserved. The probability, or cross section, for this process – also called inverse bremsstrahlung – depends on the distribution of electrons in velocity space. If there are $dn_e(v)$ electrons with velocities between v and $v + dv$ (relative to the third particle), then the differential cross section for hydrogen-like atoms or ions per atom (or ion) is

$$d\sigma_{ff}(v, \nu) = \frac{e^6 Z^2 g_{ff}(v, \nu)}{48\sqrt{3}\pi^2 \varepsilon_0^3 chvm_e^2 \nu^3} dn_e(v) . \tag{2.87}$$

As we have seen in Sect. 2.3.4, partial degeneracy of electrons plays only a minor role in the Sun. We may therefore use the Maxwellian distribution

$$dn_e = n_e \left(\frac{m_e}{2\pi kT}\right)^{3/2} \exp\left(-\frac{m_e v^2}{2kT}\right) 4\pi v^2 dv . \tag{2.88}$$

The total cross section is then obtained as the integral over (2.87), cf. Problem 2.18:

$$\sigma_{ff}(\nu) = \frac{e^6 Z^2 n_e \overline{g_{ff}}}{24\pi^2 \varepsilon_0^3 chm_e^{3/2} (6\pi kT)^{1/2} \nu^3} . \tag{2.89}$$

The main task is again the calculation of the wave functions, the transition probability, and then the Gaunt factor g_{ff}. Free-free processes are possible at any photon energy; hence the cross section is a continuous function of frequency.

Problem 2.18. How is the average Gaunt factor $\overline{g_{ff}}$ in (2.89) defined?

Scattering by Electrons. Thomson scattering of photons by free electrons is independent of photon energy. This process can be classically treated: the electron is set into oscillatory motion by the electric field of the incident wave. The cross section is then calculated as the ratio of radiation emitted per unit time and incoming energy flux density. The result (per electron) is the constant

$$\sigma_s = \frac{e^4}{6\pi \varepsilon_0^2 m_e^2 c^4} \approx 6.65 \times 10^{-29}\,\text{m}^2 , \tag{2.90}$$

or $\sigma_s = 8\pi r_0^2/3$, where $r_0 = e^2/4\pi\varepsilon_0 m_e c^2 \approx 2.818 \times 10^{-15}$ m is the "classical electron radius".

An important correction to (2.90) arises from collective effects; i.e., the electrostatic interaction of ions and electrons already discussed in the context of the equation of state. For photons with a wavelength comparable or larger than the Debye radius (2.40) the probability to be scattered becomes smaller. The total opacity reduction in the Sun's core, where electron scattering is most important, may then reach a few per cent (Bahcall et al. 1982,

Bahcall and Ulrich 1988). This may appear little; but, according to (2.19), the opacity governs the temperature gradient, and thus a reduction of κ leads to a smaller central temperature. As we shall see in Sect. 2.4.2 (Problem 2.19), the branching of the pp chains and hence the production of energetic neutrinos by ^{8}B decay is extremely sensitive to changes of temperature.

Calculation of the Rosseland Mean. Before the integral (2.20) can be calculated we must multiply, for each frequency ν , the cross sections for the various processes by the number density of the respective absorbers, and add all the results. Thus, for bound-bound, bound-free, and free-free absorption, we need the number density n_j of atoms (or ions) in the state from which the transition to a higher state (or to the continuum) occurs. These number densities are provided by the Boltzmann and Saha equations. For scattering by electrons, we need the electron density, n_e, which can be approximated by the expression (2.45) for full ionization, because electron scattering is important only in the deep interior. For the three processes involving an atom (or ion), a correction for stimulated emission, i.e., a factor $1 - \exp(-h\nu/kT)$, must also be applied. An example for the total absorption coefficient obtained in this way is shown in Fig. 2.7. The 20 most abundant elements, cf. Sect. 4.4, have been taken into account for this calculation.

Let us substitute $h\nu/kT = u$ in the integral (2.20). The *radiative opacity*, or Rosseland mean absorption coefficient, per unit mass is then given by

$$\frac{1}{\kappa} = \rho \int\limits_0^\infty \frac{15u^4 e^u/4\pi^4(e^u - 1)^2}{(1 - e^{-u})\sum_j \sigma_j n_j + \sigma_s n_e} du \, , \tag{2.91}$$

where the sum is over all bound-bound, bound-free, and free-free processes, with the corresponding cross sections σ_j and initial-state number densities n_j.

Most extensive opacity calculations have been performed by groups at the Los Alamos Scientific Laboratory (Huebner et al. 1977) and at the Lawrence Livermore National Laboratory (Iglesias and Rogers 1996). The OPAL tables of the latter group contain Rosseland mean opacities that explicitly include the contributions from 19 of the most abundant elements, in addition to hydrogen and helium. Since hydrogen is converted into helium during the solar main-sequence evolution, it is necessary in principle to calculate such an opacity table for each depth in the Sun, and for each solar age. Usually a small number of tables, and linear interpolation between them, is used. The Los Alamos tables extend down to $kT \approx 1\,\mathrm{eV}$ (11600 K); the OPAL tables of Iglesias and Rogers (1996) reach further, down to $\approx 5600\,\mathrm{K}$. Complementary tables for the range of lower temperature have been prepared by Kurucz (1979) and by Alexander and Ferguson (1994). For the present Sun, Fig. 2.8 shows the opacity as a function of depth, together with its temperature derivative.

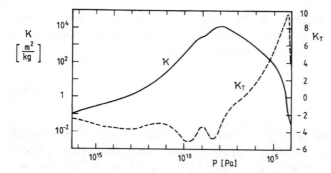

Fig. 2.8. Opacity, κ, and its double-logarithmic temperature derivative, κ_T, in the solar interior. The pressure P is used as a depth coordinate (center at left)

2.3.8 Boundary Conditions and Method of Solution

Let us first collect the equations which describe our model of the solar interior. We have four first-order differential equations:

$$\frac{\partial r}{\partial m} = \frac{1}{4\pi\rho r^2} \; , \tag{2.8}$$

$$\frac{\partial P}{\partial m} = -\frac{Gm}{4\pi r^4} \; , \tag{2.9}$$

$$\frac{\partial L}{\partial m} = \varepsilon - T\frac{\partial S}{\partial t} \; , \tag{2.12}$$

$$\frac{\partial T}{\partial m} = \begin{cases} -\dfrac{3\kappa L}{256\pi^2\sigma r^4 T^3} & \text{(in a stable layer)} \; , \tag{2.19} \\[2mm] (\partial T/\partial m)_\mathrm{C} & \text{(in an unstable layer)} \; , \tag{2.21} \end{cases}$$

and four constitutive equations:

$$\rho = \rho(P,T) \; , \qquad dS = dS(P,T) \; ,$$

$$\varepsilon = \varepsilon(\rho,T) \; , \qquad \kappa = \kappa(\rho,T) \; .$$

The four constitutive relations have been treated in the preceding sections, and are here written in symbolic form only. For the differential equations we must apply four boundary conditions.

Two of the four boundary conditions will be imposed at the center, $m = 0$. These conditions are

$$r(0) = 0 \quad \text{and} \quad L(0) = 0 \; . \tag{2.92}$$

Outer Boundary Conditions. The remaining two boundary conditions are applied at the surface. We place this surface (index s) at optical depth $\tau(r_s) = 2/3$; the optical depth is defined through

$$\tau(r) = \int\limits_{r}^{\infty} \kappa\rho\, dr' \ . \tag{2.93}$$

The boundary values adopted by the solutions of the four differential equations are r_s, P_s, L_s, and T_s. We do not know these values beforehand (except for the *present* Sun), but we shall be able to derive 2 relationships between them; these relationships are the desired boundary conditions.

For this purpose we must consider the atmosphere, here defined as the layer $r > r_s$, or $\tau < 2/3$. As the energy sources lie deep in the interior, we have $L \equiv L_s$ in the whole atmosphere. Moreover, we assume (or know from observation) that the atmosphere is *geometrically thin*, so that $r \equiv r_s$. Finally, the atmosphere contains very little mass, so that $m \approx m_\odot$. In this case the use of m as an independent coordinate would lead to very inaccurate results. We therefore replace m by a more rapidly varying quantity such as the optical depth introduced in (2.93) and set $m = m_\odot$ whenever it occurs as a coefficient.

Being consistent with the simplifications made, we discard the equations for r and L, (2.8) and (2.12). The pressure equation, (2.9), is transformed into

$$\frac{\partial P}{\partial \tau} = \frac{Gm_\odot}{r_s^2 \kappa} \ , \tag{2.94}$$

and, integrated, leads to the first of our desired surface conditions:

$$P_s = \frac{Gm_\odot}{r_s^2} \int\limits_{0}^{2/3} \frac{1}{\kappa}\, d\tau \tag{2.95}$$

(strictly, we have used the condition $P(0) = 0$).

In order to obtain the second condition, we make further assumptions in the atmosphere: we adopt an absorption coefficient κ independent of frequency ("grey"), integrate the equation of radiative transfer over all frequencies, and assume the expansion

$$\int\limits_{0}^{\infty} I_\nu d_\nu \equiv I(\tau,\theta) = I_0(\tau) + \cos\theta I_1(\tau) \ . \tag{2.96}$$

For the study of the atmosphere itself this model, called the Eddington approximation, is much too coarse. But for the present purpose, the derivation of a boundary condition for the interior model, it is quite sufficient.

We can determine the functions $I_0(\tau)$ and $I_1(\tau)$ in the following way: define the momenta

$$J = \frac{1}{4\pi} \int I \, d\Omega \, ,$$

$$F = \int I \cos\theta \, d\Omega \, , \tag{2.97}$$

$$K = \frac{1}{4\pi} \int I \cos^2\theta \, d\Omega \, ,$$

where the integrals are over the full solid angle Ω. The equation of transfer

$$\cos\theta \frac{dI}{d\tau} = I - B \tag{2.98}$$

then immediately leads to

$$\frac{dF}{d\tau} = 4\pi(J - B) \, , \qquad \frac{dK}{d\tau} = \frac{F}{4\pi} \, . \tag{2.99}$$

If (2.96) is substituted into (2.97) we find $J = 3K = I_0$ and $F = 4\pi I_1/3$. Since F is the constant energy flux, the first of (2.99) yields $J = B$, while the second leads to

$$I_0 = \frac{3}{4\pi} F\tau + b \, .$$

The constant of integration b is now obtained from the "radiation condition", namely from the requirement that, at $\tau - 0$, the net radiation into the upper half-space is the total flux F. Equivalently, we could require that there is no net radiation into the lower half-space. In both cases we find $b = F/2\pi$. Since $I_0 = B = \sigma T^4/\pi$, we have the final result

$$T^4 = \frac{L_s}{4\pi\sigma r_s^2}(3\tau/4 + 1/2) \tag{2.100}$$

for the atmospheric temperature; in (2.100) we have expressed the flux in terms of the luminosity and radius. This relation is used to evaluate the integral (2.95), where we need $\kappa(\tau)$ instead of $\kappa(T)$; taken at $\tau = 2/3$, (2.100) is the second boundary condition at the surface.

We see from (2.100) why $\tau = 2/3$ has been chosen as the "surface" of our interior model: At this depth the temperature equals the effective temperature,

$$T_{\text{eff}} = (L_s/4\pi\sigma r_s^2)^{1/4} \, .$$

Iteration Scheme. The integration of the model equations is usually carried out by means of an iteration scheme. For example, we may begin with a first guess at r_s and L_s, use (2.95) and (2.100) to calculate P_s and T_s, and then integrate inwards. At the center the two conditions (2.92) generally will not be satisfied, but we may improve the first guess of r_s and L_s and iterate until this is the case. The technique used to obtain most of the results shown in this chapter essentially follows this idea, but is described in much more detail by Kippenhahn et al. (1967).

2.4 Results for a Standard Solar Model

2.4.1 General Evolution

In the preceding sections we have collected almost all the information needed to calculate the evolution of the Sun from an initial homogeneous star to the present state. The only missing detail is the temperature gradient in the convectively unstable layers; this will be supplemented in Chap. 6, where we shall treat the mixing-length theory of convection. However, the results presented in the present chapter already employ this theory.

Figure 2.9 shows the Sun's evolution in the familiar Hertzsprung–Russell diagram. The approach from the Hayashi line to the zero-age main sequence took $\approx 5 \times 10^7$ years, which is of the order of the Kelvin–Helmholtz time. During its main-sequence life of 4.57×10^9 years the Sun has changed very little in effective temperature and luminosity. We may conclude that such

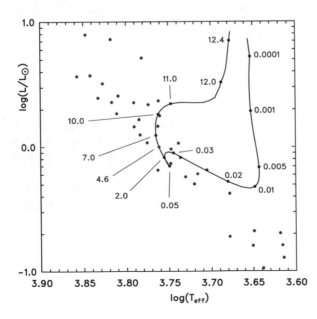

Fig. 2.9. Hertzsprung–Russell diagram with main-sequence stars (*asterisks*), and an evolutionary track of the Sun. The *numbers* on the track are ages in 10^9 years, counted from an initial state with $L = 7L_\odot$

evolutionary effects can explain only part of the spread of the stars within the main-sequence band (another reason for the finite width of the main sequence is that the initial chemical composition is not exactly the same for all stars). The asterisks in Fig. 2.9 represent visual binary stars for which T_{eff} and L have been determined with the best possible accuracy [taken from Audouze and Israël (1985, p. 252)].

For the Sun itself, and even more so for the Earth, the change of the solar luminosity during its main-sequence life is, however, substantial. The increase from about $0.7L_\odot$ to the present value often has been considered as a great puzzle. If a reduction of the "solar constant" by a factor 0.7 happened today, it would lead to dramatic cooling and very probably to a complete ice cover of the Earth. But if the Earth was completely iced over once, it would remain so even if the incoming radiation was increased to the present level – a consequence of the increased albedo. Yet the present Earth is not ice-covered, and geological evidence suggests that it never was.

The solution to this puzzle probably lies in the evolution of the Earth's atmosphere. For example, if the early concentrations of ammonia and (or) of carbon dioxide were higher than today, the greenhouse effect in the infrared could have compensated for the lower solar luminosity (e.g., Wigley 1981).

As shown in Fig. 2.10, the major part of the luminosity increase is due to an increase of the radius from about $0.87r_\odot$ to the present radius. During this evolution the effective temperature changed by only about 125 K, and hence contributed less to the luminosity change (the neutrinos, also shown in Fig. 2.10, will be considered in the following section).

A model of the present Sun, of age 4.57×10^9 years, is compiled in Table 2.4. It is a standard model, based on OPAL tables for the equation of state and opacity, on nuclear reaction rates as listed in Table 2.3, and on the local mixing-length theory as described in Chap. 6; it includes diffusion of helium as outlined in Sect. 2.3.3. The levels in the table are not equidistant in any of the variables; they are chosen so that both the outer layers and the central region are reasonably resolved. The surface (the first line) is at

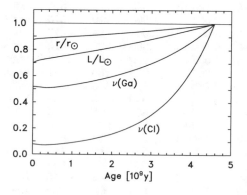

Fig. 2.10. Evolution of the solar radius and luminosity, and of the neutrino rates for the ^{37}Cl and ^{71}Ga experiments

Table 2.4. A model of the present Sun ($Z + n$ means $Z \times 10^n$)

$m/m_\odot$	$r/r_\odot$	P [Pa]	T [K]	ρ [kg/m^3]	$L/L_\odot$	X	μ	Γ_1
1.0000	1.0000	9.550+03	5.778+3	2.512−4	1.000	.735	1.247	1.636
1.0000	.9999	1.726+04	8.241+3	3.159−4	1.000	.735	1.243	1.373
1.0000	.9997	3.121+04	1.032+4	4.380−4	1.000	.735	1.195	1.215
1.0000	.9994	5.641+04	1.163+4	6.704−4	1.000	.735	1.139	1.194
1.0000	.9991	1.020+05	1.274+4	1.056−3	1.000	.735	1.089	1.191
1.0000	.9988	1.843+05	1.382+4	1.689−3	1.000	.735	1.044	1.196
1.0000	.9985	3.331+05	1.493+4	2.721−3	1.000	.735	1.004	1.203
1.0000	.9981	6.022+05	1.612+4	4.402−3	1.000	.735	.967	1.212
1.0000	.9976	1.089+06	1.743+4	7.103−3	1.000	.735	.932	1.223
1.0000	.9971	1.968+06	1.890+4	1.144−2	1.000	.735	.898	1.236
1.0000	.9965	3.557+06	2.059+4	1.836−2	1.000	.735	.865	1.252
1.0000	.9959	6.429+06	2.257+4	2.928−2	1.000	.735	.832	1.275
1.0000	.9952	1.162+07	2.495+4	4.624−2	1.000	.735	.800	1.299
1.0000	.9944	2.101+07	2.783+4	7.272−2	1.000	.735	.769	1.327
1.0000	.9934	3.797+07	3.141+4	1.127−1	1.000	.735	.741	1.363
1.0000	.9923	6.864+07	3.589+4	1.727−1	1.000	.735	.715	1.404
1.0000	.9910	1.241+08	4.166+4	2.610−1	1.000	.735	.692	1.461
1.0000	.9894	2.243+08	4.929+4	3.881−1	1.000	.735	.674	1.528
1.0000	.9874	4.054+08	5.953+4	5.669−1	1.000	.735	.660	1.585
1.0000	.9850	7.328+08	7.288+4	8.210−1	1.000	.735	.652	1.601
1.0000	.9820	1.325+09	8.938+4	1.191+0	1.000	.735	.644	1.583
1.0000	.9783	2.394+09	1.093+5	1.732+0	1.000	.735	.637	1.581
1.0000	.9737	4.328+09	1.342+5	2.513+0	1.000	.735	.630	1.604
.9999	.9681	7.823+09	1.661+5	3.619+0	1.000	.735	.624	1.633
.9999	.9611	1.414+10	2.073+5	5.187+0	1.000	.735	.620	1.654
.9998	.9523	2.556+10	2.601+5	7.413+0	1.000	.735	.618	1.662
.9996	.9415	4.620+10	3.273+5	1.058+1	1.000	.735	.616	1.666
.9993	.9281	8.352+10	4.127+5	1.509+1	1.000	.735	.615	1.667
.9987	.9117	1.510+11	5.209+5	2.152+1	1.000	.735	.614	1.668
.9978	.8917	2.729+11	6.580+5	3.069+1	1.000	.735	.613	1.668
.9963	.8676	4.933+11	8.315+5	4.378+1	1.000	.735	.613	1.667
.9940	.8387	8.916+11	1.051+6	6.245+1	1.000	.735	.612	1.666
.9904	.8057	1.586+12	1.320+6	8.826+1	1.000	.735	.612	1.665
.9746	.7101	5.959+12	2.230+6	1.955+2	1.000	.735	.611	1.665
.9423	.5988	2.200+13	3.153+6	5.180+2	1.000	.714	.614	1.665
.9019	.5193	5.750+13	3.822+6	1.120+3	1.000	.712	.616	1.666
.8569	.4611	1.205+14	4.403+6	2.039+3	1.000	.710	.616	1.667
.7872	.3991	2.754+14	5.158+6	3.984+3	1.000	.708	.617	1.667
.7173	.3539	5.144+14	5.831+6	6.589+3	1.000	.706	.618	1.668
.6342	.3117	9.321+14	6.583+6	1.058+4	.999	.704	.619	1.668
.5445	.2741	1.585+15	7.382+6	1.607+4	.995	.700	.621	1.668
.4452	.2376	2.627+15	8.296+6	2.378+4	.980	.694	.624	1.668
.3388	.2013	4.255+15	9.362+6	3.444+4	.940	.680	.631	1.668
.2430	.1686	6.401+15	1.046+7	4.720+4	.856	.655	.643	1.668
.1629	.1391	8.995+15	1.156+7	6.186+4	.721	.617	.663	1.668
.0897	.1072	1.254+16	1.280+7	8.193+4	.503	.554	.699	1.668
.0310	.0705	1.730+16	1.419+7	1.108+5	.219	.460	.761	1.668
.0034	.0321	2.183+16	1.530+7	1.417+5	.028	.369	.833	1.668
.0002	.0127	2.313+16	1.559+7	1.513+5	.002	.343	.855	1.668
.0000	.0000	2.338+16	1.565+7	1.533+5	.000	.338	.860	1.668

optical depth $\tau = 2/3$. The pressure and the density change very rapidly in the outer layers, while the luminosity and the hydrogen content vary only in the deep interior. The mean molecular weight μ first decreases inwards due to ionization of the abundant elements hydrogen and helium, but increases again near the center, where half of the hydrogen has already been converted into helium.

The solar mass is strongly concentrated towards the center. The outer convection zone, which extends from immediately below the surface down to $r/r_\odot \approx 0.71$, contains only about 2.5% of the total mass. And only about 10% of the mass lies outside $r/r_\odot = 0.5$, although 7/8 of the volume is there. – There is no interior convection zone in this model.

A number of other properties of the present Sun's interior have already been discussed in earlier sections, and are illustrated in Figs. 2.2 to 2.5, and 2.8. Here we only add the distribution of ^{3}He. Consistent with the lifetime variation of this isotope inside the Sun, there has been an accumulation near $r = 2 \times 10^8$ m (Fig. 2.11). The height of this peak is almost 100 times the original uniform abundance $Y_{30} = 4 \times 10^{-5}$ adopted for this model.

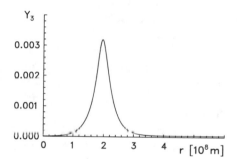

Fig. 2.11. Mass fraction Y_3 of ^{3}He in the present Sun

2.4.2 Neutrinos

Since about 1970 the main motivation for the calculation of solar interior models has been the persistent discrepancy between the predicted and measured flux of neutrinos from the Sun. We must therefore treat the solar neutrino production in some detail.

First it must be mentioned that, in addition to the reactions already discussed in Sect. 2.3.6, an important source of neutrinos is the "pep reaction"

$$p(pe^-,\nu)d , \qquad Q_\nu = 1.442 \, \text{Mev} . \tag{2.101}$$

As a three-particle reaction, the pep reaction occurs rather infrequently, only at about 0.25% of $p(p,e^+\nu)d$ (which we abbreviate by p+p). For the energy production the pep reaction is therefore unimportant; in fact it constitutes a small loss because the *total* energy goes into the neutrino. Nevertheless the

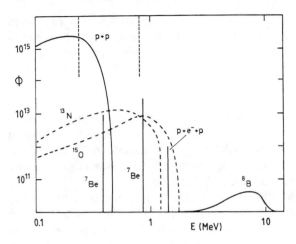

Fig. 2.12. Energy spectrum of the solar neutrino flux at 1 astronomical unit, after Bahcall (1979). Units are neutrinos/(m^{2}s MeV) for the continua, and neutrinos/(m^{2}s) for the lines. The *dotted vertical lines* mark the thresholds at 233 keV and 814 keV for the ^{71}Ga and ^{37}Cl experiments, respectively. The threshold of the water detector Super-Kamiokande is at ≈ 5 MeV

pep reaction has some importance for the neutrino experiments since Q_ν is markedly above the threshold of 814 keV of the ^{37}Cl detector (Sect. 3.6.2).

Figure 2.12 shows an energy spectrum of solar neutrinos, as predicted by a standard solar model for the distance of 1 astronomical unit. The continuous neutrino spectra shown in this figure are the usual β-decay spectra, which account for the conservation of energy and angular momentum. On the other hand the three-particle pep reaction and the electron capture of ^{7}Be produce monoenergetic neutrinos. The ^{7}Be neutrinos occur in two lines at 862 keV (89.7%) and 384 keV (10.3%), because the resulting ^{7}Li nucleus can be either in the ground state or in its first excited state.

The spectral distribution of the solar neutrinos strongly depends on temperature T, as a consequence of the dependence of the pp branching on T (Problem 2.19). This is illustrated in the evolution of the neutrino flux that would be detectable by ^{37}Cl (Fig. 2.10). As the central temperature evolves from 1.35×10^7 K to 1.56×10^7 K, this flux, which largely arises from ppIII, increases much more than the luminosity, which almost exclusively is generated in ppI and ppII. In contrast to this, the flux of neutrinos that would be detectable by a gallium experiment follows much more closely the evolution of the solar luminosity. The threshold of ^{71}Ga at 233 keV is low enough to allow the capture of a substantial fraction of neutrinos from the p+p reaction which is largely responsible for the energy generation.

Of course the *total* neutrino flux can be predicted without detailed knowledge of the Sun's interior: since two neutrinos are produced along with each ^{4}He nucleus, i.e., on the average, for every 25 MeV of the Sun's radiative output, the total flux can directly be obtained from the luminosity.

Problem 2.19. Write the nuclear reaction rates listed in Table 2.3 in the form $r = r_0 T^\eta$ and calculate r_0 and η at various temperatures between 10^7 K and 1.6×10^7 K. Determine the branching ratios for the pp chains for the solar center (use Table 2.4).

The neutrino fluxes from the various sources can be calculated as integrals over the reaction rates, r_j, as given by (2.75):

$$\Phi_j = \frac{1}{4\pi A^2} \int_0^{m_\odot} r_j \, dm \; ; \tag{2.102}$$

the subscript j runs over the neutrino-producing reactions. Here A is the astronomical unit; hence Φ_j is the mean flux at the Earth's orbit. The results for the solar model used in the present text are presented in Table 2.5. Also given are the cross sections σ_j, averaged over the respective energy spectra, for ^{37}Cl and ^{71}Ga. The product, $\Phi_j\sigma_j$, is the number of neutrinos captured per unit time and per atom. It is given in "solar neutrino units", or snu, defined as:

1 snu = 1 capture per second and per 10^{36} target atoms.

Table 2.5. Neutrino fluxes, cross sections (from Bahcall and Ulrich 1988, Bahcall et al. 1996, and Bahcall 1997), and predicted capture rates for the chlorine and gallium experiments

Reaction	Φ_j [10^{14}/m^{2}s]	σ_j [10^{-50}m^2]		$\Phi_j\sigma_j$ [snu]	
		^{37}Cl	^{71}Ga	^{37}Cl	^{71}Ga
p + p	5.99	0	11.72	0	70.2
pep	0.0136	16	204	0.22	2.8
^{7}Be	0.48	2.4	71.7	1.14	34.2
^{8}B	0.00062	11400	24000	7.06	14.9
^{13}N	0.045	1.7	60.4	0.08	2.7
^{15}O	0.039	6.8	113.7	0.27	4.4
Total				8.76	129.2

Other standard models predict capture rates that are similar to those listed in Table 2.5. The uncertainty partly comes from uncertainties in the opacity and in the heavy element abundance, which affect the temperature and hence the relative frequency of the pp chains. Another source of possible errors are the reaction rates themselves (Bahcall et al. 1998). For the flux of ^{7}Be neutrinos the largest uncertainty comes from the ^{3}He-^{4}He reaction, mainly due to the need of extrapolating the S factor down to low energy, cf.

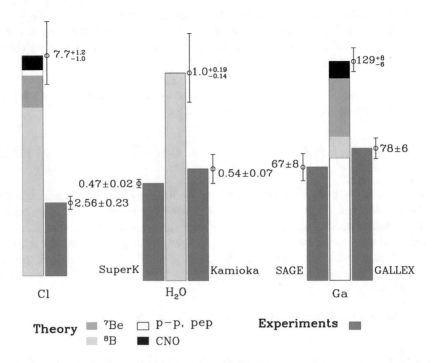

Fig. 2.13. Solar neutrinos: Prediction from the standard solar model BP98 of J. N. Bahcall and M. H. Pinsonneault (*high columns*) and experimental results (*lower columns*), for the ^{37}Cl (*left*), water (*middle*), and ^{71}Ga detectors (*right*). The *shading* indicates the contributions to the theoretical prediction from the diverse nuclear reactions in the Sun, as indicated at the bottom

Sect. 2.3.6; for the ^{8}B neutrinos the largest uncertainty lies in the S factor of the ^{7}Be + p reaction.

The neutrino fluxes from the decays of ^{13}N and ^{15}O listed in the last two lines of Table 2.5 are not the same because the CNO cycle has not yet reached equilibrium. Their total contribution to both the ^{37}Cl and ^{71}Ga experiments is, however, so small that the error would be insignificant if equilibrium had been assumed.

The experimental results for ^{37}Cl and ^{71}Ga, as well as for the water Čerenkov detectors (Sect. 3.6.2), are shown in Fig. 2.13, together with the theoretical predictions. Three main conclusions can be drawn from these results: The first is that solar neutrinos have indeed been detected. This directly confirms the concept of nuclear energy generation in the Sun. The second is that the experimental result is significantly below the theoretical expectation based on the standard solar model. This is the often quoted neutrino discrepancy, or the "solar neutrino problem". The third is that the neutrino discrepancy depends on neutrino energy; this conclusion is derived from the different energy thresholds of the three detector types.

Two ways of resolving the neutrino discrepancy have been proposed. The first is to modify the solar model in such a manner that the relative frequency of the pp reaction chains is modified, especially so that the ppIII chain is reduced and hence less ^{37}Cl captures and less neutrino-electron scattering events in the water detectors are predicted. This has been attempted with non-standard solar models some of which will be presented in Sect. 2.5 below. However, this "astrophysical" solution of the solar neutrino problem appears to fail in a quantitative test. The total flux from the ppIII branch is $\propto T_{c}^{18}$, where T_c is the temperature at the Sun's center (Bahcall 1989). A model with a, say, 3% reduction of T_c would therefore correctly predict the results of Kamiokande and Super-Kamiokande. On the other hand, the yield of the ppII branch is $\propto T_c^8$, which means that the same temperature reduction would *not* sufficiently reduce the flux of ^{7}Be neutrinos for a correct prediction of the gallium results (Fig. 2.13).

The alternative is to reconsider the *physics of the neutrino*. Already in 1957 B. Pontecorvo discussed the possibility of *neutrino oscillations*, that is the change from one neutrino flavor into another. In the solar case this could mean that, during their transit from the Sun to Earth, the neutrinos, which all are generated as electron neutrinos, partially change their flavor and become μ or τ neutrinos. Such neutrino oscillations are not part of the standard model of elementary particles; on the other hand, they are not forbidden by any fundamental principle. A necessary condition is that at least one type of neutrino must have a finite rest mass, which would be consistent with measurements at nuclear reactors. Even with a very small rest mass ($m_\nu c^2 \ll 1\,\text{eV}$) the MSW effect, so named after the work of Mikheyev and Smirnow (1985) and Wolfenstein (1978), would permit efficient neutrino oscillations in the presence of other matter, i.e., in the deep interior of the Sun. In either case, both μ and τ neutrinos would escape detection in the present experiments, and the capture rate would be greatly reduced (see e.g., the discussion of Bethe 1986).

First results from the heavy-water detector at the Sudbury Neutrino Observatory (Sect. 3.6.2) indeed indicate that neutrino flavor transformation is the answer. The charged-current reaction (3.97), which is sensitive only to electron neutrinos, yields a significantly smaller flux than the elastic-scattering reaction (3.96) which, in addition to being sensitive to ν_e, has a small sensitivity to the two other flavors (Ahmad et al. 2001).

2.5 Non-Standard Models

Models of the Sun that substantially deviate in one or more aspects from the standard model are called non-standard models. These models all have been constructed *ad hoc* in order to alleviate or even remove the neutrino discrepancy. As explained above, this "astrophysical solution" of the neutrino problem probably will not work. It is nevertheless instructive to consider some of the non-standard models because they shed light on the internal constitution of a star like the Sun.

2.5.1 The Low-Z Model

Via the various absorption processes enumerated in Sect. 2.3.7, and as electron donators in general, heavy elements contribute much to the opacity – disproportionately so as far as their small abundance is concerned. Lowering Z therefore lowers κ. This leads to a smaller temperature gradient according to (2.19), and hence to a smaller central temperature. In such a model the pp branching ratios change away from the ppIII chain, thus reducing the ^{8}B neutrinos. The energy release at the center is also reduced, but the luminosity remains the same, as the energy generating region is slightly expanded. An abundance $Z = 0.001$, which would be typical for a population II star, is sufficiently small to bring the predicted ^{37}Cl capture rate down to the observed 2.5 snu.

In the low-Z model the surface value $Z \approx 0.02$ is explained by the assumption that the Sun has passed through dense clouds of interstellar matter, where large amounts of dust have been accreted. Since the Sun is well-mixed in the outer convection zone, enough dust must have been collected during such passages that this whole layer has obtained the large Z abundance. Because of its supposed dust accretion the low-Z model has been called the "dirty solar model" by Christensen-Dalsgaard et al. (1979).

In addition to the above-mentioned problem of the ^{7}Be neutrinos, at least two more difficulties arise with the low-Z model. One is that the initial helium mass fraction Y_0, which (together with the mixing-length) is adjusted to yield the luminosity and radius of the present Sun, is found to be 0.20, or even less. This is difficult to reconcile with the primordial (big bang) value of about 0.23. The other difficulty is that the eigenfrequencies, calculated according to the prescription outlined in Chap. 5, match the observed solar oscillation frequencies less accurately than the eigenfrequencies of the standard model.

2.5.2 Rapidly Rotating Core

The centrifugal acceleration of a rapidly rotating star helps to support the star's mass against gravity. In such a model the thermal pressure (which is proportional to T) can be reduced. Thus T can be reduced, and again the branching of the pp chains is shifted into the desired direction. An estimate of the required rate of rotation is obtained by equating the centrifugal potential with, say, one tenth of the pressure, so that the rotational influence is a 10% effect:

$$\rho r^2 \Omega^2 = P/10 . \tag{2.103}$$

Hence $\Omega = (P/10\rho)^{1/2}/r$. At a level close to the center, say $r = 0.1 r_\odot$, this yields $\Omega \approx 1.5 \times 10^{-3}\,\mathrm{s}^{-1}$ (use Table 2.4) which is about 500 times faster than the rotation of the solar surface! Detailed calculations of solar models with a depth-dependent angular velocity (Bartenwerfer 1973) confirm this estimate.

A solar model with a such a rapidly rotating core is excluded by observations. One point is the oblateness of the core, which of course is invisible but which adds a quadrupole moment to the gravitational field. The outer layers of the Sun would be deformed, and the solar surface would show an oblateness in excess of the observed $\approx 10^{-5}$ that is due to the (slow) *surface rotation* (cf. Chap. 7). The other point is the observed rotational splitting of solar oscillation frequencies which only allows for a small core that rotates at most twice as rapidly as the solar surface (Chaplin et al. 1996 b, 1999; Lazrek et al. 1996; Charbonneau et al. 1998).

2.5.3 Internal Magnetic Field

The model with a strong internal magnetic field is formally related to the model with the rapid internal rotation. Assumed is a randomly oriented, small-scale field which, as a net effect, applies pressure to the material which it permeates. This *magnetic pressure* (cf. Chap. 8) acts in complete analogy to the before-mentioned centrifugal potential. For a 10% effect the necessary field strength follows from

$$B^2/2\mu = P/10 . \tag{2.104}$$

With a central pressure of 10^{16} Pa (Table 2.4) this means that the required field strength for a noticeable effect is of order 10^9 G. Again detailed calculations of Bartenwerfer (1973) confirm this estimate.

The problem here is that a small-scale field would not survive for 4.6×10^9 years. Due to ohmic dissipation the electric current that supports a field of scale l decays at a rate proportional to l^{-2}. If $l \ll r_\odot$ the lifetime of the field would be much smaller than the Sun's age (cf. Chap. 8).

2.5.4 The Internally Mixed Model

Mixing in the solar core would replace some of the helium accumulated near the center by hydrogen and thereby reduce the mean molecular weight μ. A lower temperature is then sufficient to give the same thermal pressure $P \propto T/\mu$, and this eventually leads to lower neutrino detection rates in the ^{37}Cl and water experiments, by the same reasoning as in the case of the other non-standard models. The mixing has been proposed both as a statistically stationary process (Schatzman et al. 1981) and as a time-dependent, intermittent process (Dilke and Gough 1972). In the latter case, an instability of the ^{3}He accumulated near $r = 2 \times 10^8$ m (Fig. 2.11) would cause the mixing. The hydrostatic equilibrium would immediately adjust itself to the changed μ, but the thermal adjustment to the new stratification would take a Kelvin–Helmholtz time (2.6), of order 3×10^7 years. In such a time-dependent model the low neutrino rate is explained by a real minimum in energy production. The corresponding luminosity minimum is, however, phase shifted because of the long thermal relaxation time of the Sun, and hence would not be observed at present. As for the low-Z model, the matching of observed p-mode oscillation frequencies with eigenfrequencies of the mixed model is worse than with standard model frequencies. Another general difficulty is that mixing the core requires *energy:* the heavy helium must be lifted in order to be replaced by the lighter hydrogen. In other words, the inwards directed gradient of μ strongly stabilizes the stratification (Sect. 6.1).

2.6 Bibliographical Notes

Whether or not a supernova has triggered the collapse of the pre-solar cloud has been discussed by Clayton (1979). The alternative to the hypothesis of freshly synthesized ^{26}Al is that isotopic anomalies such as the ^{26}Mg anomaly are residuals of pre-solar interstellar grains. According to Wielen et al. (1996) the slightly higher iron content of the Sun as compared to nearby stars of the same age suggests that the Sun was formed at ≈ 6.6 kpc from the galactic center, somewhat less than the present distance of ≈ 8.5 kpc. For the ages of meteorites a comprehensive source is the monograph of Dodd (1981); cf. also the discussion of Wasserburg (1995).

Appenzeller (1982) gives an account of the processes of star formation and pre-main-sequence evolution. He describes in detail the mechanisms responsible for the fragmentation of the original collapsing cloud, and of the loss of angular momentum and magnetic flux. An evolutionary track in the Hertzsprung–Russell diagram, which closely resembles Fig. 2.1, has been calculated by Appenzeller and Tscharnuter (1975). Bodenheimer (1983) reviews the techniques used for such calculations. Magnetic braking of an interstellar cloud has been studied numerically by Dorfi (1989), numerical results for a fragment of $1\,m_\odot$ are described by Bodenheimer et al. (1990) and Yorke et al. (1993). Bodenheimer (1995) reviews the relevant observations and physical processes; the magnetohydrodynamic aspects are discussed in several chapters of the volume edited by Tsinganos (1996).

The equations of stellar structure are described in a number of standard texts, e.g., Schwarzschild (1958), Clayton (1968), Cox and Giuli (1968), or Kippenhahn and Weigert (1990). While the derivation of the four differential equations is straight forward, the treatment of convection and diffusion as well as the four constitutive equations for ρ, dS, ε, and κ require physical assumptions.

Braginsky (1965) and Burgers (1969) describe the foundations of element diffusion, for which first ideas had been advanced by Chapman (1917) and Biermann (1937). Solar models including element diffusion have also been calculated by Wambsganss (1988), Cox et al. (1989), Proffitt and Michaud (1991), and Bahcall and Pinsonneault (1992, 1995).

An analytical version of the minimization method for the free energy has been obtained by Däppen (1980), who also quotes earlier authors who employed (2.30). In particular, Däppen introduces an analytical approximation for the partition functions which appear in the Saha equations (2.37); this approximation is valid in the framework of the Debye–Hückel treatment.

More detailed studies of the electrostatic interaction between the charged particles in a stellar plasma have of course been made. Here I mention only Stewart and Pyatt's (1966) treatment of the lowering of ionization potentials, with its relation to the concept of "pressure ionization", and the general discussion of Rogers (1984) of the ionization equilibrium and its effect on the equation of state in the solar interior. More references can be found in this latter paper. The influence on solar models has been investigated by Ulrich (1982). An attempt beyond the Debye–Hückel approximation has been made by Meister et al. (1999).

As far as the thermodynamic relations are concerned the present text follows Baker and Kippenhahn (1962). As a general text in thermodynamics I found the monograph of Callen (1985) most useful. In the context of the equation of state, thermodynamic quantities have been treated by Däppen et al. (1988) and Mihalas et al. (1990).

Nuclear reaction rates are most thoroughly described in the book of Clayton (1968). In particular the exponential factor arising from the penetration of the Coulomb barrier is discussed in detail. Rates at energies close to the Gamow peak have been measured by Junker et al. (1998). Adelberger et al. (1998) give a critical discussion. Electron screening, first treated by Salpeter (1954), is described in Chap. 17 of Cox and Giuli (1968). Attempts to reduce the solar neutrino discrepancy also include modifications in the treatment of electron screening (Dzitko et al. 1995, Dar and Shaviv 1996), but Gruzinov and Bahcall (1998) find a slight *increase* of the neutrino rates.

For the calculation of opacities I refer to the review of Huebner (1986) and to the results of the *Opacity Project* (Seaton 1995, Berrington 1997). A collection of Los Alamos tables has been compiled by Weiss et al. (1990). As far as "collective effects" on electron scattering are concerned, the original estimate by Diesendorf (1970) has been revised by Boercker (1987). The effects of opacity changes on the solar model have been pointed out by Kim et al. (1991), Faulkner and Swenson (1992), Charbonnel and Lebreton (1993), and by Tripathy and Christensen-Dalsgaard (1998).

An extensive description of the program used here to calculate the solar main sequence evolution has been given by Kippenhahn et al. (1967). In one form or another, this code has been used by many different groups. When results are compared one should be aware of this. The iteration used essentially is a multi-dimensional Newton–Raphson method, and is also called the *Henyey method* because in the context of stellar evolution it was first employed by Henyey et al. (1959). Many related codes have been developed since then. The programs of, e.g., Eggleton (1971, 1972), Christensen-Dalsgaard

(1982), D. B. Guenther [quoted by Demarque et al. (1986)], or Reiter et al. (1995) confirm the results obtained for the earlier standard models, e.g., those of Abraham and Iben (1971) or Demarque et al. (1973).

A good account of the solar standard model is the review of Castellani et al. (1997). These authors study in particular the consequences that the diverse observational and theoretical uncertainties have for the predicted neutrino flux. Kayser (1981) treats the quantum mechanics of neutrino oscillations, the monographs of Bahcall (1989), Klapdor-Kleingrothaus and Zuber (1997), and Schmitz (1997) give comprehensive introductions to neutrino physics and neutrino astrophysics. Current results and literature can be found on J. B. Bahcall's home page, www.sns.ias.edu/~jnb/. Other standard model calculations include those of Berthomieu et al. (1993), Turck-Chièze and Lopes (1993), Morel et al. (1997), Schlattl et al. (1997), and – with a view of the Sun's past and future – the series of Sackmann et al. (1990, 1993) and Boothroyd et al. (1991).

The standard solar models all have their original helium content and their mixing length adjusted so that the present solar luminosity and radius are obtained. The combined heavy element abundance Z is considered as given and kept constant. It should be kept in mind, however, that not Z itself, but the ratio Z/X is the observed quantity (cf. Chap.4). Therefore Z/X should also be adjusted, as has been pointed out (and done) by Bahcall et al. (1982).

Another non-standard model which has not been mentioned in Sect. 2.5 has been proposed by Faulkner and Gilliland (1985) and Gilliland et al. (1986). This model employs WIMP's, i.e., weakly interacting massive particles. Supposedly these particles have survived from the big bang up to the present day, and are gravitationally held in the solar interior because of their large mass. Due to their weak interaction, the WIMP's have a long mean-free-path and thus very effectively contribute to the energy transport. A reduced temperature gradient is the consequence, and the further reasoning is as in the low-Z model: smaller central temperature, less ppIII, fewer ^{8}B neutrinos.

3. Tools for Solar Observation

The Sun is a main-sequence star of average mass, luminosity, and effective temperature – one of about 10^{11} in our galaxy. The only distinction is that the Sun is much closer to us. Next to the Sun the nearest star, α Centauri with its companions, is more than 10^5 times farther away. This has two consequences: First, the Sun is the only star where details on the surface can be spatially resolved; and second, the flux of radiation from the Sun is so intense that many studies which cannot be undertaken with other stars are possible, in particular those which simultaneously require very high spectral resolution *and* high resolution in space and time.

However, even for solar observations there are limitations, which have their origin mostly in the Earth's atmosphere, but sometimes also in the instrumentation itself. In this chapter we shall first treat these limitations, then describe the telescopes used for high spatial resolution and the spectroscopic and polarimetric devices attached to these telescopes, and finally mention a number of special instruments that have been used for particular problems in solar physics.

3.1 Limitations

3.1.1 General Difficulties

Problem 3.1. An area of $0.3'' \times 0.3''$ with average intensity (Fig. 1.4) on the solar surface is observed with a 60-cm telescope and a high-resolution spectrograph. The desired exposure time is 1 ms, and the bandwidth is 10 mÅ at $\lambda = 5000$ Å. How many photons are detected if the efficiency of the entire telescope-spectrograph-detector system is 1%?

The answer is: 400 photons, i.e. the *photon noise* is 1/20 or 5%. But often the requirements are still more stringent: One would like to observe in a dark sunspot, where the intensity is an order of magnitude less, or in the center of a strong spectral line with, e.g., one tenth residual intensity, or both at the same time. Or one would like to observe polarization of, say, 1% or less! These examples illustrate that even in solar physics the number of available photons can be a problem.

The methods to overcome this problem are the same as in general astronomy: First, a larger aperture is used to collect more photons. Currently the largest solar telescope in operation is the McMath–Pierce telescope of the National Solar Observatory at Kitt Peak, USA, with a clear aperture of 1.52 m; the largest presently in planning is ATST, the Advanced Technology Solar Telescope, with a 4 m aperture. Second, the least possible fraction of photons should be lost. This is achieved especially by the use of image intensifiers or highly efficient detectors. The most common detector is the CCD or "Charge-Coupled Device", a detector based on the internal photoelectric effect in semiconductors. Such a detector has a quantum efficiency of up to 90%, which must be compared to the 5–25% (depending on wavelength) of a photomultiplier, or the $\approx 1\%$ of a photographic emulsion!

Another limitation that solar physics has in common with general astronomy is *atmospheric extinction*. Of course this can only be overcome by observing from above the atmosphere. Numerous solar experiments have been carried out with the help of rockets or satellites since about 1950. Most notable are a series of Orbiting Solar Observatories (OSO 1 – OSO 8) between 1962 and 1975, the Apollo Telescope Mount (ATM) on the Skylab mission (May 1973 to January 1974), the Solar Maximum Mission (SMM, 1980 – 1989), and Spacelab 2 (1985). More recent space missions are:

Ulysses. The Ulysses mission is the first spacecraft that explores the interplanetary space at high solar latitudes. It was launched on October 6, 1990 and, through a close fly-by at Jupiter in February 1992, entered a 6-year eccentric orbit that is highly inclined with respect to the ecliptic. The maximum southern latitude of $-80°$ with respect to the solar equator was first reached in September 1994, the maximum northern latitude of $80°$ in September 1995. Ulysses carries a number of instruments to study *in situ* the solar wind and the interplanetary magnetic field at high solar latitude: composition, particle energies, plasma waves and radio waves, etc.. Other instruments observe dust grains in the interplanetary space, or monitor bursts of solar x-rays and cosmic γ-rays. The Ulysses mission is described in Astron. Astrophys. Suppl. Ser., Vol. 92 (1992).

Yohkoh. The Yohkoh satellite is an observatory for the study of x-rays and γ-rays from the Sun. It was launched on August 31, 1991 into a near-Earth orbit, and carries a telescope for soft x-rays, a telescope for hard x-rays (up to 100 keV), and spectrometers for a wide range of wavelengths from soft x-rays up to γ-rays of 100 MeV. Its main purpose is the investigation of flares and of transient phenomena in the solar corona. The Yohkoh instruments are described in 6 articles in *Solar Physics*, Vol. 136 (1991).

SOHO. The Solar and Heliospheric Observatory was launched on December 2, 1995 and inserted into its orbit around the Lagrangian Point L1 on March 20, 1996. There, ca. 1.5×10^6 km sunward from the Earth, its instruments have uninterrupted sunlight, which is especially important for the three instruments that record the global oscillations of the Sun. Five instruments on the SOHO satellite take images and spectra in diverse wavelength bands from the ultraviolet to soft x-rays, whereby the transition layer between the chromosphere and the corona, and the corona itself, can be studied; in addition, the corona is monitored out to a distance of more than $10r_\odot$ by a triple-coronagraph. Three further instruments investigate the solar wind *in situ*, in particular the energy distribution of its constituents. The SOHO mission is described in detail in a volume edited by Fleck et al. (1995).

TRACE. The Transition Region And Coronal Explorer, launched on April 2, 1998, carries a 30-cm normal-incidence telescope that images fields of $8.5' \times 8.5'$ on the Sun. Diverse wavelength bands between 28 nm and 250 nm are used, so that the continuum as well as the emission of highly ionized atoms can be studied, characteristic for temperature ranges from 6000 K to 1 000 000 K. The images have a spatial resolution of better than $1''$; images of the whole Sun are composed in a mosaic. Since the orbit is Sun-synchronous, long uninterrupted observations of the transition region and the corona are possible. Handy et al. (1999) describe the TRACE mission.

3.1.2 Seeing: Description and Definitions

The second serious limitation to solar observations, which also has its origin in the Earth's atmosphere, is *seeing*. Seeing is the degradation of image quality by fluctuations of the refractive index in the light path. As such it is of course a general astronomical problem. However, in the context of solar observations the difficulties are aggravated by the Sun itself. The Sun causes thermal convection in the entire troposphere, it heats the ground around the observatory and thus initiates local convection, it heats the observatory building, in particular the dome, and finally the telescope into which it must shine. Any heated surface is a source of air instability. Since the viscosity of air is small, the resulting convective motion will have a large Reynolds number and therefore be turbulent. The concomitant temperature fluctuations, up to 0.1 K in the free atmosphere but frequently much larger in and around the building and telescope, affect the refractive index and hence generate wave-front aberrations. The refractive index varies according to

$$n - 1 = 2.79 \times 10^{-4} \frac{P/P_0}{T/T_0} , \tag{3.1}$$

where $P_0 = 1$ bar and $T_0 = 273$ K. Since atmospheric turbulence is nearly in pressure equilibrium, it is predominantly the temperature fluctuation that causes the seeing.

Image degradation by seeing is a very complicated process, but often three different aspects can be identified. *Blurring* is the defocusing effect of air schlieren with an index of refraction that varies from place to place. The whole image looses its sharpness. If the image essentially remains sharp but is rapidly shifted back and forth we speak of *image motion*. If substantial parts of the image remain sharp but are shifted relative to each other then we have *image distortion*. The frequency spectrum of image motion typically reaches up to $100\,\mathrm{Hz}$ (e.g., Brandt 1969). Exposures of $10^{-2}\,\mathrm{s}$ and faster can therefore substantially reduce this aspect of solar seeing.

Point Spread Function and Modulation Transfer Function. Let us consider a real image formed in the x, y-plane by a solar telescope. At each point the intensity $I(x, y)$ has contributions which "really" belong to neighboring points of the (non-existing) perfect image. Introducing the *point spread function* $\mathrm{PSF}(x, y; \xi, \eta)$ we have

$$I(x, y) = \int\!\!\!\int\limits_{-\infty}^{\infty} I_0(\xi, \eta)\, \mathrm{PSF}(x, y; \xi, \eta)\, d\xi d\eta \ . \tag{3.2}$$

The convolution (3.2) is often abbreviated in the form $I = I_0 * \mathrm{PSF}$.

A simplification occurs when the point spread function depends only on the distances, $\xi - x$ and $\eta - y$. This is the case if there is uniform blurring over the whole image. It is then most practical to consider the Fourier-transformed image. Because of the convolution theorem, the Fourier transform $F_0(\boldsymbol{q})$ of I_0 is simply multiplied by the Fourier transform $S(\boldsymbol{q})$ of PSF; here $\boldsymbol{q}$ is the spatial frequency vector (q_x, q_y). The function $S(\boldsymbol{q})$ is the *transfer function*, its modulus is the *modulation transfer function*, or MTF:

$$S(\boldsymbol{q}) = \int\!\!\!\int\limits_{-\infty}^{\infty} \mathrm{PSF}(x, y)\, \exp[-2\pi\mathrm{i}(q_x x + q_y y)]\, dx dy \ , \tag{3.3}$$

and $\mathrm{MTF}(\boldsymbol{q}) = |S(\boldsymbol{q})|$. The advantage of the Fourier representation is that several degrading effects simply appear as a sequence of factors in the MTF. For example, we may separate the optical imperfections of the telescope from the seeing:

$$\mathrm{MTF}_{\mathrm{total}} = \mathrm{MTF}_{\mathrm{telescope}} \cdot \mathrm{MTF}_{\mathrm{seeing}} \tag{3.4}$$

If the point spread function is rotationally symmetric in the x, y-plane, i.e., if $\mathrm{PSF}(x, y) = \mathrm{PSF}(r)$, where $r = (x^2 + y^2)^{1/2}$, then the modulation transfer function is rotationally symmetric in the q_x, q_y-plane, and is given by the Hankel transform:

$$\mathrm{MTF}(q) = 2\pi \int\limits_{0}^{\infty} r\, \mathrm{PSF}(r) J_0(2\pi q r)\, dr \ , \tag{3.5}$$

where $q = (q_x^2 + q_y^2)^{1/2}$ and J_0 is the Bessel function of order 0.

Problem 3.2. Derive (3.5) from (3.3).

Problem 3.3. Calculate the *line spread function*, LSF, as the image of a line $I_0 = \delta(\xi - \xi_0)$. Assume that PSF is rotationally symmetric and show that LSF is an Abel transform of PSF.

Diffraction-Limited Telescopes. An example of a rotationally symmetric point spread function is the image ("Airy image") of a point source formed by a diffraction-limited telescope with a circular unobstructed aperture

$$\mathrm{PSF_D}(r) = \frac{1}{\pi}[J_1(br)/r]^2 \ , \tag{3.6}$$

where J_1 is the Bessel function of order 1 and $b = D\pi/\lambda f$; D and f are the aperture and focal length of the imaging system, and λ is the wavelength of the light. The first zero of $\mathrm{PSF_D}$ is at $r = r_1 = 3.832/b$, the radius of the "central Airy disc". The angle $\alpha_1 = r_1/f$ corresponding to this distance is usually considered as the "resolution" of a diffraction-limited telescope:

$$\alpha_1 = 1.22\lambda/D \ . \tag{3.7}$$

When (3.6) is transformed according to (3.5) we obtain

$$\mathrm{MTF_D}(q) = \frac{2}{\pi}\left[\arccos(q/q_m) - \frac{q}{q_m}\sqrt{1 - (q/q_m)^2}\right] \ . \tag{3.8}$$

Spatial frequencies larger than $q_m = b/\pi$ are not transmitted by the telescope; $\mathrm{MTF_D} = 0$ for $q \geq q_m$.

The solid curve of Fig. 3.1a shows $\mathrm{MTF_D}$. According to its definition, the modulation transfer function is the factor by which a signal is reduced at each spatial frequency q. The spatial frequency corresponding to the "resolution" (3.7) is $q = 1/r_1 = q_m/1.22$, and $\mathrm{MTF_D} = 0.0894$ for this q (the black dot in Fig. 3.1a). This means that of the original, say, 15% intensity contrast of the solar granulation, little more than 1% is left in the image if a telescope is used which according to the criterion (3.7) just "resolves" the granular structure. And the effects of optical aberration and seeing are not yet considered!

Problem 3.4. Prove (3.8). – What aperture needs a diffraction-limited solar telescope in order to transmit, at $\lambda = 500\,\mathrm{nm}$, 60% of the contrast of a $1''$ feature?

Diffraction and Seeing. Averaged over periods of one second or longer (i.e., for "long" exposure) the seeing point spread function is often well represented by a rotationally symmetric (normalized) Gaussian

$$\mathrm{PSF_S}(r) = \frac{1}{2\pi s_0^2} \exp(-r^2/2s_0^2) \ . \tag{3.9}$$

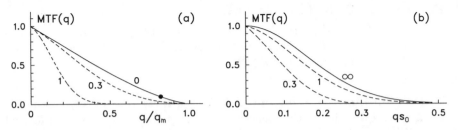

Fig. 3.1. Modulation transfer function for diffraction and seeing: (**a**) constant q_m, or given telescope aperture, as a function of seeing; (**b**) constant s_0, or given seeing, as a function of aperture. The label on the curves is $s_0 q_m$ in both cases

The parameter s_0 is a quantitative measure of the seeing. In angular distance, $s_0 = 1''$ is generally considered as good seeing; $s_0 = 0.5''$ is excellent. The Fourier transform of a Gaussian is another Gaussian; the MTF that belongs to the "long-exposure" PSF (3.9) is

$$\mathrm{MTF_S}(q) = \exp(-2\pi^2 s_0^2 q^2) \ . \tag{3.10}$$

This is the solid curve of Fig. 3.1b. The total MTF is the product of (3.8) and (3.10), and is shown in form of the dashed curves in Figs. 3.1a and 3.1b. The former shows how the image formed by a certain telescope is degraded as a function of the seeing parameter s_0. The latter demonstrates how large a telescope should be in order to make the best possible use of the seeing conditions at a certain observatory (raising q_m above $q_m s_0 = 1$ still increases the contrast, but there is little gain beyond $q_m s_0 = 2$). An observation with an optically good small telescope ($D \leq 20\,\mathrm{cm}$, say) is usually diffraction limited, while even perfect large telescopes, with aperture $D > 50\,\mathrm{cm}$, mostly allow only seeing-limited work.

As another quantitative measure of the seeing, we may define the parameter r_0 introduced by Fried (1966). For this purpose we consider the coherence function $B(\boldsymbol{\xi})$ of the complex wave amplitude which, at ground level, is the result of the phase changes caused by the random variation of the refractive index along the light path through the Earth's atmosphere. To a good approximation the atmospheric turbulence is governed by Kolmogorov's laws. In this case the variation of the refractive index n at height h obeys

$$\langle [n(\boldsymbol{x}) - n(\boldsymbol{x} + \boldsymbol{\xi})]^2 \rangle = C_n^2(h)\xi^{5/3} \ , \tag{3.11}$$

where $\boldsymbol{\xi}$ is the distance between two points, ξ its absolute magnitude, and C_n^2 is the *refractive index structure coefficient* at height h. As explained, e.g., by Roddier (1981), the coherence function is then

$$B(\boldsymbol{\xi}) = \exp\left(-1.455 k^2 \xi^{5/3} \int C_n^2(h) dh / \cos z\right) \ . \tag{3.12}$$

Here z is the zenith angle and k is the wave number of the light. The factor k^2 arises from the fact that the phase change is proportional to that wave number, and that a second-order moment of the phase variation is involved. The structure coefficient C_n^2 strongly depends on height; often the main contributions to the integral come from a small number of turbulent layers in the atmosphere, and in particular from a layer near the ground.

The Fried parameter is the characteristic decay length of the exponential (3.12). More specifically, we may write

$$r_0 = \left(0.423k^2 \int C_n^2(h)dh/\cos z\right)^{-3/5},$$
(3.13)

where the numerical factor originates from Fried's comparison of the seeing-limited resolution with the diffraction limit of a circular aperture. Thus, the Fried parameter r_0 is the aperture of an imaginary diffraction-limited telescope which has the same resolving power as a large, optically perfect telescope used under the conditions of the actual seeing. As the seeing is variable, r_0 depends on the observatory site and on time; according to (3.13) its dependence on wavelength is $r_0 \propto \lambda^{6/5}$.

Problem 3.5. Draw the modulation transfer function for the case of equal half-widths of the diffraction and seeing point spread functions. What is the value of $s_0 q_m$?

3.1.3 Seeing: How to Live with It

Site Selection. A substantial part of the seeing originates in the free atmosphere and is outside the astronomer's control. The first step in the construction of a solar observatory, therefore, is to select an appropriate site where the effects of the atmospheric turbulence are tolerable.

The search for a site may include *in-situ* measurements in the free atmosphere. For example, K. O. Kiepenheuer used an aircraft equipped with a micro-thermal sensor to measure temperature fluctuations of down to 0.01 K with a spatial resolution of 10 cm. Flights over and around the Canary Islands (Fig. 3.2) revealed that, above the inversion layer, an intermediate wind speed, say 5 m/s to 15 m/s, is favorable: Not violent enough to shake the observatory building, it blows the locally generated turbulence off the mountain.

Specially equipped radiosondes have also been used for *in-situ* measurements of atmospheric temperature fluctuations. The results can be converted into a value of the Fried parameter r_0 as long as the distance between two points of measurement falls into the *inertial range* of the atmospheric turbulence, i.e., lies between the large *outer scale* of energy input and the small *inner scale* of energy dissipation (in the Earth's atmosphere this means roughly between 30 m and 3 mm). For the Canary Islands the radiosonde results (Fig. 3.3) show that most often 20 cm < r_0 < 30 cm. These values

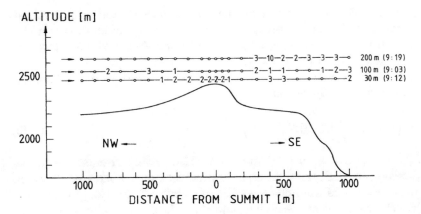

Fig. 3.2. Aircraft measurements of temperature fluctuations above the Roque de los Muchachos (La Palma, Canary Islands) on June 23, 1973. Heights above the site (and local time) are given on the *right*. Numbers in the scans are peak-to-peak values in units of 0.01 K. *Open circles* mean $\Delta T < 0.01$ K. After Kiepenheuer and Schellenberger (1973)

are rather large, but the measurements do not include the seeing caused by ground-level turbulence.

A method that includes the effect of *all* layers is the *spectral-ratio* method proposed by von der Lühe (1984). It is based on a large number of short-exposure images of the same object. If this object is structured on scales below the seeing limit, such structure will be manifest in the Fourier transform, e.g., for the ith image

$$F_i(\boldsymbol{q}) = F_0(\boldsymbol{q})S_i(\boldsymbol{q}) \;, \tag{3.14}$$

where F_0 is the transform of the object. The average $\langle |S_i|^2 \rangle$ over all images will still be non-zero at spatial frequencies q for which the average transfer

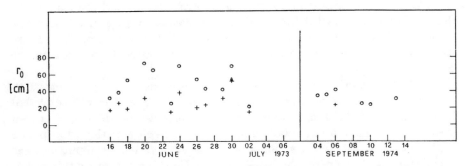

Fig. 3.3. The Fried parameter r_0 deduced from radiosonde measurements of temperature fluctuations. *Circles* include measurements up to 10 km, *crosses* up to 17 km. After Barletti et al. (1977)

function $\langle S_i(\boldsymbol{q}) \rangle$ has already dropped to an insignificant value. The spectral ratio

$$\varepsilon(\boldsymbol{q}) = \frac{|\langle F_i(\boldsymbol{q}) \rangle|^2}{\langle |F_i(\boldsymbol{q})|^2 \rangle} = \frac{|\langle S_i(\boldsymbol{q}) \rangle|^2}{\langle |S_i(\boldsymbol{q})|^2 \rangle} \tag{3.15}$$

therefore has a cutoff at a certain spatial frequency that is related to the Fried parameter r_0. Results based on this method are shown in Fig. 3.4. The values are markedly smaller than those of Fig. 3.3, typically 10 cm.

The Fried parameter is a statistical property and therefore relevant for long-exposure observation (i.e., long compared to the time scale of the seeing). Occasionally the conditions might be more homogeneous than indicated by the above results; the resolution of short-exposure images will be much better during such moments.

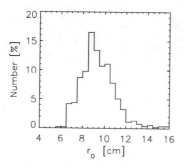

Fig. 3.4. Distribution of the Fried parameter, measured at 550 nm by the spectral-ratio method during good seeing conditions at the Vacuum Tower Telescope, Izaña, Tenerife. After Denker (1996)

Another convenient ground-based method to assess the seeing is the photoelectric *seeing monitor* (Brandt 1970). This device is attached to a telescope and rapidly scans the intensity profile across the limb of the solar image. Image motion is thus eliminated, and the gradient of the profile, represented in form of a Gaussian line spread function $\exp(-x^2/2\sigma^2)$, provides the blurring parameter σ. If, in addition, the total light falling into a narrow window (positioned across the solar limb) is measured as a function of time, one may also deduce an image motion parameter, e.g., the rms value of the fluctuation of this signal. For the R. B. Dunn Telescope at Sacramento Peak Observatory Brandt et al. (1987) derived a Fried parameter from such measurements, with a distribution similar to that shown in Fig. 3.4.

The variation of the refractive index n not only causes a variation in direction of the light, but also in intensity. This effect, known as *scintillation*, is small for an extended source like the Sun because $\Delta I / \langle I \rangle \simeq N^{-1/2}$, where N, the number of air parcels with different n that cover the source, is large. Nevertheless the solar scintillation can be measured, and Seykora (1993) found its close correlation to the seeing as measured by the directional effect. A theoretical treatment of Beckers (1993) showed that predominantly near-ground layers contribute to the solar ΔI, opposite to the scintillation of the stars.

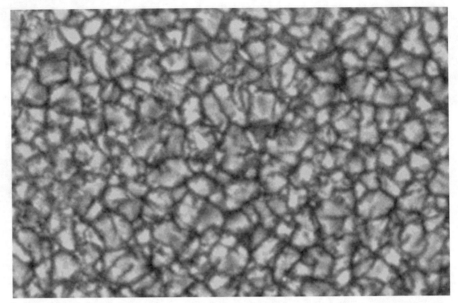

Fig. 3.5. Solar granulation, in a field of 30 000 km × 20 000 km on the Sun. The image has been taken at the Swedish Vacuum Solar Telescope on the Roque de los Muchachos, La Palma, Canary Islands. Exposure 13 ms in the blue spectral region, 468 ± 5 nm. Courtesy P. N. Brandt, G. W. Simon, and G. Scharmer

From many test measurements it now appears that the most favorable site for a solar observatory is on a high mountain (above the inversion layer) on an island surrounded by a large water surface, which makes the incoming air mass as homogeneous as possible. The Canary Islands and Hawaii seem to satisfy these conditions very well and at the same time have enough sunshine to justify the construction of large observatories. Images of the solar granulation (Fig. 3.5) confirm this conclusion. Of course, photographs of comparable quality have been obtained for some time at other observatories, notably at Pic du Midi. But more persistent seeing conditions, which are required for evolutionary studies of the small-scale photospheric structure, are met at the island sites.

Local Turbulence. Beyond a careful selection of a good observatory site nothing can be done about the seeing originating in the free atmosphere. But the effects of ground-level turbulence can considerably be reduced by the use of a high building. *Tower telescopes* were first constructed by G. E. Hale at Mt. Wilson, California; his 20 m and 50 m solar tower telescopes, completed in 1908 and 1911, were pioneering instruments. Since then these telescopes provided important data, predominantly of velocity and magnetic fields on the solar surface, and they are still in use today.

Local turbulence develops during the morning hours when the Sun heats the ground around the observatory. This effect is reduced if the observatory

Fig. 3.6. Big Bear Solar Observatory of the New Jersey Institute of Technology. Courtesy H. Zirin

is surrounded by water, which has a high thermal inertia. Big Bear Solar Observatory, situated on an artificial island in Big Bear Lake, California, USA (Fig. 3.6), is an example. The scintillation measured at this site is particularly small. Other lake sites are the Udaipur Solar Observatory in Rajasthan, India, the Huairou Solar Station of the Beijing Observatory in China, and the San Fernando Observatory at Northridge, California, USA.

Open Structure and Domeless Design. The design of the observatory building is another important matter. We have already seen that a wind of intermediate strength is welcome because it clears the site from local turbulence. Exposing the coelostat (Sect. 3.2.2) to the open air may therefore be better than protecting it by a dome. Examples are the the 60-cm tower telescope at Kitt Peak, Arizona, and the very similar 70-cm telescope (Fig. 3.7) at Izaña, Tenerife, which have a cylindrical dome that can be completely retracted, but also partially opened to act as a *wind screen*. An additional

Fig. 3.7. The German Vacuum Tower Telescope at the Observatorio del Teide, Izaña, Tenerife. The dome is partially retracted. Courtesy D. Soltau

possibility to prevent the build-up of heated, unstable air masses is an air suction system (installed, e.g., at the refractor at Culgoora, Australia).

The tower of the Dutch Open Telescope (Fig. 3.8) has a completely open structure, so that there is very little perturbation of the incoming uniform wind. The framework of this tower is designed in such a way that even when a deformation occurs the platform remains horizontal and the Sun is seen under the same angle.

As an alternative to the retractable dome various domeless designs also minimize the perturbation of the onstreaming uniform wind. The solar telescopes at Capri, Italy and at Hida, Japan (Fig. 3.19) are important examples. A special optical arrangement is the *turret* (Sect. 3.2.2), first installed at the R. B. Dunn Solar Telescope of Sacramento Peak Observatory (Fig. 3.9), and also used for the Swedish Solar Observatory on La Palma, Canary Islands.

Telescope Evacuation. Solar telescopes with their large focal length are particularly vulnerable to internal seeing, i.e., turbulence generated inside the telescope by solar heating, especially heating of the air in front of the telescope mirrors. A most effective measure against this is to enclose the optical path

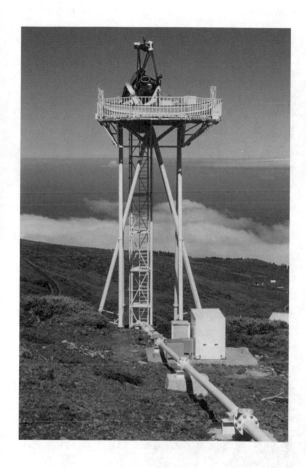

Fig. 3.8. The Dutch Open Solar Telescope, on the Roque de los Muchachos (La Palma, Canary Islands). Design R. H. Hammerschlag, Astronomical Institute Utrecht

by an evacuated tank. Experience shows that a vacuum of a few mbar is sufficient. A disadvantage is the necessity of entrance and exit windows which must withstand the atmospheric pressure; their thickness should therefore be of the order of one-tenth of the diameter. But even a thick window is affected by mechanical stress; in addition, the window absorbs a fraction of the sunlight (cf. Fig. 3.17) and therefore suffers thermal effects. In spite of these problems most modern solar telescopes have an evacuated light path. Some, e.g., the Domeless Solar Tower Telescope at Hida Observatory, Japan, even have both the telescope and the spectrograph evacuated.

If a turret is used, the vacuum can include the first and second telescope mirrors (Fig. 3.15). An elegant solution is to use the telescope objective itself as entrance window. The Swedish Solar Telescope on La Palma is of this type.

Instead of the evacuation a helium-filled tank can be used. For very large entrance windows this may be a necessity. The refractive index n of helium is much closer to 1 than that of any other gas: $n = 1.000036$, as compared to $n = 1.000293$ for air. In addition, temperature fluctuations are smoothed

Fig. 3.9. The Richard B. Dunn Telescope at Sacramento Peak, New Mexico, USA. Courtesy National Solar Observatory, AURA, Inc.

more rapidly in helium than in air because of the larger thermal conductivity. Both effects reduce the internal seeing.

Other Passive Measures. White paint on the whole building reflects the sunlight and so helps to avoid unwanted heating. Mostly TiO_2 is used, which has a high reflectivity in the visible, but is nearly black in the far infrared and hence radiates the building's own heat away.

Thermal problems often are worst at the edges of optical elements. These problems can therefore be alleviated by using oversize optical elements, i.e., by larger windows, lenses, mirrors, etc., than would correspond to the actual aperture of the telescope. The aperture is then defined by an additional diaphragm ("stop") behind (or in front of) the main focusing objective or mirror. A white annular diaphragm just outside the main lens has been used with great success at several observatories.

3.1.4 Adaptive Optics

Adaptive optics comprises the measurement of wave-front perturbations, the extraction of a correction signal from these measurements, and the application of this signal to an *active*, i.e., tiltable or deformable, optical element, in order to restore the wave front. The active element is generally an additional mirror, and is often divided into a large number of independently steerable elements.

The first successful adaptive optics devices used for solar observation were *image motion compensators*. Image motion is caused by a uniform, although time-dependent, tilt of the wave front. To correct for such a perturbation is relatively easy. Figure 3.10 illustrates an image motion compensation system which uses a dark solar feature (a spot or a pore) to generate the correction signal. A glass slide in the main beam diverts part of the light to form an ancillary image of the spot onto a quadrant photocell. Differences in the signals from the four quadrants measure the displacement of the beam in two perpendicular coordinates. Converted into steering signals, these displacements directly control three piezoelectric transducers (the *actuators*) mounted at the back of a tiltable mirror in the main beam. The system works well with image-motion frequencies of up to several hundred hertz. An important application is the production of motion pictures for the study of time-dependent solar phenomena.

In the case of image distortion it is not possible to stabilize the whole image with a single mirror. Nevertheless the system permits the stabilization of the spot or pore used to generate the correction signal, and of its

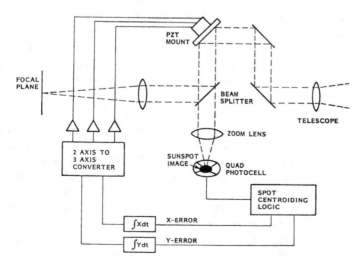

Fig. 3.10. Image motion compensation by three piezoelectric transducers (PZT) mounted on the back of the correcting mirror (Tarbell and Smithson 1981). Courtesy Lockheed Palo Alto Research Laboratories

neighborhood. The angular diameter of the stabilized field (the *isoplanatic patch*) is typically $5''$ if visible light is used, and up to $15''$ in the infrared. The limitation arises mainly from turbulence in higher layers of the Earth's atmosphere.

In order to study time-dependent solar fine structure even in the absence of dark features, the *correlation tracker* has been developed (von der Lühe 1983) and installed at several observatories. This device employs an array detector to take pictures of the solar granulation (Fig. 3.5), separated by short intervals of time. Two consecutive pictures are shifted (in the computer) relative to each other until the maximum correlation is found. The shift is then interpreted as image motion and accordingly compensated in real time. Thus the correlation tracker can tranquillize an image during the one or several seconds of exposure needed for a narrow-band filtergram or a highly resolved spectrum. Of course, the success of this system depends on the separation of the time scales of image motion on the one hand and of true granular evolution on the other hand; typically the latter is one minute or longer, while the former is one second or shorter.

The scheme shown in Fig. 3.10 can be generalized. A wave-front sensor replaces the quadrant photocell. Since the Fried parameter r_0 is the length over which the wave front of the incident light has an approximately uniform tilt, an aperture of diameter D must be divided into $\simeq (D/r_0)^2$ sub-apertures. For this purpose an image of the entrance pupil is formed onto an array of lenslets, as first proposed by Shack and Platt (1971) as a variant of the Hartmann test, where a perforated screen is used. Each lenslet forms a separate image of the object, slightly displaced because of the local tilt of the wave front. The displacement is measured, e.g., by means of a correlation tracker, and the measurement is used to control an active mirror. This mirror, either segmented or deformable, must have sufficient degrees of freedom so that the local wave-front tilts can be corrected. Adaptive optics systems using Shack-Hartmann wave-front sensing are under development at several solar observatories. First results, e.g., concerning the sub-arcsecond structure of the photospheric magnetic field outside sunspots (Sigwarth 2001 b), have been obtained at the Dunn Solar Telescope at Sacramento Peak.

If we include the observer, then we may say that adaptive work has been practiced for a long time. As the quality of seeing rapidly varies in time it has been common to take large numbers of photographs with short exposure, and then select the best ones. With a fast electronic camera and a large data storage capacity this method has been automatized: the images are continuously taken and compared to each other, and only the best of a given time interval is retained; this procedure is known as *frame selection*. In order to improve the chance for good pictures the above-mentioned seeing monitor can be used: this instrument, or similar ones measuring for instance the granular contrast, beforehand selects the right moments for exposure in real time. All such conventional measures against seeing require short exposure. Image intensifiers can reduce the exposure time by a factor of 10 or more, and thus help in an indirect way. Figure 8.9 shows a spectrum taken with the aid of an image intensifier.

3.1.5 Image Restoration

There are several methods of finding an estimate of the best image of an object from degraded images. We briefly mention two of these image-restoration methods, and show examples of solar applications.

Speckle Interferometry. Speckles are seen as a grainy interference pattern in images formed by laser light reflected from a diffusive surface, or in images of astronomical objects taken with short exposure, say 10 ms. Labeyrie (1970) first suggested that information beyond the seeing limit may be obtained from such short-exposure images. Taking the absolute magnitude of (3.14) and averaging we obtain the power spectrum of the object,

$$|F_0(\boldsymbol{q})|^2 = \frac{\langle |F_i(\boldsymbol{q})|^2 \rangle}{\langle |S_i(\boldsymbol{q})|^2 \rangle} \; . \tag{3.16}$$

In this expression the phase of the object Fourier transform has been lost. Nevertheless it is possible to recover essential information about certain objects, e.g., the angular separation of two stars, if $\langle |S_i(\boldsymbol{q})|^2 \rangle$, the *speckle transfer function*, is known. In the stellar case this function can be obtained as the power spectrum of a point source if such is available within the isoplanatic angle. The solar case is more complicated because there are no point sources on the Sun. Therefore models have been derived (e.g. Roddier 1981) that depend on the statistical properties of the atmosphere as well as on the diffraction of the actual telescope.

In order to restore the image of an extended object, such as an area on the solar surface, the Fourier *phase* is required in addition to the Fourier amplitude. For this purpose Knox and Thompson (1974) have proposed to use the autocorrelation of the image transform (3.14), or the *cross spectrum*

$$\langle F_i(\boldsymbol{q}) F_i^*(\boldsymbol{q} + \Delta\boldsymbol{q}) \rangle = F_0(\boldsymbol{q}) F_0^*(\boldsymbol{q} + \Delta\boldsymbol{q}) \cdot \langle S_i(\boldsymbol{q}) S_i^*(\boldsymbol{q} + \Delta\boldsymbol{q}) \rangle \; , \tag{3.17}$$

where the left-hand side is an average over a large number of short-exposure image transforms, while the right-hand side contains the cross spectrum of the object and the corresponding signal transfer function $\langle S_i(\boldsymbol{q}) S_i^*(\boldsymbol{q} + \Delta\boldsymbol{q}) \rangle$. Again models have been calculated that include atmospheric and telescopic properties (von der Lühe 1988). A most important result is that the signal transfer function is real. Therefore the phase difference occuring in (3.17), $\phi(\boldsymbol{q}) - \phi(\boldsymbol{q} + \Delta\boldsymbol{q})$, is due to the object cross spectrum alone. An integration then yields the Fourier phase of the object. The amplitude is determined as above, and an inverse Fourier transform restores the image.

A further extension of the Knox-Thompson method is *speckle masking*, so named because it corresponds to the application of certain masks to speckle

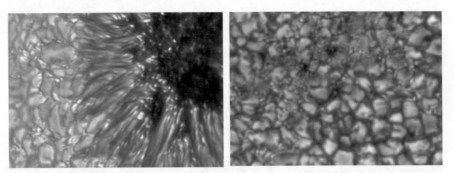

Fig. 3.11. Image restoration by speckle masking. Sunspot (*left*) and anomalous granulation (*right*). Speckle images taken at $\lambda \approx 550\,\mathrm{nm}$ by C. R. de Boer and C. Denker with the 70-cm Vacuum Tower Telescope at Izaña, Tenerife. Both pictures cover a field of $24'' \times 17''$. Courtesy F. Kneer

images, as originally introduced by Weigelt (1977). Here one calculates the triple correlation of short-exposure images or, equivalently, the mean *bispectrum*

$$\langle F_i(\boldsymbol{q})F_i(\boldsymbol{p})F_i^*(\boldsymbol{q}+\boldsymbol{p})\rangle \;, \tag{3.18}$$

which, by (3.14), is the product of the object bispectrum and a transfer function that is again real and can be constructed on the base of an atmospheric model and the telescopic diffraction (von der Lühe 1985). Hence the Fourier phase of the object can be recovered. Figure 3.11 shows results obtained with the speckle-masking technique.

Phase Diversity. In its simplest version the phase-diversity method uses two images taken simultaneously of the same object to derive two unknown functions, namely the desired intensity distribution I_0 and the distorted wave front in the entrance pupil of the telescope. We consider two convolutions of type (3.2), each with additive noise N_i, i.e.,

$$I_i = I_0 * \mathrm{PSF}_i + N_i \;, \qquad i = 1, 2 \;. \tag{3.19}$$

The two point spread functions PSF_i contain information from the *same* atmospheric degradation. They differ from each other because the phase of the complex wave field is changed ("diversified") in a well-defined manner for the formation of the two images. One way to achieve this is to take one image in the focal plane and the other slightly out of focus, at a distance Δz, say; this was first proposed by Gonsalves and Chidlaw (1979), the application to solar images has been described by Löfdahl and Scharmer (1994). The equivalent phase difference at the edge of the entrance pupil is

$$\Delta\phi = \frac{\pi\Delta z}{4\lambda}\left(\frac{D}{f}\right)^2 \;, \tag{3.20}$$

Fig. 3.12. Image restoration of a sunspot section, $7.7'' \times 7.7''$, by phase diversity. Original images, focussed (*left*) and defocussed (*middle*), and restored image (*right*). Exposure 48 ms, Vacuum Tower Telescope, Izaña, Tenerife; bandpass 2.6 nm at $\lambda = 569.5$ nm. Courtesy A. Tritschler

as can be shown, e.g., by means of spherical waves that originate from the focus and from a point displaced from the focus by Δz, and propagate towards the entrance pupil. In practical work Δz should be chosen such that the phase difference at the edge of the aperture is at least of order π, say, so that the two point spread functions differ in a significant manner, and a solution of the system (3.19) can be found. In the presence of noise this solution is obtained by a simultaneous minimization of $|I_i - I_0 * \mathrm{PSF}_i|^2$, for $i = 1, 2$.

The phase-diversity method relies on *isoplanatic* degradation, i.e. on the uniformity of the point spread function over the field of view. In reality this condition is satisfied only over small sections of a few arcseconds angular extent. Larger images must therefore be composed of sections that are restored separately. Figure 3.12 illustrates the method. The phase-diversity method can be generalized to several instead of two images, and to different means of diversifying the phases. The method has also been combined with speckle interferometry (Paxman and Seldin 1993).

3.2 High-Resolution Telescopes

3.2.1 Image Scale

The image of the Sun formed in the telescope must be recorded in some way, for example photographically or with a photoelectric detector. Let us take an array detector with 60 detector elements per millimeter. We want to project 50 km of the solar surface onto such a picture element (short *pixel*). Then the scale of the projection must be chosen in such a way that the whole solar image has a diameter of $d = 46.5$ cm. This size of an image is typical for a large solar telescope, although often the field of view is limited to a small fraction of the whole solar image. The diameter of the image in the prime focal plane is

$$d = 2fr_\odot/A \, ,\tag{3.21}$$

where f is the focal length, $r_\odot$ the solar radius and A the distance of the Sun. If $d = 46.5$ cm as in the example above, then f must be 50 meters. Such a long focal length is a major characteristic of many solar telescopes. This is in contrast to common astronomical telescopes for which the light-gathering power [proportional to $(D/f)^2$ for extended objects and to D^2 for point sources] is of greater importance.

A large focal length makes a freely steerable solar telescope inconvenient. Special mirror arrangements are therefore used in order to guide the sunlight into a fixed direction: the heliostat, the coelostat, and the turret, which we shall describe in the following section. For a tower observatory the fixed direction is mostly vertical, and the height of the tower is often adapted to the desired focal length.

Another possibility for obtaining a solar image of order 0.5 m in diameter is to use a smaller primary focal length, and then to form a secondary image that is large enough for high-resolution work. The diameter d' of the secondary image then defines the *effective* focal length, f_{eff}, via (3.21). A great number of solar telescopes now in use are of this type. Generally these telescopes are steerable; however, in order to feed a spectrograph or other heavy post-focus instruments a fixed beam is still necessary. Mostly a Coudé arrangement is chosen where the light is directed into the polar axis.

In all cases the required scale of the solar image essentially influences the design of the telescope. The development of detectors with finer resolution, of order a few microns, is a step forward. Not only the telescope itself but also, for example, the accompanying spectrograph can thus be more compact. For space experiments this is a particular advantage.

The design of solar telescopes is somewhat facilitated by the fact that only a part of the sky must be within reach: the Sun always stays in the declination range $\pm 23.5°$ with respect to the celestial equator.

3.2.2 Mirrors for Fixed Telescopes

The Heliostat. In principle a single flat mirror is sufficient to reflect the sunlight into a fixed direction. If this direction is the polar axis, the mirror is called a (polar) heliostat. Figure 3.13 is a schematic diagram of the 2 m-heliostat installed at the McMath–Pierce Solar Telescope at Kitt Peak. The daily rotation of the Earth is compensated by a corresponding rotation of the fork which holds the mirror. The mirror itself is tilted about the declination axis according to the yearly variation of the Sun's declination.

A variant, designed by J. B. Foucault around 1869, is the siderostat. The siderostat mirror directs the sunlight into the horizontal direction. For this purpose it must turn simultaneously about two axes.

A disadvantage of the heliostat (and also of the siderostat) is that the solar image *rotates* (by $15°/$h) as the Sun moves about the sky. Except for single short-exposure images this rotation must be compensated, either by an extra optical element (the Dove prism) or by rotating the whole post-focus

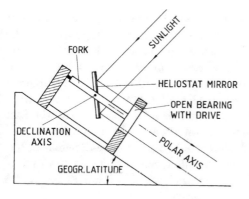

Fig. 3.13. Heliostat

instrumentation to match the rotation of the solar image. For polarimetric studies it is a comfort that (at constant declination) the angle of incidence, and hence the instrumental polarization, is constant, although this angle can be quite large.

The Coelostat. Image rotation is avoided by the following two-mirror arrangement, called the coelostat. As with the heliostat, the primary mirror (the "coelostat mirror") is mounted on the polar axis, but this time *parallel* to this axis and fixed to it. A secondary mirror (the "second flat") then

Fig. 3.14. Coelostat of the German Vacuum Tower Telescope at the Observatorio del Teide, Izaña, Tenerife. *Right*, primary mirror, mounted on the polar axis; *upper left*, secondary mirror, at the variable mast. Courtesy W. Schmidt

reflects the light into the desired direction, which is horizontal in some, but vertical in most cases. The vertical coelostat has an advantage in that the thermal stratification is perpendicular to the light path; it also naturally fits into the tower observatory. The height of the secondary mirror remains fixed during the day, but must be adjusted over the year to the Sun's varying declination. Figure 3.14 shows a rather low position (summer). In order to avoid the second mirror's shade to fall at a particular time of the day onto the first mirror, the latter can be moved sidewards on a track.

The angles of incidence on the coelostat mirrors are time-dependent, but in general smaller than for the heliostat. At any given declination and time of the day there is a coelostat position on the track which minimizes the angles of incidence and thus the instrumental polarization.

The coelostat mirrors do not rotate the image. For this reason today coelostats are much more in use than heliostats, even though there is an additional reflection.

The Turret. The turret, also called the "sunseeker", is an altazimuth mirror arrangement which is very compact and thus minimizes the local turbulence generated by the tower top. It consists of two spheres, each containing a flat mirror. One moves about a vertical axis in the azimuthal direction and carries with it the other, which seeks the Sun's elevation (Fig. 3.15). Altazimuth telescope mounts are more complicated than equatorial mounts in that both

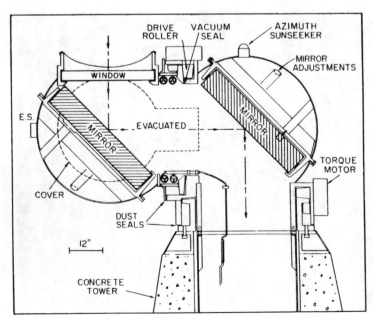

Fig. 3.15. Turret of the Richard B. Dunn Telescope at Sacramento Peak, New Mexico, USA (Dunn 1969). E.S. is the elevation sunseeker. Courtesy National Solar Observatory, AURA, Inc.

axes must be turned simultaneously in order to compensate for the Earth's rotation. But all modern telescopes are computer controlled in any case. Hence there is no difficulty, except that the solar image rotates; in contrast to the heliostat this rotation varies in time. To compensate for image rotation the R. B. Dunn Telescope at Sacramento Peak is rotated as a whole.

3.2.3 Telescopes with Long Primary Focus

We now discuss the solar telescopes which have a focal length f large enough for high-resolution work in the primary focus, with presently available detectors. The typical aperture, D, of a solar telescope is between 0.5 m and 1.5 m, hence large f also means large focal ratio, $N = f/D$. The long-primary-f telescopes are fixed in the observatory building; they are fed by steerable mirrors as explained in the preceding section.

Let us, as a typical example, consider the 60-cm Vacuum Tower Telescope at Kitt Peak. This telescope will be replaced in the near future by SOLIS, a specialized tool for Synoptic Optical Long-term Investigations of the Sun, including a 54 cm telescope with a vector-spectromagnetograph. Still, the Vacuum Telescope at Kitt Peak is well-suited for demonstration because of its clear-cut design and good documentation (Livingston et al. 1976 a); moreover, since its construction in 1973 it has been of invaluable service to many solar astronomers through its magnetograms, which were regularly recorded and distributed, e.g., in the *Solar Geophysical Data*.

A north-south section across the building and telescope is shown in Fig. 3.16. The tower is located at the southeast edge of Kitt Peak and has a completely retractable wind screen, so that the coelostat is well exposed to the wind. The tower platform is kept rather small by the use of a curved track for the coelostat mirror (the "Zeiss arrangement").

The light enters the vacuum tank through a 86 cm diameter, 10 cm thick fused-quartz window. Inside the tank the beam is folded, first by the focusing main mirror, and then by another flat mirror. The main mirror is spherical, with $f = 36$ m, stopped to an aperture of 60 cm. The f ratio, $N = 60$, is typical, cf. the solar telescopes listed in Table 3.1. The inclination of the mirror is 1.25°.

Spherical Aberration. Thanks to the large focal ratio, spherical aberration is negligible for this telescope. In principle, spherical aberration limits the usable aperture D, because off-axis rays have a shorter focus than rays close to the axis (the *paraxial* rays). Hence with increasing D the point spread function widens.

In the following we shall apply *Strehl's criterion* for the quality of the image. This criterion demands that the central intensity of the point spread function is at least 80% of the central intensity if there was pure diffraction (i.e., that the *Strehl ratio* is at least 0.8). In the case of spherical aberration this condition coincides with the *Rayleigh $\lambda/4$-criterion*, namely that the

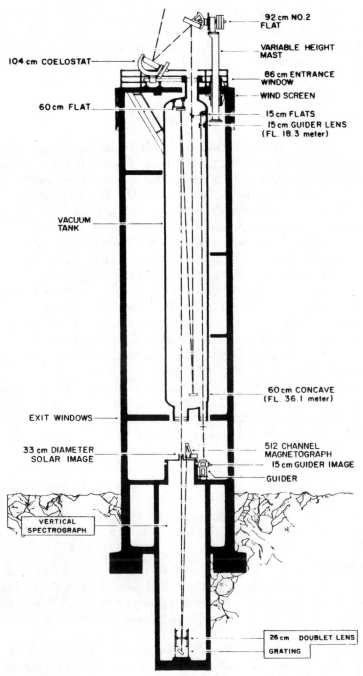

Fig. 3.16. The Kitt Peak 60-cm Vacuum Tower Telescope (Livingston et al. 1976 a).
Courtesy National Solar Observatory, AURA, Inc.

wave-front aberration should be no more than one quarter wavelength. It is therefore also called the "Rayleigh-Strehl criterion". The detailed theory of aberration shows that the criterion requires

$$D \leq 512\lambda N^3 \; . \tag{3.22}$$

For $D = 60\,\mathrm{cm}$, $N = 60$, and visible light, (3.22) is amply satisfied. For this reason spherical main mirrors are commonly used for solar telescopes with large N. These mirrors are easier to manufacture, and therefore cheaper, than parabolic mirrors.

Coma. Also still tolerable is the coma. Increasing with angular distance from the center of the field, the coma appears as a radially elongated asymmetric point spread function, and hence limits the usable field of view. If we apply again Strehl's criterion (which for the coma is equivalent to wave-front aberrations of no more than 0.6λ) and use the results of optical aberration theory, we find the angular radius of the usable field (in radians) to be

$$w_0 = 19.2\lambda N^2/D \; , \tag{3.23}$$

or about $3°$ for the Kitt Peak tower telescope at $\lambda = 500\,\mathrm{nm}$. Thus the criterion is satisfied, even if we take the $1.25°$ inclination of the main mirror into account, although the margin is not as wide as for the criterion (3.22) for spherical aberration.

Astigmatism. The severest optical aberration of our example telescope is astigmatism. Astigmatism could be avoided by the use of an off-center piece cut from a parabolic mirror instead of an inclined sphere. With the inclined sphere the telescope is a "schiefspiegler"; astigmatism occurs because even for an axial incident beam the mirror appears no longer rotationally symmetric, but rather as an ellipsoid. Rays reflected from the "major" (and "minor") axes (perpendicular and parallel to the direction of inclination) are focussed in two elliptical patches slightly outside (inside) the mean focal plane. Like the coma, astigmatism generally limits the usable field of view. For a spherical mirror Strehl's criterion (now equivalent to wave-front aberrations of no more than 0.17λ) determines the angular radius of the usable field (in radians):

$$w_0 = (2.7\lambda N/D)^{1/2} \; . \tag{3.24}$$

In our case $w_0 = 0.67°$ at $\lambda = 500\,\mathrm{nm}$, i.e., w_0 is little more than half the inclination of the main mirror!

At the Kitt Peak tower telescope the main mirror's astigmatism is largely compensated by cylindrical bending of the flat mirror in the vacuum tank which folds the beam down towards the exit window. The bending amounts to a 1.2λ deformation of this mirror.

Telescope MTF and Transmission. The modulation transfer function of the entire telescope has been measured (Livingston et al. 1976 a). It lies markedly below the diffraction limited $\mathrm{MTF_D}$ (3.8). Approximately we have $\mathrm{MTF}(k) \approx \mathrm{MTF_D}(3k)$, i.e., the spatial resolution is only about one-third of the optically perfect telescope!

Each optical surface of the instrument absorbs a certain fraction of the light, as illustrated in Fig. 3.17. The exact absorption depends on wavelength, but by and large the intensity is reduced by a factor 0.5 due to the telescope, a factor 0.2 due to the spectrograph, and again by a factor 0.2 due to the magnetograph. Altogether, the transmission is only a few percent, and this is similar with many other telescopes.

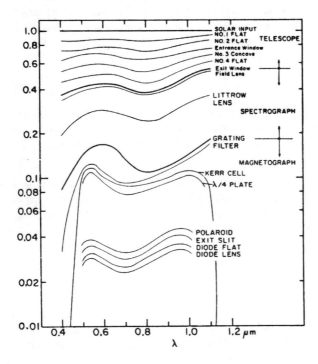

Fig. 3.17. Measured transmission of the 60-cm Kitt Peak Vacuum Tower Telescope, and of the accompanying spectrograph and magnetograph (Livingston et al. 1976 a). Courtesy National Solar Observatory, AURA, Inc.

Guiding. An important feature of solar telescopes is the guiding system. Guiding is necessary because normally only a small part of the solar image, namely the narrow strip falling onto the entrance slit of the spectrograph, is investigated. During the time of measurement one must be sure that this strip remains the same, or that the solar image is scanned in a controlled way, if so desired.

In our example telescope a 15 cm × 22 cm flat mirror, located just below the entrance window and inclined by 45°, directs the central part of the beam to an objective with $f = 18$ m, forming an auxiliary image of 15 cm diameter

Table 3.1. Examples of high-resolution solar telescopes with large primary focal length. Coel coelostat (V vertical, H horizontal), D aperture (L Lens, M mirror), f focal length

Telescope	Year	Type	D [cm]		f [m]	$N = f/D$
Mt. Wilson, 150 ft Tower	1911	Coel, V	L	30	45	150
Einstein-Turm, Potsdam	1924	Coel, V	L	60	14	23
Kitt Peak, McMath-Pierce	1962	Heliostat	M	160	82	51
Meudon	1968	Coel, V	M	60	45	75
Sac. Peak, R.B. Dunn Tel.	1969	Turret	M	76	55	72
Crimea	1973	Coel, V	M	90	50	56
Kitt Peak, Tower	1973	Coel, V	M	60	36	60
Baikal, Russia	1978	Siderostat	L	76	40	53
Sayan, USSR	1979	Coel, II	M	80	20	25
German Tower, Tenerife	1987	Coel, V	M	70	46	66
Swedish Tel., La Palma	2002	Turret	L	97	20.8	21

(Fig. 3.16). The position of this image is tracked automatically. In this way signals are generated which are used to correct the tilt (in two directions) of the second flat coelostat mirror. Because thermal expansion can cause the guide and telescope optics to drift apart, a laser system monitors and compensates for departure.

Variants. A number of other high-resolution telescopes with long primary focus are listed in Table 3.1. They are all fixed telescopes, but with variations. For example, the German Vacuum Tower Telescope on Tenerife, which is very similar to the Kitt Peak vacuum tower, has a double building. An inner tower carries the coelostat, the telescope, the guiding system, and the spectrograph. An outer, completely separated tower houses ancillary equipment, the elevator, etc.; its main purpose, however, is to protect the telescope from being shaken by the wind, which is generally stronger on the Canary Islands than at Kitt Peak. Another difference in detail is that the astigmatism originating from the slight inclination of the main mirror is corrected by a cylindrical bending of the main mirror itself.

Although Table 3.1 is by no means complete, it is representative in that the vertical coelostat is the most popular configuration. There are essentially two reasons for this. First, coelostats have no image rotation. Second, the vertical arrangement is less vulnerable to the seeing conditions: it avoids ground-level turbulence and also has the advantage that the light path is perpendicular to the stratification of air layers with different indices of refraction.

Problem 3.6. Write the field limitations (3.23) and (3.24) in seconds of arc for $\lambda = 500\,\text{nm}$ and for D given in centimeters. Apply the result to the telescopes of Table 3.1; consider in particular the 0.84° tilt of the main mirror of the Vacuum Tower Telescope at Izaña.

3.2.4 Telescopes with Short Primary Focus

We shall now consider a second class of high-resolution solar telescopes. The design of these telescopes widely varies, but they all have in common a primary focal length which is much smaller than that of the instruments described in the preceding section.

The main advantage of a short primary focus is that the whole telescope can be compact and therefore can be built as a freely steerable instrument. This makes it possible to have a solar image before any reflection at inclined surfaces, which is important for polarimetry.

A complication is the necessity of secondary optical elements to form an image with a sufficiently large scale for high-resolution work. Also, because of the weight and size of diverse post-focus equipment, a stationary image is desirable in any case. Usually a Coudé system provides such an image. But even then polarimetry is safer with this type of instrument (in comparison to a coelostat system): apart from the small dependence on declination, δ, the angles of incidence on the diverse mirrors remain constant.

We shall again consider a typical example, the 45-cm evacuated Gregory Coudé Telescope at Izaña, Tenerife. This telescope will be replaced by a larger instrument (GREGOR, cf. Table 3.2), but an almost identical copy is in operation at the Istituto Ricerche Solari Locarno. The optical scheme is illustrated in Fig. 3.18. With $f = 2.48\,\mathrm{m}$ the focal ratio is $N = 5.5$. The

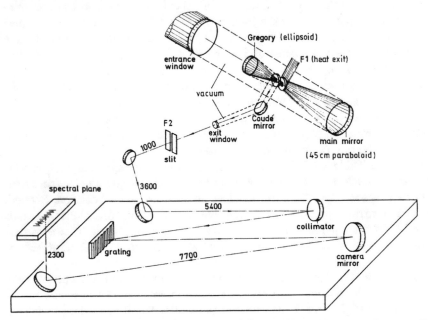

Fig. 3.18. Optical arrangement of the German 45-cm Vacuum Gregory-Coudé Telescope at the Observatorio del Teide, Izaña, Tenerife

Fig. 3.19. The Zeiss 60-cm Domeless Solar Tower Telescope at Hida Observatory, Japan. Courtesy S. UeNo

main mirror is a paraboloid; a sphere would not satisfy the Rayleigh-Strehl criterion (3.22).

Excess Heat and Scattered Light. A solar image 2.5 cm in diameter is formed at a water-cooled copper disc in the prime focus, F 1. A 2.5-mm hole in this disc selects 1/10 of the solar diameter for further enlargement. Thus, 99% of the sunlight is deflected out of the telescope. This is important, because less scattered light can be produced further in the telescope, and unwanted heat is removed. In case of the 45-cm telescope the power is only about 200 W, but for a 4-m telescope it would be near 15 kW!

The possibility to remove unwanted heat and light is an essential advantage of the Gregory system as compared to the Cassegrain system. The latter has a convex mirror *between* the main mirror and F 1, and therefore no real primary image where the heat trap could be located. In addition, the Cassegrain system requires a special baffle that prevents light from outside the wanted field from reaching the (secondary) focus and detector; this is important for extended sources such as the Sun.

The limitation of the field of view is meaningful also because the coma, which is the dominant aberration, in any case would permit only a slightly larger field, cf. (3.23).

 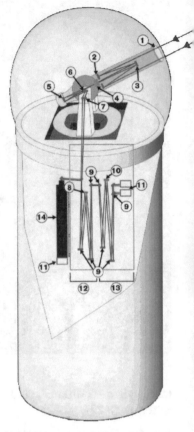

Fig. 3.20. The French-Italian solar telescope THEMIS at the Observatorio del Teide, Izaña, Tenerife. 1 Entrance window, 2–3 Ritchey-Chrétien system, 4 exit window, 5–7 secondary optics, 8–13 spectrograph with camera, 14 universal birefringent filter

Alignment. Special care must be applied to the alignment of the primary and secondary optical elements. In contrast to the spherical surface, the parabolic main as well as the elliptic secondary mirror each have a vertex and only a single axis of symmetry. These axes must coincide. Moreover, the two foci must coincide. For the focus tolerance we apply again the Strehl criterion of at least 80% intensity at the center of the point spread function. As in the case of spherical aberration, such a criterion for the focus tolerance is equivalent to Rayleigh's $\lambda/4$ criterion, and yields an upper bound for the allowable defocusing Δ:

$$|\Delta| \leq 2N^2\lambda . \tag{3.25}$$

While this upper bound is rather generous for the telescopes with large N described in the preceding section, we find (at $\lambda = 500\,\mathrm{nm}$) $|\Delta| \leq 30\,\mu\mathrm{m}$ for the example Gregory telescope!

A flat mirror behind the heat trap reflects the light into the declination axis. The position of this mirror must be stable in spite of the near-by heat concentration. A second flat mirror (the Coudé mirror) deflects the light into the polar axis. These two mirrors must be carefully aligned in order to avoid a circular wandering of the solar image. At zero declination the two mirrors are oriented perpendicular to each other. In this case the instrumental polarization is minimal.

Guiding of the Gregory Coudé Telescope is achieved with the help of an auxiliary telescope, rigidly attached to the main tubus, which allows the observation of the solar limb. According to the momentary limb position, signals are generated which control the telescope drives at the declination and polar axes.

Variants. Examples of other solar telescopes with short primary focal length are given in Table 3.2. Figure 3.19 shows the domeless solar telescope at Hida, Japan, designed by Zeiss with the aim to avoid dome-generated seeing. The Hida telescope has a Gregory system similar to the one described above. The mounting is altazimuthal. A 45° flat mirror, as in a Newton telescope, reflects the light into the elevation axis, which at its end has the Gregory mirror. THEMIS, the Télescope Héliographique pour l'Etude du Magnétisme et des Instabilités Solaires, is a solar telescope especially designed for polarimetry (Fig. 3.20). The polarization modulator (Sect. 3.5.4) is located on the main

Table 3.2. Examples of high-resolution solar telescopes with short primary focal length. E equatorial, A altazimuth; D aperture (L Lens, M mirror); f_p, f_eff primary and effective focal length. The ATST is in the study phase

Telescope	Year	Mount	D [cm]		f_p [m]	f_eff [m]
Okayama Coudé	1968	E	M	65	6.0	37
Pic du Midi	1972	E	L	50	6.45	35
Big Bear	1973	E	M	65	3.5	50
Domeless Coudé, Hida	1979	A	M	60	3.15	32.2
Multi-Channel, Huairou	1984	E	M	60	1.5	6/12
Gregory Coudé, Izaña	1985	E	M	45	2.48	25
THEMIS	1996	A	M	90	3.15	15
Dutch Open Telescope	1997	E	M	45	2.0	44
SOLIS VSM [a]	2002	E	M	54	0.8	3.3
GREGOR [b]	2005	A	M	150	2.50	60
ATST [c d]	2008	A	M	400	10–20	80–120
LEST design study [e]	1990	A	M	240	5.5	16.8/163.4

[a] The Vector-Spectromagnetograph (Keller 2001), [b] von der Lühe et al. (2001), [c] Beckers (1995), [d] Keil et al. (2001), [e] Engvold and Andersen (1990)

optical axis, so that for this particular purpose the configuration is completely axisymmetric and hence free of instrumental polarization. With an entrance window of 1.1 m diameter and 6.6 cm thickness, THEMIS is not evacuated but has a helium filling.

New Projects. The fine structure of the magnetic field at the solar surface requires polarimetric investigations at smaller spatial scale than that resolved by existing solar telescopes. Large telescopes, with apertures up to 4 m, have therefore been proposed. A detailed study (Engvold and Andersen 1990) has been made for LEST, a Large Earth-Based Solar Telescope. The aperture is 2.4 m and, according to (3.7), the diffraction limit is 0.05″ at $\lambda = 500$ nm. It is a modified Gregorian system, with a second and a third focus as listed in Tab. 3.2, and with an adaptive optics system. The alignment requirements are rather strict for this instrument. Coma and defocusing are the major sources for optical image degradation. All mirrors are equipped with an active alignment system. In addition, the two ellipsoidal mirrors are especially shaped to correct the optical aberration of the primary. This can be demonstrated by the calculation of *spot diagrams*. The spot diagram shows where rays emanating at various angles from an object point hit the image plane. As illustrated in Fig. 3.21, the coma at F 1 becomes intolerably large even as little as 3″ off the optical axis, but is substantially reduced in the secondary and tertiary images.

The LEST study employs a helium-filled telescope tubus. Even the smaller SOLIS telescope will have a helium filling, which permits an entrance window of only 6 mm thickness. The alternative is an open structure (Fig. 3.8):

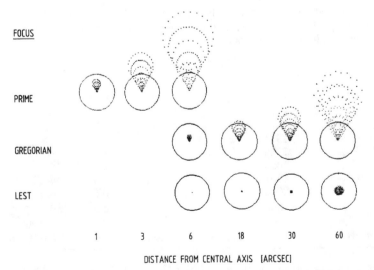

Fig. 3.21. Spot diagrams calculated for the three foci of LEST (Andersen et al. 1984). Courtesy LEST Foundation

the German telescope GREGOR for the Observatorio del Teide, which will replace the 45 cm evacuated Gregory telescope described in this section, and the proposed ATST (Advanced Technology Solar Telescope) will be open solar telescopes.

Problem 3.7. Calculate the field limitation due to coma and astigmatism for the primary images of the telescopes of Table 3.2. What are the tolerances for the primary and secondary foci?

Problem 3.8. Suppose there are dark and bright filaments with 20% intensity contrast and about 0.25″ apart in a sunspot penumbra. How much of this contrast would be transmitted by a 2.4-m telescope under perfect seeing conditions?

3.3 Spectrographs and Spectrometers

3.3.1 The Grating Spectrograph

The solar telescope projects the image of the Sun into its focal plane. This is also the location of the *entrance slit* of the spectrograph. A light cone of width $D/f = 1/N$ enters, is made parallel by the *collimator*, and falls onto the *grating*. In order to make the best possible use of the available light and of the optical elements, precisely the whole cone should be collimated. It follows that the focal ratio of the collimator should be the same as that of the telescope:

$$f_S/D_S = f/D \ . \tag{3.26}$$

Most modern solar spectrographs have reflection gratings. The dispersed reflected light is refocussed either by the collimator itself as in the *Littrow* arrangement (Fig. 3.16), or by a separate camera as in the *Czerny–Turner* system (Fig. 3.18). In the latter case the collimator and camera both operate slightly off-axis and therefore introduce coma and astigmatism. In a symmetric configuration the coma is cancelled; thus generally collimator and camera have equal focal lengths, f_S, the "focal length of the spectrograph".

Dispersion and Spectral Resolution. Let us consider a reflection grating, and let a be the distance between two grooves (Fig. 3.22), the *grating constant*. If α is the angle of incidence, and $\beta(\lambda)$ the angle of reflection, then the directions β of maximum intensity are given by the grating equation

$$m\lambda = a(\sin \alpha + \sin \beta) \ , \tag{3.27}$$

where m is an integer, the *order* of the spectrum. Differentiation of (3.27) yields the angular *dispersion*

$$\frac{d\beta}{d\lambda} = \frac{m}{a \cos \beta} \ . \tag{3.28}$$

The linear dispersion in the spectral plane is $dx/d\lambda = f_S d\beta/d\lambda$; hence, with $\beta \simeq \alpha$,

$$f_S = \frac{dx}{d\lambda}\frac{a\cos\alpha}{m} \ . \tag{3.29}$$

Available gratings have, e.g., 632 grooves/mm[1], and are typically used in the 5th order at $\alpha \approx 60°$. The required dispersion $dx/d\lambda$ is $4\,\mathrm{mm/\AA}$ if, for example, we want to resolve $5\,\mathrm{m\AA}$ and have a detector in the spectral plane with pixels of $20\,\mu\mathrm{m}$. With these numbers (3.29) gives $f_S = 6.3\,\mathrm{m}$. On the other hand, D_S is determined by the dimensions of the grating. A $15\,\mathrm{cm} \times 30\,\mathrm{cm}$ grating is common, and yields a projected aperture of $15\,\mathrm{cm} \times 15\,\mathrm{cm}$ ($\alpha = 60°$). The focal ratio of the spectrograph is thus fixed and, by (3.26), the focal ratio of the example telescope is equally fixed:

$$N = 42 \ .$$

Indeed, the telescopes listed in Tables 3.1 and 3.2 all have (effective) focal ratios of this order of magnitude. We now see that this is a consequence of the available detectors and spectrograph gratings. Detectors with smaller pixels permit smaller focal ratios, and hence more compact instruments.

Problem 3.9. Show that the dispersion of a grating depends only on λ and the angle of incidence (use $\beta \simeq \alpha$).

Problem 3.10. An aperture D_S permits a diffraction-limited angular resolution of order λ/D_S, cf. (3.7). Using this, show that the resolving power $\lambda/\Delta\lambda = nm$, where m is the order of the spectrum and n is the total number of grooves in the grating. What is $\lambda/\Delta\lambda$ for the above example?

Problem 3.11. The entrance slit has a finite width, and its image must be folded with the spectrum. How narrow a slit must be used in order to obtain the $5\,\mathrm{m\AA}$ resolution of the example considered in the text? If the seeing is $1''$ and the telescope has a focal length of $30\,\mathrm{m}$, by how much should the entrance slit be opened so that no light is wasted? What is the spectral resolution in this case?

Problem 3.12. The various orders m overlap each other in the spectrum. Show that a wavelength interval which does not overlap with itself is limited by $\Delta\lambda < \lambda/m$.

The coma is compensated in the symmetric Czerny–Turner configuration because it is itself *antisymmetric* in the angle of inclination. On the other hand, astigmatism is *symmetric* and therefore remains. It limits the field of view in the spectral plane, i.e., the spectral bandwidth that can be used at the same time. Using (3.24), we find in our example spectrograph a band of

---

[1] In this case the separation of grooves is exactly 2.5 times the wavelength of the He-Ne laser used for the interferometric control of the grating manufacture.

$f_S w_0 (dx/d\lambda)^{-1} \approx 37\,\text{Å}$. If a larger bandwidth is necessary, one may correct for the astigmatism by an additional lens or by bending the camera mirror. For example, at the Domeless Solar Tower Telescope of Hida Observatory the correction is achieved by a cylindrical shape of the entrance window of the evacuated vertical spectrograph.

Echelle Spectrograph. There are gratings which allow spectroscopic work at high orders ($m = 50$, say). The grooves must have a special profile for this purpose, strongly *blazed* – cf. Fig. 3.22 –, and the reflectivity of the groove surfaces must be high so that there is enough intensity at high orders. Such a grating is usually called an *echelle* or *echelle grating*. Its density of grooves is low (e.g., 79 per mm, in which case the grating constant is 20 times the wavelength of the He-Ne laser, cf. the footnote on p. 100).

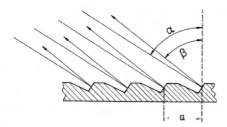

Fig. 3.22. Blazed reflection grating

The advantage of the overlapping spectra is that lines from different regions of the solar spectrum can be recorded simultaneously in a single camera. At the same time, of course, much unwanted light from various orders is superimposed onto the desired spectrum. For this reason the echelle spectrograph has a *predisperser*. Working at low spectral resolution, the predisperser images the spectrum either perpendicular or parallel to the main direction of dispersion. In both cases the wanted and unwanted spectral parts can be separated by masks or filters.

3.3.2 The Fourier Transform Spectrometer

The spectrograph grating essentially is an interferometric device; this is evident from the grating equation (3.27), which is the condition of constructive interference. Other types of interferometers have been used in solar physics, too. Of these, we shall treat here the *Fourier transform spectrometer*.

This instrument essentially is a Michelson interferometer which allows for a *large* variation of the optical path difference. The principle is illustrated in Fig. 3.23. Consider the "balanced" output (as shown in the figure) where each of the two beams undergoes one external reflection at the beamsplitter or recombiner, and each is once transmitted. If the two retroreflectors are identical, with amplitude reflection coefficient r_c, and if the beamsplitter

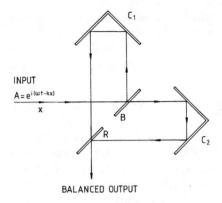

Fig. 3.23. Fourier transform spectrometer. C_1, C_2 retroreflectors, B beamsplitter, R recombiner. Adapted from Brault (1985). Courtesy National Solar Observatory, AURA, Inc.

and recombiner are identical glass plates, with external reflection and transmission coefficients r_e and τ, then a monochromatic input emerges with an amplitude

$$A = r_c r_e \tau e^{i\omega t}(e^{-ikx_1} + e^{-ikx_2}) , \qquad (3.30)$$

where $k = 2\pi/\lambda$; x_1 and x_2 are the lengths of the two paths, measured from some arbitrary common origin before the beamsplitter. The emergent intensity is

$$I = AA^* = \frac{\eta}{2}[1 + \cos k(x_2 - x_1)] , \qquad (3.31)$$

where $\eta = 4r_c^2 r_e^2 \tau^2$ is the efficiency of the instrument.

In general the input to the spectrometer is not monochromatic, but has an intensity distribution $B(k)$. We use $x = x_2 - x_1$ for the path difference, and denote the term that is independent of x by I_0; then

$$I(x) = I_0 + \frac{1}{2}\int_0^\infty \eta B(k) \cos kx \, dk . \qquad (3.32)$$

Hence, we can recover $B(k)$ by a measurement of $I(x)$ and subsequent computation of the cosine transform of $I(x) - I_0$. In practice, one or both of the mirrors are translated with a uniform velocity v, so that $x = vt$, and I is a function of time t. Of course, great care must be applied to monitor the position of the mirrors. Auxiliary interferometers, working with laser beams, are used for this purpose.

The Fourier transform technique yields the *absolute* wavelengths of spectral lines, limited in accuracy only by the accuracy of the laser used in the auxiliary interferometer. Presently absolute wavelengths can be measured to a precision of order $1\,\text{mÅ}$.

The spectral resolution of the Fourier transform spectrometer is determined by the total length L of the signal $I(x)$, i.e., by the range of possible path differences: $\Delta k = 2\pi/L$. The wavelength resolving power, therefore, is given by $\lambda/\Delta\lambda = L/\lambda$. For example, the instrument installed at Kitt Peak Observatory permits $L = 1\,\mathrm{m}$; hence $\lambda/\Delta\lambda = 2 \times 10^6$ at $\lambda = 500\,\mathrm{nm}$.

The lowest wave number which can be determined equals Δk and thus is of no concern. On the other hand, the largest possible k is limited by the *sampling* process: if Δx is the distance between two successive measurements, then $\lambda \geq 2\Delta x$ or $k \leq \pi/\Delta x \equiv k_{\mathrm{Ny}}$; k_{Ny} is the *Nyquist* wave number. The necessity of precise sampling is the reason why Fourier transform spectrometers traditionally have been used in the infrared, where this condition is less stringent. But today's instruments cover the whole visible spectrum.

A major advantage of the interferometer technique is that high spectral resolution is obtained without the need of a narrow entrance slit. Therefore, much more light can enter (high throughput), and the signal-to-noise ratio may be improved; the gain is up to a factor 100 in comparison to a conventional spectrograph (of course, simultaneous high resolution in spectrum *and* space has the same limitations as before).

A wide spectral range can simultaneously be observed with the Fourier transform spectrometer. Also, the efficiency η varies very smoothly with k, so that rather accurate *absolute* intensity spectra can be obtained by calibration of selected bands.

Problem 3.13. The Doppler Compensator. Show that a glass plate of thickness d and refractive index n laterally translates a beam by an amount

$$\Delta x = d\sin\alpha \left(1 - \sqrt{\frac{1 - \sin^2\alpha}{n^2 - \sin^2\alpha}}\right), \tag{3.33}$$

where α is the angle between the plate normal and the incident light. This device has been used to compensate for an unwanted Doppler shift of a spectral line; at the same time, that shift is determined by a measurement of α. Calculate the tilt of a 4-mm glass plate which compensates the Doppler shift caused by $v = 10\,\mathrm{m/s}$. Take the Fe I line at 557.6 nm in a spectrum of 10 mm/Å dispersion, and use $n = 1.5$ for the refractive index of the glass.

3.3.3 The Measurement of Line Shifts

Two examples of spectral line shifts which are of interest in solar physics are the gravitational redshift, $\Delta\lambda_{\mathrm{G}} = \lambda G m_{\odot}/r_{\odot}c^2$, and the Doppler shift, $\Delta\lambda_{\mathrm{D}} = \lambda v/c$. In particular the Doppler shift, which occurs whenever the distance between the source of radiation and the observer changes (with relative velocity v), has been an important object of solar research since about 1960, and special instruments have been designed for its measurement. These instruments all work with the solar spectrum, or at least with a small part of it, and hence belong to the section on spectroscopy; because of their common purpose we treat them here in a separate subsection.

Of course, the whole profile of a spectral line may be recorded and subsequently be used to determine the exact position of the line. However, often this position is the only information of interest. Its direct measurement can save much labor in data collection and data analysis. Nevertheless we must keep in mind that sometimes knowledge of the whole profile is indeed necessary, especially when the line profile is not symmetric.

The Resonance-Scattering Spectrometer. A very accurate method to measure line shifts is the resonant-scattering solar spectrometer. Figure 3.24 illustrates the principle (e.g., Brookes et al. 1978). The instrument works with a selected single spectral line; the sodium line $D1$ at 589.6 nm and the potassium line at 769.9 nm have mostly been used so far.

In order to improve the signal-to-noise ratio and to reduce unwanted heat, a filter first blocks all the sunlight except a band of, say, 1 nm around the line under investigation (Fig. 3.24 a). Next the light is linearly polarized, and in this form enters the electro-optic modulator. This is a special crystal which is uniaxial in the absence of an electric field. Mostly potassium-dihydrogen-phosphate (KH_2PO_4 – "KDP") or ammonium-dihydrogen-phosphate ($NH_4H_2PO_4$ – "ADP") crystals are used. Brookes et al. use a deuterated KDP. These crystals can be made biaxial by means of an electric field applied parallel to the optical axis. If the electric field strength is properly chosen (a voltage of several kV is typically required), the wave polarized parallel to the slow axis is retarded by $\lambda/4$ so that the combined light emerges in circular polarization. If the electric field is reversed the fast and slow axes change their roles, and the circular polarization is of opposite sense. Application of an AC voltage therefore yields alternating right-handed and left-handed circular polarization.

The heart of the resonance spectrometer is the vapor cell. An external magnetic field parallel to the incident beam splits the states of the atoms in

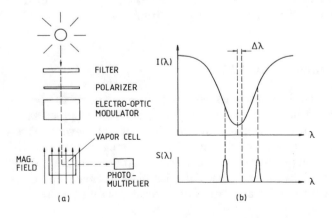

Fig. 3.24. Determination of spectral line shift with a resonance-scattering spectrometer

the cell so that unpolarized light would show the longitudinal Zeeman effect, visible in the resonant-scattered light. For the circular polarized light, however, only one of the two shifted components (i.e., one of the σ components) is present in the scattered light, because only the properly polarized atoms respond. Now, the spectral lines of the metal vapor in the cell are much narrower than the broad line originating in the (much hotter) solar atmosphere, and the scattering cross sections – $S(\lambda)$ in Fig. 3.24 b – can therefore be located at the two sides of the solar line profile; it is only necessary to choose the proper strength of the external magnetic field. If the solar line is shifted, the intensity of the light scattered in the two components is not the same. The difference can be determined by separating the photomultiplier signals (currents or photon counts) measured during the two phases (left and right) of the incident polarized light. For small shifts we may assume that the solar line profile is linear; if, in addition, the line is symmetric, then the velocity is

$$v = k\frac{N_\mathrm{l} - N_\mathrm{r}}{N_\mathrm{l} + N_\mathrm{r}} \,, \tag{3.34}$$

where N_l and N_r are the two photomultiplier signals, and k is a constant. The denominator in (3.34) corrects for intensity variations.

The resonance-scattering spectrometer illustrated in Fig. 3.24 employs a magnetic field *parallel* to the incident beam. As an alternative, a *perpendicular* field can be used. The scattered light must then be observed through right and left circular analyzers (Grec et al. 1976).

As with the interferometer, no slit is necessary for the resonance-scattering spectrometer. A large entrance beam is available and thus the signal-to-noise ratio is high. Since, in contrast to the Doppler compensator, the instrument works without an imaging system, it is particularly suited for the measurement of Doppler shifts in the *mean* solar radiation.

Indeed, since 1976 G. R. Isaak and his collaborators have been able to observe the spherically symmetric pulsations of the Sun (together with the pulsations of low degree, cf. Chap. 5) with a resonance-scattering spectrometer – one of the most important discoveries in recent solar physics (Claverie et al. 1979). Since then, several of these instruments have been installed at various observatories around the world, and the SOHO satellite also carries one with the GOLF experiment.

The wavelength measured in the resonance spectrometer is *absolute*, because of the direct comparison to atomic states. After correction for the Earth's orbital and rotational motions, absolute solar wavelengths can be determined to an accuracy of about 1 mÅ. However, the amplitudes of the solar pulsations can be determined without knowledge of the absolute line position. Because they have specific frequencies, overtones with less than 1 cm/s amplitude have now been identified – this corresponds to a Doppler shift of less than 1 µÅ!

Problem 3.14. Determine the constant k in (3.34) from the known line profile.

Problem 3.15. The widths of the solar $Na\,D\,1$ line at $\lambda = 589.6\,nm$ and of the solar potassium line at $\lambda = 769.9\,nm$ are $0.05\,nm$ and $0.02\,nm$, respectively, at the steepest points of the profile. Calculate the magnetic field strength required to place the Zeeman σ components at these points (for the Zeeman splitting, see Sect. 3.5.1).

3.4 Filters and Monochromators

Broad-band filters, with a bandwidth of order $100\,nm$, essentially are colored glass plates and have already been mentioned in Chap. 1 in the context of the UBV system of stellar colors. We do not discuss these filters further because they are quite generally used in astronomy. Neither shall we discuss the so-called interference filters, which consist of a multitude of thin dielectric layers with alternate high and low index of refraction, and have transmission windows typically 0.5 to $20\,nm$ wide (such filters are often used in order to preselect a certain spectral band). Instead, we shall concentrate on the filters with very narrow bandwidths, $0.1\,nm$ and less, which are exclusively used for solar observation because only here is sufficient light available.

3.4.1 The Lyot Filter

A filter with a very narrow bandwidth was first described by Lyot (1933) and, independently, by Öhman (1938). It is called, variously, the polarization-interference monochromator, the birefringent filter, or just the Lyot filter, and consists of an alternating sequence of polarizers and birefringent crystals, cf. Fig. 3.25.

All the polarizers are mounted parallel to each other. The crystals are uniaxial, cut parallel to their optical axis and positioned in such a way that

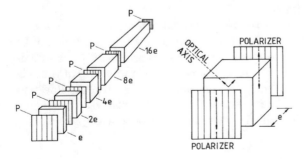

Fig. 3.25. Optical components of the Lyot Filter

this axis makes an angle of 45° with the axis of the polarizers. The thickness of the nth crystal plate is

$$e_n = 2^{n-1}e \ , \tag{3.35}$$

where e is the thickness of the first plate ($n = 1$).

Let us first consider the effect of a single crystal and its two enclosing polarizers. An incident wave, of amplitude A, becomes linear polarized, and in the crystal can be decomposed into two perpendicular components, each with amplitude $A/\sqrt{2}$. One has its electric vibration parallel to the optical axis; this component propagates with velocity c/n_e of the extraordinary ray. The other vibrates in the perpendicular direction, and propagates with velocity c/n_o of the ordinary ray. After traveling across the first plate of thickness e there is a phase difference $\delta = 2\pi e(n_o - n_e)/\lambda$, where λ is the vacuum wavelength. The difference $J = n_o - n_e$ is called *the birefringence* of the material.

Of the two components the second polarizer transmits only the respective parts that are parallel to its own direction; these parts each have an amplitude $A/2$. Let ϕ be a common phase; the interference of the two waves leaving the crystal then yields the amplitude

$$\frac{A}{2}\cos(\phi + \delta) + \frac{A}{2}\cos\phi = A\cos(\delta/2)\cos(\phi + \delta/2) \ . \tag{3.36}$$

That is, we have again a vibration with phase ϕ (and shift $\delta/2$); but its amplitude $A' = A\cos(\delta/2)$ is now modulated in dependence of λ. The intensity

$$I = A^2\cos^2(\delta/2) \ , \tag{3.37}$$

has maxima at $\delta = 2k\pi$, and is zero at $\delta = (2k-1)\pi$, where k is any integer. The wavelengths of the maxima are $\lambda = eJ/k$; for large k, the distance of one maximum to the next is $\simeq eJ/k^2$.

Now take N plates, with thicknesses (3.35), between $N+1$ polarizers. After passage of the last polarizer the intensity is

$$I = A^2\cos^2(\pi eJ/\lambda)\cos^2(2\pi eJ/\lambda)\cos^2(4\pi eJ/\lambda)$$
$$\ldots \cos^2(2^{N-1}\pi eJ/\lambda) \ . \tag{3.38}$$

The individual factors have maxima at

$$\lambda = eJ/k \ , \quad \lambda = 2eJ/k \ , \quad \lambda = 4eJ/k \ , \quad \ldots \lambda = 2^{N-1}eJ/k \ . \tag{3.39}$$

The product (3.38) is nearly zero over most of the interval between two maxima of the first factor. These primary maxima are strongly narrowed by the other factors, but secondary maxima remain, cf. Problem 3.17. The various factors are shown in Fig. 3.26 for the case of 6 crystals. Figure 3.27 demonstrates the transmittance for the individual plates, and also for the various combinations.

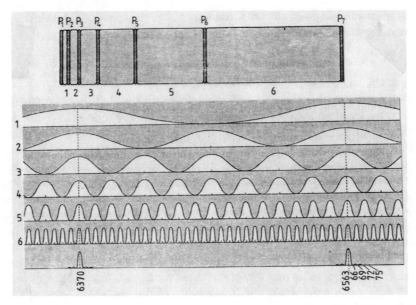

Fig. 3.26. Intensity transmitted by the elements of a Lyot filter. After Lyot (1944)

The width of the transmitted band is determined by the *thickest* crystal plate. Let λ be the desired central wavelength of that band, and $\Delta\lambda$ the desired width. Then we must have

$$\lambda = 2^{N-1}eJ/k \tag{3.40}$$

for some integer k, and

$$\Delta\lambda = 2^{N-1}eJ\left(\frac{1}{k-1/2} - \frac{1}{k+1/2}\right) \tag{3.41}$$

for that same k. As k is large, we have approximately $\Delta\lambda \simeq 2^{N-1}eJ/k^2$; we may eliminate k and obtain

$$2^{N-1}e = \lambda^2/J\Delta\lambda \ . \tag{3.42}$$

On the left of (3.42) we have the thickness of the thickest plate. We see that this thickness is entirely determined by λ, $\Delta\lambda$ and the birefringence J of the material chosen.

The distance in wavelength between the windows of a filter is called the *free spectral range*. Any one of the windows of the Lyot filter can be selected by a broad-band interference filter that closes the other windows.

The birefringent filter is temperature sensitive. At the wavelength of Hα there is a shift of $-0.04\,\mathrm{nm/K}$ for calcite and of $-0.07\,\mathrm{nm/K}$ for quartz. Figure 3.27 illustrates how Lyot was able to employ this effect in order to shift the transmitted band over a few nanometers. Nevertheless, the adjustment

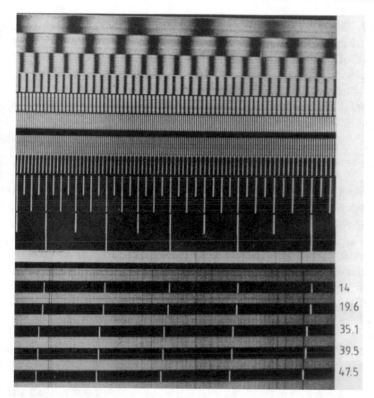

Fig. 3.27. *Upper part*: transmitted intensity for 6 individual birefringent crystals, each between two polarizers. *Middle*: combined transmittance for the filter elements No. 6, 6+5, 6+5+4, ... *Lower part*: transmittance of the complete filter, for various temperatures (in centigrade). For comparison a solar spectrum, ranging from about Na D on the left to Hα on the right. After Lyot (1944)

("tuning") of the filter to a specific wavelength (e.g., an absorption line in the solar spectrum) by way of temperature variation is impractical if rapid changes are desired. Moreover, because of the different temperature coefficients, it would introduce an asymmetric window function for filters which are composed of calcite (the thick ones) and quartz crystals (the thin ones). The whole filter therefore is usually enclosed inside a thermostat, which keeps temperature fluctuations smaller than, say, 0.05 K.

Problem 3.16. The birefringences of quartz and calcite are 0.0092 and −0.172, respectively. What is the required thickness of the thickest plate if $\lambda = 656.3$ nm (Hα), and $\Delta\lambda = 0.05$ nm is desired? How many plates are necessary if the nearest spectral windows of the filter should be no closer than ±25 nm?

Problem 3.17. Calculate the intensity (3.38) transmitted by a Lyot filter composed of 9 crystal plates, with a transmission width of 0.05 nm at Hα, as a function of distance from the window center. What are the heights and the wavelengths of the secondary maxima? (Even a height of only a few percent introduces much unwanted light if the primary maximum is in a strong absorption line and the secondary maxima lie in the neighboring continuum! Cf. the study of an Hα filter by Krafft 1968).

3.4.2 Tuning: the Universal Filter

A better way of tuning the Lyot filter is the addition of a *quarterwave plate* to the birefringent crystal, together with the possibility to rotate the subsequent polarizer. The quarterwave plate is itself a birefringent (uniaxial) crystal. As the main crystal, it is cut parallel to its optical axis and mounted in such a way that this axis is perpendicular to the main axis of the filter. The difference is that the quarterwave plate has its optical axis at 45° relative to the optical axis of the crystal, and that it retards the two rays relative to each other exactly by $\lambda/4$. If we now rotate the exit polarizer relative to the entrance polarizer (which remains fixed to the two crystals) by an angle α, then the intensity of the emergent linear polarized light will depend on λ (because of the λ-dependent retardation in the main crystal) and on α (because of the rotation). Instead of (3.37) we obtain (Problem 3.18)

$$I = A^2 \cos^2(\delta/2 + \alpha) , \tag{3.43}$$

where $\delta = 2\pi e J/\lambda$, as before. Maxima of the intensity occur at

$$\delta/2 + \alpha = k\pi . \tag{3.44}$$

Hence, a rotation by an angle $\alpha = \pi$ shifts an intensity maximum to the same position where the adjacent maximum was at $\alpha = 0$, and rotation angles $\alpha < \pi$ suffice to cover the whole λ range between two maxima. For a given wavelength λ, the required angle of rotation is therefore

$$\alpha = -\frac{\pi e J}{\lambda} \bmod(\pi) . \tag{3.45}$$

If only the thickest element of the Lyot filter is tunable then only the narrowest maxima (line No. 6 of Fig. 3.26) can be shifted. The wavelength range over which the filter can be used is then comparable to the width of the filter window itself. Obviously, tuning over a wide range of wavelengths is possible if the whole filter is subdivided into single tuning elements, so that the maxima in each line of Fig. 3.26 can be shifted to any desired wavelength. In such an assembly, a quarterwave plate must be added to each crystal, and each exit polarizer (together with the following crystal and quarterwave plate) must be made rotatable.

The problem is that simple quarterwave plates work for one specific wavelength only. This is satisfactory if we are only interested in the narrow spectral region around a single solar line, which can be covered by tuning of, say, the two thickest filter elements.

A successful "universal birefringent filter" became possible only after the development of *achromatic quarterwave plates* (e.g., Beckers 1971 b). These plates are themselves combinations of different materials, e.g., quartz and magnesium fluoride, which compensate their respective spectral variations of the retardation. The compensation is achieved in a similar way as the compensation of the chromatic aberration in an achromatic objective.

The spectral region over which the filter can be tuned is limited to the range in which the quarterwave plates can be made achromatic. The Zeiss universal filter has 9 tuning elements, is tunable within the range 450 nm to 700 nm, and has a transmission window 0.0095 nm to 0.025 nm wide, depending on wavelength. A complementary tuning element narrows the window by another factor $1/2$ (Beckers et al. 1975).

It is a complicated problem to find, for a given wavelength, the exact rotation angles of all the tuning elements, and to actually rotate the elements into their precise positions. November and Stauffer (1984) describe a computer-controlled fully tunable 9-element filter.

Problem 3.18. For a single tuning element write down the vibrations of the electric vector upon entrance to, an emergence from, the birefringent crystal and the quarterwave plate. Prove relation (3.43).

3.4.3 A Tunable Michelson Interferometer

The 6 stations of GONG, the Global Oscillation Network Group, as well as MDI, the Michelson Doppler Imager on board of the SOHO satellite, employ tunable wide-field Michelson interferometers as filters. The optical scheme of this filter is shown in Fig. 3.28.

The use of a wide field, and hence an image of the whole Sun, is made possible by inserting two glass blocks, with thicknesses d_1 and d_2 and refraction indices n_1 and n_2, into the two interferometer arms (one of the blocks may be replaced by air or empty space). If $d_1/n_1 = d_2/n_2$, the two arms are optically equivalent, and the angle of incidence does not affect the separation of the two rays.

The path difference of the rays traversing the two interferometer arms is

$$\Delta = 2(n_1 d_1 - n_2 d_2) = 2\frac{n_1^2 - n_2^2}{n_1}d_1 , \tag{3.46}$$

and the phase difference is $\delta = 2\pi\Delta/\lambda$. This closely resembles the phase difference $2\pi e J/\lambda$ between the ordinary and extraordinary rays passing through a Lyot filter element (Sect. 3.4.1), except that the latter have orthogonal polarizations. Since such orthogonal polarization allows for tuning of the filter

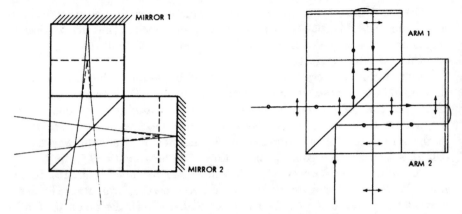

Fig. 3.28. Tunable Michelson interferometer. *Left*: Wide-field function, with two glass blocks between the beam splitter and the mirrors of the two arms. *Right*: Polarizing interferometer, with quarterwave plates placed before the mirrors, and orthogonally polarized output beams. After Title and Ramsey (1980)

element (Sect. 3.4.2), this feature has been added to the Michelson interferometer as well: The input beam is polarized at 45° with respect to the vertical, and the beam splitter is made *polarizing* by means of a multi-layer coating in the splitting plane so that it transmits and reflects, respectively, the components parallel and perpendicular to the plane of incidence. A quarterwave plate is placed in front of each mirror and, since it is traversed twice, rotates the direction of polarization by 90°. As a result one has two orthogonally polarized output beams, as shown in Fig. 3.28. Tuning is then possible with an additional rotatable half-wave retarder. The temperature of the interferometer is controlled by a thermostat.

The Michelson interferometer has periodic transmission maxima. The maxima can be narrowed by a combination of two such interferometers, with free spectral ranges in the ratio 1:2. A prefilter selects the spectral region around a certain solar line; both GONG and MDI use the nickel line at 676.8 nm. The interferometer itself is tuned to several positions within the profile of that line, so that its Doppler shift can be determined.

3.4.4 The Fabry–Perot Interferometer

The interferometer that is named after its inventors (Perot and Fabry 1899) has been used since about 1970 for narrow-band solar spectroscopy. The core of this interferometer consists of two plates with parallel surfaces of reflectance R, enclosing a layer of width d with a medium of refractive index n (Fig. 3.29). The device can be used to define a wavelength standard and is called *etalon* for this reason.

An incident beam is multiply reflected between the two parallel surfaces, but with each reflection a fraction T of the intensity is transmitted, and

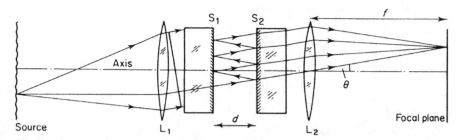

Fig. 3.29. Fabry–Perot interferometer. S_1 and S_2 are partially-reflecting parallel surfaces; L_1 and L_2 are lenses for collimation and imaging, respectively. After Vaughan (1989)

all these transmitted fractions interfere in the outgoing beam. If the angle of incidence is θ, then the path difference between two successive beam fractions is $\Delta = 2nd\cos\theta$, and the phase difference is

$$\delta = 2\pi\Delta/\lambda = 4\pi nd\cos\theta/\lambda \ . \tag{3.47}$$

For an incoming wave of form $\exp(i\omega t)$ the transmitted and reflected (absolute) wave amplitudes are $T^{1/2}$ and $R^{1/2}$, respectively; thus, counting all the reflections and the concomitant phase differences, the outgoing wave is a geometric series,

$$A\mathrm{e}^{\mathrm{i}\omega t} = T\mathrm{e}^{\mathrm{i}\omega t} + TR\mathrm{e}^{\mathrm{i}(\omega t+\delta)} + TR^2\mathrm{e}^{\mathrm{i}(\omega t+2\delta)} + \ \ldots \tag{3.48}$$

or

$$A = T(1 + R\mathrm{e}^{\mathrm{i}\delta} + R^2\mathrm{e}^{2\mathrm{i}\delta} + \ \ldots \ ) = \frac{T}{1 - R\mathrm{e}^{\mathrm{i}\delta}} \ . \tag{3.49}$$

The intensity of the outgoing beam is, therefore,

$$I = AA^* = \frac{T^2}{1 - 2R\cos\delta + R^2} = \frac{T^2}{(1 - R)^2 + 4R\sin^2(\delta/2)} \tag{3.50}$$

or

$$I = I_{\max}\frac{1}{1 + \frac{4R}{(1-R)^2}\sin^2\frac{\delta}{2}} \ , \tag{3.51}$$

where $I_{\max} = T^2/(1 - R)^2$. This result was first obtained by G. B. Airy in 1831. For two values of R the transmitted intensity as a function of phase δ is shown in Fig. 3.30. The transmission is periodic; the mth maximum is at $\delta = 2m\pi$ or, according to (3.47), at wavelength $\lambda = 2nd\cos\theta/m$. The free spectral range, or distance between two successive maxima, is $\mathrm{FSR} = \lambda/m = \lambda^2/2nd\cos\theta$.

The transmission peak becomes narrower as R approaches 1. If the width is small in comparison to the free spectral range, we may expand the $\sin(\delta/2)$ in (3.51) and obtain for the full width $\Delta\lambda$ at half-maximum transmission

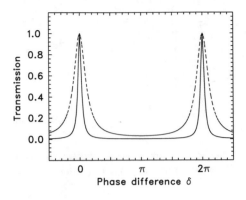

Fig. 3.30. Transmission of a Fabry–Perot interferometer, for reflectivities $R = 0.9$ (*solid*) and $R = 0.7$ (*dashed*)

$$\Delta\lambda = \text{FSR}/\mathcal{F} , \tag{3.52}$$

where

$$\mathcal{F} = \frac{\pi\sqrt{R}}{1 - R} \tag{3.53}$$

is the *finesse* of the interferometer. The finesse is the ratio of the free spectral range to the spectral resolution.

In the absence of absorption we have $T = 1-R$; in this case $I_{max} = 1$, i.e., the intensity at the transmission maxima is equal to the incident intensity, independently of the value of R. If a fraction A of the intensity is absorbed with each reflection, then $T = 1 - R - A$, and $I_{max} = [1 - A/(1 - R)]^2$.

The free spectral range of a Fabry–Perot interferometer is rather small. For $\lambda = 500\,\text{nm}$, $d = 1\,\text{mm}$ we have FSR $\approx 0.13\,\text{nm}$. Therefore, further filters are necessary for the selection of a particular transmission maximum. Cavallini et al. (1987) combine a Fabry–Perot interferometer with a grating spectrograph, Bendlin et al. (1992) describe a combination with a universal birefringent filter. Because of the high maximum transmission of the Fabry–Perot (in comparison to other filters or spectral devices) the combination of two or three such interferometers is most advantageous. An assembly of three Fabry–Perot interferometers for the solar observatory at Culgoora, Australia, has been described by Ramsay et al. (1970). Since the transmission maxima and free spectral ranges of the etalons are essentially determined by the plate separations d, the ratios of these separations must be chosen such that the unwanted transmission windows are closed. Darvann and Owner-Petersen (1994) have studied this problem in detail. Prefilters must still be applied in order to select a certain spectral band, but their pass bands may be wider, with better transmission. Technically, the challenge is to control the parallelism of the etalon plates and the plate-separation ratios with high precision. Kentischer et al. (1998) have built a two-etalon instrument and describe its upgrade to a triple system (which has been realized in 2001). By variation of the plate separation their spectrometer can be used in the wavelength range

420–750 nm, with a spectral resolution of 250 000 or 150 000, for field-of-view diameters 50″ and 100″, respectively.

3.4.5 A Magneto-Optical Filter

The magneto-optical filters are related to the resonance-scattering spectrometer (Sect. 3.3.3). The principle of such a filter is illustrated in Fig. 3.31. The two crossed polarizers block all light except at a wavelength where the state of polarization is changed in the first vapor cell between the polarizers. This is the case for $\lambda = \lambda_L$, λ_R, the wavelengths of the two σ components (Sect. 3.5.1) of a spectral line that is Zeeman-split in a longitudinal magnetic field. The linear polarized light entering the cell can be decomposed into right- and left-handed circular polarized components. Of these, the atoms in the cell will scatter, and so effectively absorb, the component that corresponds to their own state of energy and polarization; the other component passes through the cell, and a fraction of it passes, as linear polarized light, the second polarizer. The state of linear polarization is then again converted to circular polarization by a quarterwave plate. Hence, the second vapor cell will transmit only light in one of the two σ components.

A variant of the scheme shown in Fig. 3.31 has the second vapor cell immediately behind the second polarizer (Cimino et al. 1970). In this case both σ components are transmitted with opposite circular polarizations. The two components can then be separated by a birefringent optical element and recorded independently. An example for this arrangement is the LOWL experiment (Tomczyk et al. 1995 a), installed at Mauna Loa, Hawaii, and Izaña, Tenerife, for the measurement of solar oscillations with low degree l. The potassium line at 769.9 nm is used; a magnetic field of strength 0.3 T places the two σ components into the wings of the solar absorption profile. The width of the two passbands is ≈ 2 pm, one-tenth of the solar line width.

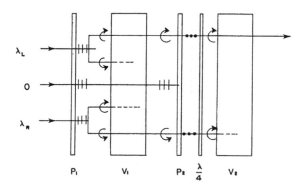

Fig. 3.31. A magneto-optical filter. P_1 and P_2 are crossed linear polarizers, V_1 and V_2 are vapor cells; $\lambda/4$ marks the quarterwave plate, λ_L and λ_R are the wavelengths of the Zeeman-σ components, and O symbolizes the light at other wavelengths. After Cimino et al. (1968)

3.4.6 A Double Monochromator

We have already, in Chap. 1, discussed the absolute measurement of the solar spectral irradiance by D. Labs and H. Neckel. In order to select the 20-Å
bands a double monochromator as shown in Fig. 3.32 was used. It consists
of two identical spectrographs of the Czerny–Turner type. The two gratings
are $52 \times 52\,\mathrm{mm}^2$ in size and have 600 grooves/mm; they are fixed on a single
axis and therefore are rotated together. The resolution is $\lambda/\Delta\lambda \approx 50\,000$;
the entrance slit is $10\,\mu\mathrm{m}$ wide, so that the width of the instrumental profile
is a small fraction of an ångstrøm (cf. Problem 3.11). This makes the 20-Å
window almost rectangular, as shown in Fig. 3.33.

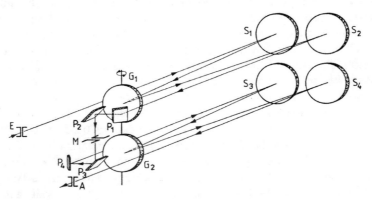

Fig. 3.32. Double monochromator. E entrance slit, M middle slit, A exit slit; S_1,
S_3 collimators, S_2, S_4 camera mirrors; G_1, G_2 gratings; P_1–P_4 plane mirrors. After
Labs and Neckel (1962)

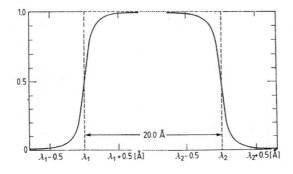

Fig. 3.33. Window function
of the double monochroma-
tor. After Labs and Neckel
(1962)

Problem 3.19. Calculate the width of the exit slit for a spectral band of 20 Å.

3.4.7 The Spectroheliograph

The spectroheliograph is a monochromator for solar imaging which historically preceded the diverse filters described above. It was developed by H. Des-
landres and by G. E. Hale around 1890. Like the monochromator of Labs and
Neckel, the spectroheliograph is a solar spectrograph with an exit slit in the
spectral plane. The solar image is moved across the entrance slit, and simultaneously the photographic plate is moved along behind the exit slit. As an
alternative, the Sun's image and the photographic plate may remain fixed,
but the whole spectrograph with its two slits is moved. In either way, if the
exit slit is placed at a wavelength λ in the spectrum, a monochromatic solar
image in that wavelength λ is generated on the photographic plate.

As compared to the monochromatic filter, the spectroheliograph has the
disadvantage that only a narrow strip of the Sun is imaged at any one time,
so the generation of an image of an extended solar region may take several
minutes of time. Nevertheless, important contributions to solar research have
been made by the use of the spectroheliograph. An example is the discovery
of the solar 5-minute oscillations (Leighton et al. 1962). In this case it was
even possible to take advantage of the time delay between different parts of
the image (Sect. 5.1.1)!

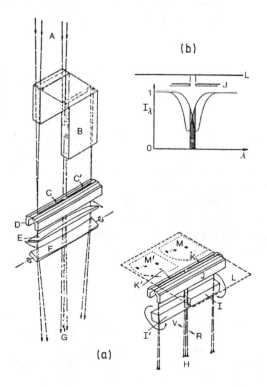

Fig. 3.34. Doppler Spectroheliograph, after Leighton et al.
(1962). (a) Optical scheme, (b)
the shifted line profiles

The principle of Leighton's spectroheliograph is illustrated in Fig. 3.34. It is a modified version, specifically designed for the measurement of Doppler shifts of solar spectral lines.

A beam splitter (B) divides the light beam (A) into two parts, so that duplicate images (C, C') are formed at the entrance slit (D). Lenses (E, F) correct the tilt and the curvature of the spectral lines, and at the same time serve as field lenses for the spectrograph. The split light beam (G) proceeds toward the collimator, and returns from the camera in dispersed form (H). Before the exit slit (J) the two beams pass two plane-parallel glass blocks (I, I') whose equal but opposite tilt slightly shifts the beams toward the red (R) or violet (V), just as the Doppler compensator described in Sect. 3.3.3. As the spectroheliograph scans the solar image, latent images (M, M') are built up on the photographic plate (L) of the successive image strips (K, K'). The crosshatched sections on Fig. 3.34 b represent the portions of the two line profiles admitted to the photographic plate.

The two spectroheliograms which are simultaneously obtained in the two wings of the spectral line are subsequently subtracted. The effect of the line *shift* on the photographic plate is doubled in this way, while *intensity* variations are cancelled.

A variant of the spectroheliograph described here includes means to measure the circular polarization of spectral line radiation. With a proper calibration, a solar *magnetogram* is thus produced as the instrument scans across the solar image (e.g., Fig. 8.11).

3.5 Polarimetry

The presence of a magnetic field can be deduced from the Zeeman splitting of spectral lines. However, this is possible only for a sufficiently strong field. If the field is weak, we must resort to the fact that the Zeeman components are polarized. We have already seen how the resonance-scattering spectrometer and the magneto-optic filter utilize this polarization.

In the present section we shall describe the splitting and polarization of Zeeman components in the solar spectrum, and how the Zeeman effect can be used to extract information about the magnetic field in the region on the Sun where the spectral line has been formed.

3.5.1 Zeeman Splitting

We recall that in the case of a weak magnetic field (weak in the sense that LS – or Russell–Saunders – coupling is the appropriate description) L, S, J, and M_J are the quantum numbers that define the state of an atom. L characterizes the total orbital angular momentum of the electrons (more precisely: of the outer electrons which are treated explicitly), S is the analogue for the spin, and J for the total angular momentum; M_J is the "magnetic" quantum number that determines the component of the total angular momentum in any one direction (z, say), and adopts the values $-J$, $-J+1$, ... J, while J itself can have the values $|L - S|$, $|L - S| + 1$, ... $L + S$.

When the field strength, B, is zero all M_J-states have the same energy. This degeneracy is removed if $B \neq 0$; the z-direction is then the direction of the field vector $\boldsymbol{B}$. For the energy shift and the resulting Zeeman displacement of a spectral line the Landé factor of each state,

$$g = 1 + \frac{J(J+1) + S(S+1) - L(L+1)}{2J(J+1)} \, , \tag{3.54}$$

is of importance. Let g, g' and M, M' be the Landé factors and magnetic quantum numbers of the lower and upper states of the transition considered. Then the displacement of the line from its original position λ_0 is

$$\lambda - \lambda_0 = \frac{e}{4\pi c m_e} g^* \lambda^2 B \, , \tag{3.55}$$

where

$$g^* = gM - g'M' \; ; \tag{3.56}$$

g^* is the g factor for the transition. When λ is measured in cm and B in gauss, (3.55) takes the form $\lambda - \lambda_0 = 4.67 \times 10^{-5} g^* \lambda^2 B$. In the following we shall use the notation $\Delta\lambda_B = \lambda - \lambda_0$.

States with $S = 0$ have $g = 1$. If both states of a transition have $S = 0$ then the selection rule for M_J, namely $\Delta M_J = -1$, 0, $+1$, together with (3.56) yields $g^* = -1$, 0, $+1$. The result is the *normal Zeeman splitting*, or Lorentz triplet. In some exceptional cases a triplet is also possible for transitions between states with $S \neq 0$, but the splitting may then differ from the normal case.

In general there is a Zeeman *multiplet*, or *anomalous* splitting. Since solar lines are broad, the components of such a multiplet normally are not resolved. For a weak magnetic field it is possible to treat the multiplet as if it was a triplet. An *effective* g factor is then calculated from the g^* values of the contributing components; each component is given a weight according to its intensity. A table of Zeeman multiplets and effective g factors has been compiled by Beckers (1969 a), and Harvey (1973) gives a list of lines with large Zeeman splitting.

The Zeeman triplet consists of the two shifted σ *components* and the unshifted π *component*. When the line of sight is in the direction of the magnetic field (the longitudinal Zeeman effect), the observer sees only the σ components, which have circular polarization of opposite sense. When the observer looks in a direction perpendicular to $\boldsymbol{B}$ (the transverse Zeeman effect), he sees all three components: the π component, linearly polarized perpendicular to $\boldsymbol{B}$, and the σ components, linearly polarized parallel to $\boldsymbol{B}$.

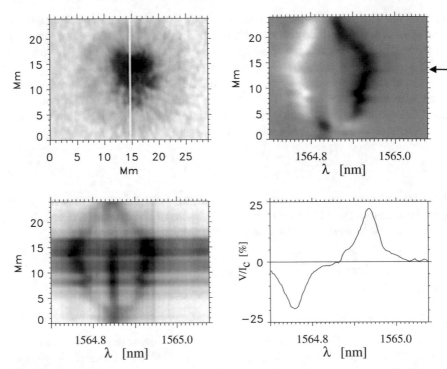

Fig. 3.35. Zeeman splitting and circular polarization of the FeI line at 1564.8 nm. The correlation tracker is used to scan the sunspot image (*upper left*) across the entrance of the spectrograph; the white line is the slit position for the spectrum shown in the *lower left*. The *upper right* panel shows the circular polarization, with an arrow marking the position of the $V(\lambda)$ profile in the *lower right*. Observation on 9 Nov 1999 at the German Vacuum Tower Telescope, Tenerife, with the infrared polarimeter TIP (Schlichenmaier and Collados 2002). Courtesy R. Schlichenmaier

These rules for the polarization of the normal Zeeman triplet are valid for an *absorption* line originating in an optically thin layer. For a line in *emission*, the sense of circular polarization is reversed, and (for the transverse Zeeman effect) "perpendicular" and "parallel" must be exchanged. The reason is of course that only emission lines have intensity by themselves, while what is seen in an absorption line is merely the residual intensity.

The polarization rules apply also in the more general (anomalous) case where $S \neq 0$ and where g^* depends on the quantum numbers: for $\Delta M_J = \pm 1$ there is σ radiation, for $\Delta M_J = 0$ we have π radiation.

When the observer is neither in the direction of $\boldsymbol{B}$ nor in a direction perpendicular to $\boldsymbol{B}$ all the Zeeman components are seen. We shall treat this case below; we shall then also drop the restriction that the spectral line originates in an optically thin layer.

Solar spectral lines are Doppler and collision broadened (cf. Sect. 2.3.7). This has the consequence that the magnetic field B must be of order 0.15 T or larger before the Zeeman splitting can be recognized. Indeed, G. E. Hale in 1908 discovered the magnetic field of sunspots by means of Zeeman splitting. According to (3.55) Zeeman splitting increases with increasing wavelength, $\Delta\lambda_B \propto \lambda^2$, while $\Delta\lambda_D \propto \lambda$; hence the separation of the Zeeman components is seen more clearly in the infrared (Fig. 3.35) than in the visible. Sunspots generally have fields up to $\approx 0.3\,\mathrm{T}$. For most spectral lines this is still weak in the sense of the LS coupling, so that (3.54) to (3.56) are applicable.

If B is substantially smaller than in sunspots, or if B is comparable with the sunspot field strength, but the field covers only a small fraction of the resolved area on the Sun, then the Zeeman effect is at best discernable as line broadening. However, the Zeeman components are polarized regardless how small the splitting is. H. W. Babcock and K. O. Kiepenheuer around 1950 first investigated the possibility of separating the Zeeman components by means of polarimetric methods. Below we shall describe their polarimeters, or *magnetographs*. When we use this latter name we must however keep in mind that rather involved theoretical interpretation lies between the polarimeter output and the solar magnetic field.

3.5.2 Polarized Light

Let us first consider a single monochromatic wave train of electromagnetic radiation, propagating in the z direction. The electric vector lies in the x, y-plane:

$$E_x = \xi_x \cos\phi , \quad E_y = \xi_y \cos(\phi + \varepsilon) , \tag{3.57}$$

where $\phi = \omega t - kz$; ε is the phase difference between E_x and E_y; ξ_x and ξ_y are constant amplitudes which, together with ε, describe the state of polarization of the wave, which is elliptic in general. Equivalent to ξ_x, ξ_y, and ε are the parameters introduced in 1852 by G. G. Stokes,

$$I = \xi_x^2 + \xi_y^2 , \qquad Q = \xi_x^2 - \xi_y^2 ,$$

$$U = 2\xi_x\xi_y \cos\varepsilon , \quad V = 2\xi_x\xi_y \sin\varepsilon . \tag{3.58}$$

The Stokes parameters have the advantage of being measurable by polarimetry. In the form (3.58) they obey $I^2 = Q^2 + U^2 + V^2$, an equation which characterizes *complete polarization*; of course, a single wave train is always completely polarized.

Natural light never is perfectly monochromatic, but possesses a finite bandwidth. Let us first consider a beam with a bandwidth small in comparison to the width of a solar spectral line. Still, such a beam must be viewed as a wave packet, i.e., as a superposition of a large number of independent wave trains. These wave trains all have their own amplitudes and phases, and the Stokes parameters are the respective averages:

$$I = \langle \xi_x^2 + \xi_y^2 \rangle , \qquad Q = \langle \xi_x^2 - \xi_y^2 \rangle ,$$

$$U = 2\langle \xi_x \xi_y \cos \varepsilon \rangle , \quad V = 2\langle \xi_x \xi_y \sin \varepsilon \rangle . \tag{3.59}$$

The averages can also be thought of as time averages, taken over a period of observation that is sufficiently long so that the result is independent of it.

For unpolarized light all transverse directions are equivalent, and all phases ε between 0 and 2π occur with the same probability (and uncorrelated to the product $\xi_x \xi_y$), so obviously $Q = U = V = 0$. At the other extreme, completely polarized light is characterized by ε and the ratio ξ_x/ξ_y being the same for *all* constituent wave trains. In this case, we indeed find $I^2 = Q^2 + U^2 + V^2$. In general, however, there is *partial* polarization, with the degree of polarization defined by

$$P = \left(\frac{Q^2 + U^2 + V^2}{I^2} \right)^{1/2} . \tag{3.60}$$

The parameters Q and U describe the linearly polarized part of a beam, while V stands for the circular polarization.

Problem 3.20. Show that $P \leq 1$ always holds. What is the state of polarization if only Q, or only U, or only V does not vanish? What is the meaning of the *sign* of Q, U, and V?

Stokes Profiles. So far we have considered monochromatic light – monochromatic in the sense that the bandwidth was small compared to the width of solar spectral lines. In a solar line, the radiation at each wavelength λ has a certain state of polarization, i.e., the Stokes parameters (3.59) are functions of λ. The aim, then, is first to measure the profiles $I(\lambda), Q(\lambda), U(\lambda), V(\lambda)$, and second to relate this measurement to the solar magnetic field. We shall consider the second step first, thereby restricting the discussion to the special case of the normal Zeeman effect [Mathys and Stenflo (1987) describe how anomalous Zeeman patterns can be treated in a similar manner].

To start with the simplest case, consider a longitudinal Zeeman triplet in a layer of small optical thickness τ, illuminated from one side with the continuous intensity I_C, unpolarized and independent of λ. At the other side of the layer we observe two shifted absorption-line profiles superimposed on the uniform attenuation due to the layer:

$$I = I_C(1 - \tau) - I_C \tau (\eta^+ + \eta^-)/2 , \tag{3.61}$$

where

$$\eta^\pm = \eta(\lambda \pm \Delta\lambda_B) , \tag{3.62}$$

and $\eta(\lambda) = \kappa_l(\lambda)/\kappa_C$ is the ratio of the line absorption coefficient $\kappa_l(\lambda)$, to the absorption coefficient of the continuum, κ_C, in the vicinity of the line. Since the Doppler and collision broadening is the same in the magnetic and non-magnetic cases, we may use the same absorption profile $\eta(\lambda)$. The second term of (3.61) describes the absorption in the two σ components. This absorption renders the emergent radiation circularly polarized, and may therefore be described in terms of the Stokes parameter V. The polarization is antisymmetric with respect to the line center, as expressed by the sign of $V(\lambda)$:

$$V(\lambda) = -I_C \tau (\eta^+ - \eta^-)/2 \ . \tag{3.63}$$

There is no linear polarization, hence $Q = U = 0$.

Next we consider the normal transverse Zeeman effect in an optically thin layer. We may choose the x, y coordinates such that $U \equiv 0$ (i.e., x is parallel to the magnetic field, and y is perpendicular). We know that the intensities of the $(\sigma^+, \pi, \sigma^-)$ triplet are in the ratios $1/4{:}1/2{:}1/4$ and that the linear polarizations of the σ and π components are at right angles, and hence contribute with different signs to Q. Therefore the radiation emerging from the layer can be described by

$$I = I_C(1 - \tau) - I_C \tau \left(\frac{1}{2}\eta + \frac{1}{4}(\eta^+ + \eta^-) \right) \ , \tag{3.64}$$

$$Q = -I_C \tau \left(\frac{1}{2}\eta - \frac{1}{4}(\eta^+ + \eta^-) \right) \ , \tag{3.65}$$

while $U = V = 0$.

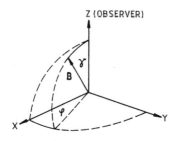

Fig. 3.36. Geometry of the magnetic field

3.5.3 Unno's Equations

In general the magnetic field has an inclination, γ, with respect to the line of sight, and an azimuth, ϕ, cf. Fig. 3.36. In this case all four Stokes profiles will be affected. Let us once again consider first an optically thin absorption layer, but let the incident light, of intensity I, be already polarized, with Stokes parameters Q, U, and V. We define the Stokes vector

$$I = \begin{pmatrix} I \\ Q \\ U \\ V \end{pmatrix} . \tag{3.66}$$

Across a *thin* layer the total change, ΔI, ΔQ ..., in each of the four parameters must be a linear combination of these variables themselves. Thus

$$\Delta I = -\tau I - \tau \eta I . \tag{3.67}$$

The first term on the right of this equation is the continuous absorption, the second describes the absorption due to the line. The elements of the line absorption matrix η depend on the magnitude of the magnetic field via the shifted profiles (3.62); in addition, they explicitly depend on the field angles γ and ϕ. We shall not go through the the detailed derivation of η here, but rely on the work of Unno (1956).

Unno follows the classical theory of H. A. Lorentz and treats the absorbing electrons as linear oscillators. In this picture the atomic state with its Zeeman-split energy levels is represented by a precession of the linear oscillators around the magnetic lines of force. The frequency shift, as well as the directional characteristics of the line absorption, is then a consequence of this precession. Unno's result is

$$\eta = \begin{pmatrix} \eta_I & \eta_Q & \eta_U & \eta_V \\ \eta_Q & \eta_I & 0 & 0 \\ \eta_U & 0 & \eta_I & 0 \\ \eta_V & 0 & 0 & \eta_I \end{pmatrix} , \tag{3.68}$$

where

$$\eta_I = \frac{1}{2}\eta \sin^2 \gamma + \frac{1}{4}(\eta^+ + \eta^-)(1 + \cos^2 \gamma) ,$$

$$\eta_Q = \left(\frac{1}{2}\eta - \frac{1}{4}(\eta^+ + \eta^-) \right) \sin^2 \gamma \cos 2\phi ,$$

$$\eta_U = \left(\frac{1}{2}\eta - \frac{1}{4}(\eta^+ + \eta^-) \right) \sin^2 \gamma \sin 2\phi , \tag{3.69}$$

$$\eta_V = \frac{1}{2}(\eta^+ - \eta^-) \cos \gamma .$$

Although we have not deduced the absorption matrix, we can see a number of its properties. In the diagonal we have always η_I, i.e., the amount of absorption proportional to the respective parameter itself is the same for all four Stokes parameters. The reason is that they all are intensities and behave as such. Outside the diagonal all entries are zero, except for the first line and column; that is, of the three polarization parameters Q, U, and V each changes only in proportion to itself and to the intensity, while the intensity has changes in proportion to all four parameters. Additional entries to the absorption matrix, due to magneto-optic effects, will be mentioned below.

For $\gamma = 0$ we recover the special case of the longitudinal Zeeman effect (3.61) and (3.63), for $\gamma = \pi/2$ and $\phi = 0$ we find the transverse effect represented by (3.64) and (3.65).

Transfer of Polarized Radiation. The solar atmosphere is not an optically thin layer. Moreover, there is not only absorption, but also emission and scattering of radiation. The general theory of radiative transfer in an atmosphere permeated by a magnetic field is complicated; here we shall again follow Unno and assume local thermodynamic equilibrium. For clarity, we shall give an index λ to the Kirchhoff–Planck function $B(T, \lambda)$ in order to distinguish it from the magnetic field B.

We write the equation of transfer in vectorial form,

$$\cos\theta \frac{d\boldsymbol{I}}{d\tau} = (\mathbf{1} + \boldsymbol{\eta})(\boldsymbol{I} - \boldsymbol{B}_\lambda) \ , \tag{3.70}$$

where $\boldsymbol{I}$ is the four-dimensional Stokes vector (3.66), and $\boldsymbol{B}_\lambda \equiv (B_\lambda, 0, 0, 0)$; τ is the optical depth in the neighboring continuum, and θ is the angle to the local vertical direction. The absorption on the right of (3.70) immediately follows from (3.67) because the atmosphere can be envisaged as a sequence of thin layers. The emission, on the other hand, follows from the assumption of local thermodynamic equilibrium: in each of the four equations (3.70) there must be a term formally corresponding to the I term, but with I replaced by $-B_\lambda$. Of course, (3.70) is a generalization of the one-dimensional equation of transfer (2.15), which we have already used in Sect. 2.3.2.

There is a number of special cases in which Unno's equations (3.70) can be solved analytically, or at least can be reduced to the problem of evaluating integrals. One such case is the optically thin absorption layer, which yields (3.67). Further cases have been considered by Mattig (1966). In all these cases an atmospheric solar model, i.e., $T(\tau)$ and therewith $B_\lambda(\tau)$, is a necessary ingredient. In addition, the absorption profile in the absence of a magnetic field, $\eta(\lambda)$, must be known, as described in Chap. 4.

Longitudinal Magnetic Field. In this case we have $\gamma = 0$; hence $\eta_Q = \eta_U = 0$, and

$$\eta_I = \frac{1}{2}(\eta^+ + \eta^-) \ , \quad \eta_V = \frac{1}{2}(\eta^+ - \eta^-) \ . \tag{3.71}$$

The system (3.70) then becomes (with $\mu = \cos\theta$)

$$\mu\frac{dI}{d\tau} = (1+\eta_I)(I-B_\lambda) + \eta_V V \;,$$

$$\mu\frac{dQ}{d\tau} = (1+\eta_I)Q \;,$$

$$\mu\frac{dU}{d\tau} = (1+\eta_I)U \;,$$

(3.72)

$$\mu\frac{dV}{d\tau} = (1+\eta_I)V + \eta_V(I-B_\lambda) \;.$$

We may set $Q = U = 0$; if there is linear polarization at large optical depth, we easily find

$$Q(0,\mu) = Q(\infty,\mu)\exp\left(-\int_0^\infty \frac{1+\eta_I}{\mu}\,d\tau\right)$$

(3.73)

and an analogue expression for U. In any case, the interesting variables are I and V. We define $X(\tau,\mu) = I+V$ and $Y(\tau,\mu) = I-V$, and obtain

$$\mu\frac{dX}{d\tau} = (1+\eta^+)(X-B_\lambda) \;,$$

(3.74)

$$\mu\frac{dY}{d\tau} = (1+\eta^-)(Y-B_\lambda) \;.$$

Each of these two equations is exactly of the type described in Chap. 4 in the context of radiative transfer in a spectral line. We may therefore immediately take the solution (4.7) derived there:

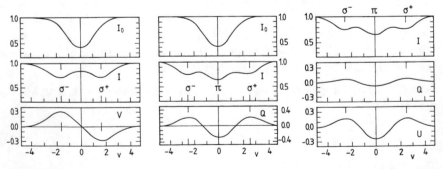

Fig. 3.37. Calculated Stokes parameters, as functions of $v = \Delta\lambda/\Delta\lambda_\mathrm{D}$. *Left*: longitudinal magnetic field, with $v_\mathrm{B} = -1.7$; *middle*: transverse magnetic field, with constant azimuth and $v_\mathrm{B} = 2.5$; *right*: transverse field, with varying azimuth, $\phi = \exp(-10\tau)$, and $v_\mathrm{B} = 2.5$. Adapted from Beckers (1969 b)

$$X(0,\mu) = \int\limits_{0}^{\infty} \frac{1+\eta^+}{\mu} B_\lambda \exp\left(-\int\limits_{0}^{\tau} \frac{1+\eta^+}{\mu}\, d\tau'\right) d\tau .$$ (3.75)

The other variable, Y, is obtained if we replace η^+ by η^-. Since η^+ and η^- are the shifted line absorption coefficients, X and Y are the shifted line profiles:

$$X = I_0(\lambda + \Delta\lambda_B) , \quad Y = I_0(\lambda - \Delta\lambda_B) ,$$ (3.76)

where $I_0(\lambda)$ is the line profile in the absence of the magnetic field. Thus, according to the definition of X and Y, we find

$$I = \frac{1}{2}\left[I_0(\lambda + \Delta\lambda_B) + I_0(\lambda - \Delta\lambda_B)\right] ,$$ (3.77)

$$V = \frac{1}{2}\left[I_0(\lambda + \Delta\lambda_B) - I_0(\lambda - \Delta\lambda_B)\right] .$$ (3.78)

Figure 3.37 shows I and V profiles, together with $I_0(\lambda)$, for a typical value of the ratio $v_B = \Delta\lambda_B/\Delta\lambda_D$ of the Zeeman displacement to the Doppler width of the line.

Transverse Magnetic Field. Here we have $\gamma = 90°$. Let us additionally assume $\phi = 0$. This corresponds to a special choice of the system of coordinates which describe the transverse field components; this choice is possible if the azimuth ϕ does not vary with optical depth. The absorption coefficients are now $\eta_U = \eta_V = 0$, and

$$\eta_I = \frac{1}{2}\eta + \frac{1}{4}(\eta^+ + \eta^-) ,$$

$$\eta_Q = \frac{1}{2}\eta - \frac{1}{4}(\eta^+ + \eta^-) .$$ (3.79)

The equation of transfer, (3.70), is reduced to

$$\mu\frac{dI}{d\tau} = (1+\eta_I)(I - B_\lambda) + \eta_Q Q ,$$

$$\mu\frac{dQ}{d\tau} = (1+\eta_I)Q + \eta_Q(I - B_\lambda) ,$$ (3.80)

$$\mu\frac{dV}{d\tau} = (1+\eta_I)V ,$$

while $U \equiv 0$ due to our choice of coordinates. We may also set $V \equiv 0$ unless there is circular polarization deep in the atmosphere – in which case V is of the form (3.73).

The equations for I and Q are treated in the same manner as the equations for I and V in the previous case. Introducing $X(\tau, \mu) = I + Q$ and $Y(\tau, \mu) = I - Q$ we find that $X(0, \mu)$ is identical to the original line profile, I_0, and that

$$Y(0, \mu) = \int\limits_0^\infty \frac{2 + \eta^+ + \eta^-}{2\mu} B_\lambda \exp\left(-\int\limits_0^\tau \frac{2 + \eta^+ + \eta^-}{2\mu} d\tau'\right) d\tau . \qquad (3.81)$$

Hence the Stokes profiles are

$$I(0, \mu) = \frac{1}{2}\left[I_0(0, \mu) + Y(0, \mu)\right] ,$$

$$Q(0, \mu) = \frac{1}{2}\left[I_0(0, \mu) - Y(0, \mu)\right] . \qquad (3.82)$$

Typical profiles of I and Q are shown in Fig. 3.37 (middle).

Weak Magnetic Field. We now consider the case where the magnetic field is weak, in the sense that $\Delta\lambda_B \ll \Delta\lambda_D$, where $\Delta\lambda_D = (2RT/A)^{1/2}\lambda/c$ is the Doppler width of the original line profile (cf. Chap. 4). The ratio $v_B = \Delta\lambda_B/\Delta\lambda_D$ is then small, and an expansion of the Stokes parameters in powers of v_B is possible. In this expansion I, Q, and U have only even terms, while V has only odd terms. This becomes evident if we consider the symmetry of η_I, η_Q, and η_U with respect to v_B, and the antisymmetry of η_V, and if we inspect Unno's equations under the transformation $v_B \rightarrow -v_B$. Using $v = \Delta\lambda/\Delta\lambda_D$ we obtain the absorption coefficients

$$\eta_I = \eta \qquad\qquad\qquad\qquad +O(v_B^2) ,$$

$$\eta_Q = -\frac{1}{4}\frac{\partial^2 \eta}{\partial v^2} v_B^2 \sin^2 \gamma \cos 2\phi \quad +O(v_B^4) ,$$

$$\eta_U = -\frac{1}{4}\frac{\partial^2 \eta}{\partial v^2} v_B^2 \sin^2 \gamma \sin 2\phi \quad +O(v_B^4) , \qquad (3.83)$$

$$\eta_V = \frac{\partial \eta}{\partial v} v_B \cos \gamma \qquad\qquad\quad +O(v_B^3) ,$$

and the expansion

$$I = I_0 \qquad\quad + \quad v_B^2 I_2 + \dots ,$$

$$Q = \qquad\qquad\quad v_B^2 Q_2 + \dots ,$$

$$U = \qquad\qquad\quad v_B^2 U_2 + \dots , \qquad (3.84)$$

$$V = \quad v_B V_1 + \dots .$$

The transfer equations, to the lowest (zero) order, give

$$\mu \frac{dI_0}{d\tau} = (1+\eta)(I_0 - B_\lambda) ;$$ (3.85)

that is, I_0 is our original line profile.

To the first order we obtain the circular polarization:

$$\mu \frac{dV_1}{d\tau} = (1+\eta)V_1 + \frac{\partial \eta}{\partial v} \cos\gamma (I_0 - B_\lambda) ,$$ (3.86)

which can be solved by an integral of the form which we have had before. Without writing down this solution we note that for a weak magnetic field the leading term of the Stokes parameter V is proportional to $B\cos\gamma$, i.e., to the longitudinal (line-of-sight) component of the (possibly inclined) field vector.

The second order yields the linear polarization:

$$\mu \frac{dQ_2}{d\tau} = (1+\eta)Q_2 - \frac{1}{4}\frac{\partial^2 \eta}{\partial v^2} \sin^2\gamma \cos 2\phi (I_0 - B_\lambda) ,$$

$$\mu \frac{dU_2}{d\tau} = (1+\eta)U_2 - \frac{1}{4}\frac{\partial^2 \eta}{\partial v^2} \sin^2\gamma \sin 2\phi (I_0 - B_\lambda) .$$ (3.87)

Again the solutions can be written in form of our now well-known integral. We see that, to leading order, the Stokes parameters Q and U are proportional to $B^2 \sin^2\gamma$, i.e., to the *square* of the transverse component of the (possibly inclined) magnetic field.

Example calculations of Jefferies et al. (1989) show that the weak-field approximation yields reasonable results if $v_B \leq 0.5$.

Magneto-Optic Terms. There are several effects that cause further terms in the matrix η, Eq. (3.68). The coupling between the Stokes parameters arising from those terms must be taken into account if high accuracy is desired. One of those effects is the *anomalous dispersion* of an electromagnetic wave propagating through a magnetized plasma, in the present context first treated by Rachkovsky (1962 a); in a longitudinal field, for example, there is Faraday rotation of the direction of linear polarization. A second effect is *radiative scattering*: for scattered light the state of polarization generally depends on the angles of the incident and scattered rays relative to the direction of B.

General Solution. For an arbitrary depth-dependence of the absorption matrix Unno's equations, supplemented by the magneto-optic terms, must be integrated numerically. Diverse algorithms to do this have been described; for the interpretation of observed Stokes parameters, however, an *inverse* problem must be solved: The profiles $I(\lambda)$, $V(\lambda)$ ... are given, and the atmospheric structure and motion, the magnetic field, etc., are wanted. One way is to guess the magnitude and direction of the magnetic field in the solar atmosphere, to calculate the elements of the absorption matrix, and then to solve (3.70). If the emergent Stokes profiles agree with the observed profiles we may say that the field has been determined, otherwise we must iterate.

In this context the *response function* is a useful concept. For any physical variable x, say the atmospheric temperature, the magnetic field strength, or a velocity component, the response function $\boldsymbol{R}_x$ is an integral kernel that determines how the Stokes profiles respond to a change of x, viz.

$$\delta \boldsymbol{I}(\lambda) = \int\limits_0^\infty \boldsymbol{R}_x(\lambda, \tau)\delta_x(\tau)d\tau \ . \tag{3.88}$$

Real Stokes profiles show a great variety of forms, and often differ substantially from the schematic cases of Fig. 3.37. For example, the two wings of $V(\lambda)$ may be asymmetric, in amplitude as well as in area (Fig. 3.38, first row). These asymmetries, δa and δA, are usually defined as the difference in amplitude and area, respectively, between the blue and red wings, divided by the sum. Another quantity of interest is the wavelength at which $V(\lambda) = 0$ between the two wings. Interpreted in terms of the Doppler effect this *zero-crossing velocity* may amount to several km/s, cf. the examples in the second row of the figure. If the resolution element of the observed area contains an unresolved inhomogeneous magnetic field, e.g., a tube of magnetic flux in a field-free environment, then the zero-crossing velocity refers to the magnetic part alone. The third row of Fig. 3.38 shows more irregular V profiles; some resemble the theoretical profiles of Q and U, possibly due to a superposition of V profiles from both magnetic polarities, Doppler-shifted relative to each other.

If the circular polarization $V(\lambda)$ is integrated over a spectral band including lines with asymmetric V profiles, then a non-zero net polarization is obtained. Illing et al. (1975) have measured such *broad-band polarization*, and first suggested a model containing layers of different velocity and magnetic field strength as an explanation.

For the numerical treatment of Unno's equations it makes no difference whether or not the field magnitude B and the angles γ and ϕ are functions of depth. The inversion method could be extended even to the case of an atmospheric and magnetic structure that is horizontally inhomogeneous. This would require to solve the transfer equations in three dimensions, instead in terms of optical depth τ only. A great difficulty, however, is the limited resolution of the polarimetric observations, which often precludes the retrieval of detailed information from the inversion. We must keep in mind that a certain amount of circular polarization can be generated in two different ways: either by a weak uniform field, or by a stronger field that occupies only a fraction of the resolved area at the solar surface. We postpone the discussion on how to discriminate between these two cases to Chap. 8.

An example of Stokes parameters computed for a transverse field and depth-dependent field azimuth ϕ is shown in Fig. 3.37. In Fig. 3.39 the maximum of the V profile is shown as a function of B and $\cos\gamma$. An atmosphere $T(\tau)$ appropriate to a sunspot umbra (Hénoux 1969) has been employed in

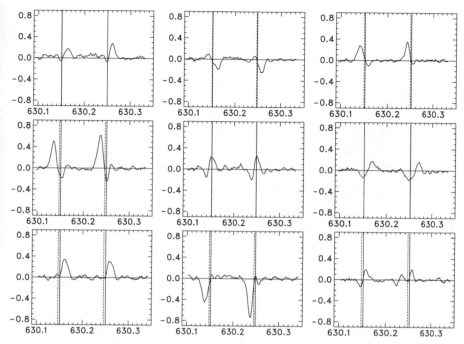

Fig. 3.38. Circular polarization $V(\lambda)$ of two Fe I lines (at 630.15 nm and 630.25 nm), in % of the intensity I_C of the continuum near the lines. The *upper row* shows large amplitude and area asymmetries, the *middle row* large shifts v_{ZC} of the zero-crossing velocity, and the *lower row* shows "irregular" profiles. Wavelength λ is in nm; *solid* vertical lines mark the rest positions of the two iron lines, *dashed* lines mark the cores of the corresponding $I(\lambda)$ profiles. From measurements of small-scale magnetic flux concentrations in the photosphere, made with the Advanced Stokes Polarimeter at Sacramento Peak Observatory. Adapted from Sigwarth (1999)

this latter calculation. The close proportionality of V to $B\cos\gamma$ at small field strength, and the saturation at large field strength, is evident from the result.

Problem 3.21. Find (3.67) as a special solution of (3.70). Derive an equation for the degree, P, of polarization; show that $dP/d\tau$ depends only on P, I, B, $dI/d\tau$, and η_I.

3.5.4 Solar Polarimeters

A solar polarimeter is a spectrograph with additional optical elements which allow discrimination between the various kinds of polarization. The most important type is the photoelectric V-polarimeter. Kiepenheuer (1953) and Babcock (1953) first built such instruments. Since then, circular polarization has regularly been measured at many solar observatories. A number of polarimeters that determine the whole Stokes vector also have been built, but linear polarization is more difficult to measure. One reason is that V is linear

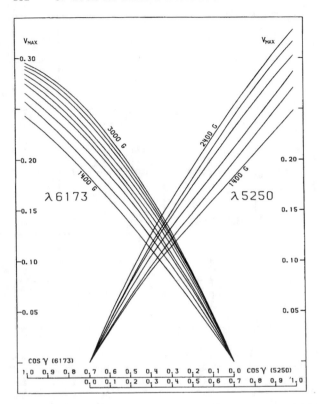

Fig. 3.39. Maximum amplitude of the V parameter for two spectral lines, as a function of field strength and field inclination. After Wittmann (1974 a)

in B, while Q and U are of second order, as we have seen. Another reason is that solar telescopes with inclined mirrors introduce linear polarization which must be corrected for.

Figure 3.40 illustrates the principle of a photoelectric V-polarimeter. Behind the entrance slit is an electro-optic light modulator (EOLM), which we have already encountered in the context of the resonance-scattering spectrometer (Sect. 3.3.3). The purpose here, however, is opposite: the orientation of the crystal and the applied AC voltage are chosen in such a way that the circularly polarized part of the light becomes linearly polarized. Thus, an intensity signal which alternately originates from the right and left circular polarization passes the polarizer and enters the spectrograph. A servo mechanism, e.g., the Doppler compensator of Sect. 3.3.3, centers a magnetically sensitive line between two exit windows in the spectral plane. An electronic device measures the difference between the two signals and, because of the modulation, recognizes the part which originally was circularly polarized. This part yields the desired V signal.

In solar physics a polarimeter is often called a *magnetograph*, in particular if – e.g., by means of the weak-field approximation or by a fit of the

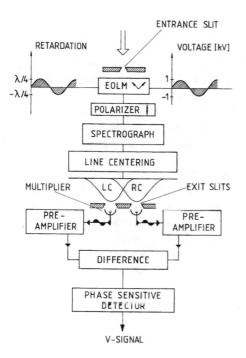

Fig. 3.40. V-polarimeter. Adapted from Schröter (1973)

line profile to some specified form – the output signal is directly converted to a magnetic field strength. Magnetograms have been obtained regularly since 1974 with the V-polarimeter installed at the Kitt Peak Vacuum Tower Telescope, which was taken as an example telescope in the present text (Livingston et al. 1976 a, b, Jones et al. 1992). This magnetograph employs a Kerr cell in combination with a $\lambda/4$ plate instead of an electro-optical crystal, but otherwise very much follows the principles described above. Another alternative of modulating the polarization is the rotating $\lambda/4$ plate first used by Kiepenheuer (1953). The modulation must be either sufficiently fast so that the two intensities that are subtracted from each other are measured under the same atmospheric seeing distortion, or the measurement must be strictly simultaneous by means of a beam-splitting device. If the full Stokes vector is desired, the diverse states of polarization must all be encoded separately by appropriate phase retarders within each modulation cycle. Another important ingredient is a polarizer that can be mounted in front of the polarimeter; by introducing a well-defined amount and kind of polarization the whole apparatus can be calibrated.

Polarimeters belong to the standard equipment of most solar observatories. The review of Beckers (1971 a) lists 22 magnetographs, half of which are vector magnetographs. Examples of more recent instruments are the Advanced Stokes Polarimeter (ASP; Elmore et al. 1992) at Sacramento Peak Observatory, as well as the La Palma Stokes Polarimeter (LPSP) and the

Tenerife Infrared Polarimeter (TIP), both described by Martínez Pillet et al. (1999). The Zürich Imaging Stokes Polarimeters ZIMPOL I and ZIMPOL II (Stenflo et al. 1992, Povel et al. 1994, Gandorfer and Povel 1997) reach a polarimetric accuracy better than 10^{-4}. ZIMPOL II uses a single CCD camera to record four independent modulated polarization signals and so allows to recover the complete Stokes vector. In this case the camera chip itself is used as a demodulator: within each modulation period one out of each four rows is exposed successively to the four signals; the other three rows are masked and serve as buffers before readout.

Particularly fine V spectra have been obtained with a polarimeter used in combination with the Fourier transform spectrometer at Kitt Peak Observatory. Figure 3.41 shows an example that contains the Fe I line at 525.02 nm used in the discussion below. With this polarimeter no phase-sensitive detector is necessary to sort out the modulated part of the signal. The polarized and unpolarized spectra can automatically be separated when the Fourier transform of the recorded intensity (3.32) is computed.

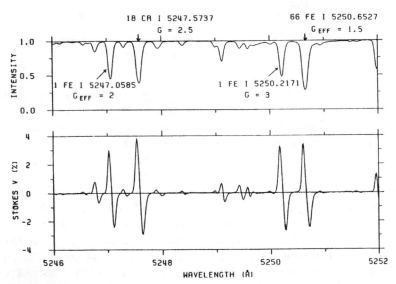

Fig. 3.41. Stokes I and V spectra obtained with a Fourier transform spectrometer. From Stenflo et al. (1984)

Let us use the result (3.78) for an estimate of the expected circular polarization for a given (weak) longitudinal magnetic field (following Schröter 1973). For small v_B (3.78) yields $V = v_B dI/dv = -v_B I_C dr/dv$, where I_C is the continuum intensity, and $r = 1 - I/I_C$ is the relative line depression. Let us assume a Gaussian profile with central depression r_0, i.e., $r = r_0 \exp(-v^2)$. And let us place the two windows at the steepest part of this profile, that is at $v = \pm 2^{-1/2}$. At these positions we have $r = 0.607 r_0$ and $dr/dv = \pm 0.86 r_0$.

The difference of the two V signals, divided by the sum of the two I signals, yields the degree of circular polarization:

$$\frac{V}{I} = v_B \frac{0.86 r_0}{1 - 0.607 r_0} \ .$$

(3.89)

A frequently used line for polarimetry is the Fe I line at $\lambda = 525.02\,\text{nm}$ (although this line poses problems because of its temperature sensitivity), with $g^* = 3$, $\Delta\lambda_D = 42\,\text{mÅ}$, and $r_0 \approx 0.7$. We substitute (3.55) for $\Delta\lambda_B$ and obtain

$$\frac{V}{I} \approx 9.6 \times 10^{-4} B \ ,$$

(3.90)

where B must be measured in gauss. We see that a field of 10 G still produces about 1% circular polarization, which can easily be measured with a good polarimeter.

Since we have already seen that for a weak magnetic field $V \propto B \cos\gamma$ we may generalize (3.90) for an inclined weak field to $V/I \approx 9.6 \times 10^{-4} B \cos\gamma$.

A rough estimate how much *linear* polarization we may expect in solar spectral lines can be obtained as follows. We set again $\phi = 0$, and compare (3.87) to (3.86). In combination with (3.84) we see that we obtain Q instead of V if we make the replacement

$$v_B \frac{\partial\eta}{\partial v} \cos\gamma \rightarrow -\frac{1}{4} v_B^2 \frac{\partial^2\eta}{\partial v^2} \sin^2\gamma \ .$$

If, in addition, we take the line profile $r(v)$ in place of the absorption profile [which is permitted only for the optically thin layer, cf. (3.67)], then

$$\frac{Q}{I} \simeq \frac{1}{4} \frac{I_C}{I} \frac{d^2 r}{dv^2} v_B^2 \sin^2\gamma \ .$$

Let us use the Gaussian profile as above. The maximum of $|d^2 r/dv^2|$ occurs at $v = 0$, and has the value $2 r_0$. At $v = 0$ we have $I_C/I = (1 - r_0)^{-1}$, so that the degree of linear polarization becomes

$$\frac{Q}{I} \simeq \frac{r_0}{2(1 - r_0)} v_B^2 \sin^2\gamma \ .$$

(3.91)

For the Fe I λ 525.02 nm line this finally gives (with B in gauss)

$$\frac{Q}{I} \approx 10^{-6} B^2 \sin^2\gamma \ .$$

(3.92)

This rough estimate indicates that linear polarization originating from a weak magnetic field is very difficult to measure. A transverse field with a strength of 100 G yields ca. 1% linear polarization; by comparison, the same amount of circular polarization is obtained from a longitudinal field of 10 G, according to the estimate (3.90).

3.5.5 Scattering Polarization and Hanle Effect

Polarization does not necessarily require a magnetic field in the region where the light originates. Polarized radiation is also produced by scattering, as is well-known from the blue light of the sky. A necessary condition for scattering polarization is the anisotropic illumination of the scattering particles. Thomson scattering by free electrons in the corona is the most prominent solar example (Sect. 9.1.3). But also at lower levels in the atmosphere one can observe the effect near the limb. The polarization is linear; in the continuum it is due to Rayleigh scattering by neutral hydrogen atoms and to Thomson scattering by electrons; in spectral lines it is due to coherent scattering in bound-bound transitions of atoms.

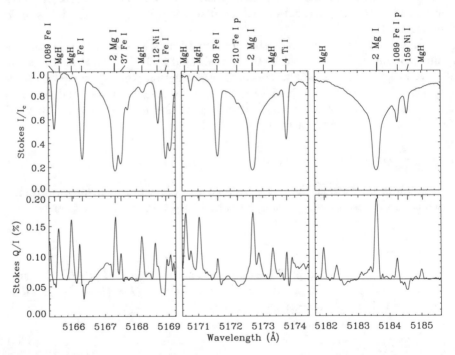

Fig. 3.42. Spectral regions around three Mg I lines, from records made near the northern solar limb. Notice especially the lines of MgH that are rather weak in the intensity spectrum (*top*), but very conspicuous in the linear polarization Q/I (*lower spectrum*). From Stenflo et al. (2000 b)

Polarization through scattering in the solar atmosphere is a small effect, with Q/I of order 10^{-3} and smaller. The largest Q signals have been measured in resonance lines (transitions involving the ground state); the first results were obtained by Brückner (1963) who used the Ca I line at 422.67 nm. More recent observations have been made with the ZIMPOL instruments (Stenflo and Keller 1997); Gandorfer (2000) has published a spectral atlas that displays Q/I for the wavelength range $462.5\,\mathrm{nm} \leq \lambda \leq 699.5\,\mathrm{nm}$. The differences to the intensity spectrum are so remarkable that one may speak of the "second solar spectrum" (Fig. 3.42), with molecular lines, e.g., from C_2 and MgH, and other unusual features (Stenflo et al. 2000 ab).

The *Hanle effect* causes a decrease of the scattering polarization and a rotation of the plane of polarization, and occurs when a magnetic field is present in the observed part of the atmosphere. This effect was first investigated in the laboratory by Hanle (1924) in the context of resonant fluorescence. One can understand the Hanle effect if one realizes that the anisotropic radiation field leads to the polarized scattered light by means of polarizing the states of the scattering atoms. An atomic state is polarized if there is a non-equilibrium population of its magnetic sublevels with well-defined phase relations between these sublevels (quantum interference). In general, both the upper and the lower state will be polarized. A magnetic field perturbs the polarization of the atomic states (in the classical picture, it causes a precession of the oscillating dipoles). Even a quite weak magnetic field may lead to depolarization via the Hanle effect, and there is no cancellation of the contributions of unresolved magnetic dipoles such as occurs with the circular polarization caused by the Zeeman effect. Thus, the Hanle effect bears new possibilities to investigate the solar magnetic field (e.g., Stenflo 1982; Landolfi and Landi degl'Innocenti 1986; Bianda et al. 1998, 1999).

3.6 Special-Purpose Instruments

3.6.1 The Pyrheliometer

The Pyrheliometer is an instrument which measures the total solar flux of radiation. Early instruments, designed by C. G. Abbot from 1908 onwards, essentially consisted of a black chamber which traps all entering radiation by virtue of its form. The walls of the chamber are continually bathed by flowing water, and the water temperature is measured before it gets in contact with the walls, and again immediately afterwards. This water-flow pyrheliometer could determine the solar constant with an uncertainty of about 1%.

A modern instrument is the Active Cavity Radiometer Irradiance Monitor (ACRIM), which first has been installed in the Solar Maximum Mission. Figure 3.43 illustrates the principle (Willson 1979; Willson and Hudson 1981). The central part is a conical cavity which, via a thermal impedance, is connected to a heat sink. A shutter alternately admits the solar radiation to, or blocks it from, the cavity. At the base of the cone (where it is in contact to the thermal impedance) a temperature sensor is bonded; near the tip of the cone (where – in the interior – the solar radiation is absorbed) there is a winding for electrical heating. Both the sensor and the heater are at the outside of the cone, while its inside is coated with specular black paint.

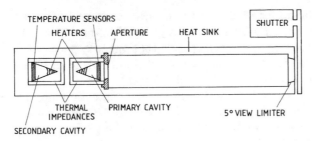

Fig. 3.43. Active cavity radiometer. Adapted from Willson (1979)

The active cavity radiometer operates in such a way that a constant temperature difference of about 1 K is maintained between the cavity and the heat sink. This is achieved by a servo loop which uses the signal of the temperature sensor. Thus the heating is automatically increased when the shutter closes the cavity. Let P_r and P_o be the electrical power in the reference (closed) and observational (open) phases. The incident radiation flux density, S, is then obtained from

$$SA_c(\alpha_c + \rho\rho_c) = P_r - P_o + C \ . \tag{3.93}$$

Here A_c is the area of the aperture, α_c and ρ_c are the absorptance and reflectance of the cavity for solar radiation, and ρ is the fraction of the radiation not initially absorbed which is reflected back into the cavity. The additional term C comprises several small corrections, e.g., heat conducted to, or away from, the cone by air or by the electrical lead, or radiation from the cone to the surroundings.

The specular black coating in the cavity has an absorptance of $\alpha_s = 0.9 \pm 5\%$. As the cone angle is $30°$, there are six internal reflections of an axial ray before reflection out of the aperture. Theoretically, the absorptance of the cavity is therefore

$$\alpha_c = 1 - (1 - \alpha_s)^6$$

or $0.999\,999$ with an uncertainty of 0.0003%. In practice, an absorptance of 0.99943 ± 0.00002 has been measured. Hence, the term $\rho\rho_c$ is also a small correction.

Because of a possible temperature drift a second identical cavity and thermal impedance is connected to the heat sink. Its temperature is allowed to drift passively. This cavity always views the heat sink, just as the primary cavity in the shutter-closed phase. It can therefore be used to eliminate the effects of the heat sink temperature drift.

The ACRIM flown on the Solar Maximum Mission consisted of three identical pyrheliometers of the kind just described. They were operated with different shutter frequencies, and thus the effect of cavity surface degradation could be controlled to some extent. The instrument measures the absolute irradiance with an uncertainty of order 0.2%, but relative variations much smaller than this have been recorded.

3.6.2 Neutrino Detectors

The Earth is essentially transparent for neutrinos, and the same is of course true for all terrestrial materials from which neutrino detectors could be built. Fortunately, the flux of neutrinos from the Sun is intense. Two neutrinos are produced per synthesized α particle, i.e., for roughly every 25 MeV generated in the Sun. At the Earth we thus have approximately 6.5×10^{14} neutrinos/m^{2}s, cf. Table 2.5. Also fortunately, there are a few nuclei which have a sufficiently large cross section to capture one or the other neutrino out of the solar flux; the yield is about one neutrino in every 10^{20}. Among these nuclei are the ^{37}Cl and the ^{71}Ga isotopes. Although other target atoms have been investigated, these two are the most important ones used so far for real solar neutrino detectors. A ^{37}Cl experiment has been in operation since 1968, and results from two gallium experiments have been obtained since 1991. In addition to the neutrino capture by nuclei, neutrino-electron scattering has been used successfully in two water Čerenkov detectors since 1987.

All solar neutrino experiments are located in underground laboratories. This is necessary because muons, produced in the Earth' atmosphere by cosmic rays, would cause unwanted signals in the detector and therefore must be shielded off. Only at a depth of 1 to 2 km, equivalent to 3 to 6 km water, the muon flux is sufficiently attenuated. The results of all solar neutrino experiments have been discussed in Scct. 2.4.2.

The ^{37}Cl Experiment. The basic reaction of this experiment is

$$\nu + {}^{37}\text{Cl} \underset{\text{capture}}{\overset{\text{decay}}{\rightleftharpoons}} {}^{37}\text{Ar} + e^- \ . \tag{3.94}$$

The energy threshold for this reaction is 814 keV. Above this energy the cross section of ^{37}Cl for neutrino absorption increases by several orders of magnitude, cf. Fig. 3.44, or the values listed in Table 2.5 (for the continuous solar neutrino spectra the cross sections in this table are averages over energy).

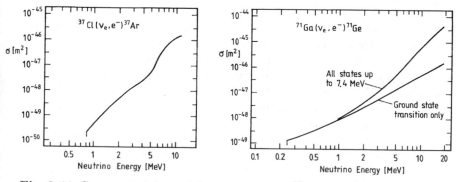

Fig. 3.44. Cross section for neutrino capture. *Left:* ^{37}Cl, after Kitchin (1984); *right:* ^{71}Ga, after Hampel (1986)

The reason is that the ^{37}Ar nucleus has a large number of excited states into which the target atom can go if the neutrino energy is sufficiently high.

The ^{37}Cl experiment, carried out by R. Davis, measured the flux of solar neutrinos for the first time in 1968 (e.g., Bahcall and Davis 1976). The target consists of ca. 615 t of tetrachloroethene, C_2Cl_4, which is an inexpensive cleaning fluid. The fluid is contained in a 400 m^3 tank, 1500 m underground in the Homestake gold mine near Lead, South Dakota, USA.

A typical run goes as follows. The tank is left alone for about 100 days. As the half-life of ^{37}Ar is 35 days, the number of ^{37}Ar atoms has then almost reached the saturation level. The argon atoms can be assumed to be freely dissolved in the liquid, because the neutrino provides ample energy to break up the original C_2Cl_4 molecule. Helium is then pumped through the tank; argon, a noble gas like helium, is carried along, and subsequently collected by a charcoal trap at liquid nitrogen temperature. As a check a small amount of a different Ar isotope, say 3×10^{19} atoms of ^{36}Ar, is added to the tank prior to the helium pumping. Since the charcoal trap recovers the majority of these "carrier atoms" ($\geq 90\%$), and since all isotopes of an element chemically behave in the same way, it is reasonable to conclude that all the ^{37}Ar atoms have been recovered as well.

The argon is placed into a miniature proportional counter where the decay is monitored. The ^{37}Ar nucleus captures an inner electron, – the right-to-left reaction of (3.94). At the same time a 2.8 keV electron is emitted; by this signature the ^{37}Ar decay is distinguished from other events in the counter.

Besides the test with the ^{36}Ar carrier gas, two other tests confirm that ^{37}Ar is being recovered efficiently. One is to irradiate the tank with a calibrated neutron source, the other to directly inject 500 atoms of ^{37}Ar. In both cases the recovery is satisfactory.

The ^{71}Ga Experiment. If the neutrino energy exceeds the threshold of 233 keV, the neutrino can be captured according to

$$\nu + {}^{71}\text{Ga} \underset{\text{capture}}{\overset{\text{decay}}{\Longleftrightarrow}} {}^{71}\text{Ge} + e^- . \tag{3.95}$$

As Fig. 3.44 illustrates, the cross section for this reaction also increases with increasing neutrino energy. But because of the relatively low energy threshold the main solar contribution comes from the pp reaction, cf. Table 2.5, where again energy-averaged cross sections are listed for the various solar spectra. The isotope ^{71}Ga has a half-life of 11.4 days. The gallium target is therefore exposed for about 3 weeks; within this period the number of ^{71}Ge atoms reaches ca. 70 % of the saturation level.

One of the gallium experiments, the Russian–American SAGE in the Baksan underground laboratory in the Caucasus, uses 57 t of metallic liquid Ga (gallium has a melting temperature of 29.8°). The other experiment, the collaboration GALLEX of groups in France, Germany, Israel, and Italy, is performed in the Gran Sasso underground laboratory in Italy; its target is a tank with 101 t of GaCl$_3$/HCl solution, containing 30.3 t of gallium. As in the ^{37}Cl

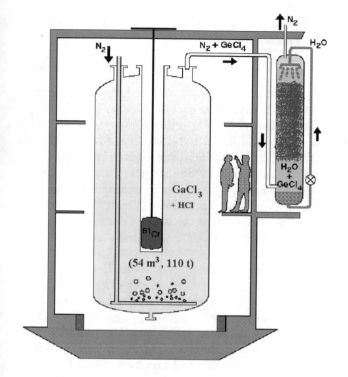

Fig. 3.45. The ^{71}Ga neutrino detector of the GALLEX experiment in the Gran Sasso Underground Laboratory, Italy. The picture shows the ^{51}Cr calibration source in the central shaft. Courtesy MPI für Kernphysik, Heidelberg

experiment, the ^{71}Ge atoms are extracted from the tank. Subsequently their decay according to (3.95) is measured in a miniature proportional counter. Again the typical signature of the ^{71}Ge decay in the counter is used to discriminate other events.

In order to ensure that the extraction is quantitative, a stable Ge isotope is added as a "carrier" to each experimental run of GALLEX. As a further test a known quantity of ^{71}As has been introduced into the tank; this arsenic isotope decays with a half-life of 2.9 days into ^{71}Ge, which has been fully recovered. Both gallium detectors have been tested successfully by calibrated artificial ^{51}Cr neutrino sources, consisting of chromium irradiated with neutrons (Fig. 3.45). The isotope ^{51}Cr decays by electron capture and thereby emits neutrinos of 430 keV and 750 keV, which is in the low-energy part of the solar neutrino spectrum.

Problem 3.22. The abundances of the ^{37}Cl and ^{71}Ga isotopes in Cl and Ga are 24.23% and 40%, respectively. With the numbers given in the text, calculate the number of target atoms in the two neutrino detectors. What is the number of ^{37}Ar atoms after 100 days, and the number of ^{71}Ge atoms after 14 days? What are the saturation levels?

Water Čerenkov Experiments. The reaction used in the water detectors is elastic neutrino-electron scattering

$$\nu_x + e^- \;\rightarrow\; e^- + \nu_x \;. \tag{3.96}$$

This reaction occurs with all three neutrino flavors, although with greatly reduced sensitivity to ν_μ and ν_τ. The scattered electron emits Čerenkov light if its velocity v exceeds the velocity of light in water, c/n, where $n = 1.34$ is the refraction index of water. The light is emitted along a circular cone around the path of the electron. The half-angle θ of the cone is given by $\cos\theta = c/nv$ and is $41.9°$ in the relativistic limit $v \simeq c$. Photomultiplier tubes at the walls of the detector monitor the Čerenkov light so that the trajectory and the timing of the electron can be determined. This allows to infer the energy and the direction of the incoming neutrino. Thus the water Čerenkov detector is a *neutrino telescope*, in contrast to a radiochemical experiment that has no directional information.

Solar neutrinos have been measured with two water Čerenkov detectors, Kamiokande since 1987 and Super-Kamiokande since 1996, both in the Mozumi mine near Kamioka, Japan. The detectors consist of cylindrical tanks filled with 4500 t and 50 000 t water, respectively, and a large number of photomultiplier tubes mounted at the walls, over 11 000 in the case of Super-Kamiokande. The detector volume is divided into an inner part, in which the neutrino-electron scattering is observed, and an outer ("anti-detector") part to discriminate events originating from outside. The time resolution is a few ns; the angular resolution for neutrinos of the solar energy range is ca. $28°$.

The energy threshold is $\approx 7.5 \, \mathrm{MeV}$ for Kamiokande, and $\approx 5 \, \mathrm{MeV}$ for Super-Kamiokande. For smaller neutrino energy the measurement is spoiled by events arising from radioactive impurities dissolved in the water. Hence the two water detectors are sensitive only to the $^8\mathrm{B}$ neutrinos from the solar reaction chain ppIII.

Kamiokande was originally built for the search of nucleon decay. Such has not been discovered, but by improving the water purity the energy threshold was lowered so that it could be used as a detector for solar neutrinos. The name may stand for both endeavors: Kamioka Nucleon Decay Experiment, or Kamioka Neutrino Detection Experiment.

The Sudbury Neutrino Observatory. A Čerenkov detector using *heavy water* instead of ordinary water has been built 2000 m underground in a nickel mine in Sudbury, Ontario, Canada. The detector is a spherical acrylic (transparent) vessel of 12 m diameter, containing 1000 t of D_2O. Surrounding this vessel is an outer container filled with 7000 t of H_2O that acts as shielding. The walls of this outer container hold the 9500 photomultiplier tubes for the detection of the reaction products.

In addition to neutrino-electron scattering, the following two deuteron reactions can be monitored in the heavy-water tank:

$$\nu_e + d \; \rightarrow \; p + p + e^- , \tag{3.97}$$

$$\nu_x + d \; \rightarrow \; p + n + \nu_x . \tag{3.98}$$

The first is a charged-current reaction of electron neutrinos ν_e, the second a neutral-current reaction that can occur with *any* neutrino flavor, ν_x. Due to this distinction the heavy-water detector is able to measure the ratio $\Phi(\nu_e)/\Phi(\nu_x)$ of the electron neutrino flux to the total neutrino flux (all flavors); a ratio smaller than 1 is an indication of *neutrino oscillations*, i.e., the transmutation of electron neutrinos emitted in the Sun into other neutrino flavors that could not be detected in the previous experiments.

The energy threshold of the heavy-water detector is at 5 MeV, as for the ordinary water detector.

Borexino. A detector consisting of a spherical container filled with 300 t organic scintillator material, with 2200 photomultiplier tubes mounted at the walls of an external water tank, will be built in the underground laboratory in the Gran Sasso, Italy. This detector monitors scintillation photons caused by electrons arising from neutrino-electron scattering. The energy threshold is at 250 keV, substantially lower than the threshold of the water Čerenkov detector. Thus low-energy solar neutrinos can be monitored. The line at 862 keV arising from ^{7}Be neutrinos is a prominent feature in the solar neutrino spectrum (Fig. 2.12); the energy resolution of Borexino will be sufficiently high to identify this feature. Hence the flux of neutrinos from the ppII chain of the Sun's nuclear reactions can be measured.

3.6.3 The Coronagraph

In the visible part of the spectrum the solar corona radiates 10^5 to 10^6 times less intensely than the solar disc. Under normal circumstances this faint radiation is buried below stray light (originating from the disc) which is generated mainly in the Earth's atmosphere, but also in the telescope.

We may observe the corona at the occasion of a solar eclipse, because then the bright solar disc is occulted by the moon *outside the atmosphere* and so both the atmospheric and the telescopic contributions to the stray light are largely eliminated.

But even without an eclipse it is possible to see the solar corona under special conditions: the observatory site must have a sky as dark as possible, i.e., as little atmospheric scattered light as possible, (high mountain sites often are "coronal" in this sense), and a specialized telescope must be used – the coronagraph invented by B. Lyot around 1930.

The essence of the coronagraph is to produce an artificial eclipse by means of an occulting disc inside the telescope, and to take special precautions to reduce the *instrumental* scattered (and diffracted) light. Figure 3.46 illustrates the principle of Lyot's original coronagraph.

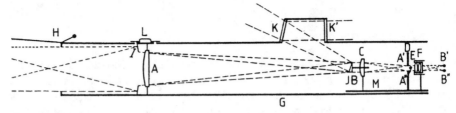

Fig. 3.46. Coronagraph. After Lyot (1932)

The light enters the telescope through an elongated tube (H) which protects the objective (A) from dust. The objective itself is made of a single lens so that the number of scattering surfaces is minimized. The primary image is formed at B. From here the central part (the solar disc) is reflected at an inclined surface (J) through a window (K) out of the telescope – reflections from the window itself leave through a second window (K'). Behind the occulting surface (J) there is a field lens (C) which images the objective at A'A'' on a diaphragm D. Light scattered at the objective and diffracted at its edges is blocked in this way. The center of the diaphragm is occupied by a small screen (E) which stops the solar image formed by secondary reflections at the surfaces of the objective. The objective lens itself must be cleaned frequently, and for that purpose can be taken out together with a side cover (L), to which it is mounted. The setting (I) of this lens has a concave surface, so that all light falling on it is also reflected out of the tube. The final image of the corona is produced at B'B'' by an achromatic objective (F).

Many variants of Lyot's coronagraph have been built. Of course, it is most advantageous not only to reduce the stray light in the instrument, but also to eliminate the atmospheric stray light. Coronagraphs have therefore been flown on space missions. Numerous coronal photographs have been taken with an instrument flown with Skylab (May 1973 to January 1974). The Large-Angle Spectroscopic Coronagraph (LASCO) on the SOHO satellite consists of three instruments with occulting discs of different size so that the corona can be imaged at various distances, namely 1.1–3, 1.5–6, and 3.7–30 solar radii, respectively (Brueckner et al. 1995). The two instruments for the outer corona have an external occulting disc in front of the entrance aperture, similar to the Skylab coronagraph; the one for the inner corona is internally occulted as in Fig 3.46. Instead of an occulting disc an (external and/or internal) occulting edge may be used for the investigation of a segment of the corona. The UV Coronagraph Spectrometer (UVCS) on board of SOHO is of this type (Kohl et al. 1995).

3.7 Bibliographical Notes

Whenever *Fourier transforms* or related transforms occur, the monograph of Bracewell (1965) is most useful. An example is the line spread function, and the Abel transform (Problem 3.3).

Tatarski (1961, 1971) deals with wave propagation through a turbulent medium, especially for the case of isotropic turbulence. Roddier (1981) comprehensively treats the atmospheric origin of *solar seeing*; in particular, he discusses the relation between the fluctuations of temperature and refractive index, and the Fried parameter. The *Correlation Tracker* is treated by von der Lühe et al. (1989); Schmidt and Kentischer (1995) describe a modern version. Adaptive optics and solar image restoration are the topics of the volume edited by Radick (1993) and the book of Roggemann and Welsh (1996). Paxman et al. (1996) describe two implementations of the phase diversity technique.

In addition to the seeing arising from atmospheric or local turbulence, solar observations (and not only of the corona!) often are severely hampered by *stray light*. Mattig (1983) describes stray light measurements and separates the two contributions from the Earth's atmosphere and from the solar telescope.

Many developments in *solar instrumentation* have been discussed during conferences at Sacramento Peak Observatory (Dunn 1981, von der Lühe 1989, November 1991, Kuhn and Penn 1995, Rimmele et al. 1999).

Solar *telescopes* for high resolution are divided by Dunn (1985) into fixed and steerable telescopes, a division which roughly corresponds to the distinction made in this text between instruments with long and short primary focal length. Details of telescopes in general, in particular about the various typical aberrations, can be found in the Handbuch article of Bahner (1967). King (1979) gives an historical account.

The optical principles that determine the angular resolution, etc., of telescopes used in the visible are of course equally valid in other spectral regions. Rohlfs and Wilson (2000) describe the tools of radio astronomy. At wavelengths in the extreme UV and soft x-ray spectral regions optical surfaces coated with aluminum or silver do not reflect normally incident light. Therefore, *grazing-incidence* optics has been used, e.g., for the Skylab and Yohkoh soft-x-ray *Wolter Telescopes* (Zombeck et al. 1978, Tsuneta et al. 1991), and for the SOHO coronal diagnostic spectrometer, CDS (Harrison et al. 1995). A more recent alternative is the use of optical surfaces with *multilayer coating*. Layers with two different indices of refraction that are periodically stacked yield a surface that reflects at particular wavelengths. A normal-incidence x-ray telescope, NIXT, was first used by Golub et al. (1990) to observe the solar corona. The extreme-UV telescope EIT on board of the SOHO mission (Delaboudinière et al. 1995) and the TRACE soft-x-ray telescope (Handy et al. 1999) are of this type. The XUV Doppler telescope XDT described by Sakao et al. (1999) is a variant. It uses the selective reflectivities of two different multilayer coatings to take images in two narrow spectral bands on either side of a coronal emission line; a Doppler shift of the line thus manifests itself in different intensities of the two images.

I have put no emphasis on light *detectors*, except in the context of the scale of solar images. General texts on astronomical instruments, e.g., Kitchin (1984), or the monographs of Holst (1998) and Howell (2000) treat this subject.

Diffraction gratings are extensively treated by Stroke (1967). An introductory monograph on spectroscopy has been written by Kitchin (1995). Howard (1974) describes a variant of the *Doppler compensator*: instead of tilting a glass plate, the pair of exit slits is moved until each of the two slits gets an equal amount of light; the velocity signal is then taken from the shaft which drives the slit motion.

The theory of birefringent filters is outlined in the classical papers of Lyot (1944) and Evans (1949). Harvey et al. (1988) and Scherrer et al. (1995) treat the GONG and MDI versions of the tunable Michelson interferometer; earlier versions, also called "Fourier Tachometers", have been described by Beckers and Brown (1978), Brown (1981), and Evans (1981). For the Fabry–Perot interferometer the monographs by Tolansky (1948) and Vaughan (1989) provide introductory as well as special material. For the solar application the Fabry–Perots are mounted either in the collimated beam, as shown in Fig. 3.29, or in the telecentric beam near the image plane; Darvann and Owner-Petersen (1994) and Kentischer et al. (1998) discuss the specific advantages of these two mountings.

A magneto-optical filter that is related to the Lyot filter has been investigated first by Öhman (1956). Beckers (1970) reviews the diverse types of magneto-optical filters.

For the study of polarized light it seems necessary to distinguish between strictly monochromatic light, and light with a bandwidth small compared to the width of solar lines. As Shurcliff (1962) remarks, the definition of the Stokes parameters in terms of electromagnetic theory "is awkward in that one must assume that the light is sufficiently monochromatic that, at any time, a definable phase angle γ exists between the instantaneous scalar components ... of the electric field, yet the light must be sufficiently polychromatic that the unpolarized state is not precluded".

It is also awkward, of course, that the classical theory of atoms is still used to derive the absorption coefficients (3.69) for polarized light in the presence of a magnetic field. Unno's result, which was independently derived by Stepanov (1958), rests on formulae which were originally presented by Sears (1913). Fortunately, a quantum-mechanical treatment (Landi degl'Innocenti and Landi degl'Innocenti 1972) confirms the classical result.

The monograph by Stenflo (1994) covers the theory of polarized radiation as well as the application to the Sun, earlier reviews were written by Evans (1966), Beckers (1971 a), and Stenflo (1971). The volume edited by Trujillo-

Bueno et al. (2002) covers solar polarimetry, but also other astrophysical applications. The inversion of polarimetric data has been treated by Wittmann (1974 a), Landolfi et al. (1984), Skumanich and Lites (1987), Rees et al. (1989, 2000), Ruiz Cobo and del Toro Iniesta (1992), and Bellot Rubio et al. (2000). The magneto-optic effects are included in these treatments; their significance has been demonstrated by Landolfi and Landi degl'Innocenti (1982), although earlier results of Rachkovsky (1962 b) and Beckers (1969 b) had indicated that the solar Stokes profiles are not largely altered. The volumes edited by November (1991), Rutten and Schrijver (1994), and Stenflo and Nagendra (1996) contain reviews and detailed contributions.

A major difficulty in the interpretation of polarimetric data arises from the *inhomogeneity* of the solar magnetic field. Stenflo (1971) has discussed the problem. Solanki (1993) reviews the effects of the small-scale flux concentrations that will be treated in Chap 8, and Sigwarth et al. (1999) present results with a spatial resolution of $1''$. Grossmann-Doerth et al. (2000), Martínez Pillet (2000), and Steiner (2000) demonstrate that magnetic discontinuities such as occur in penumbral channels, or in a *canopy*-like transition from a non-magnetic to a magnetic layer in the atmosphere, can cause diverse line profiles such as shown in Fig. 3.38. On the other hand, magnetic structure on scales much smaller than the photon free path has been proposed as the origin of the "irregular" line profiles of polarized radiation (Sánchez Almeida et al. 1996). Keller (1995) shows how the speckle technique can be combined with polarimetry in order to measure the solar magnetic field with diffraction-limited resolution.

The measurement of *vector* fields has been documented in the proceedings edited by Hagyard (1985). Mickey et al. (1996) and LaBonte et al. (1999) describe an imaging vector magnetograph based on a tunable Fabry–Perot filter. The instrumental polarization introduced by coelostat mirrors has been considered by Capitani et al. (1989). Stenflo (1982, 2001) and Landolfi and Landi degl'Innocenti (1986) discuss the Hanle effect and its possible use to measure the magnetic field in prominences and other places of the outer solar atmosphere.

A description of the water-flow pyrheliometer and results obtained from it can be found in Aldrich and Hoover (1954).

A number of neutrino detectors other than those described in Sect. 3.6.2 have been proposed. The proceedings of a Brookhaven conference (Friedlander 1978) and the books of Bahcall (1989), Klapdor-Kleingrothaus and Zuber (1997), and Schmitz (1997) provide detailed introductions to this field.

4. The Atmosphere

We now turn to a more detailed study of the solar atmosphere, which, in the present chapter, will be understood as the combination of the *photosphere* and the *chromosphere*.

The *photosphere* is a layer of little more that one hundred kilometers thickness where – going inwards – the solar gas changes from almost completely transparent to completely opaque. Virtually all the light which we receive from the Sun originates in the photosphere. Therefore, most of the information which we have about the Sun stems from this layer.

Fortunately the transition from optically thin to optically thick depends on position on the Sun, and on wavelength – in a most pronounced way in the spectral lines. This circumstance allows us to study the thermodynamic state of the solar atmosphere, its chemical composition, and a variety of hydrodynamic and hydromagnetic phenomena which abound on the Sun.

In fact, in a number of strong spectral lines the transition to the optically thick state occurs high in the atmosphere. When we observe the Sun in such a line we therefore see the upper part of the solar atmosphere, the *chromosphere*. This layer has obtained its name from the colorful appearance which it presents at the time of a solar eclipse.

In the present chapter we shall concentrate on results concerning the *mean* atmosphere, i.e., averaged over the horizontal coordinates on the solar surface. Examples are the run of temperature and pressure with height in the atmosphere, or the abundances of the elements. In the subsequent chapters, we shall then treat spatially resolved and local phenomena.

4.1 Radiative Transfer
– Local Thermodynamic Equilibrium

4.1.1 The Equation of Transfer

In order to derive detailed atmospheric models we must return to the equation of radiative transfer, (2.15). Let us introduce the optical depth at frequency ν as the independent coordinate

$$d\tau_\nu = -\kappa_\nu \rho dr .$$

(4.1)

Hence

$$\mu \frac{dI_\nu}{d\tau_\nu} = I_\nu - S_\nu \tag{4.2}$$

describes the variation of the intensity with depth; S_ν is the source function, i.e., the ratio between emission and the absorption coefficient, and $\mu = \cos\theta$ (the intensity depends on τ_ν, μ, and ν; as for the other variables, κ_ν, S_ν ..., the argument ν is often written as an index).

A formal solution of the radiative transfer equation, i.e., a reduction to the problem of evaluating an integral, can be obtained in the following way: Multiply (4.2) by $\mu^{-1} \exp(-\tau_\nu/\mu)$ and integrate; the result is

$$I_\nu(\tau_\nu, \mu) = I_\nu(\tau_{0\nu}, \mu) \exp[-(\tau_{0\nu} - \tau_\nu)/\mu]$$
$$+ \frac{1}{\mu} \int_{\tau_\nu}^{\tau_{0\nu}} S(\tau_\nu') \exp[-(\tau_\nu' - \tau_\nu)/\mu] d\tau_\nu' , \tag{4.3}$$

where $\tau_{0\nu}$ is the optical depth at some level of reference. If we integrate from $\tau_\nu = 0$, where we observe, to $\tau_{0\nu} = \infty$, i.e., deep into the star, then we have the total emergent intensity

$$I_\nu(0, \mu) = \frac{1}{\mu} \int_0^\infty S_\nu(\tau_\nu) \exp(-\tau_\nu/\mu) \, d\tau_\nu . \tag{4.4}$$

Equation (4.4) can be used in a variety of ways. For instance, we may assume a known source function and predict the emergent *intensity*. Or, which is more interesting in the context of solar atmospheric models, we may derive the *source function* from the (absolute) intensity, if the latter is measured as a function of ν and (or) μ. This requires an inversion of the integral (4.4).

Problem 4.1. Assume that there are neither absorbers nor emitters between two points of a radiation field. Show that the intensity, according to its definition in Sect. 1.5.1, is the same at these two points; convince yourself that this is consistent with the equation of transfer, (4.2).

4.1.2 Various Equilibria

Before we present results of an inversion of the emergent intensity integral (4.4) we must clarify our assumptions concerning the thermodynamic state of the solar atmosphere.

Thermodynamic equilibrium prevails when a single value T of the temperature is sufficient to describe the thermodynamic state everywhere. The particles then have Maxwellian velocity distributions for that T; the states of ionization and excitation of the atoms are distributed according to the Saha and Boltzmann equations for that same T; and the radiation field has the homogeneous and isotropic black-body form given by the Kirchhoff-Planck function, again for the same T:

$$B_\nu(T) = \frac{2h\nu^3}{c^2} \frac{1}{e^{h\nu/kT} - 1} \ . \tag{4.5}$$

No temperature gradient exists in thermodynamic equilibrium, and it is obvious that this situation is realized virtually nowhere.

Very often, however, the conditions of *local thermodynamic equilibrium* (LTE) are very closely satisfied. That is, at a certain place, a single temperature T does suffice to describe the statistical particle velocities, the population of the atomic states, and the local ratio of emission to absorption of radiation. In LTE the most important simplification of the radiative transfer problem is the relation $S_\nu = B_\nu(T)$.

Whether or not LTE can be assumed depends on the *thermalization length*. This is the distance over which a particle or photon emitted in a certain collision or transition has undergone sufficient further collisions or absorption/emission processes so that it can no longer be distinguished within the respective distribution. In LTE the thermalization length must be shorter than the distance over which the temperature of the gas changes markedly.

The preceding definition of the local thermodynamic equilibrium makes clear that, at the same place in the atmosphere, LTE may be a good assumption for one particular process or species, but may be completely wrong for another. Consequently, the substitution $S_\nu = B_\nu$ may be allowed when the formation of a certain spectral line is considered, but may fail for other lines. Often the difference becomes apparent only after departures from LTE are explicitly taken into account. A crude rule of thumb is that the continuum in the visible and infrared, the wings of most spectral lines, and the entire profiles of weak lines are formed in LTE, while possible departures from LTE must be considered for the line cores, and for strong lines.

Departures from LTE occur in particular when radiative interactions are too rare to establish the distributions enumerated above. The thermalization length is large, and the distributions are influenced by photons coming from great distances, where different conditions (e.g., different velocity or atomic state distributions) might prevail. Thus the state is "non-local", and often called "non-LTE". There might still be an equilibrium everywhere. However, the equilibrium is no longer characterized by a single temperature. In the solar atmosphere a characteristic situation is that the electrons satisfy a Maxwell distribution, with *electron temperature* T_e, because collisions are still frequent. On the other hand, the population of atomic levels depends on radiative processes, which become rare in a rarefied gas. These populations therefore must be described by *statistical equations*. Therefore, instead of non-LTE more descriptive names are SE, for *statistical equilibrium*, or KE, for *kinetic equilibrium*.

4.1.3 Absorption Lines in LTE

In principle we could use expression (4.4) for the emergent intensity regardless whether we consider a frequency ν in the continuum or in an absorption line. However it is often advantageous to realize that in an absorption line τ_ν is the optical depth calculated from the continuum *and* line absorption coefficients, i.e., $d\tau_\nu = d\tau_C + d\tau_l$, or

$$d\tau_\nu = (1 + \eta_\nu)d\tau_C \;, \tag{4.6}$$

where $\eta_\nu \equiv \eta(\nu) = \kappa_l(\nu)/\kappa_C$ was already defined in Sect. 3.5.2 (although there we used wavelength λ instead of frequency ν as the argument). Using (4.6) we rewrite the intensity integral (4.4). At disc center ($\mu = 1$), and under the assumption of LTE ($S_\nu = B_\nu$), we obtain

$$I_\nu(0,1) = \int\limits_0^\infty (1 + \eta_\nu) B_\nu \exp\left(-\int\limits_0^\tau (1 + \eta_\nu)\, d\tau'\right) d\tau \;, \tag{4.7}$$

where τ is now the optical depth in the continuum. As we saw, (4.7) is most useful in the context of radiative transfer in magnetically sensitive lines.

It remains to determine the line absorption coefficient. We have already, in Sect. 2.3.7, written the cross section *per atom* as a function of frequency. We restrict our attention to the two mechanisms of Doppler and collisional line broadening, as expressed by profiles (2.84) and (2.85). At each frequency of the collisional profile we must apply the Doppler-broadening formula (or vice versa), i.e., we must fold the two expressions and obtain

$$\phi(\nu) = \frac{\gamma}{\sqrt{\pi}\Delta\nu_D} \int\limits_{-\infty}^\infty \frac{\exp(-(\nu - \nu')^2/\Delta\nu_D^2)}{[2\pi(\nu' - \nu_0)]^2 + \gamma^2/4}\, d\nu' \;. \tag{4.8}$$

With the definitions

$$y = \frac{\nu - \nu'}{\Delta\nu_D} \;, \tag{4.9}$$

$$v = \frac{\nu - \nu_0}{\Delta\nu_D} \;, \quad \text{and} \tag{4.10}$$

$$a = \frac{\gamma}{4\pi\Delta\nu_D} \tag{4.11}$$

we find

$$\phi(\nu) = \frac{1}{\sqrt{\pi}\Delta\nu_D} H(a, v) \;, \tag{4.12}$$

where we have introduced the *Voigt function*

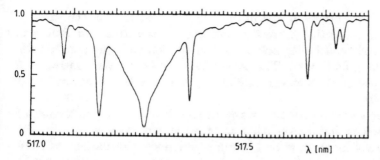

Fig. 4.1. Solar spectrum, with Mg b_2 line. From Beckers et al. (1976)

$$H(a, v) = \frac{a}{\pi} \int\limits_{-\infty}^{\infty} \frac{e^{-y^2}}{(v-y)^2 + a^2} \, dy \, . \tag{4.13}$$

The Voigt function is normalized such that for small a its value at the line center is 1, i.e.,

$$H(a, 0) = 1 + O(a) \, , \tag{4.14}$$

while $\phi(\nu)$ is normalized so that the integral over all frequencies is 1.

For solar spectral lines we generally have $a \ll 1$. Inspection of (4.13) shows that in this case the Doppler profile dominates near the line center ν_0. On the other hand, the Gaussian rapidly drops with distance from ν_0, and hence the line wings are shaped according to the collisional, or "damping", part of the profile. An example where the two parts of the line profile can clearly be distinguished is shown in Fig. 4.1 (although a strong line such as Mg b_2 cannot very accurately be treated under the assumption of LTE!).

Problem 4.2. Show that the integral over v of $H(a, v)$ is $\sqrt{\pi}$. Prove (4.14).

The cross section per atom, (2.83), is now

$$\sigma = \frac{e^2 f}{4\varepsilon_0 m_e c \sqrt{\pi} \Delta\nu_D} H(a, v) \, . \tag{4.15}$$

In order to obtain the line absorption coefficient, i.e., the cross section per unit mass, we must multiply σ by the number of absorbing particles per unit mass. This number is obtained as the number n_i/ρ of particles per unit mass of the species i considered, multiplied by the fraction n_{ij}/n_i of particles of that species in ionization state j, and finally multiplied by the fraction of these particles which are in excitation state k, i.e., by n_{ijk}/n_{ij}. Using $\rho = \mu m_H \sum n_i$ we therefore have

$$\kappa_l = \frac{\sigma}{\mu m_H \sum n_i} \frac{n_i}{n_i} \frac{n_{ij}}{n_i} \frac{n_{ijk}}{n_{ij}} \, . \tag{4.16}$$

The various terms are separated in this way because, in LTE, the third and fourth factors, respectively, are determined from the Saha and Boltzmann equations, (4.59) and (4.64) below, in combination with the normalization conditions (4.60) and (4.65). The second factor of (4.16), the relative (number) abundance of the element considered, will be taken as known for the present purpose.

The emergent line intensity can now be calculated if two atomic constants are known: the oscillator strength f and the damping constant γ.

Laboratory experiments, or quantum-mechanical calculations of transition probabilities, yield the f values. Standard tables have been prepared by Wiese et al. (1966) and Wiese et al. (1969), but new results appear constantly. The damping constant γ generally depends on density and temperature, as well as on atomic properties. We do not discuss this matter further, since this is done in extenso in many texts on spectroscopy. Also, we shall not treat line-broadening mechanisms other than the two considered above, although some can be important (e.g. Stark-effect broadening for hydrogen lines).

We must however add a complementary remark on the Doppler width, $\Delta\nu_D$. The thermal random motion of atoms normally does not suffice to explain quantitatively the "Doppler core" of spectral line profiles. Therefore, the effect of a turbulent gas motion is added, and we write

$$\Delta\nu_D = \frac{\nu_0}{c}\sqrt{\frac{2\mathcal{R}T}{A} + \xi_t^2} \, . \tag{4.17}$$

The parameter ξ_t is called "microturbulence", or "microvelocity" (as opposed to the "macroturbulence" which will be discussed in Sect. 4.3.3 below).

4.2 Radiative Transfer – Statistical Equilibrium

4.2.1 Model Assumptions

In the higher part of the solar atmosphere we must consider departures from local thermodynamic equilibrium. We shall treat such departures here for a single species only. An example is silicon, which is an important absorber in the ultraviolet, cf. Fig. 4.5. In general both the continuum and the line absorption will be affected, because the population of the atomic levels from which bound electrons are excited (either to higher bound levels or to the "free", or continuum, state) may no longer obey Boltzmann's law.

We shall however still assume that collisions between particles occur with sufficient frequency for the velocity distributions to become Maxwellian. In the atmosphere, our present concern, this is a reasonable assumption, but we must keep in mind that in more tenuous environments, such as the corona or the solar wind, it will certainly be violated. In the present treatment the various particles all will have Maxwellian velocity distributions with the same temperature; since collisions with *electrons* play a dominant role for the atomic populations discussed below, this temperature will also be called the *electron Temperature*, T_e. In the following we mean this parameter whenever we speak of temperature (a term which strictly would be reserved for LTE).

Vernazza et al. (1973) have outlined the equations describing the statistical equilibrium in the context of solar atmospheric models. Here I shall follow their work, but also draw from Schleicher (1976). We shall consider an atom with an electron for which N discrete bound states, plus a continuum state (i.e., the state of ionization) are possible. For brevity we shall omit the indices denoting the chemical element and the state of ionization. To begin with, we enumerate the various transitions between the atomic states and their characteristic constants.

4.2.2 Line Radiation and Einstein Coefficients

As far as *line radiation* is concerned, let us concentrate on a particular pair among the N bound states. We denote the lower and upper level of this pair by L and U, respectively. If E_L and E_U are the energies of the two levels, then $h\nu = E_U - E_L$ defines the frequency ν of the line under consideration. Three radiative transitions must be distinguished: spontaneous emission, induced emission, and absorption.

Spontaneous emission of a photon can occur if the electron initially is in the upper level. The number of spontaneous emissions per unit time is proportional to the number n_U of atoms per unit volume in that level; the constant of proportionality is a characteristic number for the atom and the levels considered, and will be denoted by A_{UL}. Thus, $n_U A_{UL}$ photons are spontaneously emitted per unit time and unit volume. If we recall that the energy levels of an atom have a finite spread, and that the emitting atoms are in random motion, we see that the emitted photons must have a frequency distribution, say $\chi(\nu)$, around the frequency ν_0 of the line center. Then, $n_U A_{UL} \chi(\nu)$ is the number of photons spontaneously emitted per unit time, per unit volume, and per frequency interval around ν. Since the photons are spontaneously emitted with equal probability into all directions we obtain

$$n_U A_{UL} \chi(\nu)/4\pi \tag{4.18}$$

as the number of such emissions per unit time, volume, frequency interval, and solid angle.

The atomic constant A_{UL} is called the Einstein coefficient for spontaneous emission. Its dimension is 1/time, and A_{UL} can be interpreted as the inverse of the lifetime of the upper state against spontaneous radiative transition into the lower state. Lifetimes of 10^{-8} s are typical, but different values (by orders of magnitude) may occur as well.

Induced emission, also called stimulated emission, also requires an atom with an electron initially in the upper state. If such an atom is exposed to radiation of frequency $\nu = (E_U - E_L)/h$, a transition to the lower state may be initiated. This time the number of emitted photons is again proportional to the number of atoms in the upper state, but also to the intensity of the incident radiation. As for the spontaneous emission, we must allow for a finite width of the emitted line, for which we introduce the profile $\psi(\nu)$. In analogy to (4.18) we write the constant of proportionality in the form $B_{UL}/4\pi$. Since the induced emission is in the same direction as the inducing radiation, and since the intensity of the latter is already related to unit solid angle, we have

$$n_U B_{UL} I_\nu \psi(\nu)/4\pi \tag{4.19}$$

as the number of induced emissions per unit time, volume, frequency interval, and solid angle. Again, B_{UL} is an atomic constant, called the Einstein coefficient of induced emission. Notice that $B_{UL} I_\nu$ has the same dimension as A_{UL} and can be interpreted as an inverse lifetime.

The third radiative transition is *absorption*, also called radiative excitation. It is the opposite of induced emission. Initially the atom is in the lower state. When exposed to radiation of frequency ν, the atom takes a photon out of the radiation field, and thereby is excited into the upper state. The number of these radiative excitations will be proportional to the number n_L of atoms in the lower state, and to the intensity I_ν. As before, we must allow for the finite width in frequency of the absorption process, so that

$$n_L B_{LU} I_\nu \phi(\nu)/4\pi \tag{4.20}$$

is the number of absorbed photons per unit time, volume, frequency interval, and solid angle; B_{LU} is the Einstein coefficient for radiative excitation.

The profile $\phi(\nu)$ is of course the absorption profile which we have already considered in Sects. 2.3.7 and 4.1.3.

Problem 4.3. Relate the Einstein coefficient B_{LU} to the oscillator strength defined in (2.83).

Problem 4.4. Relations between Einstein coefficients. Consider the special case of thermodynamic equilibrium, where the levels U and L are populated according to (4.64), where $I_\nu = B_\nu(T)$, and where the *principle of detailed balance* is valid, i.e., the number of upward transitions must be equal to the number of downward transitions. Show that under these circumstances the following relations must hold:

$$g_U B_{UL} = g_L B_{LU} \tag{4.21}$$

$$\frac{g_U}{g_L} A_{UL} = \frac{2h\nu^3}{c^2} B_{LU} . \tag{4.22}$$

Notice that, because the Einstein coefficients are atomic constants, relations (4.21) and (4.22) are generally valid, and not only in the special case of thermodynamic equilibrium!

The Einstein coefficients A_{UL}, B_{UL}, and B_{LU} can be calculated theoretically as transition probabilities, or may be determined by laboratory experiments. For the present text we shall take them as known quantities.

4.2.3 Continuum Radiation

Next we consider transitions to and from the continuum state.

Photoionization occurs if the energy of an incident photon exceeds the energy of ionization (counted from the energy of the level occupied by the target atom). The number of photoionizations must be proportional to the intensity of the incident radiation. For each level j we introduce a photoionization cross section, $\alpha_j(\nu)$, so that $\alpha_j(\nu)I_\nu$ is the absorbed energy per time, frequency interval, target atom, and solid angle; hence, if n_j is the number density of atoms in the jth bound state, then

$$n_j\alpha_j(\nu)I_\nu/h\nu \tag{4.23}$$

is the number of photoionizations from level j per time, volume, frequency interval, and solid angle.

The cross section $\alpha_j(\nu)$ must be determined by measurements in the laboratory, or by quantum-mechanical calculation. An example, valid for hydrogen-like atoms, was already presented in Sect. 2.3.7.

The inverse of photoionization is *radiative recombination*. As for the case of line radiation, we must allow for an induced contribution, which is proportional to the intensity, and an independent term. Both contributions must be proportional to the number density, n_{N+1}, of atoms in the continuum state. To be concise we use C instead of $N+1$ for the index of the continuum state, and write

$$n_C(\gamma_j(\nu) + \beta_j(\nu)I_\nu)/h\nu \tag{4.24}$$

for the number of radiative recombinations to level j per time, volume, frequency interval, and solid angle. The frequency interval corresponds to a certain velocity interval of the electron distribution.

Problem 4.5. In analogy to Problem 4.4, use the special case of local thermodynamic equilibrium (where the principle of detailed balance is valid) to show that (4.24) can generally be replaced by the expression

$$\alpha_j(\nu)n_C\frac{n_j^*}{n_C^*}\exp\left(-\frac{h\nu}{kT}\right)\left(\frac{2h\nu^3}{c^2} + I_\nu\right)\bigg/h\nu\,, \tag{4.25}$$

where

$$\frac{n_j^*}{n_C^*} = \left(\frac{h^2}{2\pi m_e kT}\right)^{3/2}\frac{n_e g_j}{2u_C}\exp\left(\frac{E_C - E_j}{kT}\right) \tag{4.26}$$

is the LTE ratio of the populations of levels j and C (E_C is the ionization energy, u_C is the partition function of the ionized state, cf. Sect. 2.3.4).

4.2.4 Collisions

In addition to radiative transitions, we must consider *collisional transitions* between the levels of our model atom. Collisions have no direct influence on the radiation field. Nevertheless, they affect the populations of the atomic levels, and must therefore be included into the statistical equations. What we need in these equations are the collisional transition *rates* C_{ij}, i.e., the number of transitions per unit time and per atom from state i to state j (bound or continuum).

In principle, these rates are determined in complete analogy to the procedure outlined in Sect. 2.3.6 in the context of nuclear reaction rates. We must separate the effect arising from the distribution of the colliding particles in velocity space from the effect of intrinsic atomic properties. An integral of the form (2.77) is written down; as we are mainly interested in collisions of atoms with electrons, the relative velocity that appears in the integrand is essentially the electron velocity (which is much larger due to the much smaller electron mass).

We shall not go into further detail concerning the rates C_{ij}. However, it should be mentioned that, once C_{ij} has been calculated, and once one assumes (as we do) that the velocities are distributed as in LTE, the inverse rate, C_{ji}, is also known. For, the collision rates C_{ij} and C_{ji}, and hence their ratio C_{ij}/C_{ji}, depend only on atomic properties and on the particle distributions. Since the former are constant and the latter are in LTE, the ratio C_{ij}/C_{ji} must have its LTE value, i.e.,

$$C_{ij}/C_{ji} = n_j^*/n_i^* . \tag{4.27}$$

This relation follows from the principle of detailed balance which holds in local thermodynamic equilibrium.

For collisional transitions between two bound states, L and U, it follows that

$$C_{\mathrm{UL}} = \frac{g_{\mathrm{L}}}{g_{\mathrm{U}}} \exp\left(\frac{E_{\mathrm{U}} - E_{\mathrm{L}}}{kT}\right) \cdot C_{\mathrm{LU}} . \tag{4.28}$$

On the other hand, for transitions to and from the continuum we can apply (4.26) and obtain

$$C_{\mathrm{C}j} = \left(\frac{h^2}{2\pi m_{\mathrm{e}} kT}\right)^{3/2} \frac{n_{\mathrm{e}} g_j}{2u_{\mathrm{C}}} \exp\left(\frac{E_{\mathrm{C}} - E_j}{kT}\right) \cdot C_{j\mathrm{C}} . \tag{4.29}$$

4.2.5 The Source Function

We now return to the equation of transfer. We shall consider this equation at each frequency ν, although the argument (or index) ν will mostly be suppressed. Let κ be the total coefficient of absorption (defined, as before, in terms of area per unit mass). If r is the outward radial coordinate, then

$$\mu \frac{dI_\nu}{dr} = -\kappa \rho I_\nu + \varepsilon \ . \tag{4.30}$$

According to the diverse processes enumerated above, we must distinguish between line absorption and continuum absorption; that is, we write

$$\kappa = \kappa_l + \kappa_C \ , \tag{4.31}$$

where κ_l is the line absorption coefficient, and κ_C is the continuum absorption coefficient. We also divide the source term of (4.30), ε, into line emission ε_l and continuum emission ε_C:

$$\varepsilon = \varepsilon_l + \varepsilon_C \ . \tag{4.32}$$

Let us first consider *line* radiation. Knowing the Einstein coefficients for the radiative transitions we can write down the net gain of the intensity at frequency ν through these transitions. For the particular transition between levels L and U the number of involved photons is given by expressions (4.18) to (4.20). Each photon carries an energy $h\nu$; hence

$$\frac{h\nu}{4\pi}[n_U(A_{UL}\chi + B_{UL}I_\nu\psi) - n_L B_{LU}I_\nu\phi] \tag{4.33}$$

is the change of energy, at frequency ν, per unit time, volume, frequency interval, and solid angle. Instead to the volume we may as well relate this change to area times length; in this way we recognize that (4.33) is just the right-hand side of the transfer equation – as far as line radiation is concerned – i.e., identical to

$$-\rho \kappa_l I_\nu + \varepsilon_l \ . \tag{4.34}$$

We replace ε_l by a line source function S_l, defined by $S_l = \varepsilon_l / \rho \kappa_l$, compare expressions (4.33) and (4.34), and find

$$\kappa_l = \frac{h\nu}{4\pi\rho}(n_L B_{LU}\phi - n_U B_{UL}\psi) \ , \tag{4.35}$$

$$S_l = \frac{n_U A_{UL}\chi}{n_L B_{LU}\phi - n_U B_{UL}\psi} \ . \tag{4.36}$$

Expression (4.35) makes clear why stimulated emission is also called "negative absorption" (e.g., Aller 1963): it simply enters as a negative contribution to the line absorption coefficient.

At this point we introduce another simplifying concept, known as *complete redistribution*. This concept consists of the assumption that the frequency profiles of the three radiative processes are all equal:

$$\chi(\nu) = \psi(\nu) = \phi(\nu) \ . \tag{4.37}$$

This assumption is valid only under special circumstances; for example in the extreme case of exact *coherence*, where a photon is emitted at precisely the same frequency at which it was absorbed; or in the other extreme case of complete *incoherence* where the frequency of emission is completely independent of the frequency of absorption. In general these conditions are not satisfied, and we must deal with the individual profiles; this case is called *partial redistribution*. Here we shall be content with (4.37), which makes the line source function (4.36) independent of frequency:

$$S_l = \frac{n_U A_{UL}}{n_L B_{LU} - n_U B_{UL}} \ . \tag{4.38}$$

Under the approximation of complete redistribution the line absorption coefficient (4.35) is proportional to the function $\phi(\nu)$, which we shall consider as given in form of a Voigt function (4.13).

Problem 4.6. Use (4.21) and (4.22) to eliminate the Einstein coefficients from the line source function. Derive the line source function for the case where several lines are treated simultaneously.

Next we consider the continuum. With expressions (4.23) and (4.25) above we have already prepared what we need for the continuum part on the right-hand side of the transfer equation. Therefore

$$-\rho \kappa_{Cj} I_\nu + \varepsilon_{Cj} \equiv \alpha_j(\nu) \left[n_C e^{-h\nu/kT} \frac{n_j^*}{n_C^*} \left(\frac{2h\nu^3}{c^2} + I_\nu \right) - n_j I_\nu \right] \ . \tag{4.39}$$

This defines the continuum absorption coefficient κ_{Cj} and the continuum source function $S_{Cj} = \varepsilon_{Cj}/\rho \kappa_{Cj}$ for transitions to and from level j:

$$\kappa_{Cj} = \rho^{-1} \alpha_j(\nu) \left(n_j - n_C e^{-(h\nu/kT)} \frac{n_j^*}{n_C^*} \right) \ , \tag{4.40}$$

$$S_{Cj} = \frac{2h\nu^3}{c^2} \frac{1}{(n_j/n_C)(n_C^*/n_j^*) \exp(h\nu/kT) - 1} \ . \tag{4.41}$$

The ratio n_j^*/n_C^* is again given by (4.26); obviously $S_{Cj} = B_\nu$ in the case of LTE.

Problem 4.7. Derive the continuum absorption coefficient κ_C and the continuum source function S_C for the case where transitions to and from several bound states are considered simultaneously.

We may now treat line and continuum radiation together. While the absorption coefficient is simply the sum of the diverse κ's, the source function is the weighted sum:

$$S = \frac{\kappa_l S_l + \kappa_C S_C}{\kappa_l + \kappa_C} \ . \tag{4.42}$$

4.2.6 The Equations of Statistical Equilibrium

In order to evaluate the source function we must know the populations n_L and n_U of the lower and upper levels of the transition under consideration. If more than one transition is to be studied, we must know the populations n_j of all the respective levels.

In statistical equilibrium the total number of transitions which populate a certain level must be equal to the total number of depopulating transitions. If there are only the two levels L and U then we obtain these numbers by integrating expressions (4.18) to (4.20) for the three radiative transitions over frequency ν and solid angle Ω, and by adding the respective numbers of collisions. This yields the balance

$$n_L(B_{LU}\bar{J}_{LU} + C_{LU}) = n_U(A_{UL} + B_{UL}\bar{J}_{UL} + C_{UL}) , \tag{4.43}$$

where

$$\bar{J}_{LU} = \bar{J}_{UL} = \frac{1}{4\pi} \int I_\nu \phi(\nu) \, d\nu d\Omega \tag{4.44}$$

is the intensity averaged over frequency *and* over solid angle [we assume complete redistribution, i.e., use (4.37)]. If ε is the abundance of the considered chemical element relative to hydrogen (by number), and n_H is the number density of hydrogen, then

$$n_L + n_U = \varepsilon n_H . \tag{4.45}$$

Now we generalize the statistical equations to the model atom with $N+1$ levels. The population of the jth level is governed by

$$n_j \sum_{i \neq j} R_{ji} = \sum_{i \neq j} n_i R_{ij} . \tag{4.46}$$

On the left of this equation we have the transitions from the jth level to other levels, on the right we have the transitions from those other levels to level j (the sums include the bound states as well as the continuum).

For bound levels the *rates* R_{ji} are given in terms of the Einstein coefficients and the collisional rates:

$$R_{ji} = A_{ji} + B_{ji}\bar{J}_{ji} + C_{ji} ; \tag{4.47}$$

the term A_{ji} on the right must be deleted whenever level i is higher than level j.

For transitions from level j to the continuum (level C) the rate follows from an integration of (4.23) over all frequencies above the frequency ν_{jC} needed for ionization, integration over solid angle, and addition of the collisional rate:

$$R_{jC} = 4\pi \int_{\nu_{jC}}^{\infty} \frac{\alpha_j(\nu)}{h\nu} J_\nu d\nu + C_{jC} , \tag{4.48}$$

where J_ν is the mean intensity, defined as

$$J_\nu = \frac{1}{4\pi} \int I_\nu \, d\Omega \ . \tag{4.49}$$

In the same manner, we integrate (4.25) and add the corresponding collisional rate to obtain the rate for recombinations (to level j):

$$R_{Cj} = 4\pi \frac{n_j^*}{n_C^*} \int\limits_{\nu_{jC}}^{\infty} \frac{\alpha_j(\nu)}{h\nu} e^{-(h\nu/kT)} \left(\frac{2h\nu^3}{c^2} + J_\nu \right) d\nu + C_{Cj} \ . \tag{4.50}$$

Of the $N+1$ linear equations (4.46) only N are independent. The system is complemented by the generalization of (4.45):

$$\sum_j n_j = \varepsilon n_H \ . \tag{4.51}$$

Equations (4.46) and (4.51) determine the number densities that are needed for the radiative transfer in statistical equilibrium. Instead of these number densities it is illustrative and common to determine the *departure coefficients* defined by

$$b_j = \frac{n_j/n_j^*}{n_C/n_C^*} \ , \quad b_C = 1 \ . \tag{4.52}$$

It is important to realize that the number densities which are obtained by solving the statistical equations themselves depend on the intensity (via $\bar{J}_{ji}$); that is, the absorption coefficient and the source function depend on intensity. The equation of transfer is therefore generally non-linear, and is solved by an iterative method.

Problem 4.8. Transform (4.46) into a system of equations for the departure coefficients.

4.3 Atmospheric Models

4.3.1 Limb Darkening

The variation of temperature T and pressure P with height in the atmosphere is often called an "atmospheric model", or simply a "solar model" – not to be confused with the interior solar models which we have treated in Chap. 2. We shall now describe such atmospheric models, and thereby neglect horizontal variations of T and P.

A first direct evidence for the existence of a temperature gradient in the solar atmosphere is the *limb darkening*, cf. Fig. 4.2. Near its center, the solar disc appears brighter, and hence hotter, than near the limb because we see

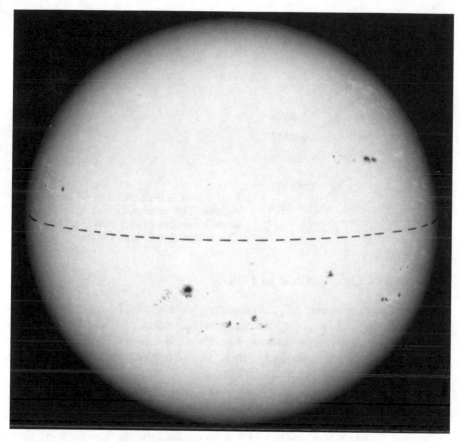

Fig. 4.2. The Sun on 11 August 1958, with limb darkening and several sunspot groups. The *dashed curve* is the equator. Photograph Einsteinturm, Astrophysical Observatory Potsdam, Germany

into deeper layers when we observe normal to the surface, and into shallower depth near the limb where the line of sight is nearly tangential.

In fact we have already been acquainted with an atmospheric model which *predicts* the limb darkening: the Eddington approximation of Sect. 2.3.8. At $\tau = 0$ this approximation yields an intensity variation which is linear in μ:

$$I(0,\mu)/I(0,1) = (2 + 3\mu)/5 \tag{4.53}$$

(where $\mu = \cos\theta$). The comparison of this result with observations (Fig. 4.3) shows that the model is quite satisfactory near the disc center, i.e., at great depth, but fails near the limb.

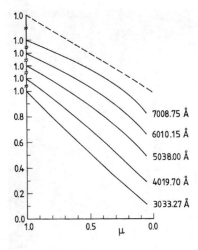

Fig. 4.3. Limb darkening at various wavelengths. The data are fitted with polynomials of 5th degree (Pierce and Slaughter 1977). The *dashed* line follows (4.53)

4.3.2 Model Calculations in LTE

We return to expression (4.4) for the emergent intensity. Let us now use wavelength λ instead of frequency for the spectral coordinate, and let us assume local thermodynamic equilibrium, i.e., $S_\lambda = B_\lambda$. Let us also, for the moment, restrict ourselves to the continuum, and neglect lines. The absorption coefficient then varies in a smooth fashion.

We have already seen in Chap. 1 that the absolute intensity has been observed over a wide range of the spectrum. Using this result we may in principle derive both the temperature as a function of τ_λ and the variation of κ_λ with λ. To see this consider first a fixed wavelength, say $\lambda = 500$ nm, where the continuum can clearly be distinguished (Fig. 4.4) and which is often taken as a reference wavelength. Due to the exponential factor in the integrand of (4.4) the contribution at any given τ_λ is more strongly attenuated near the limb (small μ) than near the center ($\mu = 1$). Hence, taking observations at various μ, the function $B_\lambda(T(\tau_\lambda))$ is weighted in different ways, and may thus

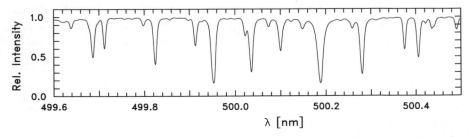

Fig. 4.4. Normalized intensity of the solar spectrum near 500 nm. Data from the *Liège Atlas* of Delbouille et al. (1973)

be determined from the center-to-limb variation. Problem 4.9 is an explicit, although approximative, illustration. From B_λ we then calculate $T(\tau_\lambda)$.

Next we can apply the same procedure for two different values of λ. We find, say, $T(\tau_\lambda)$, in addition to $T(\tau_{500})$ Then

$$\frac{dT}{d\tau_\lambda} \Big/ \frac{dT}{d\tau_{500}} = \kappa_{500}/\kappa_\lambda \;, \tag{4.54}$$

which serves to derive the absorption coefficient as a function of wavelength.

Problem 4.9. The Eddington–Barbier approximation. Expand the source function $S(\tau_\lambda)$ around some optical depth τ_λ^*. Find the intensity $I_\lambda(0, \mu)$ and show that, up to terms of second order in $\tau_\lambda - \tau_\lambda^*$, we have $S_\lambda(\tau_\lambda^*) = I_\lambda(0, \mu)$ for $\tau_\lambda^* = \mu$.

At the extreme limb the resolution in μ of the observed absolute intensity is not satisfactory. As a consequence, the temperature $T(\tau_\lambda)$ can be obtained from the center-to-limb variation with sufficient accuracy only for $\tau_{500} > 0.1$ [recall that for small τ_λ the inversion of (4.4) requires good observational data at small μ]. There, in the deep part of the atmosphere, the center-to-limb variation of the intensity yields relatively safe results: The temperatures obtained in a variety of studies are rather consistent at large τ_{500}, cf. Fig. 4.6.

Photospheric models which include layers of smaller optical depth may be obtained from the wavelength dependence of I_λ if the absorption coefficient is known. We have already seen in Sect. 2.3.7 that $\kappa(\lambda)$ can be calculated theoretically as a function of density and temperature.

For typical conditions in the solar photosphere, Fig. 4.5 shows the continuous absorption coefficient $\kappa_C(\lambda)$. In the visible and infrared, the H$^-$ ion is the dominant absorber, as recognized by R. Wildt in 1938. Its bound-free continuum extends to $\lambda = 1.645\,\mu$m, corresponding to the ionization energy of $0.754\,$eV for H$^-$. For larger λ, there is a free-free continuum, with strongly increasing absorption towards the far infrared.

Even more than in the infrared, the continuum absorption coefficient increases in the ultraviolet. Several bound-free continua are clearly discernable in Fig. 4.5 by means of their respective ionization edges.

In a solar photospheric model the absorption coefficient is a function of depth. In order to construct a complete model we may proceed as follows. We start with a first guess at the temperature. Let us use τ_{500}, the optical depth at $\lambda = 500\,$nm, in place of the geometrical depth. Thus $T(\tau_{500})$ is assumed to be a known function. If we use LTE and restrict our attention to the disc center the integral (4.4) becomes

$$I_\lambda(0, 1) = \int_0^\infty B_\lambda(T) \exp(-\tau_\lambda)\, d\tau_\lambda \;. \tag{4.55}$$

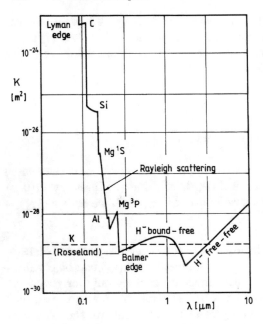

Fig. 4.5. Continuum absorption coefficient (per particle) in the solar atmosphere, at $\tau_{500} = 0.1$. Adapted from Unsöld and Baschek (1999)

In order to evaluate this integral at a wavelength other than 500 nm we need $\kappa_C(\lambda)$, cf. (4.66) below; hence we need, in addition to T, the diverse particle densities so that we can calculate κ_C as explained in Sect. 2.3.7. To this end we use the pressure equation

$$\frac{dP}{d\tau_{500}} = \frac{g_\odot}{\kappa_{500}} \; , \tag{4.56}$$

and thereby ensure that our model atmosphere will be in hydrostatic equilibrium. The integral of (4.56) is

$$P = \int_0^{\tau_{500}} \frac{g_\odot}{\kappa_{500}} \, d\tau'_{500} \; . \tag{4.57}$$

Thus we know T and P at each step of integration. The particle densities now follow from the equation of state and from the equilibria of ionization and excitation. The electron density is

$$n_e = \sum_{ij} j n_{ij} \; , \tag{4.58}$$

where n_{ij} is the number density of species i in the jth state of ionization. The ratios of the n_{ij} are determined by Saha equations, cf. (2.37), which we write in abbreviated form

$$n_{i,j+1}/n_{ij} = f_{ij}(P_e, T) \; ; \tag{4.59}$$

their sum gives the total density of a species:

$$n_i = \sum_j n_{ij} = \rho X_i / m_i \ . \tag{4.60}$$

The X_i are the mass fractions of the chemical elements (denoted earlier by $X, Y, \dots$) and are considered as given for the present purpose. The system (4.57) to (4.60) is then closed by the relations for the electron pressure

$$P_{\mathrm{e}} = n_{\mathrm{e}} kT \ , \tag{4.61}$$

the density

$$\rho = P\mu/\mathcal{R}T \ , \tag{4.62}$$

and the mean molecular weight

$$\mu = \sum_i n_i A_i / (n_{\mathrm{e}} + \sum_i n_i) \tag{4.63}$$

(not to be confused with $\mu = \cos\theta$!). Equation (4.62) implies that only the gas pressure is taken into account, which may suffice for the present purpose. Finally, for $\kappa_C(\lambda)$ we also need the number n_{ijk} of each atom (or ion) in the diverse excited levels (ith species, jth ionization state, kth excited level). Since we have assumed LTE, these numbers are given by Boltzmann distributions,

$$n_{ijk} = n_{ij0}(g_{ijk}/g_{ij0}) \exp(-E_{ijk}/kT) \ , \tag{4.64}$$

and by

$$\sum_k n_{ijk} = n_{ij} \ . \tag{4.65}$$

When the atmospheric integration is completed we may calculate the emergent intensity (4.55) over as many wavelength bands as we wish, or rather as our available observational data allow us to compare with. For each band it is only necessary to calculate the τ_λ scale according to

$$\tau_\lambda = \int_0^{\tau_{500}} \frac{\kappa_\lambda}{\kappa_{500}} \, d\tau'_{500} \ . \tag{4.66}$$

Of course, the first guess of $T(\tau_{500})$ may not provide the best fit to the observed intensities. Then we must iterate.

As a practical matter we may integrate (4.56) in the inverted form, because when $\tau_{500} \ll 1$, equidistant steps in pressure make more sense than equidistant steps in τ_{500}; or we may use $\ln \tau_{500}$ or $\ln P$ as an independent variable; or we may transform (4.56) into a differential equation for the electron pressure P_{e}. But in principle the scheme is always as outlined by (4.55) to (4.66).

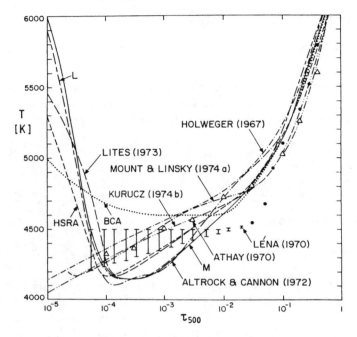

Fig. 4.6. Model atmospheres of the Sun. From Vernazza et al. (1976)

It is also important to realize that often not all the information is needed, and that only the dominant absorbers must be treated with care. In the infrared, for example, only the H^- ion must be taken into account.

A solar model that has been obtained on the basis of continuum intensities alone is the "Bilderberg Continuum Atmosphere", or BCA, so named after a 1967 conference hotel near Arnhem, Holland (Gingerich and de Jager 1968). The BCA is depicted in Fig. 4.6 as a dotted curve. Although this model disagrees in detail with the later results, also shown in Fig. 4.6, it does show a temperature minimum and a temperature rise for $\tau_{500} < 10^{-4}$. The information concerning these higher layers comes from both the ultraviolet and the infrared. Only in these spectral regions the atmosphere is sufficiently opaque that the higher layers can be observed in the continuum. But in the infrared, near 1.6 μm, is also the spectral region where the solar atmosphere is most transparent, i.e., where we see the deepest of all observable layers.

Problem 4.10. Calculate $d\ln B_\lambda/d\ln T$ as a function of T and λ. If the intensity changes by an order of magnitude, how much does the brightness temperature change, at $T = 5000\,\text{K}$, for the two wavelengths $\lambda = 150\,\text{nm}$ and $\lambda = 150\,\text{μm}$?

A Model Including Lines. In the continuum, high layers of the solar atmosphere are opaque, and therefore observable, only in the ultraviolet and infrared. In the visible, these layers are accessible to observation only if we use the large opacity in spectral lines. The information contained in spectral lines has been exploited to construct models of the atmosphere. In the simplest case, this is done again under the assumption of local thermodynamic equilibrium, LTE. The model of Holweger (1967), also shown in Fig. 4.6, is an example.

When we construct a solar model atmosphere according to the recipe outlined above we may assume that the turbulence parameter ξ_t introduced in (4.17) varies as a function of depth. In fact $\xi_t(\tau)$ is an additional function which can be adapted to improve the agreement between observations and model. If measurements at various distances from the disc center (i.e., various μ) are exploited, it is even possible to distinguish between vertical and horizontal turbulent motions. The latter are seen near the limb, the former near the disc center. Holweger, who investigated 900 lines of the visible solar spectrum under the assumption of LTE, found that the vertical component decreases from $2.5\,\mathrm{km/s}$ to $1\,\mathrm{km/s}$ between $\tau = 1$ and $\tau = 10^{-3}$, and that the horizontal component decreases from $3\,\mathrm{km/s}$ to $1\,\mathrm{km/s}$ in the same range of optical depth.

4.3.3 Models with Departures from LTE

The Bilderberg Continuum Atmosphere had a temperature which was not well determined near its minimum. Intensities in the UV and in the far IR which emerge from the coolest layer were not known with sufficient precision. In the infrared there is the additional difficulty that the intensity is not very sensitive to changes of temperature (Problem 4.10) so that the latter cannot easily be determined accurately.

A lower temperature minimum, of $T \approx 4200\,\mathrm{K}$ near $\tau_{500} = 10^{-4}$, is one of the main improvements of later models, such as the Harvard-Smithsonian Reference Atmosphere, or HSRA (Gingerich et al. 1971). The main support for such lower minimum temperature comes from rocket observations of the ultraviolet intensity around $\lambda = 165\,\mathrm{nm}$, but also from infrared observations near $\lambda = 300\,\mathrm{\mu m}$.

In addition to new data, the later atmospheric models incorporated departures from local thermodynamic equilibrium. Near the temperature minimum a typical effect of the statistical equilibrium is to *underpopulate* (in comparison to LTE) the states of the neutral atoms. For example, for the ground state of Si I the departure coefficient reaches ≈ 0.2 at $500\,\mathrm{km}$ above the level where $\tau_{500} = 1$ (Fig. 4.8). The reason is that at the low temperature of that height photoionization dominates over radiative recombination. As a consequence of the lower population of the neutrals one may explain the observational data with a lower temperature minimum than would have been possible under the assumption of LTE. The situation is opposite in the layers lying above the temperature minimum: here radiative recombination is dominant and the neutrals are strongly overpopulated, at the expense of the ionized state.

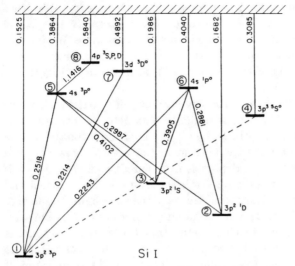

Fig. 4.7. Level diagram for a Si I model atom. Vertical distances are in energy; numbers on the *solid lines* (the radiative transitions) are wavelengths in μm. From Vernazza et al. (1976)

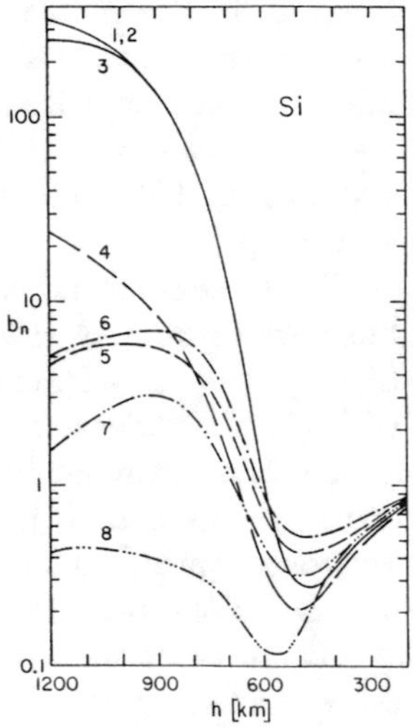

Fig. 4.8. Departure coefficients for the 8-level Si I atom. From Vernazza et al. (1981)

Most extensive calculations in statistical equilibrium have been presented by Vernazza et al. (1976, 1981). An example of a model atom as it was explained in Sect. 4.2 is shown in Fig. 4.7. This Si I atom consists of 8 bound levels and a continuum; level no. 8 artificially combines three levels that lie close together. Figure 4.8 shows the departure coefficients that result from model calculations as outlined in Sects. 4.1 and 4.2. A similar treatment (with similar results) has been applied to hydrogen, carbon, iron, and a number of other absorbers. It thus became possible to make detailed comparisons of model predictions with the observational data. A temperature model and the various spectral features which helped to establish it is shown in Fig. 4.9.

Both continuum and line data can be fitted with the model. The continuum in the EUV is shown in Fig. 4.10 as an example. The four distinct peaks in the spectral range shown are, from left to right, the Lyman α line and the continua following the ionization edges of the ground states of C I (at 109.8 nm), H I (at 91.2 nm – the "Lyman continuum"), and He I (at 50.4 nm). As an example of a line fit, Fig. 4.11 shows the Hα profile. This figure also demonstrates how the agreement between observation and model is improved as more levels are included into the statistical equilibrium calculations.

Even when the departures from LTE are taken into account, there may still be a discrepancy between observation and model. The profiles of strong lines such as Hα are an example, cf. Fig. 4.11. Because of their double-peaked contribution functions (see below) these lines are particularly sensitive to er-

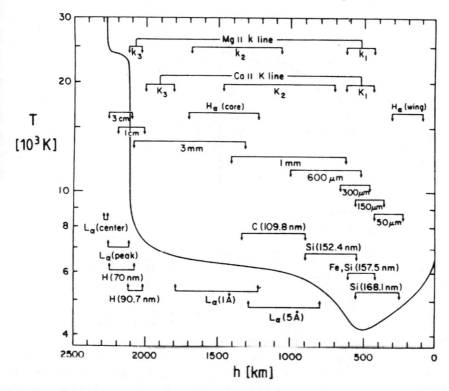

Fig. 4.9. Temperature as a function of depth in the solar atmosphere, and approximate depths where diverse continua and lines originate. From Vernazza et al. (1981)

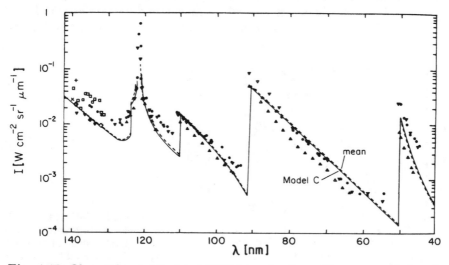

Fig. 4.10. Observed and calculated EUV intensity. From Vernazza et al. (1981)

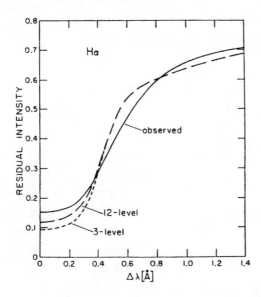

Fig. 4.11. Observed and calculated Hα profiles. From Vernazza et al. (1981)

rors in the atmospheric model. Lines of hydrogenic atoms are also complicated due to their broadening by the linear Stark effect.

In order to reconcile the model with the observations it may be necessary to introduce a "macroturbulence" or "macrovelocity". In contrast to the microturbulence mentioned earlier, which broadens the Doppler core of the absorption coefficient and so affects κ_l and S_l, the macroturbulence consists of a normalized distribution (mostly a Gaussian is chosen) of width $\Delta\lambda_{\mathrm{mac}} = \lambda V_{\mathrm{mac}}/c$, which is folded with the line profile after the calculation is completed. It adds no net absorption, i.e., the total area between the con-

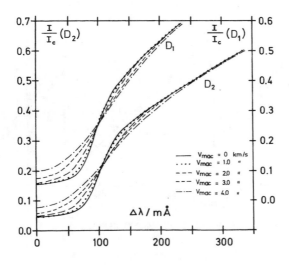

Fig. 4.12. Calculated NaD profiles, as functions of the macroturbulence parameter, V_{mac}. The best fit to observed profiles is obtained with $V_{\mathrm{mac}} = 2\,\mathrm{km/s}$. From Schleicher (1976)

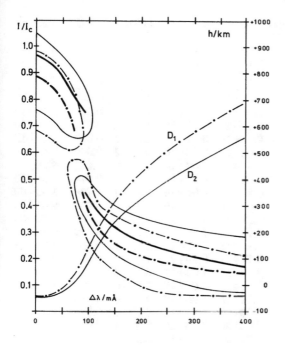

Fig. 4.13. Profiles of Na D lines (*left scale*) and height of line formation (*right scale*). The *heavy curves* mark the maximum of the contribution to each wavelength, the *light curves* enclose the area where the contribution function exceeds 1/3 of the maximum. From Schleicher (1976)

tinuum and the line profile remains unchanged (so does the *equivalent width*, which is that area divided by the continuum intensity). For the two Na D lines Fig. 4.12 demonstrates how the profiles are influenced; in this case the best agreement with observed profiles is obtained with $V_{\mathrm{mac}} = 2\,\mathrm{km/s}$. The name macroturbulence is derived from the idea that in the solar atmosphere large moving parcels of gas introduce local Doppler shifts which are not resolved observationally.

It is often useful to know which layers of the solar atmosphere contribute to the emergent intensity in a certain line profile. For this purpose we must evaluate not only the *total* emergent intensity, i.e., the integral over optical depth, but also the integrand itself. The latter is also called the *contribution function*. Figure 4.13 is an example: for the two Na D lines it shows, at each wavelength, where the contribution exceeds 1/3 of its maximum, and the height of the maximum itself. It is remarkable that around 80 mÅ from the line center the contribution to these lines is more or less divided into two layers, with a clear gap at 500 km. The reason is the gradient of temperature in the solar atmosphere, and the concomitant gradient in the width of the Doppler core. Hence the absorption profile is broad in the highest layer because the Doppler core is broad, it is narrow in the intermediate layers where the Doppler core is narrow and collisions are still unimportant, and it is again broad in the deepest layer because there collisional "damping" generates the broad wings.

Table 4.1. Atmospheric parameters for the quiet Sun. From model C of Vernazza et al. (1981). $Z + n$ means $Z \times 10^n$

h [km]	τ_{500}	T [K]	ξ_t [km/s]	n_H [m^{-3}]	n_e [m^{-3}]	P [Pa]	P_G/P	ρ [kg/m^3]
2080	3.51$-$7	8180	8.55	6.54+16	3.78+16	1.80$-$2	0.689	1.53$-$10
2070	3.77$-$7	7940	8.50	6.96+16	3.78+16	1.84$-$2	0.681	1.63$-$10
2050	4.30$-$7	7660	8.42	7.71+16	3.79+16	1.94$-$2	0.670	1.80$-$10
2016	5.20$-$7	7360	8.22	9.08+16	3.81+16	2.12$-$2	0.662	2.12$-$10
1990	5.90$-$7	7160	8.01	1.03+17	3.86+16	2.28$-$2	0.660	2.42$-$10
1925	7.72$-$7	6940	7.63	1.38+17	4.03+16	2.78$-$2	0.662	3.23$-$10
1785	1.21$-$6	6630	6.92	2.60+17	4.77+16	4.51$-$2	0.677	6.08$-$10
1605	1.96$-$6	6440	5.85	6.39+17	6.01+16	9.33$-$2	0.726	1.49$-$09
1515	2.42$-$6	6370	5.26	1.05+18	6.46+16	1.41$-$1	0.760	2.45$-$09
1380	3.29$-$6	6280	4.51	2.27+18	7.60+16	2.77$-$1	0.805	5.32$-$09
1280	4.08$-$6	6220	3.92	4.20+18	7.49+16	4.79$-$1	0.842	9.82$-$09
1180	5.08$-$6	6150	3.48	7.87+18	8.11+16	8.53$-$1	0.869	1.84$-$08
1065	6.86$-$6	6040	2.73	1.71+19	9.35+16	1.73+0	0.914	4.00$-$08
980	9.15$-$6	5925	2.14	3.15+19	1.04+17	3.01+0	0.944	7.36$-$08
905	1.24$-$5	5755	1.70	5.55+19	1.05+17	5.04+0	0.963	1.30$-$07
855	1.55$-$5	5650	1.53	8.14+19	1.06+17	7.21+0	0.969	1.90$-$07
755	2.54$-$5	5280	1.23	1.86+20	8.84+16	1.53+1	0.978	4.36$-$07
705	3.29$-$5	5030	1.09	2.94+20	7.66+16	2.28+1	0.982	6.86$-$07
655	4.45$-$5	4730	0.96	4.79+20	8.09+16	3.50+1	0.985	1.12$-$06
605	7.02$-$5	4420	0.83	8.12+20	1.11+17	5.52+1	0.988	1.90$-$06
555	1.46$-$4	4230	0.70	1.38+21	1.73+17	8.96+1	0.991	3.23$-$06
515	3.01$-$4	4170	0.60	2.10+21	2.50+17	1.34+2	0.993	4.90$-$06
450	1.02$-$3	4220	0.53	3.99+21	4.52+17	2.57+2	0.995	9.33$-$06
350	5.27$-$3	4465	0.52	9.98+21	1.11+18	6.80+2	0.995	2.33$-$05
250	2.67$-$2	4780	0.63	2.32+22	2.67+18	1.69+3	0.994	5.41$-$05
150	1.12$-$1	5180	1.00	4.92+22	6.48+18	3.93+3	0.985	1.15$-$04
100	2.20$-$1	5455	1.20	6.87+22	1.07+19	5.80+3	0.980	1.61$-$04
50	4.40$-$1	5840	1.40	9.20+22	2.12+19	8.27+3	0.975	2.15$-$04
0	9.95$-$1	6420	1.60	1.17+23	6.43+19	1.17+4	0.970	2.73$-$04
$-$25	1.68+0	6910	1.70	1.26+23	1.55+20	1.37+4	0.969	2.95$-$04
$-$50	3.34+0	7610	1.76	1.32+23	4.65+20	1.58+4	0.970	3.08$-$04
$-$75	7.45+0	8320	1.80	1.37+23	1.20+21	1.79+4	0.971	3.19$-$04

Variations in time and in the horizontal coordinates have been neglected in our discussion of model atmospheres. Time variations are negligible as long as we are only interested in the steady state or in slow phenomena, i.e., those which allow the continuous adjustment of the population and ionization equilibria. However, such adjustment may take several hundred seconds in the chromosphere, and may thus bear on the propagation and dissipation of *waves* in this region.

Horizontal fluctuations are indeed a problem. In particular towards the limb, the observer may average over an area of variable temperature, and this is probably one reason why some results obtained from center-to-limb studies disagree with results from the disc center alone.

As long as the horizontal scale of the fluctuations is large in comparison to the scale of vertical stratification, it is meaningful to construct individual atmospheric models for the various features. Vernazza et al. (1981) have calculated such models. Their model C, representative for the quiet Sun, is listed in Table 4.1. This table does not include the transition layer to the corona. There the more recent models of Fontenla et al. (1990, 1993) should be consulted; these models, based on additional information from helium lines, do not need the temperature plateau at ≈ 2120–2260 km (Fig. 4.9) to account for the Lyman α emission (cf. Sect. 9.1.2).

Models with different temperature profiles have also been "mixed". That is, as a crude way to consider the horizontal variation, the various emergent spectral intensities have been added, with the weights chosen so that the result gives a better match to observational data. An example is the "mean model", which, in addition to model C, is shown in Fig. 4.10. We shall return to such multi-component model atmospheres when we discuss – in the context of the granular velocity field – asymmetries of solar spectral lines.

The atmospheric models presented in this section are called *semi-empirical* because their temperature is adapted in order to reproduce the observed spectral intensity. The law of conservation of energy has *not* been used. If we had done that, we would have been able to *predict* the temperature as a function of height, i.e., to calculate a *theoretical model*. In Sect. 9.4 we shall address this much more difficult problem.

Problem 4.11. What is the difference between the contribution function for the emergent *line intensity* and the contribution function for the *line depression*?

4.4 The Chemical Composition of the Sun

4.4.1 Spectrum Synthesis

For the treatment of model atmospheres we have so far pretended that the abundance of the chemical elements were known. For example, we could have used the abundances measured elsewhere in the solar system, e.g., in meteorites. But the assumption that those are valid for the Sun itself of course needs justification. Therefore, the solar abundances must be determined directly in an analysis of the solar spectrum.

The most common approach is *spectrum synthesis*. That is, the number densities n_i [in (4.16) or (4.60)] are treated as adjustable free parameters. Together with the other quantities (temperature, turbulence parameter, etc.) they must be chosen in such a way that the intensity of as many spectral lines as possible is correctly modeled.

In the measurement as well as in the synthesized spectrum the line intensity is determined *relative* to the nearby continuum. Because the continuum absorption essentially is due to hydrogen and depends on its abundance n_H, while the line absorption depends on the abundance n_i of the element in question, the result of the spectrum synthesis is an *abundance relative to hydrogen*, n_i/n_H. It is common to list

$$\log A = 12 + \log(n_i/n_H) \ , \tag{4.67}$$

i.e., the logarithmic abundance normalized to $n_H = 10^{12}$ particles per unit volume. Table 4.2, taken from Grevesse and Sauval (1998), is a rather complete example.

Often an exact fit to all observed lines of an element is not possible, and often a particular line may be matched in a synthetic spectrum for the disc center, but the same line (with the same atmospheric model) may show a discrepancy near the limb. Such a discrepancy may be used to assess the uncertainty of the determined abundance. Mostly this uncertainty is between 0.02 and 0.1 in $\log A$, and in some cases up to 0.3, cf. Table 4.2. The reason for the uncertainty is not always known, although the horizontal inhomogeneity may play a role. Smaller errors, of order 0.01 to 0.02, arise from uncertainties about the correct atmospheric model.

An important source of uncertainty is the oscillator strength f, or transition probability, which in the absorption coefficient always appears in combination with the abundance n_i. A prominent example is iron. The iron abundance "increased" by a factor 10 around 1968 when it became clear that the f values used previously were too large by such an amount! Another example is lutetium; a new measurement of transition probabilities for Lu II lines resulted in a significantly smaller abundance, such that there is no longer a discrepancy between the solar and meteoritic values (Bord et al. 1998, Den Hartog et al. 1998). Generally the uncertainties arising from not well-known oscillator strengths contribute significantly to the errors given in Table 4.2.

The radioactive elements Tc, Pm, and those beyond Bi have no entries in Table 4.2. Others have no lines originating in the photosphere that are particularly suited for the quantitative chemical analysis. Thus, the abundances of F and Cl are obtained from sunspots, while the abundances of Ne and Ar are determined from coronal lines and from direct measurements in the solar wind. Of course helium has also been measured in the solar wind, but its solar abundance is best determined by helioseismology, see below. For comparison, meteoritic abundance values are included in the table; these are predominantly derived from carbonaceous chondrites.

4.4.2 The Light Elements Lithium, Beryllium, and Boron

In contrast to earlier results the solar abundance of beryllium shows no significant difference to the meteoritic value. Boron appears slightly depleted on

Table 4.2. Abundance of chemical elements, $\log(n/n_H) + 12$, in the photosphere and in meteorites. The value for He is derived by helioseismology, the values for F, Ne, Cl, and Ar are from sunspots, the corona, or solar wind particles. After Grevesse and Sauval (1998)

Element	Photosphere	Meteorites	Element	Photosphere	Meteorites
1 H	12.00	–	42 Mo	1.92 ± 0.05	1.97 ± 0.02
2 He	10.93 ± 0.004	–	44 Ru	1.84 ± 0.07	1.83 ± 0.04
3 Li	1.10 ± 0.10	3.31 ± 0.04	45 Rh	1.12 ± 0.12	1.10 ± 0.04
4 Be	1.40 ± 0.09	1.42 ± 0.04	46 Pd	1.69 ± 0.04	1.70 ± 0.04
5 B	2.55 ± 0.30	2.79 ± 0.05	47 Ag	0.94 ± 0.25	1.24 ± 0.04
6 C	8.52 ± 0.06	–	48 Cd	1.77 ± 0.11	1.76 ± 0.04
7 N	7.92 ± 0.06	–	49 In	1.66 ± 0.15	0.82 ± 0.04
8 O	8.83 ± 0.06	–	50 Sn	2.0 ± 0.3	2.14 ± 0.04
9 F	4.56 ± 0.3	4.48 ± 0.06	51 Sb	1.0 ± 0.3	1.03 ± 0.07
10 Ne	8.08 ± 0.06	–	52 Te	–	2.24 ± 0.04
11 Na	6.33 ± 0.03	6.32 ± 0.02	53 I	–	1.51 ± 0.08
12 Mg	7.58 ± 0.05	7.58 ± 0.01	54 Xe	–	2.17 ± 0.08
13 Al	6.47 ± 0.07	6.49 ± 0.01	55 Cs	–	1.13 ± 0.02
14 Si	7.55 ± 0.05	7.56 ± 0.01	56 Ba	2.13 ± 0.05	2.22 ± 0.02
15 P	5.45 ± 0.04	5.56 ± 0.06	57 La	1.17 ± 0.07	1.22 ± 0.02
16 S	7.33 ± 0.11	7.20 ± 0.06	58 Ce	1.58 ± 0.09	1.63 ± 0.02
17 Cl	5.5 ± 0.3	5.28 ± 0.06	59 Pr	0.71 ± 0.08	0.80 ± 0.02
18 Ar	6.40 ± 0.06	–	60 Nd	1.50 ± 0.06	1.49 ± 0.02
19 K	5.12 ± 0.13	5.13 ± 0.02	62 Sm	1.01 ± 0.06	0.98 ± 0.02
20 Ca	6.36 ± 0.02	6.35 ± 0.01	63 Eu	0.51 ± 0.08	0.55 ± 0.02
21 Sc	3.17 ± 0.10	3.10 ± 0.01	64 Gd	$1.12 + 0.04$	1.09 ± 0.02
22 Ti	5.02 ± 0.06	4.94 ± 0.02	65 Tb	-0.1 ± 0.3	0.35 ± 0.02
23 V	4.00 ± 0.02	4.02 ± 0.02	66 Dy	1.14 ± 0.08	1.17 ± 0.02
24 Cr	5.67 ± 0.03	5.69 ± 0.01	67 Ho	0.26 ± 0.16	0.51 ± 0.02
25 Mn	5.39 ± 0.03	$5.53 + 0.01$	68 Er	0.93 ± 0.06	$0.97 + 0.02$
26 Fe	7.50 ± 0.05	7.50 ± 0.01	69 Tm	0.00 ± 0.15	0.15 ± 0.02
27 Co	4.92 ± 0.04	4.91 ± 0.01	70 Yb	1.08 ± 0.15	0.96 ± 0.02
28 Ni	6.25 ± 0.04	6.25 ± 0.01	71 Lu	0.06 ± 0.10	0.13 ± 0.02
29 Cu	4.21 ± 0.04	4.29 ± 0.04	72 Hf	0.88 ± 0.08	0.75 ± 0.02
30 Zn	4.60 ± 0.08	4.67 ± 0.04	73 Ta	–	-0.13 ± 0.02
31 Ga	2.88 ± 0.10	3.13 ± 0.02	74 W	1.11 ± 0.15	0.69 ± 0.03
32 Ge	3.41 ± 0.14	3.63 ± 0.04	75 Re	–	0.28 ± 0.03
33 As	–	2.37 ± 0.02	76 Os	1.45 ± 0.10	1.39 ± 0.02
34 Se	–	3.41 ± 0.03	77 Ir	1.35 ± 0.10	1.37 ± 0.02
35 Br	–	2.63 ± 0.04	78 Pt	1.8 ± 0.3	1.69 ± 0.04
36 Kr	–	3.31 ± 0.08	79 Au	1.01 ± 0.15	0.85 ± 0.04
37 Rb	2.60 ± 0.15	2.41 ± 0.02	80 Hg	–	1.13 ± 0.08
38 Sr	2.97 ± 0.07	2.92 ± 0.02	81 Tl	0.9 ± 0.2	0.83 ± 0.04
39 Y	2.24 ± 0.03	2.23 ± 0.02	82 Pb	1.95 ± 0.08	2.06 ± 0.04
40 Zr	2.60 ± 0.02	2.61 ± 0.02	83 Bi	–	0.71 ± 0.04
41 Nb	1.42 ± 0.06	1.40 ± 0.02	90 Th	–	0.09 ± 0.02
			92 U	< -0.47	-0.50 ± 0.04

the Sun, but the error margin is large. On the other hand, the photospheric depletion of lithium by about two orders of magnitude, first recognized by Greenstein and Richardson (1951), is firmly established.

The light elements Li, Be, and B can be destroyed by nuclear reactions with protons. In addition to those already listed in Sect. 2.3.6, the actual reactions are

$$^6\text{Li}(\text{p},^3\text{He})\alpha \,, \quad ^9\text{Be}(\text{p},\alpha)^6\text{Li} \,,$$
$$^{10}\text{B}(\text{p},\alpha)^7\text{Be} \,, \quad ^{11}\text{B}(\text{p},\gamma)3\alpha \,.$$

(4.68)

A case of special interest is the "burning" of lithium and beryllium. The reaction rates for these two elements become significant (in view of the solar age) at temperatures of 2.5×10^6 K and 3×10^6 K, respectively. Thus the strong (but not total) depletion of Li and the absence of Be depletion can provide a hint to the depth of the outer convection zone of the Sun: Convection (including convective overshooting into the subadiabatic range below the proper convection zone) and the concomitant mixing should reach the layer where $T \approx 2.5 \times 10^6$ K, but essentially avoid the layer where $T \approx 3 \times 10^6$ K. We may conclude from this argument that the solar convection zone should have a depth of $\approx 2 \times 10^8$ m, in rough agreement with model calculations which show that the stratification becomes subadiabatic, and hence stable against convection, at about this depth (Sect. 6.2), and in agreement with the helioseismic result (Sect. 5.3).

However, a detailed calculation shows that the problem of Li depletion is not fully solved: During the $\approx 5 \times 10^7$ years of the Sun's evolution from the Hayashi line toward the zero-age main sequence the outer convection zone may have reached a depth where $T \approx 3.5 \times 10^6$ K, and *all* the lithium may have been lost (e.g., Ahrens et al. 1992, Schlattl and Weiss 1999). But the Sun is not completely depleted of lithium; moreover, there is evidence that Sun-like stars, younger than the Sun but also on the main sequence, have less lithium depletion or even the "normal" abundance. Hence it appears that Li depletion takes place during the main-sequence evolution. Since in this phase the outer convection zone has not been sufficiently deep for the necessary mixing, some additional mixing *below* the convection zone should occur. – In terms of *energy generation* the contribution of lithium burning is small and of no concern.

4.4.3 Helium

Helium deserves a special comment. Although a line of this element had been discovered as early as 1868 by N. Lockyer (and ascribed to "helium" because at the time it was unknown on Earth) the spectroscopic determination of its abundance has remained rather inaccurate, with an error of order 0.2 in $\log A$. The reason is that all important helium lines fall into the ultraviolet or infrared parts of the spectrum, and that these lines are produced in the chromosphere and corona under conditions which largely deviate from LTE.

In view of the spectroscopic difficulties it is fortunate that a most accurate method to determine the solar photospheric helium abundance is helioseismology, as first proposed by Gough (1984 c). In a zone where an abundant element is partially ionized the specific heats c_P and c_V both increase because of the energy required for ionization, while their difference remains nearly unchanged. Therefore the ratio of those specific heats, which is approximately equal to the adiabatic exponent Γ_1, becomes closer to 1. Since the velocity of sound is proportional to $\Gamma_1^{1/2}$, it is clear that the frequencies of oscillation must be influenced by a change of the abundance. The second ionization of helium occurs sufficiently deep in the Sun so that its signature on the eigenfrequencies is not spoiled by the uncertainty of near-surface convection, and is clearly discernable. Thus, the value $\log A = 10.93 \pm 0.004$ listed in Table 4.2 is based on helioseismic inversion (Sect. 5.3), which yields the mass fraction $Y_s = 0.248 \pm 0.002$ of helium at the solar surface, and thus

$$n_{He}/n_H = Y_s/4X_s \,, \tag{4.69}$$

where X_s is the surface mass fraction of hydrogen.

The value $Y_s = 0.248$ is very close to the result $Y_s = 0.245$ obtained for the theoretical solar model described in Sect. 2.4, where the helium abundance has been calibrated by observed properties of the Sun, and where the process of He diffusion towards the solar center (Sect. 2.3.3) has been taken into account. Without such diffusion the calibration would yield $Y_s \approx 0.28$. This demonstrates that it is appropriate to include that process into the standard solar model.

4.5 Bibliographical Notes

There are a large number of monographs on the physics of stellar atmospheres. The Sun, being the star which has been investigated in most detail, is treated extensively in these texts. Unsöld (1955), Aller (1963), Jefferies (1968), and Athay (1972) have written classical textbooks. Particularly useful as far as departures from local thermodynamic equilibrium are concerned, are the books of Mihalas (1978), or Cannon (1985).

Mihalas further provides chapters on *partial frequency redistribution*, a subject also treated in a number of research papers (e.g., Rutten and Milkey 1979), and on radiative transfer in *moving atmospheres*, a subject also covered in Mihalas and Mihalas (1984). Line-profile variations arising from wave motions in the solar atmosphere (Chap. 5) are treated by Severino et al. (1986) and Gomez et al. (1987). The effect of *turbulent motions* on spectral line profiles has been discussed in a number of papers edited by Gray and Linsky (1980). Computational methods are reviewed in two volumes edited by Kalkofen (1984, 1987), and by LeVeque et al. (1998).

Traditionally radiative transfer is treated in form of a one-dimensional problem, with all variables dependent on the vertical coordinate only. However, the higher we are in the solar atmosphere, the more important is the *horizontal inhomogeneity*. This should be modeled by two- or three-dimensional calculations. Mihalas et al. (1978), Kneer and Heasley (1979), and Owocki and Auer (1980) present methods to do this, and results relevant to the solar atmosphere.

The atmospheric reference models of Maltby et al. (1986, Appendix A) and Fontenla et al. (1993) slightly differ from Table 4.2, especially by a minimum temperature some 200 K higher. Neckel and Labs (1994), and Neckel (1996, 1997), provide more details on limb darkening and its wavelength dependence.

In analogy to the *contribution function* for the emergent intensity, for which an example is shown in Fig. 4.13, the contribution function for the line depression has been considered by Beckers and Milkey (1975), Magain (1986), and Grossmann-Doerth (1994). Beckers and Milkey, and earlier Mein (1971), also discuss *line response functions*, which define the response of a line profile to changes of the physical conditions, e.g., by a velocity field or a magnetic field, along the line of sight. For contribution functions in the presence of a magnetic field see also Staude (1972).

The volume edited by Fröhlich et al. (1998) contains many useful contributions on the chemical composition of the Sun. Ross and Aller (1976) and Anders and Grevesse (1989) wrote earlier reviews; Carmeron (1973) reviews the meteoritic element abundances. The best spectroscopic determination of the He abundance comes from prominences (Heasley and Milkey 1978). Lithium depletion by internal mixing, induced by rotation, has been discussed by Zahn (1992) and Charbonnel et al. (1994). Michaud and Charbonneau (1991) review what is known about lithium in stars, while Pinsonneault (1997) in general discusses possible mixing processes in the stellar interior.

5. Oscillations

A new branch of solar research has been developed since 1975, when it was discovered that the photospheric periodic motions previously known as "5-minute oscillations" have a spectrum of discrete frequencies. A large number of such *modes*, with periods in the range 2–15 minutes, have been observed. They are identified as acoustic waves, or p modes, where the pressure gradient is the main restoring force. The discrete mode pattern is a consequence of reflecting boundaries and can be used to obtain information about the Sun's interior.

Gravity provides the restoring force for a second type of oscillation, with lower frequencies. These internal gravity modes depend on a stable stratification: they are supported by the radiative interior of the Sun, with a discrete spectrum of g modes, which, however, has not yet been confirmed by observation; on the other hand, there are hints that they exist above the convection zone in the stable solar atmosphere.

5.1 Observations

5.1.1 Five-Minute Oscillations

Oscillatory motions in the solar atmosphere were discovered in 1960 by a spectroheliographic technique (Leighton et al. 1962; cf. Sect. 3.4.7): Two spectroheliograms recorded simultaneously in the red and blue wings of a spectral line are photographically subtracted. The result is a "Doppler plate"; its intensity variation has its origin in the local Doppler shift of the line used. Two such Doppler plates, obtained by scanning the Sun first in one and immediately afterwards in the opposite direction, are then again subtracted from each other. Since each scan takes a few minutes, the resulting "Doppler difference" has a variable time delay Δt between the two constituent Doppler plates: Δt is smallest at the edge where the scanning direction was reversed and increases linearly from there. A periodic velocity field on the Sun manifests itself as a periodically changing intensity contrast on such a picture. Figure 5.1 shows a nice example.

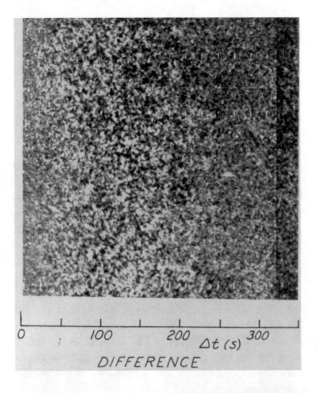

Fig. 5.1. Doppler-difference plate for the Ba II line $\lambda = 455.4\,\mathrm{nm}$. From Leighton et al. (1962)

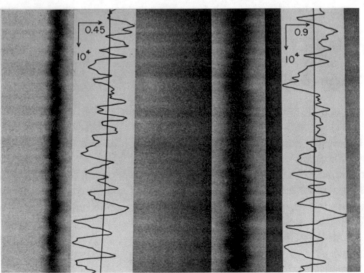

Fig. 5.2. Spectrum and velocity curves for the lines Mg b_2 and Ti I $\lambda = 517.37\,\mathrm{nm}$, obtained at Sacramento Peak Observatory. The *arrows* indicate the velocity in km/s and the distance on the Sun in km. From Evans and Michard (1962)

Problem 5.1. How does the contrast pattern of Fig. 5.1 change if the two Doppler plates are *added* instead of subtracted?

The 5-minute oscillations which are visible in the Doppler-difference plates were further investigated mainly by direct measurement of the local Doppler shift of spectral lines (Fig. 5.2). Observations made at various places between disc center and the limb show that the motion is predominantly *vertical* (e.g., Schmidt et al. 1999). The amplitude, typically 0.5 to 1 km/s, slightly *increases with height* in the atmosphere, as one can see from the analysis of lines originating in different heights. In addition to the Doppler shift a periodic *variation of the line intensity*, with relative amplitudes of a few percent, is found.

The oscillations with the largest amplitudes have frequencies 2–5 mHz, i.e., periods between 3 and 8 minutes. For this range of frequencies there is only a *small phase lag* between the velocities at different heights in the atmosphere, cf. Fig. 5.3, indicating the character of a standing wave.

The maximum amplitude in the photosphere is at a period slightly below 5 minutes; higher up in the atmosphere the maximum is shifted towards shorter periods, and 3-minute oscillations are typical in the chromosphere (observed, e.g., in Hα).

The wave numbers cover a broad range between the bounds set by the size of the observed area on the Sun and by the limited spatial resolution of the observation. Areas of 30 000 km in diameter often oscillate more or less in phase. The general appearance is that "one-third oscillates" – either of a given area at any one time, or of a given period of time at any one location on the solar surface; but this pattern changes continuously.

We shall see below that the 5-minute oscillations observed in the solar atmosphere lie in the region of *evanescent waves* of the k_h, ω-plane, with exponential rather than oscillatory variation in the vertical direction. These waves do not propagate, and this is consistent with the phase relations already mentioned.

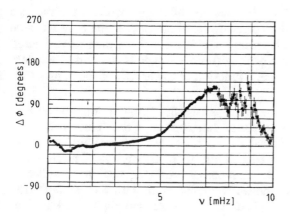

Fig. 5.3. Phase difference, $\Delta\phi$, of oscillations measured in two lines originating deep (Fe I, $\lambda = 593.0$ nm) and high (Na I, $\lambda = 589.6$ nm) in the solar atmosphere, as a function of frequency. After Staiger (1987)

5.1.2 The Spectrum of Solar Oscillations

Fourier Analysis. One of the rules of Fourier analysis is that a signal of length T allows a frequency resolution of $\Delta\omega = 2\pi/T$. That is, if we want to completely resolve two neighboring frequencies ω and $\omega + \Delta\omega$, we must observe over a time span $T = 2\pi/\Delta\omega$ after which the two oscillations acquire a phase difference of exactly 2π.

The lowest frequency visible in a Fourier-analyzed signal is also determined by the signal length, T, and is equal to the frequency resolution. In special cases still lower frequencies can be extracted by other methods, e.g., by a least-square fit of the signal to a harmonic function whose length is only a fraction of a whole period. However, in view of the large number and dense spectrum of possible modes, Fourier techniques normally must be applied. The Fast-Fourier-Transform algorithm makes such techniques very convenient.

The high-frequency limit of a spectrum is given by the time resolution, Δt, of the observed signal. This limit is called the *Nyquist frequency*, ω_{Ny}, and is given by $\omega_{Ny} = \pi/\Delta t$. Care must be taken if a signal contains oscillations with frequencies equal to or larger than ω_{Ny}: If these are not suppressed before the Fourier transformation, their spectrum will be folded into the domain below ω_{Ny}; this is called *aliasing*.

What has been said about the time and frequency domains can be repeated for the space and wave number domains. To summarize, we have

$$\Delta\omega = 2\pi/T \leq \omega < \pi/\Delta t ,$$

$$\Delta k_x = 2\pi/L_x \leq k_x < \pi/\Delta x ,$$

(5.1)

where L_x is the length of a scan in the x direction, with spacing Δx, along the solar surface, and k_x is the component of the wave vector in this direction. For observations from the ground, atmospheric seeing sets a lower bound for a meaningful Δx and, therefore, an upper bound for k_x.

Problem 5.2. Consider the Fourier transform of a signal which has data gaps as a consequence of the day/night cycle. Show that a line at ν has side lobes at $\nu \pm 11.57\,\mu$Hz. Where are the side lobes if the data are collected by a satellite in a 90-minute orbit with no continuous sunlight? Calculate the height of the side lobes as a function of the length of the gaps.

Long-Term Observation. There are essentially three options for observing without interruption over a period longer than a day. One is to use a station at high geographic latitude during the polar summer; indeed several successful campaigns have been carried out at the South Pole. The second is to combine data obtained from a number of observing stations at different geographic longitudes. Several networks of this kind, with 6 or more participating stations, have been installed, with integrated-sunlight (full-disc) as well as with imaging instruments. Examples of imaging systems are GONG (Global Oscillation Network Group; Harvey et al. 1996) that employs Michelson interferometers to measure the Doppler shift of a nickel line (Sect. 3.4.3) and TON (Taiwan Oscillation Network; Chou et al. 1995) that uses telescopes with a narrow-band filter to measure intensity oscillations in the spectral line Ca II K. Among the networks that measure the integrated sunlight are BISON (Birmingham Solar-Oscillation Network; Chaplin et al. 1996 a) and IRIS (International Research on the Interior of the Sun; Fossat 1991); both use resonance-scattering spectrometers to measure a Doppler shift, either of the potassium line at $\lambda = 769.9\,$nm (BISON), or of the sodium line at $\lambda = 589.6\,$nm (IRIS).

The third way to achieve signals of long duration is observation from space where, in addition, the degrading effect of the Earth's atmosphere is eliminated. This is done with the SOHO satellite, which orbits around the Lagrangian point L1 between the Earth and the Sun and thus has permanent sunshine (Fleck et al. 1995). SOHO carries three oscillation experiments: GOLF (Global Oscillations at Low Frequency) measures Doppler shifts in integrated sunlight with a sodium resonance-scattering spectrometer; VIRGO (Variability of solar IRradiance and Gravity Oscillations) records oscillations of the total irradiance, either spatially unresolved or with a 12-pixel imager, as well as oscillations of the irradiance in three spectral bands; and SOI (Solar Oscillation Investigation) obtains velocity images from MDI, the Michelson Doppler Imager (Sect. 3.4.3), either with moderate resolution ($4''$) for the whole disc, or with the higher resolution of $1.2''$ for a field of $10.5'' \times 10.5''$.

Power Spectrum. Let us assume that a vertical velocity signal v is known as a function of time t and position (x, y) on the solar disc. Its Fourier transform f is defined by

$$v(x, y, t) = \int f(k_x, k_y, \omega) \exp[i(k_x x + k_y y + \omega t)] \, dk_x dk_y d\omega \ . \tag{5.2}$$

In practice the integral is replaced by a sum and the rules of Fourier analysis must be observed. The *power spectrum* is

$$P(k_x, k_y, \omega) = ff^* \ . \tag{5.3}$$

If no horizontal direction is distinguished from any other, we expect a wave number dependence only on

$$k_{\rm h} = (k_x^2 + k_y^2)^{1/2} \ . \tag{5.4}$$

In any case, we may calculate from P the average

$$P(k_{\rm h}, \omega) = \frac{1}{2\pi} \int_0^{2\pi} P(k_{\rm h} \cos \phi, k_{\rm h} \sin \phi, \omega) \, d\phi \ . \tag{5.5}$$

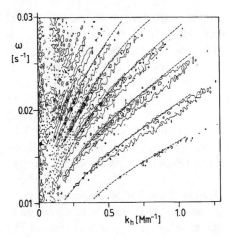

Fig. 5.4. Ridges of p and f modes in the k_h, ω-plane (contours of equal power), and eigenfrequencies of a theoretical solar model (*dotted curves*). After Deubner et al. (1979)

For the solar oscillations the important fact is that the power is not evenly distributed in the k_h, ω-plane, but instead follows certain *ridges*. As we shall see in Sect. 5.2.4, each of these ridges corresponds to a fixed number of wave nodes in the radial direction. The ridges were theoretically predicted by Ulrich (1970), and first observed by Deubner (1975) with the Domeless Coudé Telescope at Capri. Figure 5.4 shows a later example, where up to 15 ridges can be identified in the *velocity* power spectrum. The same ridge pattern is seen in the power spectrum of *intensity* variations, measured, e.g., in $\mathrm{Ca\,II}\,K$ filtergrams (Fig. 5.7).

We shall see that the *pressure perturbation* essentially provides the restoring force for the 5-minute oscillations. Hence the discrete modes into which these oscillations are decomposed, and which we shall treat in the following two sections, have been named *p modes* by Cowling (1941). The lowest ridge, called *fundamental* or *f mode*, is an exception: this mode is essentially without compression, and resembles a surface wave on deep water (Sect. 5.2.4).

In the three-dimensional k_x, k_y, ω-space the p-mode power is distributed along trumpet-shaped surfaces. On the average these trumpet surfaces are symmetric about the ω-axis, and a slice at constant ω forms a ring. However, a large-scale stationary flow that underlies the wave field causes a deviation of the rings from the circular symmetry; hence such a *ring diagram* can be used to determine large-scale flows, both at the solar surface and underneath the surface (Hill 1988).

Spherical Harmonics. Whenever the scan length becomes comparable to the solar radius, the cartesian coordinates used so far must be replaced by spherical polar coordinates (r, θ, ϕ). Instead of (5.2) we must use an expansion in terms of spherical surface harmonics, i.e.,

$$v(\theta, \phi, t) = \sum_{l=0}^{\infty} \sum_{m=-l}^{l} a_{lm}(t) Y_l^m(\theta, \phi) \ . \tag{5.6}$$

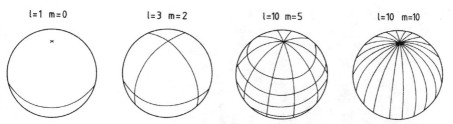

| l=1 m=0 | l=3 m=2 | l=10 m=5 | l=10 m=10 |

Fig. 5.5. Node circles of spherical harmonics. After Noyes and Rhodes (1984)

Here we shall use the complex form

$$Y_l^m(\theta, \phi) = P_l^{|m|}(\theta)e^{\mathrm{i}m\phi} , \tag{5.7}$$

where P_l^m is an associated Legendre function. Its degree, l, gives the total number of node circles on the sphere; its longitudinal order, m, is the number of node circles through the poles. Figure 5.5 illustrates a few examples.

In a spherical system which has no preferred axis of symmetry there are no *physical* poles; the mathematical poles of the polar coordinates are quite ordinary points and can be placed anywhere on the sphere. As a consequence, the oscillations and their eigenfrequencies must not depend on m (a more formal derivation of this degeneracy is given in Sect. 5.2.2 below).

On the Sun, rotation *does* distinguish a particular axis. However, the frequency of rotation is much smaller than the oscillation frequencies, and the latter are therefore *almost* independent of m. Hence we may, for the moment, neglect the m-dependence and restrict ourselves to the *zonal* harmonics, i.e., to the case $m = 0$ (the case $m \neq 0$ will be of special interest in Sect. 5.3.8, on rotational splitting).

The zonal harmonics constitute the longitudinal average over (5.6). Observationally, such an average can be obtained by means of a cylindrical lens, oriented perpendicular to the solar axis of rotation (Duvall and Harvey 1983).

The degree l of the spherical surface harmonic replaces the horizontal wave number k_h introduced above; we shall soon show that $k_\mathrm{h}r_\odot = [l(l+1)]^{1/2}$. Let us also replace the circular frequency ω by the more customary cycle frequency $\nu = \omega/2\pi$. We thus represent the power of the oscillations in the l, ν-diagram. If $\hat{a}_l(\nu)$ is the Fourier transform of the amplitude $a_{l0}(t)$, the power in the l, ν-diagram is

$$P(l, \nu) = \hat{a}_l(\nu)\hat{a}_l^*(\nu) . \tag{5.8}$$

Figure 5.6 shows an example of such a power spectrum, where more than 30 distinct ridges can be identified. The frequency range is 1.5 to 7 mHz (corresponding to periods of 2.4 to 11 minutes). The degree covers the interval $l \approx 5 \ldots 250$, thus extending the range of Fig. 5.4 ($l \approx 70 \ldots 1000$) towards smaller l. Still more detail is seen in power spectra obtained from longer data sets of the MDI, GONG, and TON experiments, with frequencies from ≈ 1 to at least 8 mHz.

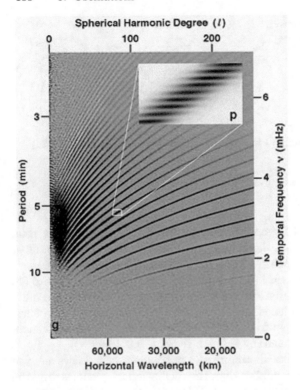

Fig. 5.6. Ridges of p modes in the l, ν-plane, calculated from a 3-day data set. The enlarged section (size $\Delta l \approx 11$, $\Delta \nu \approx 160\,\mu\mathrm{Hz}$) demonstrates the limits of resolution in degree l and frequency ν. From the GONG Image Gallery

Numerical Values of Frequencies and Amplitudes. The ridges discussed so far are *not* resolved into modes corresponding to a single degree l because the resolution given by (5.1) together with the relation $l \simeq k_h r_\odot$ would require a total solar circumference as scan length if $\Delta l = 1$ was desired. Equivalently, we recall that spherical harmonics are orthogonal on the *full surface* of the sphere, but not on the observed part, which, due to the limb darkening, is even smaller than the visible hemisphere.

The resolution of the examples shown in Figs. 5.6 and 5.7 is $\Delta l = 3 \ldots 5$. Nevertheless, single modes of oscillation can be identified at each l by means of their frequencies alone. Even with the moderate frequency resolution of Fig. 5.6 this is possible. The frequencies listed in Table 5.1 are averages over the multiplets generated by the solar rotation (Sect. 5.3.8). The uncertainties are $\approx 0.1\,\mu\mathrm{Hz}$ at intermediate frequencies (periods around 5 min), but may exceed $1\,\mu\mathrm{Hz}$ toward the edges of the recognizable frequency range. For a detailed comparison between different measurements it is also necessary to take the variation with the 11-year activity cycle into account, which amounts to several tenths of a microhertz.

The velocity amplitude of a single mode is about 30 cm/s or less. Hence the 0.5 to 1 km/s amplitude of the 5-minute oscillation and the ever-changing pattern of oscillating and non-oscillating areas on the Sun result from the in-

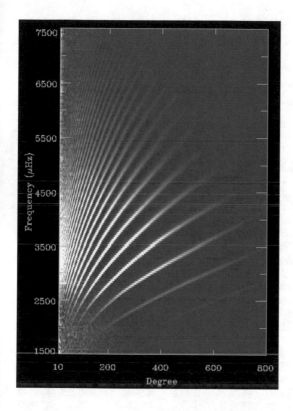

Fig. 5.7. Ridges in the l, ν-plane, calculated from Ca II K data (width 1 nm, centered at 393.4 nm) of the TON station Tenerife on 16th February 1994. The lowest (very faint) ridge is the f mode; upwards follow p 1, p 2 ... . After Chou et al. (1995)

terference of 10^7 or more single modes with randomly distributed phases. The smallest amplitudes presently measured are of order 1 mm/s (Kosovichev et al. 1997).

Problem 5.3. How many modes of oscillation lie between degrees $l = 0$ and $l = 1000$, if 10 ridges are included and all m values are counted for each l?

5.1.3 Low-Degree p Modes

Solar oscillations have also been observed as Doppler shifts in *spatially unresolved* spectra. The primary instrument used is the resonance-scattering spectrometer described in Sect. 3.3.3. Such an instrument yields a signal which is averaged over the visible solar disc. Naturally the oscillations with smaller horizontal wavelength (larger l) contribute less to this average because the effects from areas of opposite phase tend to cancel each other.

Oscillations observed in spatially unresolved sunlight are sometimes called "global" oscillations. In contrast to this definition, which would depend on the particular observational setup, we shall rather attribute the word "global" to those oscillations which can exist in certain regions extending over the *entire*

Table 5.1. Frequencies, in μHz, of low-degree solar p modes. For $l = 0$ to 3: averages from whole-disc measurements 1977–1985 (Jiménez et al. 1988). For $l = 4$ to 7: from 144 days of SOHO/MDI data taken in 1996 (Rhodes et al. 1997)

Mode	Degree l 0	1	2	3	4	5	6	7
p 7						1401.6	1443.7	1483.4
p 8					1500.3	1545.3	1587.5	1627.5
p 9					1641.0	1685.9	1727.8	1767.6
p 10					1778.0	1823.4	1866.2	1906.8
p 11			1815.4	1868.2	1914.8	1960.5	2004.2	2045.9
p 12	1823.6	1885.5	1947.5	2003.4	2051.7	2098.4	2142.6	2184.5
p 13	1957.3	2020.7	2084.1	2138.7	2188.4	2235.4	2279.9	2322.2
p 14	2093.5	2156.7	2218.4	2274.7	2324.2	2371.2	2415.6	2458.2
p 15	2228.6	2291.9	2352.2	2409.4	2458.6	2506.1	2551.1	2594.3
p 16	2362.5	2426.1	2486.8	2541.9	2593.0	2641.2	2687.0	2731.1
p 17	2496.6	2558.9	2620.3	2677.4	2728.3	2777.3	2823.7	2868.3
p 18	2629.6	2693.6	2754.5	2810.9	2864.2	2913.7	2960.6	3005.6
p 19	2764.4	2828.1	2890.1	2947.9	3000.1	3049.9	3097.2	3142.7
p 20	2899.3	2963.3	3024.1	3083.5	3135.9	3186.2	3233.9	3279.8
p 21	3033.8	3098.7	3160.0	3218.5	3271.6	3322.5	3370.8	3417.4
p 22	3168.6	3233.2	3296.1	3354.5	3408.0	3459.3	3508.2	3555.2
p 23	3304.1	3368.9	3431.2	3489.5	3544.4	3596.5	3645.9	3693.6
p 24	3439.8	3504.6	3567.2	3626.1	3681.4	3733.7	3783.7	3832.0
p 25	3576.3	3640.2	3703.3	3760.9	3818.9	3871.9	3922.1	3970.4
p 26	3711.5	3777.4	3837.8	3897.6	3956.3	4009.8	4060.7	4109.6
p 27	3847.4	3914.1	3975.5	4035.0	4093.9	4147.5	4198.9	4248.3
p 28	3984.9	4052.1	4112.9	4171.7	4231.2	4286.3	4338.1	4387.3
p 29	4121.9	4189.8	4249.3	4308.6	4371.0	4424.4	4476.5	4526.9
p 30	4257.4	4325.6	4387.3	4443.8				
p 31	4394.9	4463.6	4524.2	4583.5				
p 32	4532.3	4600.8	4656.9	4717.4				
p 33	4668.6	4738.6						

body of the Sun. We shall see that reflecting boundaries exclude some modes from some parts of the Sun, but also that it is exactly the presence of these boundaries which generates the discrete mode character of the oscillations and thus manifests the global nature.

Even in a spatially unresolved Doppler signal single modes of oscillation, with various (small) degrees l, can be distinguished. But for this purpose it is a necessity to have good resolution in frequency; according to (5.1), this means signals of sufficient duration in time. It is true that the first discovery of low-l oscillations (Claverie et al. 1979) was made with signals of one and two days length; but only the 5-day signal, obtained between 31 December 1979 and 5 January 1980 at the South Pole (Grec et al. 1983) revealed the multitude and sharpness of lines in the solar low-l oscillation spectrum. Still longer signals have been obtained since 1981 by G. R. Isaak and collaborators who combined data recorded at the Canary Islands and at Hawaii; this was the beginning of the above-mentioned BISON network. The large difference

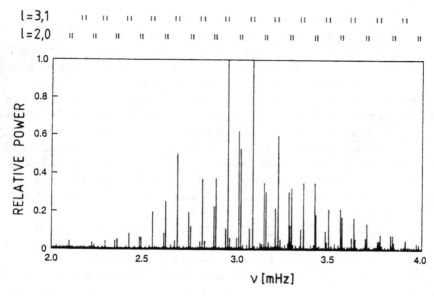

Fig. 5.8. Power spectrum of low-degree solar oscillations. From observations of G. R. Isaak and collaborators, obtained in 1981 over three months at Tenerife and Hawaii. *Top lines*: theoretical identification. From Leibacher et al. (1985)

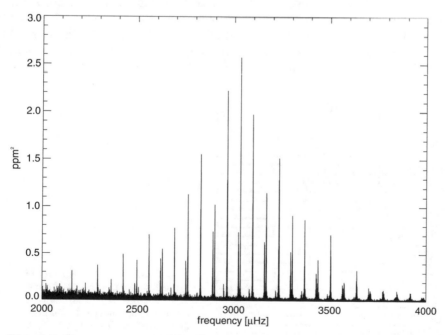

Fig. 5.9. Power spectrum of irradiance variations, measured with the VIRGO instrument on SOHO in the 402 nm band. Courtesy W. Finsterle and C. Fröhlich

in geographic longitude yielded up to 22 hours of observing time per day. A power spectrum obtained from these measurements is shown in Fig. 5.8.

The same oscillations that appear in the full-disc Doppler signal are visible in the power spectrum of the total solar irradiance variations, as well as in the irradiance in narrow spectral bands. Figure 5.9 shows an example, calculated from photometric measurements in a 5-nm band around $\lambda = 402\,\mathrm{nm}$ with the VIRGO instrument on board of the SOHO mission. The peaks in this spectrum correspond to relative variations of a few parts in 10^{-6}.

Numerical values of the oscillation frequencies seen in power spectra of the kind shown in Fig. 5.8 are listed in Table 5.1, under degrees 0 to 3. The accuracy of these frequencies is $0.1\,\mu\mathrm{Hz}$ to $1\,\mu\mathrm{Hz}$. The corresponding amplitudes lie between $3\,\mathrm{cm/s}$ and $20\,\mathrm{cm/s}$, similar to the amplitudes derived from the p-mode ridges.

The principal lines in a highly resolved oscillation spectrum are roughly equidistant in frequency, but some fine structure, in particular a splitting of almost all peaks into a pair of lines, is immediately discernable. This splitting has its theoretical interpretation (cf. Sect. 5.3.1) as indicated at the top of Fig. 5.8.

The *echelle diagram* is a convenient way to represent more details of such fine structure. It is named after the optical echelle spectrograph, where the various orders often are separated by means of a pre-disperser perpendicular to the main direction of dispersion. In the present context the echelle diagram is a superposition of spectral bands of equal length, $\Delta\nu$, for which usually twice the mean distance of the principal peaks is chosen. In this way it is possible to show both fine structure and large spectral sections at the same time in a single illustration.

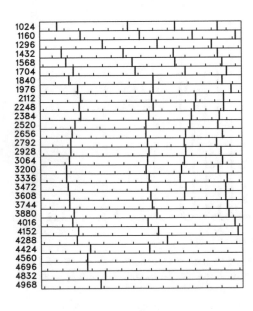

Fig. 5.10. Echelle diagram of solar p modes. Each *horizontal strip* covers $136\,\mu\mathrm{Hz}$; the columns of identified modes belong to degrees $l = 3, 4, 2$ and 5, from *left to right* (in the upper two rows the sequence is 4, 2, 5, 3). From Henning and Scherrer (1986)

Frequencies of modes with degrees $l = 2$ to $l = 5$, observed at the Wilcox Solar Observatory in Stanford, are shown as an example in Fig. 5.10. The respective mode identifications were obtained by comparison with other observations, and with theoretical results. We notice in particular the bending towards lower frequencies of the $l = $ const. columns which occurs below $\approx 2000\,\mu$Hz. Such an effect was predicted by theoretical eigenfrequency calculations.

5.1.4 Line Width and Line Asymmetry

The lines in the solar oscillation spectrum have a finite width. The line width is determined by the mode lifetime if this is shorter than the duration of the observed signal. Indeed the lifetime of a coherent oscillation is limited both because the oscillation may loose energy by some damping process and because the oscillation is continuously disturbed by the non-stationary velocity field in the convection zone. In any case the damped harmonic oscillator is an appropriate model, and we may expect a Lorentzian frequency profile as in the case of collisional broadening of the lines in an optical spectrum, cf. Eq. (2.85),

$$\phi(\nu) = \frac{\gamma}{[2\pi(\nu - \nu_0)]^2 + \gamma^2/4} \ . \tag{5.9}$$

The full width at half-maximum of the Lorentz profile is $\Delta\nu_{\rm L} = \gamma/2\pi$. The observed lines often fit very well to a Lorentz profile; the line widths obtained from such fits increase with increasing oscillation frequency, as illustrated in Fig. 5.11. The modes with the largest amplitude, with frequencies around $3\,$mHz, have a width of $\approx 1\,\mu$Hz, the lines at lower frequencies are sharper,

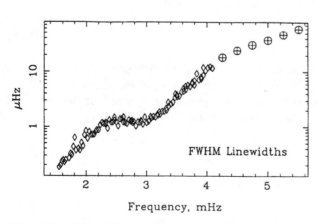

Fig. 5.11. Full width at half-maximum of p-mode lines, as a function of oscillation frequency. The *diamonds* are from fits to profiles of individual modes with $l = 19$ to 24, the *circles* are for groups of modes with $l \approx 60$. After Libbrecht (1988)

the lines at higher frequencies broader. The lifetime of a mode is given by $2/\gamma$, or the inverse of π times the full width; hence the line widths shown in the figure imply mode lifetimes between hours and months. The ratio $Q = \nu/\Delta\nu_L$ is the *quality factor* of the oscillation. Solar p modes typically have quality factors between 10^2 and 10^4.

In general the lines found in the power spectra of solar oscillations do not possess perfect Lorentzian profiles, but are slightly asymmetric. Duvall et al. (1993 a) first derived such line asymmetries from a full-disc observation of the intensity oscillations in the Ca II K line; later data, e.g. from the GONG network and from SOHO/MDI, confirmed the result with respect to velocity oscillations. Mainly the modes in the lower frequency range, $\nu < 3\,\mu\text{Hz}$, show an asymmetry, and it appears that the asymmetry is opposite in intensity and velocity power spectra. If an asymmetric line is fitted with a symmetric Lorentzian profile, a systematic error may be introduced in the determination of the mode frequency.

5.2 Linear Adiabatic Oscillations of a Non-Rotating Sun

A star like the Sun is a gaseous sphere in hydrostatic equilibrium. Perturbations of this (stable) equilibrium give rise to a restoring volume force and can therefore lead to oscillations. These oscillations will be described in the present section under simplifying assumptions.

5.2.1 Basic Equations

We shall first assume that the perturbations occur sufficiently fast so that everywhere in the star the adiabatic approximation

$$\frac{\delta P}{P_0} = \Gamma_1 \frac{\delta\rho}{\rho_0} \tag{5.10}$$

is satisfied. Here P_0 and ρ_0 are the undisturbed pressure and density, both dependent on radius r, and δP and $\delta\rho$ are the respective Lagrangian perturbations, i.e., the perturbations suffered by a fixed "parcel" or "element" of material as it moves back and forth. The adiabatic exponent,

$$\Gamma_1 = \left(\frac{d\ln P}{d\ln\rho}\right)_S , \tag{5.11}$$

is related to the adiabatic sound velocity c by

$$c^2 = \Gamma_1 P_0/\rho_0 , \tag{5.12}$$

and also depends only on depth.

It is only in the atmosphere of the Sun that the radiative exchange between places of (due to the perturbation) different temperature is sufficiently fast to render the oscillations non-adiabatic. We must consider the consequences of this when we study such details as the phase relationship between the observed velocity and temperature perturbation. Here we proceed with (5.10).

A further simplification is the neglect of rotation. Rotation is a small effect, of order Ω/ω as we shall see (Ω and ω are the frequencies of rotation and oscillation, respectively; for the Sun we have $\Omega/\omega \approx 10^{-4}$). The effect of rotation is nevertheless important; it will be discussed separately in Sect. 5.3.8 below.

Finally, we adopt the approximation of *linear* oscillations. Non-linear terms in the hydrodynamic equations are of order v/c compared to the linear terms, v is the velocity amplitude. The ratio v/c is very small almost everywhere in the Sun (again, only in the atmosphere the non-linear terms have some importance, and we must include these terms under certain circumstances).

Linearized, and in an inertial system of reference, the equation of continuity and the momentum equation can be written in the following form:

$$\rho_1 + \nabla \cdot (\rho_0 \boldsymbol{\xi}) = 0 , \tag{5.13}$$

$$\rho_0 \frac{\partial^2 \boldsymbol{\xi}}{\partial t^2} + \nabla P_1 - \frac{\rho_1}{\rho_0} \nabla P_0 + \rho_0 \nabla \Phi_1 = 0 , \tag{5.14}$$

where $\boldsymbol{\xi}$ is the vectorial distance of a gas parcel from its equilibrium position. The subscript 1 denotes Eulerian perturbations, i.e., perturbations at a fixed position in the Sun; these are related to the Lagrangian perturbations, denoted by a δ, through

$$\delta f = f_1 + \boldsymbol{\xi} \cdot \nabla f_0 . \tag{5.15}$$

The perturbation Φ_1 of the gravitational potential is related to the density perturbation through Poisson's equation

$$\Delta \Phi_1 = 4\pi G \rho_1 , \tag{5.16}$$

where G is the constant of gravitation.

After elimination of the Lagrangian perturbations with the help of transformation (5.15), relations (5.10), (5.13), (5.14), and (5.16) must be used to determine the unknowns $\boldsymbol{\xi}$, ρ_1, P_1, and Φ_1.

5.2.2 Spherical Harmonic Representation

Since the coefficients of our linear equations, which are known from the solar equilibrium model, depend only on depth, we may separate the time and angular dependencies. But first we notice, taking the curl of (5.14), that the vorticity of the perturbation $\boldsymbol{\xi}$ has no vertical component:

$$\boldsymbol{r} \cdot \text{curl} \frac{\partial^2 \boldsymbol{\xi}}{\partial t^2} = 0 \ . \tag{5.17}$$

Since we are primarily interested in *oscillations*, with $\partial/\partial t \equiv i\omega \neq 0$, this means (in spherical polar coordinates)

$$\frac{\partial}{\partial \theta}(\sin \theta \xi_\phi) - \frac{\partial \xi_\theta}{\partial \phi} = 0 \ . \tag{5.18}$$

The horizontal components, ξ_θ and ξ_ϕ, of the perturbation vector can therefore be obtained as derivatives of a single scalar function. This function, and equally the radial component of $\boldsymbol{\xi}$, will be expanded in terms of spherical harmonics. Each term has the form

$$\boldsymbol{\xi} = e^{i\omega t}\left(\xi_r(r), \xi_h(r)\frac{\partial}{\partial \theta}, \frac{\xi_h(r)}{\sin \theta}\frac{\partial}{\partial \phi}\right) Y_l^m(\theta, \phi) \ . \tag{5.19}$$

Y_l^m is the complex spherical harmonic of degree l and angular order m introduced in the preceding section. Whenever physical quantities are discussed, the real part of expression (5.19) must be taken.

The Eulerian perturbations of density, pressure, and gravitational potential are expressed in a corresponding manner:

$$(\rho_1, P_1, \Phi_1) = e^{i\omega t}[\rho_1(r), P_1(r), \Phi_1(r)] Y_l^m(\theta, \phi) \ . \tag{5.20}$$

Here, for the sake of simplicity, the same symbols are used for the perturbations and for their r-dependent factors.

It is a straight-forward matter to substitute (5.19) and (5.20) into the above equations, and to eliminate the variables ξ_h and ρ_1. We obtain

$$\frac{1}{r^2}\frac{d}{dr}(r^2 \xi_r) - \frac{\xi_r g}{c^2} + \frac{1}{\rho_0}\left(\frac{1}{c^2} - \frac{l(l+1)}{r^2 \omega^2}\right) P_1 - \frac{l(l+1)}{r^2 \omega^2}\Phi_1 = 0 \ , \tag{5.21}$$

$$\frac{1}{\rho_0}\left(\frac{d}{dr} + \frac{g}{c^2}\right) P_1 - (\omega^2 - N^2)\xi_r + \frac{d\Phi_1}{dr} = 0 \ , \tag{5.22}$$

$$\frac{1}{r^2}\frac{d}{dr}\left(r^2 \frac{d\Phi_1}{dr}\right) - \frac{l(l+1)}{r^2}\Phi_1 - \frac{4\pi G \rho_0}{g}N^2 \xi_r - \frac{4\pi G}{c^2}P_1 = 0 \ . \tag{5.23}$$

As in Chap. 2, the gravitational acceleration is $g = -\rho_0^{-1}dP_0/dr$. The velocity of sound c is defined by (5.12), and the *Brunt-Väisälä* frequency N is given by

$$N^2 = g\left(\frac{1}{\Gamma_1 P_0}\frac{dP_0}{dr} - \frac{1}{\rho_0}\frac{d\rho_0}{dr}\right) \ . \tag{5.24}$$

In the derivation of (5.21) to (5.23) the differential equation governing spherical harmonic functions has been used:

$$L^2 Y_l^m \equiv -\frac{1}{\sin \theta}\frac{\partial}{\partial \theta}\sin \theta \frac{\partial Y_l^m}{\partial \theta} - \frac{1}{\sin^2 \theta}\frac{\partial^2 Y_l^m}{\partial \phi^2} = l(l+1)Y_l^m \ . \tag{5.25}$$

Problem 5.4. Confirm (5.21) to (5.23). Derive an equation that expresses the conservation of energy.

Equations (5.21) to (5.23), together with appropriate boundary conditions (see below), must be solved numerically because of the general dependence of their coefficients on r. Only particular values of ω, the *eigenfrequencies*, will permit non-trivial solutions.

We note that the angular order m does not occur in our equations. For each l, therefore, there is a $(2l + 1)$-fold degeneracy of the eigenfrequencies. This is a consequence of the neglect of rotation and all other sources of anisotropy, and has its well-known parallel in atomic physics.

The operator L^2 defined in (5.25) is $-r^2$ times the angular part of the Laplacian, Δ. Had we, instead of spherical polar coordinates, used cartesian coordinates, with x and y pointing in two horizontal (i.e., tangential) directions perpendicular to each other, the corresponding part of Δ would be $\partial^2/\partial x^2 + \partial^2/\partial y^2$, and $-k_x^2 - k_y^2 \equiv -k_h^2$ would be the eigenvalue for an eigenfunction of form $\exp(ik_x x + ik_y y)$. For $k_h r \gg 1$, i.e., for a horizontal wavelength which is short in comparison to the radius, cartesian coordinates are a meaningful approximation. In this case the wave equation can be solved in either representation, and we can identify the eigenvalues of the Laplacian. Thus we find

$$l(l + 1) \simeq (k_h r)^2 . \tag{5.26}$$

At the solar surface, $r = r_\odot$, we have $l(l + 1) \simeq (k_h r_\odot)^2$, a relation that we have already used in the context of observed power spectra. As (5.26) is valid for large l, it may be simplified further to $l \simeq k_h r$.

5.2.3 The Cowling Approximation

We may discuss a number of properties of the oscillation equations without having a detailed solution. Usually this is done in the Cowling approximation, which consists in the neglect of the perturbation of the gravitational potential, Φ_1, in (5.21) and (5.22), and in discarding (5.23) altogether. The problem is thus reduced to two first-order differential equations:

$$\frac{1}{r^2}\frac{d}{dr}(r^2\xi_r) - \frac{\xi_r g}{c^2} + \frac{1}{\rho_0}\left(\frac{1}{c^2} - \frac{l(l+1)}{r^2\omega^2}\right)P_1 = 0 , \tag{5.27}$$

$$\frac{1}{\rho_0}\frac{dP_1}{dr} + \frac{g}{\rho_0 c^2}P_1 - (\omega^2 - N^2)\xi_r = 0 . \tag{5.28}$$

Numerical calculations of Robe (1968) and Christensen-Dalsgaard (1984) show that the eigenfrequencies obtained by solving (5.27) and (5.28) differ at most by a few percent from the eigenfrequencies of the full system. For large degree l or for very large or very small frequencies (in both limits the eigenfunctions possess a large number of nodes n in the radial direction – see below) the difference is much less than 1%. The reason is, as Cowling (1941) writes, "the large number of changes of sign" of the density perturbation ρ_1. In the solution of (5.16), that is in the Poisson integral,

$$\Phi_1(\boldsymbol{r}) = -G \int \frac{\rho_1(\boldsymbol{r}')}{|\boldsymbol{r} - \boldsymbol{r}'|}\, d\boldsymbol{r}' , \tag{5.29}$$

the contributions from different places $\boldsymbol{r}'$ in the star therefore tend to annihilate each other. The net effect is that Φ_1 becomes negligibly small.

Problem 5.5. Derive a second-order system of equations, like Eqs. (5.27) and (5.28), for the case $l = 0$, but with the variable Φ_1 retained (i.e., without making the Cowling approximation).

5.2.4 Local Treatment

The coefficients of the oscillation equations depend on depth. Nevertheless it is illustrative to consider the case where these coefficients, except for ρ_0 and P_0, are constants.

An example for which such a treatment is exact is the isothermal atmosphere, where g, c, and N all are constants, while ρ_0 and P_0 have a barometric (exponential) stratification. At any depth in a star this constitutes an acceptable approximation if g, c, and N vary much less than the perturbations themselves, i.e., if the eigenfunctions ξ_r, P_1, ... have a short vertical wavelength. In this case the Cowling approximation is appropriate, and we shall use (5.27) and (5.28). It is also appropriate to assume that the vertical wavelength is small compared to r, so that $|\xi_r/r| \ll |d\xi_r/dr|$. We define

$$S_l^2 = \frac{l(l+1)}{r^2}c^2 , \tag{5.30}$$

and search for oscillatory solutions of the following form:

$$\xi_r \sim \rho_0^{-1/2} \exp(\mathrm{i}k_r r) , \tag{5.31}$$

$$P_1 \sim \rho_0^{1/2} \exp(\mathrm{i}k_r r) . \tag{5.32}$$

It is easy to see that the $\rho_0^{1/2}$ factors take care of the (still variable) ρ_0 which occurs in (5.27) and (5.28). One must only realize that the density scale height,

$$H \equiv -\rho_0/(d\rho_0/dr) = \left(\frac{g}{c^2} + \frac{N^2}{g}\right)^{-1} , \tag{5.33}$$

is also a constant. Hence the dispersion relation is

$$k_r^2 = \frac{\omega^2 - \omega_A^2}{c^2} + S_l^2 \frac{N^2 - \omega^2}{c^2 \omega^2} \, , \tag{5.34}$$

where we have defined the acoustic cutoff frequency,

$$\omega_A = c/2H \, . \tag{5.35}$$

Problem 5.6. Confirm (5.34). Write the dispersion relation for the normalized frequency and wave number used in Fig. 5.12.

Problem 5.7. A generalization of (5.35). If the condition of an isothermal atmosphere is dropped, the wave equations for the diverse variables ξ_r, P_1, etc., have different forms. Show that the acoustic cutoff is given by

$$\omega_A^2 = \frac{c^2}{4H^2} \left(1 - 2 \frac{dH}{dr} \right) \, , \tag{5.36}$$

if div $\boldsymbol{\xi}$ is the dependent variable.

Oscillatory perturbations of type (5.31) and (5.32) require real k_r, i.e., the right-hand side of (5.34) must be positive. In order to see for which parameters, c, ω_A, etc., such oscillations can exist we set $k_r^2 = 0$, and solve the resulting algebraic equation for ω^2. Figure 5.12 shows the curves $k_r^2 = 0$ in the k_h, ω-plane, or *diagnostic diagram*, with normalized coordinates (cf. Problem 5.6). We distinguish three regions in this plane. First, at large ω, (5.34) simplifies to $\omega^2 = c^2(k_r^2 + k_h^2)$, which we recognize as the dispersion relation of ordinary acoustic waves. The restoring force for these waves is the pressure gradient; the high-frequency region is therefore called the *acoustic*, or p-mode region. Its lower boundary lies at ω_A for $l = 0$, and for large l asymptotically approaches the line $\omega = S_l \simeq c k_h$.

It is plausible that acoustic waves do not propagate at very small frequencies. Imagine that a periodic pressure perturbation is applied at the bottom of an atmosphere, and that the period is longer than the time required for a

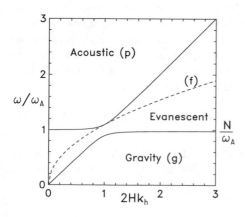

Fig. 5.12. The k_h, ω-plane, with curves $k_r = 0$ (*solid*). The *dashed* parabola marks the surface wave that corresponds to the f mode

sound wave to travel across a substantial part of the atmosphere, i.e., across a few scale heights. The whole atmosphere then has enough time to continuously adjust itself to a modified hydrostatic equilibrium, i.e., the exponential stratification is complemented by a slowly varying (in time) contribution. The vertical variation of such a perturbation is itself exponential, since $k_r^2 < 0$ according to (5.34). The corresponding intermediate frequency region of the k_h, ω-plane is called the *evanescent-wave* region.

In the evanescent region there is a particular mode for which div $\xi = 0$. In general this additional constraint is not satisfied simultaneously with (5.27) and (5.28), but for a special frequency, namely

$$\omega = \sqrt{gk_h} \tag{5.37}$$

this is possible because those two equations coincide. Equation (5.37) characterizes the dispersion of surface waves in deep water. This wave mode is marked in Fig. 5.12 by the dashed parabola. In the diagnostic diagrams (Figs. 5.4, 5.6, and 5.7) it appears as the lowest ridge, the *fundamental*, or f mode.

In the third, low-frequency, region of the k_h, ω-plane k_r^2 is again positive, so that propagating waves are possible. For small ω, and fixed non-zero l (or k_h), the dispersion relation simplifies to

$$k_r^2 = \frac{S_l^2}{c^2}\left(\frac{N^2}{\omega^2} - 1\right) \tag{5.38}$$

or [for large l, by (5.26) and (5.30)], to

$$\omega^2 = N^2 \frac{k_h^2}{k_r^2 + k_h^2} \equiv N^2 \sin^2 \Theta . \tag{5.39}$$

Waves with this dispersion relation are known as *internal gravity waves*. Their frequencies depend only on the angle Θ between the propagation vector and the vertical; they cannot propagate exactly in the vertical direction. Inspection of the dispersion relation (5.34) shows that the region of internal gravity waves, or g modes, is bounded by $\omega = N$ for large l, and by the line $\omega = S_l N/\omega_A$ for small l. The Brunt-Väisälä frequency thus turns out to be the critical frequency for internal gravity waves. In particular, it is necessary that $N^2 > 0$. We shall see in the Chap. 6 that this is also the condition of a stable stratification in the star. The larger N^2, the more solid is the stability, and the higher are the frequencies of the internal gravity waves.

Problem 5.8. Show that N^2 is proportional to the gradient, dS/dr, of the specific entropy in a star.

Problem 5.9. Calculate the diverse critical frequencies for the solar *atmosphere* and show that the observed p modes lie in the region of evanescent waves.

Already here we may see what the consequences of reflecting boundaries on acoustic or internal gravity waves are. If such waves are confined into a layer of total depth d between two reflecting boundaries, the spectrum of possible vertical wave numbers becomes discrete, and $k_r \simeq n\pi/d$ for large k_r. The respective asymptotic dispersion relations $\omega = ck_r$ and $\omega = Nk_h/k_r$ of p and g modes then imply that *the frequencies of p modes tend to infinity with equidistant frequency spacing*, comparable to the overtones of a trumpet, and that *the frequencies of g modes tend to zero, with equidistant spacing of the oscillation periods*.

That reflecting boundaries indeed exist in the Sun becomes clear if we recall that the coefficients g, c, etc., are *not* constants. The position of the dividing lines in the k_h, ω-plane between oscillatory and evanescent waves depends on depth. For a given frequency, therefore, a mode can be oscillatory at one depth, and evanescent at another. A perfect analogy to this situation is a light ray which is too oblique to enter a medium of smaller refractive index and thus gets *totally reflected*; another is a particle that has not sufficient energy to climb a potential step, and therefore changes its wave function from oscillatory to exponentially decaying.

Problem 5.10. Consider a cool layer embedded between two hot zones. Calculate the cutoff frequencies, and choose a mode which can propagate in the two hot zones, but is evanescent in the cool layer. Compare this mode to a particle which tunnels through a potential barrier.

If two evanescent zones enclose a layer of oscillatory wave behavior, the mode becomes trapped in this layer. This is the reason why the spectrum of oscillation frequencies is discrete. Figure 5.13 shows, as functions of depth in the Sun, the squared frequencies ω_A^2, N^2, and S_l^2. It is clear from these results that p modes are reflected by the outwards-increasing acoustic cutoff ω_A,

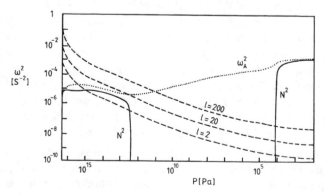

Fig. 5.13. Critical frequencies in the Sun, according to (5.24), (5.30), and (5.35). The *dashed* curves show S_l^2 for $l = 2$, 20, and 200. Courtesy M. Knölker

and by the inwards-increasing S_l. For a given degree l, p modes with higher frequencies are reflected further outward in the atmosphere, and also further inward in the core. As l increases (at fixed frequency) the inner reflection occurs at decreasing depth. For large ω (5.34) is approximately $k_r^2 = (\omega^2 - S_l^2)/c^2$. Hence $\omega^2 = S_l^2$ defines the level r_t of reflection:

$$r_t = [l(l+1)]^{1/2} c(r_t)/\omega \; . \tag{5.40}$$

A solution of this implicit equation is shown in Fig. 5.14 for a number of frequencies around $\omega/2\pi = 3\,\mathrm{mHz}$.

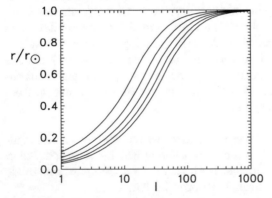

Fig. 5.14. Depth of internal reflection, according to (5.40), as a function of degree l, for p modes with oscillation frequencies 1.5, 2.25, 3, 3.75, and 4.5 mHz (*from above*). The curves are calculated for the standard solar model of Chap. 2

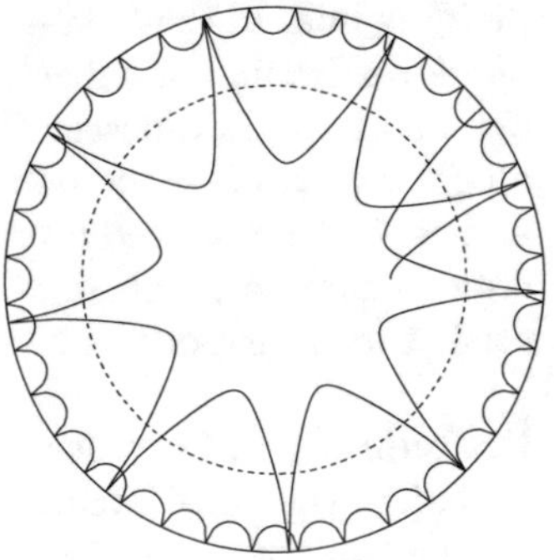

Fig. 5.15. Acoustic ray paths of two solar p modes with different depths of reflection in the solar interior. The *dashed circle* marks the base of the convection zone

The total reflection of a p mode with $l \neq 0$ at an internal evanescent region is illustrated in Fig. 5.15. Near the surface the wave motion may be essentially vertical (as observed). But as soon the wave vector is slightly inclined, it will become more so as the wave propagates downwards. The reason is that the deeper part of a surface of constant phase travels in a hotter environment, and therefore at a larger speed. Finally the ray is bent into the horizontal and then reflected. For large l this happens close to the solar surface, for smaller l at greater depth. With the upper reflection provided by the atmospheric acoustic cutoff, the waves travel in a waveguide around the Sun.

Duvall's Law. The local treatment can be used to confirm theoretically a result first obtained by Duvall (1982). He plotted, as a function of ω/k_h, the ratio $(n + \alpha)\pi/\omega$, where n is the order of the mode (number of p-mode ridge), and α is a constant. The remarkable fact is that in this representation all ridges collapse into a single curve, provided the counting of n has the right zero point and α is properly chosen, cf. Fig. 5.16.

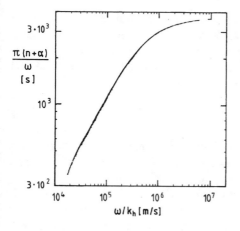

Fig. 5.16. Observed solar p modes, collapsed into a single curve, corresponding to (5.42) with $\alpha = 1.58$. Adapted from Christensen-Dalsgaard et al. (1985)

The following argument of Christensen-Dalsgaard et al. (1985) shows that such a result must be expected. The total phase difference between the internal reflection point r_t and the outer reflection point $r_\odot$ (this is sufficiently accurate for the present purpose) is

$$\Delta\psi = \int_{r_t}^{r_\odot} k_r \, dr \ . \tag{5.41}$$

For an eigensolution $\Delta\psi$ must be equal to $\pi(n + \alpha)$; the constant α occurs because the reflecting boundaries are not rigid walls, but arise from the adjacent evanescent regions. Hence, using the dispersion relation in the limit of large ω, we have

$$\pi(n + \alpha)/\omega = \int_{r_t}^{r_\odot} \left(\frac{1}{c^2} - \frac{l(l+1)}{r^2\omega^2} \right)^{1/2} dr \ . \tag{5.42}$$

Now, by (5.40), r_t depends on l and ω only in the form $l(l+1)/\omega^2$, the same combination which appears on the right of (5.42). The integral, therefore, is a function of $l(l+1)/\omega^2$. Application of (5.26), at $r = r_\odot$, then yields Duvall's law.

Regions for Internal Gravity Waves. As illustrated in Fig. 5.13, there are two regions where $N^2 > 0$ and, therefore, internal gravity waves can propagate: the atmosphere and the radiative core. In the convection zone, where $N^2 < 0$, these waves are evanescent. In the core below the convection zone internal gravity waves are trapped and thus form a discrete frequency spectrum of g modes. But the existence of g modes on the Sun has not yet been confirmed; at $\nu = 200\,\mu\text{Hz}$, upper limits are $1\,\text{cm/s}$ in velocity and 5×10^{-7} in relative intensity variation (Appourchaux et al. 2000).

On the other hand, the phase diagram, Fig. 5.3, indicates the existence of internal gravity waves in the solar atmosphere. The signature of these waves is the *negative* phase difference at low frequencies, as first observed by Schmieder (1976). To see this, we use (5.26) to rewrite the dispersion relation (5.34) for the isothermal atmosphere in the form

$$\frac{k_h^2(\omega^2 - N^2)}{\omega^2(\omega^2 - \omega_A^2)} + \frac{k_r^2}{\omega^2 - \omega_A^2} = \frac{1}{c^2} \ , \tag{5.43}$$

which, for constant ω^2, is a quadric surface in k space. The vector of phase propagation, k, is the radius vector. The group velocity is the gradient of ω in k space, and is perpendicular to the surfaces $\omega^2 = $ const.

In the region of propagating acoustic waves (Fig. 5.12) the surfaces (5.43) are oblate ellipsoids of revolution with respect to the k_r axis, because $\omega^2 > \omega_A^2$ (and $\omega^2 > N^2$). In this case the vertical components of phase velocity and group velocity have the same sign. The propagating waves can form standing waves (with almost no phase difference as in the range 2–5 mHz of Fig. 5.3), or they will have an upwards propagating phase, such as for $\nu > 5$ mHz in that figure.

By contrast, (5.43) represents a one-shell hyperboloid of revolution for internal gravity waves, where $\omega^2 < N^2$ (and $\omega^2 < \omega_A^2$). Now the r components of the phase and group velocities have *different* signs. An internal gravity wave excited from below, with upwards propagating energy, will therefore exhibit a *downwards* propagating phase. In Fig. 5.3, at frequencies around 1 mHz, such downward phase propagation is seen; these waves are therefore interpreted as internal gravity waves in the solar atmosphere.

Problem 5.11. Draw the surfaces $\omega^2 = $ const. for acoustic and internal gravity waves. Mark the vectors of the phase and group velocities.

Radiative Energy Exchange. In the solar atmosphere the propagation of waves becomes non-adiabatic because of the energy exchange by radiation. In principle this process must be described by a simultaneous solution of the equations of hydrodynamics and radiative transfer (Schmieder 1977), but many details have been discussed in terms of the radiative cooling time (Spiegel 1957)

$$\tau_R = \frac{c_V}{16\kappa\sigma T^3} \left(1 - \frac{\kappa\rho}{k} \cot^{-1}\frac{\kappa\rho}{k}\right)^{-1} \ , \tag{5.44}$$

where c_V is the specific heat at constant volume and k the wave number. In the optically thin case (the mean-free-path of photons being large compared to the wavelength, i.e., $\kappa\rho/k \ll 1$) we have $\tau_R \approx c_V/(16\kappa\sigma T^3)$, independent of k. The time τ_R changes from about a second in the photosphere to $\approx 10^3$ s in the high atmosphere, which means that an oscillation with a period of 5 min is almost isothermal in the deeper atmosphere, but nearly adiabatic in the upper layers.

The severest consequences of radiative damping occur for internal gravity waves in the solar atmosphere (Souffrin 1966; Stix 1970; Mihalas and Toomre 1981, 1982). For these waves the phase between the velocity and the various thermodynamic quantities strongly depends on τ_R. Internal gravity waves cannot even exist if τ_R is too short: the rapid radiative smoothing of temperature fluctuations destroys the restoring buoyancy force.

Below the photosphere the Sun is optically thick. Then $\tau_R \propto k^{-2}$ which, for all wavelengths of interest, is so long that the adiabatic approximation is very good everywhere in the solar interior.

5.2.5 Boundary Conditions

The Surface. The Sun has an extended atmosphere; its corona reaches far into interplanetary space. In principle it would be necessary to treat the oscillations, either propagating or evanescent, in this entire atmosphere, but this is not practical for obvious reasons. Somewhere we must place a boundary on the equilibrium model and impose boundary conditions on the perturbations. We shall see that these conditions depend on the type of boundary chosen for the solar model itself.

One possibility is to treat the atmosphere as isothermal and infinite, and to utilize the acoustic cutoff, ω_A, in the atmosphere. The atmospheric dispersion relation (5.34) gives $k_r = \pm i\alpha$, with positive α; (5.31) and (5.32) then tell us that the boundary condition must consist in the choice $k_r = +i\alpha$, so that the exponentially growing contribution is discarded. Hence it follows from (5.31) that

$$d(\xi_r \rho_0^{1/2})/dr = -\alpha \xi_r \rho_0^{1/2} \tag{5.45}$$

must be satisfied at the base of the isothermal atmosphere, which we may identify as the outer boundary of the solar model.

Condition (5.45) has been used for radial oscillations (Baker and Kippenhahn 1965) and, in combination with the Cowling approximation, for nonradial oscillations (Ando and Osaki 1975). Although this boundary condition mimics the transition into the evanescent layer in a satisfactory manner, it has its limitations for several reasons. The atmospheric reflection occurs at different levels for different frequencies, and waves with periods shorter than ca. 3.5 minutes are not reflected at all. Moreover, the temperature increases towards the upper atmosphere. The evanescent region is therefore finite, i.e., the temperature minimum constitutes a tunnel of finite length (cf. Problem 5.10). Formally this means that the exponentially growing contribution cannot be neglected entirely.

A second approach is to consider a solar model with a surface of zero density and zero pressure. Let $\rho_0 \propto (r_\odot - r)^{n_e}$ at such a surface. The hydrostatic equilibrium then requires $P_0 \propto (r_\odot - r)^{n_e+1}$. That is, P_0 and ρ_0 follow a polytrope,

$$P_0 \propto \rho_0^{(n_e+1)/n_e} \; ; \tag{5.46}$$

n_e is called the effective polytropic index. Some of the coefficients of the oscillation equations have a singularity at the surface of this model. In particular we have $N^2 \propto (r_\odot - r)^{-1}$. Therefore, the boundary condition to be imposed on the oscillations simply is that all *perturbations remain finite* at $r = r_\odot$. This approach has proved useful in analytic work, in particular in combination with the Cowling approximation. The asymptotic results to be described in Sect. 5.2.6 below are an example.

A third method is to cut off the atmosphere of the solar equilibrium model at a *finite* density. The surface of a model of this type would resemble a water surface, where the pressure falls off to zero. The boundary condition then is that the perturbed surface is also a zero-pressure surface. Hence we must impose

$$\delta P = 0 \; . \tag{5.47}$$

This boundary condition has been used in most numerical calculations.

Since the full set of (5.21) to (5.23) is normally solved in numerical work, a second outer boundary condition must be imposed. To this end we require that the internal gravitational potential smoothly matches onto the external gravitational potential. For the latter, we must select the solution of $\Delta\Phi_1 = 0$ which (for any given l) behaves like r^{-l-1} for $r \to \infty$. Thus, the boundary condition at the model surface becomes

$$r\frac{d\Phi_1}{dr} + (l+1)\Phi_1 = 0 \; . \tag{5.48}$$

We may wonder which of the three types of boundaries best represents the Sun. For the determination of eigenfrequencies alone it probably does not matter too much, if only the *place* of the boundary is chosen at the right level in the atmosphere. As test calculations demonstrate, the boundary should be placed at optical depth $\tau = 10^{-4}$, or even further outwards. The model eigenfrequencies then become more or less independent of the place and type of boundary condition. This is in particular true for the lower overtones (smaller frequencies), and the reason is of course that these are already reflected by their acoustic cutoff at some *internal* level, so that the actual boundary condition, placed in the evanescent region, has little effect: it must merely suppress the exponentially growing contribution to the eigenfunction.

The Center. Although the center of the Sun does not represent a physical surface, we must impose a boundary condition there. This is a consequence of the use of spherical polar coordinates, which introduces a formal singularity at $r = 0$ into the oscillation equations (5.21) to (5.23). As in the case of the singular surface, the boundary condition is that *all perturbations remain finite at $r = 0$*.

Problem 5.12. Expand the diverse perturbations into Taylor series around $r = 0$ and show that, for $l \neq 0$,

$$\xi_r \propto r^{l-1} , \quad P_1 \propto r^l . \tag{5.49}$$

Find the reason for $\xi_r(0) = 0$ in the case $l = 0$.

We have already seen that most p modes, in particular those of high degree l, have their point of internal total reflection r_t at a considerable distance from the center, cf. Figs. 5.14 and 5.15. For these modes it is not necessary to solve the equations all the way down to $r = 0$. It is sufficient to impose a boundary condition somewhere below $r = r_t$. The type of this condition does not matter much, if only its place is chosen well inside the evanescent region, as in the case of the outer boundary condition.

5.2.6 Asymptotic Results

When the frequencies of p modes become very large, or the frequencies of g modes become very small, approximate analytic solutions of the oscillation equations can be obtained. We have already seen that in these limiting cases the eigenfunctions have a large number of nodes along the radius, and that therefore the Cowling approximation may be employed. Tassoul (1980) has described the asymptotic methods required; the following results are taken from her work.

Let us specifically consider the p modes with large ω. For any $l \neq 0$, we have the two characteristic frequencies S_l and N, and we may therefore seek an expansion in terms of the small parameters S_l^2/ω^2 or N^2/ω^2. However, there are the singularities $S_l^2 \to \infty$ at the center, and $N^2 \to \infty$ at the surface (for the present purpose we use a stellar model having $P_0 = \rho_0 = 0$ at the surface). Therefore, neither of the expansions can be applied to the whole star, and we must divide the range $(0, r_\odot)$ into two domains. In the inner domain an expansion in terms of N^2/ω^2 is chosen, and (5.27) and (5.28) are rearranged in the form

$$\left(\frac{P_1}{\rho_0}\right)'' + (A+B)\left(\frac{P_1}{\rho_0}\right)' + \left(\frac{\omega^2 - S_l^2}{c^2} + BA + A'\right)\frac{P_1}{\rho_0}$$
$$= -\frac{N^2}{\omega^2}\left[(\omega^2\xi_r)' + \left(B + \frac{2N'}{N}\right)\omega^2\xi_r\right] , \tag{5.50}$$

and

$$\omega^2\xi_r - \left[\left(\frac{P_1}{\rho_0}\right)' + A\frac{P_1}{\rho_0}\right] = \frac{N^2}{\omega^2}(\omega^2\xi_r) , \tag{5.51}$$

where a prime denotes a derivative with respect to r, and the abbreviations

$$A = -N^2/g \, , \tag{5.52}$$

$$B = 2/r - g/c^2 \tag{5.53}$$

have been introduced.

In the outer domain the expansion parameter is S_l^2/ω^2; we derive from (5.27) and (5.28) the two equations

$$\xi_r'' + \left(A + B + \frac{2c'}{c} \right) \xi_r' + \left[\frac{\omega^2 - N^2}{c^2} + B' + B \left(A + \frac{2c'}{c} \right) \right] \xi_r$$
$$= \frac{S_l^2}{\omega^2 c^2} \left[\left(\frac{P_1}{\rho_0} \right)' + \left(A + \frac{2S_l'}{S_l} \right) \frac{P_1}{\rho_0} \right] \, , \tag{5.54}$$

and

$$\frac{P_1}{\rho_0} + c^2 (\xi_r' + B\xi_r) = \frac{S_l^2}{\omega^2} \frac{P_1}{\rho_0} \, . \tag{5.55}$$

In each domain we now have a second-order differential equation. The terms on the right are multiplied by the respective (small) expansion parameter, while the left-hand sides of (5.50) and (5.54) are rather similar to the equation that governs Bessel functions. In particular, note the singularities in the coefficients of the undifferentiated variables P_1/ρ_0 and ξ_r. Bessel functions are therefore appropriate for the expansion of P_1/ρ_0 and ξ_r, and the correct choice of their index takes care of the singularities. The leading terms of these expansions are, in the inner domain

$$\frac{P_1}{\rho_0} = k_i \left(\frac{c}{\rho_0} \right)^{1/2} \frac{v_i^{1/2}}{r} \left[J_{l+1/2}(\omega v_i) - \frac{1}{\omega} F_i(u_i) J_{l+3/2}(\omega v_i) \right] \tag{5.56}$$

and

$$\omega^2 \xi_r = -k_i \frac{v_i^{1/2}}{r(c\rho_0)^{1/2}} \left[\omega J_{l+3/2}(\omega v_i) \right.$$
$$\left. + \left(H_i(u_i) - \frac{l+1}{u_i} \right) J_{l+1/2}(\omega v_i) \right] \, , \tag{5.57}$$

where

$$u_i = v_i = \int_0^r \frac{dr'}{c} \, , \tag{5.58}$$

F_i and H_i are such that (5.50) and (5.51) are satisfied, and k_i is a constant.

In the outer domain the expansions begin with

$$\xi_r = k_o \frac{v_o^{1/2}}{r(c\rho_0)^{1/2}} \left[J_{n_e}(\omega v_o) - \frac{2}{\omega v_o} F_o(u_o) J_{n_e+1}(\omega v_o) \right] \, , \tag{5.59}$$

and

$$\frac{P_1}{\rho_0} = -k_o \left(\frac{c}{\rho_0}\right)^{1/2} \frac{v_o^{1/2}}{r} \left[\omega J_{n_e+1}(\omega v_o) \right.$$

$$\left. + \left(H_o(u_o) - \frac{n_e + 1}{2u_o}\right) \frac{v_o}{2} J_{n_e}(\omega v_o)\right] , \qquad (5.60)$$

where n_e is the polytropic index at the surface (cf. the preceding section), k_o is another constant, and u_o and v_o are defined by

$$2u_o^{1/2} - v_o = \int_r^{r_\odot} \frac{dr'}{c} . \qquad (5.61)$$

The functions F_o and H_o are determined through the differential equations (5.54) and (5.55).

Problem 5.13. Determine the functions F_i, H_i, F_o, and H_o. Show that the asymptotic expansions satisfy the boundary conditions.

The inner and outer domains overlap. In the interval that is common to both domains the two expansions (5.56), (5.57) and (5.59), (5.60) must approximate the *same* solution. This matching requirement firstly determines the ratio k_o/k_i (one of the two constants remains arbitrary since the whole problem is linear and homogeneous), and secondly yields the desired condition for the eigenfrequencies ω. For large arguments, i.e., at large distance from the respective boundaries, the Bessel functions can be represented in terms of trigonometric functions. If this is done, the matching condition is reduced to the form $\cos(q(\omega)) = 0$, so that $q = \pi(2n-1)/2$ with an integer n; excluding terms of order ω^{-2} or smaller, $q(\omega)$ is given by

$$q(\omega) = \omega(v_i + v_o) - \left(l + n_e + \frac{3}{2}\right)\frac{\pi}{2} - \frac{1}{\omega}V_{io} . \qquad (5.62)$$

We write $\omega = 2\pi\nu_{n,l}$, substitute v_i and v_o, and finally obtain

$$\nu_{n,l} = \Delta\nu\left(n + \frac{l}{2} + \frac{n_e}{2} + \frac{1}{4} + \frac{V_{io}}{2\pi^2\nu_{n,l}}\right) , \qquad (5.63)$$

where we have defined the asymptotic frequency splitting

$$\Delta\nu = \left(2\int_0^{r_\odot} \frac{dr}{c}\right)^{-1} . \qquad (5.64)$$

Formula (5.63) confirms our expectation of an (almost) equidistant frequency spacing. Ledoux (1962) first obtained this result for the special case of radial oscillations ($l = 0$), and Vandakurov (1967) first considered the general case. Evaluation of (5.64) yields

$$\Delta\nu \approx 136\,\mu\text{Hz} ,$$

(5.65)

twice the spacing of the main peaks in the power spectrum of low-l p modes. It is now clear why this value was chosen as bandwidth of the spectral sections in the echelle diagram, Fig. 5.10. The line identifications on the top of Fig. 5.8 also can be obtained from (5.63) to (5.65) (or, of course, from numerical calculations).

The close coincidence of the frequencies $\nu_{n,l}$ and $\nu_{n-1,l+2}$ seen in the observational results now turns out to be a degeneracy in the leading term of (5.63). It is worth mentioning that the asymptotic frequencies do not depend on the choice of the matching point. As Tassoul (1980) shows, this is even true for the second-order correction V_{io}. Due to this correction the degeneracy is not exact; V_{io} is related to the small separation

$$\delta\nu_{n,l} = \nu_{n,l} - \nu_{n-1,l+2} .$$

(5.66)

Various estimates have been made at $\delta\nu_{n,l}$, which all depend on the derivative of the sound velocity. Christensen-Dalsgaard (1998) gives

$$\delta\nu_{n,l} \simeq -\frac{(4l+6)\Delta\nu}{4\pi^2\nu_{n,l}} \int\limits_0^{r_\odot} \frac{1}{r}\frac{dc}{dr}dr .$$

(5.67)

Problem 5.14. Confirm (5.62).

Asymptotic g Modes. Asymptotic frequencies of g modes are obtained in an analogous manner. In an inner domain, containing the center, an expansion in terms of ω^2/S_l^2 is now possible, while ω^2/N^2 serves as expansion parameter in the outer domain, containing the surface. A complication arises as a consequence of intermediate zeros of N^2, cf. Fig. 5.13. Additional domains must therefore be considered separately, so that no more than one of the transition points (the center, zeros of N^2, and the surface) falls into each domain. We do not repeat the procedure here. For the Sun we have $N^2(r_v) = 0$, where r_v marks the lower boundary of the convection zone. The result of matching the respective expansions is, in first order

$$T_{n,l} \equiv \frac{2\pi^2(n+l/2-1/4)}{\sqrt{l(l+1)}}\left(\int\limits_0^{r_v}\frac{N}{r}dr\right)^{-1} \equiv \frac{n+l/2-1/4}{\sqrt{l(l+1)}}T_0 .$$

(5.68)

This confirms the expectation of equidistant period spacing for g modes. The time T_0 defined by the second equality in (5.68) is a constant and can be calculated for any given solar model. The standard model described in

Chap. 2 yields $T_0 = 35.4 \, \text{min}$; the model of the 1989 edition of this book gave $T_0 = 34.8 \, \text{min}$, and model 1 of Christensen-Dalsgaard (1982) [as quoted by Gough (1984 b)] $T_0 = 34.5 \, \text{min}$. The differences in the depth of the convection zone, i.e., in the value of r_v, apparently do not matter greatly because in the vicinity of r_v the Brunt-Väisälä frequency N, and therefore the contribution to the integral in (5.68), is small.

5.3 Helioseismology

The measurement and interpretation of *travel times* of earthquake signals traditionally has been used to study the Earth's interior. In addition to this classical method, the measurement of *free oscillations* has become possible since the great Chilean earthquake of May 22, 1960. Mostly this more recent branch of seismology has been applied in solar (and indeed stellar) research, although the travel-time method also has its solar analogue.

5.3.1 Direct Modeling and Inversion

The straight-forward method of helioseismology begins with the computation of a solar model according to the recipes outlined in Chap. 2. From this model we may then calculate the r-dependent coefficients of the oscillation equations, (5.21) to (5.23). Finally, we solve these equations numerically, subject to the boundary conditions discussed in Sect. 5.2.5. The result consists in *eigenfunctions* that give the radial dependence of ξ_r, P_1, etc., and of *eigenfrequencies* – the particular values of ω for which non-zero solutions exist. The approach of direct modeling mainly utilizes the latter.

Now the solar model, even the standard solar model, contains some uncertainty. The diverse corrections to the perfect-gas law in the equation of state are examples; other examples are the precise abundance of the chemical elements and the opacity. Thus, we may vary these ingredients and so try to bring as many as possible of the calculated eigenfrequencies in agreement with observed frequencies. A "best" solar model can be obtained in this way.

The observed frequencies are known with relative errors of only 10^{-4} to 10^{-3}. It is desirable to calculate the theoretical frequencies with the same precision, that is: a precision of order $1 \, \mu\text{Hz}$ or less. In general the Cowling approximation is too inaccurate for this purpose. Besides, it is necessary to apply some care in the numerical methods. For example, the results tend to depend on the grid chosen for the finite difference representation of (5.21) to (5.23). As Shibahashi and Osaki (1981) point out, an extrapolation to an infinite number of grid points must be made.

In some cases it is sufficient to compare frequency *differences*, rather than absolute frequencies. Such differences, e.g., the small separation $\delta\nu_{n,l}$ defined by (5.66), may be obtained with sufficient precision by approximate methods, for example by means of the asymptotic result (5.67).

Mode Identification. A first result of direct modeling is the identification of the ridges in the k_h, ω-plane and of the lines in the power spectrum of the full-disc oscillation signal. The p-mode ridges have been detected from degree $l = 0$ up to degree $l \approx 4000$. They are distinguished by their order n, the number of nodal surfaces in the radial direction. In the diagnostic diagram n increases upwards, with the frequency ν; orders up to $n \approx 35$ have been identified. The lowest-order ridge is the fundamental, or f mode, with dispersion relation (5.37). The lines (Figs. 5.8, 5.9) are p modes of low degree l, and intermediate and high order n.

Inversion. The iteration of a solar model with the aim of approximating a given set of observed p-mode frequencies is an inversion problem. Since the oscillation frequencies can be written as integrals over the solar volume – see (5.70) – we may characterize inversion as the art of finding a function that appears in the integrand of a given definite integral. Such is indeed possible if the integrand depends not only on the desired function, but in addition on a parameter, and if the integral itself is a function of that parameter. The determination of the atmospheric temperature profile from the observed intensity, as described in Chap. 4, was such an inversion.

In exceptional cases an analytic inversion is possible, such as (5.77) below. For the general case numerical methods are necessary. It is expedient to cast the eigenvalue problem into the variational form. Scalar multiplication of (5.14) by ξ^*, the complex conjugate of ξ, and integration over the entire volume of the Sun gives (with $\partial/\partial t = i\omega$)

$$\omega^2 \int \rho_0 \xi^* \cdot \xi \, dV = \int \xi^* \cdot \left(\nabla P_1 - \frac{\rho_1}{\rho_0} \nabla P_0 + \rho_0 \nabla \Phi_1 \right) dV \ . \tag{5.69}$$

Now P_1, ρ_1, and Φ_1 can be eliminated with the help of (5.10), (5.13), (5.15), and (5.29). Equation (5.69) can therefore be written in the form

$$\omega^2 \int \rho_0 \xi^* \cdot \xi \, dV = \int \xi^* \cdot \mathcal{L} \xi \, dV \ , \tag{5.70}$$

where $\mathcal{L}$ is a linear operator that contains the functions P_0, ρ_0, and Γ_1 (see Problem 5.15). We see that ω^2 is a weighted average over the star, and that the weighting is governed by the displacement vector ξ. Using different eigenvalues ω^2, and the corresponding eigenvectors ξ, we can obtain information from different depth ranges, in accord with the propagation and reflection properties of the various modes. Since P_0, ρ_0, and Γ_1 enter (5.70) in a non-linear manner, the determination of these functions [the "inversion" of (5.70)] is achieved by iteration of an initial guess, as described by Gough (1984 b).

Of course it is possible to take a standard solar model for the initial guess. In addition, it is possible to use some or all of the stellar structure equations as constraints for the inversion. Various seismic solar models have been constructed in this way (e.g., Dziembowski et al. 1994, 1995; Gough et al. 1996; Takata and Shibahashi 1998).

Problem 5.15. Show that the explicit form of (5.70) is

$$\omega^2 \int \rho_0 \boldsymbol{\xi}^* \cdot \boldsymbol{\xi} \, dV = \int \left[\boldsymbol{\xi}^* \cdot \nabla P_0 (\nabla \cdot \boldsymbol{\xi} + \boldsymbol{\xi} \cdot \nabla \ln \rho_0) \right.$$

$$\left. + \nabla \cdot \boldsymbol{\xi}^* (\boldsymbol{\xi} \cdot \nabla P_0 + \Gamma_1 P_0 \nabla \cdot \boldsymbol{\xi}) \right] dV \tag{5.71}$$

$$- G \iint \frac{\nabla \cdot (\rho_0 \boldsymbol{\xi}^*) \nabla' \cdot (\rho_0 \boldsymbol{\xi})}{|\boldsymbol{r} - \boldsymbol{r}'|} \, dV' dV$$

(the dash indicates the argument $\boldsymbol{r}'$). Under which boundary conditions is this correct? Show that two different eigenvectors are orthogonal to each other, and that the eigenvalues ω^2 are real.

5.3.2 Speed of Sound in the Solar Interior

As a first example we determine the speed of sound in the solar interior, following Gough (1984 a) and Christensen-Dalsgaard et al. (1985). We substitute

$$u = l(l+1)/\omega^2 , \tag{5.72}$$

$$\xi = (r/c)^2 , \tag{5.73}$$

and obtain from (5.42)

$$F(u) = \int_u^{\xi_\odot} (\xi - u)^{1/2} \frac{1}{r} \frac{dr}{d\xi} \, d\xi , \tag{5.74}$$

where $\xi_\odot = \xi(r_\odot)$. By Duvall's law, $F(u)$ is a known function. We see that u is the mentioned additional parameter (essentially the oscillation frequency, ω), while ξ contains the desired function $c(r)$. Differentiation of (5.74) with respect to u yields

$$-2 \frac{dF}{du} = \int_u^{\xi_\odot} \frac{dG/d\xi}{(\xi - u)^{1/2}} \, d\xi , \tag{5.75}$$

where $G = \ln r$. Equation (5.75) has the form of Abel's integral equation. The solution ("inversion") is

$$G(\xi) - G(\xi_\odot) = -\frac{2}{\pi} \int_{\xi_\odot}^{\xi} \frac{dF/du}{(u - \xi)^{1/2}} \, du , \tag{5.76}$$

or

$$r = r_\odot \exp\left(-\frac{2}{\pi} \int_{\xi_\odot}^{\xi} \frac{dF/du}{(u - \xi)^{1/2}} \, du \right) . \tag{5.77}$$

This is an implicit equation for $\xi(r)$ and therefore, by (5.73), for $c(r)$. The important point is that $c(r)$ can be determined from (5.77) and the known function $F(u)$ without recourse to a calculated solar model.

The result of the original inversion is shown in Fig. 5.17. In the range $0.4r_\odot < r < 0.9r_\odot$ it agrees well with the theoretical model (model 1 of Christensen-Dalsgaard 1982). Outside that range this inversion is not reliable, as was shown by inverting the eigenfrequencies of the model itself. One reason is that (5.42) was based on the local dispersion relation $\omega^2 = c^2(k_r^2 + k_h^2)$ which breaks down near the reflecting boundaries. Another reason, responsible for the uncertainty below $0.4r_\odot$, is that not enough frequencies of modes penetrating deep into the core were available.

Figure 5.18 shows a more recent result, obtained numerically according to the recipes outlined above: the difference $\delta c^2(r)$ between the squared sound velocities of a seismic and a standard solar model. Although this difference is remarkably small, there are significant excursions: a peak just below the convection zone, and a dip around $r/r_\odot = 0.2$. Both the peak and the dip occur in regions where the mean molecular weight μ varies, due to the gravitational settling of helium and to the hydrogen burning in the core, and it has been found that smoothing the gradient of μ, e.g., by some mild mixing, would reduce the excursions (Christensen-Dalsgaard 1997). In the central region, where the accuracy is least, the sound-speed deviation is still less than 1%.

The temperature of the solar center cannot be determined with the same precision as the sound speed, because of its dependence on the mean molecular weight: $T \propto \mu c^2$. Error estimates of $\approx 2\%$ have been obtained by Antia and Chitre (1995) and by Takata and Shibahashi (1998) in a combined evaluation of helioseismic results and solar model calculations. The uncertainty lies both in the abundance of the elements and in the opacity. However, a central temperature of $1.57 \times 10^7 \, \mathrm{K}$ with an uncertainty of 2% is still a sufficiently severe constraint to dismiss the non-standard solar models of Sect. 2.5.

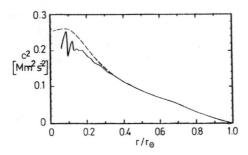

Fig. 5.17. Square of the sound speed in the Sun. *Continuous line*: inversion of the data in Fig. 5.16; *dashed*: theoretical solar model. From Christensen-Dalsgaard et al. (1985)

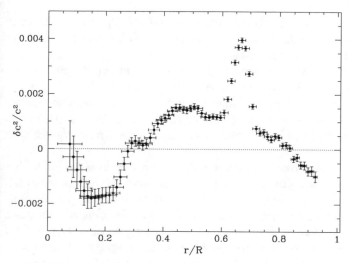

Fig. 5.18. Relative difference between the squared sound speed as inferred from 2 months of MDI data and the standard solar model of Christensen-Dalsgaard et al. (1996). *Vertical bars* are error estimates, *horizontal bars* show the resolution in depth. From Kosovichev et al. (1997)

5.3.3 Depth of the Convection Zone

As we shall see in more detail in Chap. 6, convection on the Sun is conveniently described in terms of the mixing-length theory. The crucial parameter is $\alpha = l/H_P$, the ratio of the mixing length to the pressure scale height.

An increase of α renders the convection more efficient in terms of energy transport. That is, the convection is capable of carrying the solar luminosity at a greater depth. Therefore, the depth of the Sun's outer convection zone increases with α.

Ulrich and Rhodes (1977) have demonstrated that the solar p modes of high degree l (the ridges) respond in a sensitive way to changes of α. They used solar envelope models, which is sufficient for p modes of high l in view of the relatively shallow wave guide of these modes, cf. Fig. 5.14. Their results, as well as the more refined calculations of Lubow et al. (1980) and Berthomieu et al. (1980), show that the best agreement with observed p-mode ridges is found for α between 2 and 3. (Fig. 5.4 shows the case $\alpha = 2$). The corresponding convection zones extend 200 000 and more kilometers in depth.

The more efficient the convection is in transporting energy, the smaller a difference between the actual and the adiabatic temperature gradients is required for the given energy flux, $L_\odot/4\pi r^2$. Accordingly, the excess of specific entropy over the entropy of the marginal state

$$\Delta S = \int c_P (\nabla - \nabla_a) \, d\ln P \tag{5.78}$$

decreases with increasing α. As Ulrich and Rhodes (1977) point out, it is the *entropy* which physically characterizes the convective solar envelope, while α merely serves as a technical parameter.

The entropy excess ΔS can also be viewed as the entropy "jump" across the outermost layers of the convection zone, because only there $\nabla - \nabla_a$ is substantial, cf. Table 6.1. The fact that theoretical models tend to predict too large frequencies of high-degree p modes therefore means: the theory of solar convection, as outlined in Sect. 6.2, somewhat overestimates the entropy jump ΔS.

Unfortunately, α is a free parameter only for an *envelope* model, which is not subject to central boundary conditions, but not for a full solar model. As explained in Chap. 2, α is calibrated (together with the original helium abundance, Y_0) by the requirement that the model of the present Sun must have the observed radius (and luminosity). Hence we can only vary a *third* quantity, and hope that the calibration after such a variation leads to an increased α. Partial success in this direction has come from the electrostatic correction in the equation of state, see Sect. 5.3.5 below, and – to a larger extent – from more detailed opacity calculations (the OPAL tables, Sect. 2.3.7). For example, the standard model of the present text requires $\alpha = 1.81$, as compared to $\alpha = 1.38$ for the model of the 1989 edition. Accordingly its convection zone is deeper, with $r_v/r_\odot \approx 0.708$ (compared to 0.74 of the earlier model), very close to the result (5.80) of the helioseismic inversion.

In Fig. 5.17 the base of the convection zone can be recognized by a slight change of the slope of $c^2(r)$ near $r = 0.7 r_\odot$. We differentiate $c^2(r)$, given by (5.12); in order to simplify the argument we use the perfect-gas law $P = \rho \mathcal{R} T/\mu$, with $\mu = $ const. and $\Gamma_1 = 5/3$ (ionization is unimportant near the base of the convection zone). Then

$$\frac{dc^2}{dr} = \frac{5\mathcal{R}}{3\mu}\frac{dT}{dr} = -\frac{5Gm}{3r^2}\nabla \ , \tag{5.79}$$

where $\nabla = d\ln T/d\ln P$, and where the hydrostatic equilibrium condition has been used in the second relation. Since $\nabla \simeq \nabla_a \approx 0.4$ in the deep part of the convection zone, the expression $W = (r^2/Gm)\,dc^2/dr$ is approximately $-2/3$ there. In the radiative zone the absolute magnitude of W is smaller. Hence the value of r_v can be determined as the radius in the seismic solar model where $W(r)$ reaches the constant $-2/3$. In this manner Christensen-Dalsgaard et al. (1991) and Basu and Antia (1997) obtained

$$r_v/r_\odot = 0.713 \ , \tag{5.80}$$

with errors of ± 0.003 and ± 0.001, respectively. With the smaller error, this translates into $(199\,700 \pm 700)$ km for the depth of the convection zone.

5.3.4 Chemical Constitution

We consider two examples where helioseismology has furthered the knowledge of the chemical constitution of the Sun: the abundance of helium in the solar envelope, and the content of elements heavier than helium, expressed by their mass fraction Z.

It was already pointed out in Sect. 4.4.3 that the adiabatic exponent Γ_1 decreases in a layer where an abundant element is partially ionized. As the speed of sound is $\propto \Gamma_1^{1/2}$, we must consider this quantity as a function of depth, and pay particular attention to the zones of helium ionization. Figure 5.19 shows the difference in $\Gamma_1(r)$ between two standard models with slightly different helium abundances, and the change with depth of the two ionization states of helium. The model with helium diffusion in the core has a smaller helium abundance in the convection zone, and it is this model that gives the better agreement of its eigenfrequencies with the observed ones.

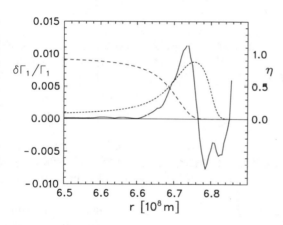

Fig. 5.19. Relative difference, $\delta\Gamma_1/\Gamma_1$, of the adiabatic exponent between solar models with and without diffusion of helium (*solid, left scale*), and the fractions η of He^+ and He^{++} particles (*dotted* resp. *dashed, right scale*) for the model with diffusion

As far as the heavy elements are concerned, helioseismology has forced us to reject the low-Z solar model described in Sect. 2.5.1. This model was designed as an attempt to remove the solar neutrino problem by virtue of its lower opacity, and therefore lower central temperature. But calculations of eigenfrequencies (Christensen-Dalsgaard and Gough, 1981) for models having interior Z values of 0.001, 0.004, and 0.02 clearly show that only the last of these three yields a satisfactory agreement with observed frequencies.

The sensitivity of low-l p modes to changes in Z is illustrated in Fig. 5.20 (the discrete sets of theoretical frequencies are connected by continuous and dashed lines merely for the sake of a clear presentation). We could conclude from this figure that still larger Z values would yield even better agreement between theory and observation. But this seems impossible: A larger Z would severely aggravate the neutrino problem, and it would contradict the spectroscopically determined abundance of heavy elements (we recall that a *smaller*

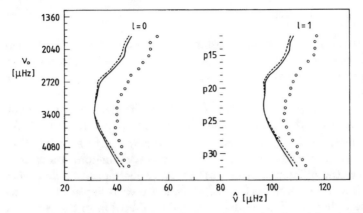

Fig. 5.20. Echelle diagram with p modes for $l = 0$ and $l = 1$. The frequencies are $\nu = \nu_0 + \hat{\nu}$. *Circles*: observations, from Jiménez et al. (1988), cf. Table 5.1; *solid* and *dashed lines*: theoretical results of Noels et al. (1984), for $Z = 0.02$ and $Z = 0.018$, respectively

Z than observed was allowed because of the possible accretion of dust at the solar surface). We must therefore find other means to repair the discrepancies of up to $\approx 10\,\mu\text{Hz}$ visible in Fig. 5.20. One is the equation of state.

5.3.5 Equation of State

In Sect. 2.3.4 an electrostatic correction to the perfect-gas law was described in the framework of the Debye–Hückel theory. The results of Noels et al. (1984), shown in Fig. 5.20, are based on a solar model which includes this effect.

The effect of the electrostatic correction upon the eigenvalue spectrum of low-degree p modes is shown in Fig. 5.21 (both calculations shown there contain the effect of partial electron degeneracy, which was also treated in Sect. 2.3.4). For the lower overtones, up to about p 23, the improvement is quite conspicuous. The higher overtones seem to be shifted towards too large frequencies; however, test calculations show that this can be repaired by modifications in the outermost layers of the solar model, e.g., a change in the atmospheric model, or a detailed account of the near-surface convection. The lower overtones are less affected by such modifications because of their deeper level of (outer) total reflection.

We conclude that the electrostatic correction is a necessity of the solar model. In fact it is now part of the standard solar model, much as the allowance for partial electron degeneracy, and the diffusion of elements in the interior.

Also at *high* degree l the electrostatic correction yields improved eigenfrequencies. As shown in Fig. 5.4, the theoretical frequencies tend to lie above

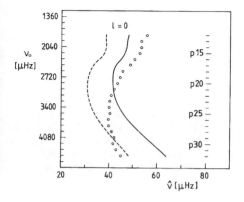

Fig. 5.21. Echelle diagram with radial ($l = 0$) p modes. *Solid* and *dashed curves*: calculated with and without the electrostatic correction in the equation of state, after Stix and Knölker (1987). *Circles*: observations of Jiménez et al. (1988)

the observed p-mode ridges. This discrepancy is considerably diminished, as shown, e.g., by Shibahashi et al. (1983) or Kaisig et al. (1984).

It must be pointed out here that, although further corrections have been incorporated, presently no solar model has eigenfrequencies that agree in all detail over the whole observed p-mode spectrum. Part of the discrepancies at low l and $n < 20$, as seen in Figs. 5.20 and 5.21, remain (see, e.g., Ulrich and Rhodes 1983), as well as some discrepancies at large l, of the kind shown in Fig. 5.4.

5.3.6 Internal Mixing

The comparison of observed and calculated oscillation frequencies *rules out the internally mixed model* of the Sun, as first noticed by Ulrich and Rhodes (1983) and confirmed in more detail by Guenther and Demarque (1997). Internal mixing evens out the gradients of the hydrogen and helium abundances and has sometimes been invoked as a means to alleviate the neutrino discrepancy. As a quantity that is most sensitive to internal mixing the small frequency separation $\delta\nu_{n,l}$, defined by (5.66), has been found. Modes with small l and large n penetrate the deepest into the Sun; hence their $\delta\nu_{n,l}$ is most significant for the physics of the core.

An example is shown in Fig. 5.22. Within the errors of order $1\,\mu\mathrm{Hz}$, there is satisfactory agreement between observations and the standard solar model. On the other hand, a model that has its inner 5% (by mass) mixed clearly has too large separations $\delta\nu_{n,1}$. Thus, the standard model seems to be confirmed, although a model of Lebreton and Maeder (1987) with mild internal mixing seems to be in tolerable agreement with the observations. We should also add that some other non-standard models yield quite consistent $\delta\nu_{n,l}$ values, e.g., a model containing weakly interacting massive particles (Faulkner et al. 1986; Däppen et al. 1986). And even the low-Z model does not substantially deviate. Thus, we cannot with certainty adopt a solar model solely on account of its $\delta\nu_{n,l}$ results.

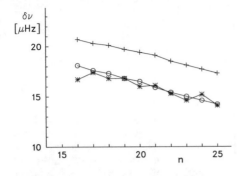

Fig. 5.22. Frequency separation $\delta\nu_{n,1}$ for $n = 16$ to 25: standard solar model (*circles*), observational results (*asterisks*; Libbrecht et al. 1990), and a model with a mixed core (*crosses*)

5.3.7 Precise Determination of the Solar Radius

The lowest ridge in the diagnostic diagram of solar oscillations is the f mode. This mode has no nodes in the radial direction. It resembles a surface wave on deep water, with dispersion according to (5.37), or $\nu = \sqrt{gk_\mathrm{h}}/2\pi$. The horizontal wave number is asymptotically related to the spherical harmonic degree l through $k_\mathrm{h} \simeq \sqrt{l(l+1)}/r$. Because of the simple dispersion relation the f-mode frequencies are essentially independent of the internal structure of the Sun, except for the radius. The mass variable $m(r)$ introduced in Chap. 2 varies very little in the observable layers, so we may use $g = Gm_\odot/r^2$ and see that $\nu \propto r^{-3/2}$. Since the frequencies can be determined with high precision, and since the product $Gm_\odot$ is known from the dynamics of the planetary system, the radius can be determined (Schou et al. 1997, Antia 1998).

In principle such a radius determination could be more precise than the common procedure where the apparent angular diameter is measured and the radius is calculated from the Sun's distance (Sect. 1.3). The problem with the f-mode method is only that the depth in the solar atmosphere that marks the radius obtained in this way must be defined precisely. It has been practical not to use the dispersion relation but instead to calculate solar models and their f-mode eigenfrequencies for different model radii. If the model radius is fixed at optical depth $\tau = 2/3$, the result is, with a conservative error estimate, $r_{\tau=2/3} = (6.957 \pm 0.001) \times 10^8$ m.

5.3.8 Internal Rotation

In spherical polar coordinates the Sun's differential rotation is represented by a velocity field

$$\boldsymbol{v}_0 = (0, 0, r\Omega\sin\theta) \equiv \boldsymbol{\Omega} \times \boldsymbol{r} , \tag{5.81}$$

where, in general, the angular velocity Ω is a function of r and θ, and

$$\boldsymbol{\Omega} = (\Omega\cos\theta, -\Omega\sin\theta, 0) . \tag{5.82}$$

Because Ω is not a constant, there is no advantage in choosing a rotating frame of reference; hence we stay in the inertial frame.

The necessary generalization of (5.70) is readily obtained from the original equation of motion. The inertia term, $\rho_0 d\boldsymbol{v}/dt$, which led to the term $\rho_0 \partial^2 \boldsymbol{\xi}/\partial t^2$ in (5.14) when there was no mean motion, now gives

$$\rho_0 \left(\frac{\partial}{\partial t} + \boldsymbol{v}_0 \cdot \nabla \right)^2 \boldsymbol{\xi} \ . \tag{5.83}$$

We consider the additional term as a small perturbation. We again set $\partial/\partial t = i\omega$, and may expect that ω differs only by a small amount from ω_α, an eigenfrequency of the unperturbed (non-rotating) oscillator:

$$\omega = \omega_\alpha + \Delta\omega_\alpha \ . \tag{5.84}$$

The index α stands for the pair of indices, l and n, which characterizes the eigenvalue, or for the triplet l, n, m whenever an additional m-dependence occurs (such as in $\Delta\omega_\alpha$), or for the pair l and m that characterizes the spherical harmonic function $Y_\alpha \equiv Y_l^m$.

Expanding (5.83), and retaining the perturbation only to first order, we obtain in place of (5.14):

$$\rho_0(\omega_\alpha^2 + 2\omega_\alpha\Delta\omega_\alpha)\boldsymbol{\xi} = \nabla P_1 - \frac{\rho_1}{\rho_0}\nabla P_0 + \rho_0\nabla\Phi_1 + 2i\omega_\alpha\rho_0(\boldsymbol{v}_0 \cdot \nabla)\boldsymbol{\xi} \ . \tag{5.85}$$

Scalar multiplication by $\boldsymbol{\xi}^*$, and integration over the solar volume, gives

$$(\omega_\alpha^2 + 2\omega_\alpha\Delta\omega_\alpha) \int \rho_0\boldsymbol{\xi}^* \cdot \boldsymbol{\xi} \, dV = \int \boldsymbol{\xi}^* \cdot \mathcal{L}\boldsymbol{\xi} \, dV$$

$$+ 2i\omega_\alpha \int \rho_0\boldsymbol{\xi}^* \cdot (\boldsymbol{v}_0 \cdot \nabla)\boldsymbol{\xi} \, dV \ . \tag{5.86}$$

Now if ω is close to ω_α, an eigenvalue, then $\boldsymbol{\xi}$ is close to $\boldsymbol{\xi}_\alpha$, the corresponding eigenvector. We may write $\boldsymbol{\xi} = \boldsymbol{\xi}_\alpha + \Delta\boldsymbol{\xi}$, but $\Delta\boldsymbol{\xi}$ is of no concern: thanks to the extremal character of the eigenvalues, first-order perturbations do not arise in the first term on the right of (5.86), and the second term is already of first order due to its $\boldsymbol{v}_0$-dependence. Using (5.70) we therefore find

$$\Delta\omega_\alpha = \frac{i \int \rho_0\boldsymbol{\xi}_\alpha^* \cdot (\boldsymbol{v}_0 \cdot \nabla)\boldsymbol{\xi}_\alpha \, dV}{\int \rho_0\boldsymbol{\xi}_\alpha^* \cdot \boldsymbol{\xi}_\alpha \, dV} \ . \tag{5.87}$$

Problem 5.16. Use (5.81) to show that (5.87) is equivalent to

$$\Delta\omega_\alpha = \frac{-m \int \rho_0\Omega\boldsymbol{\xi}_\alpha^* \cdot \boldsymbol{\xi}_\alpha \, dV + i \int \rho_0\boldsymbol{\xi}_\alpha^* \cdot (\boldsymbol{\Omega} \times \boldsymbol{\xi}_\alpha) \, dV}{\int \rho_0\boldsymbol{\xi}_\alpha^* \cdot \boldsymbol{\xi}_\alpha \, dV} \ , \tag{5.88}$$

where m is the longitudinal order of the eigenfunction. Show that the second term in the numerator also has a factor m, and that $\Delta\omega_\alpha$ is real.

The inversion problem for the angular velocity is defined by (5.88): we must find $\Omega(r, \theta)$ from the measured frequency shifts $\Delta\omega_\alpha$. This time the problem is *linear* in Ω, a consequence of our treatment of the rotational effect by perturbation theory. Thus, the shift $\Delta\omega_\alpha$ is of the same order as Ω itself. According to (5.88) there is no shift for $m = 0$; hence there is no shift for $l = 0$. But the shift is not necessarily linear in m.

For the evaluation of (5.88) we must know the eigenfunctions ξ_α of the unperturbed state. Thus, the assumption is that we possess a (non-rotating) solar model and may directly calculate its eigenfunctions. The inverse problem relates merely to $\Omega(r, \theta)$.

Radial Shear. Let us first consider the case where Ω depends only on the radial coordinate. We substitute (5.19) into (5.88) and obtain, after a few integrations by parts

$$\Delta\omega_\alpha = -m\frac{\int_0^{r_\odot} \rho_0 \Omega\{|\xi_r - \xi_h|^2 + [l(l+1) - 2]|\xi_h|^2\}r^2\,dr}{\int_0^{r_\odot} \rho_0[|\xi_r|^2 + l(l+1)|\xi_h|^2]r^2\,dr}$$

$$\equiv \int_0^{r_\odot} K_\alpha(r)\Omega(r)\,dr \tag{5.89}$$

(Gough, 1981). The rotational splitting kernel, $K_\alpha(r)$, depends on the eigenfunction with index α (which was omitted for brevity). Therefore, any given $\Delta\omega_\alpha$ samples the angular velocity in the depth range corresponding to $\boldsymbol{\xi}_\alpha$.

The integrals appearing in (5.89) are independent of m. Hence, in the case of purely radial shear, $\Delta\omega_\alpha$ is strictly linear in m. Since $-l \le m \le l$, we obtain a multiplet of $2l + 1$ frequencies, with equidistant spacing.

Problem 5.17. Write (5.89) in the form

$$\Delta\omega_\alpha = -m\Omega_0(1 - C)\ , \tag{5.90}$$

where $\Omega_0 = \Omega(r_\odot)$, and calculate C. Give an interpretation of the term $-m\Omega_0$. How long a data string is needed to resolve the rotational splitting between the frequencies of two adjacent m values if, say, $C = 0.5$?

The modes with $m = \pm l$ have the largest frequency shifts. These modes have a *sectoral* pattern, as shown in Fig. 5.5 for the case $m = l = 10$. Observationally, the sectoral modes have been isolated by Duvall and Harvey (1984) by means of the same cylindrical lens that had been used to isolate the zonal ($m = 0$) modes, cf. Sect. 5.1.2: here the lens is oriented parallel to the solar axis of rotation, so that the image is optically averaged in north-south direction. The entrance slit to the spectrograph lies parallel to

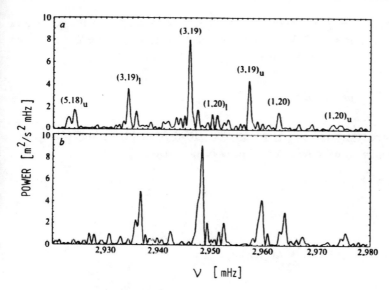

Fig. 5.23. Power spectra for p modes with $l = 3$, $m = +3$ (**a**) and $m = -3$ (**b**). The labels are (l, n); subscripts u and l mark the *upper* and *lower* side lobes arising from night gaps in the data (Duvall and Harvey 1984). Courtesy National Solar Observatory and NASA

the Sun's equator, and a sequence of north-south averages is measured along the slit. The sectoral modes with $m = -l$ and $m = +l$ represent prograde and retrograde waves propagating in longitude.

The spherical harmonic analysis of Duvall and Harvey's data, collected over 17 days at Kitt Peak, clearly shows the different rotational shifts of the two sectoral modes, cf. Fig. 5.23. We use again ν instead of ω, and – in order to compensate for the explicit l-dependence – divide the difference between the two shifts by $2l$:

$$\nu'_\alpha \equiv \frac{1}{4\pi l}(\Delta\omega_{\alpha, m=-l} - \Delta\omega_{\alpha, m=+l}) \ . \tag{5.91}$$

For rigid rotation ν'_α would be the rotation frequency, $\nu_{\rm rot}$ [up to a factor $1 - C$, cf. (5.90); but generally $C \ll 1$ in this case (Brown et al. (1986)]. The actual values of ν'_α vary between $\approx 360\,{\rm nHz}$ and $\approx 480\,{\rm nHz}$, depending on the depth ranges covered by the diverse eigenfunctions.

Problem 5.18. Show that the amplitudes of sectoral surface harmonics are very small except in a band along the equator, of width $l^{-1/2}$ in latitude. Sectoral modes with large l therefore essentially sample the *equatorial* frequency, $\nu_{\rm rot}(r, \pi/2)$, of solar rotation.

A particularly elegant method of inversion, originally used in geophysics (Backus and Gilbert, 1970), employs *optimal kernels*. These are derived as follows (Duvall et al. 1984). We write

$$\nu'_\alpha = \int_0^{r_\odot} K'_\alpha(r)\nu_{\rm rot}(r)\,dr\;,\tag{5.92}$$

where $\nu_{\rm rot} = \Omega/2\pi$, and K'_α is readily obtained from (5.89) and (5.91). For any fixed depth r_0 we now take a linear combination

$$D(r,r_0) = \sum_\alpha a_\alpha(r_0)K'_\alpha(r)\;.\tag{5.93}$$

Hence, from (5.92)

$$\int_0^{r_\odot} D(r,r_0)\nu_{\rm rot}(r)\,dr = \sum_\alpha a_\alpha(r_0)\nu'_\alpha\;.\tag{5.94}$$

The task is to find coefficients a_α which concentrate the function D of r around $r = r_0$, i.e., to make D resemble $\delta(r - r_0)$. In this case

$$\nu_{\rm rot}(r_0) \simeq \sum_\alpha a_\alpha(r_0)\nu'_\alpha\;.\tag{5.95}$$

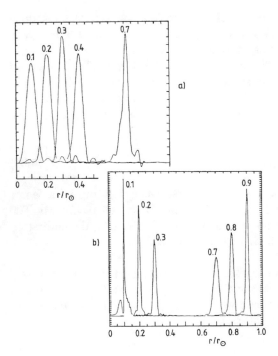

Fig. 5.24. Optimal kernels for rotational splitting: (**a**) Combinations of 140 kernels of p modes with $l = 1$ to 60; (**b**) with additional kernels of 60 g modes with $l = 1$ to 4. The labels give the maximum position. After Noyes and Rhodes (1984)

Optimal kernels for the internal solar rotation have indeed been calculated; examples are shown in Fig. 5.24. In the outer part of the Sun, the optimal kernels are predominantly combinations of p-mode kernels. In the inner part, p modes of low degree are of some value, but the inclusion of g-mode kernels would considerably improve the depth resolution.

Unfortunately the linear combinations of observed splittings on the right of (5.95) involve much cancellation, and hence an increase of relative errors. Thus, there is a tradeoff between resolution in depth and attainable accuracy.

Latitudinal Shear. So far we have treated the case of a purely depth-dependent angular velocity. This is not necessarily the solar case; in fact at the solar surface latitudinal differential rotation has long been known from the proper motion of sunspots. What has been investigated observationally is – by the selection of the two sectoral modes – the rotation frequency as a function of depth, but only at *low latitude* (Problem 5.18).

For an investigation of the full function $\Omega(r, \theta)$ the whole multiplet of $2l+1$ frequencies must be used. In particular the *deviation from equidistant spacing* within the multiplets is typical for latitudinal shear. A possible procedure is to substitute a certain form of $\Omega(r, \theta)$ into the general expression (5.88), and compare the result with observed $\Delta\omega_\alpha$ values. However, the deviation from equidistant splitting can be described by a few parameters, e.g., the coefficients a_i in the expansion

$$\Delta\nu_\alpha = L \sum_i a_i P_i(-m/L) \tag{5.96}$$

(Duvall et al. 1986). Here $\Delta\nu_\alpha = \Delta\omega_\alpha/2\pi$, P_i is the ith Legendre polynomial, and $L = [l(l+1)]^{1/2}$. An expansion in powers of m would be equivalent, but (5.96) is preferred because the P_i are orthogonal, and the coefficients a_i, therefore, independent (actually Clebsch-Gordon coefficients are used because orthogonality is required for a discrete data set, cf. Ritzwoller and Lavely 1991). As it turns out the argument m/L takes care of most of the

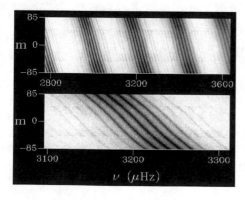

Fig. 5.25. Power of solar p modes with degree $l = 85$, as function of frequency ν and azimuthal order m: modes p 6 to p 10 (*upper panel*), and magnified mode p 8 (*lower panel*). Each mode has a number of side bands arising from adjacent values of l, as the spherical harmonics are not orthogonal over the observed portion of the solar surface. GONG data, 23 Aug 1995 to 18 Feb 1996. Courtesy F. Hill, National Solar Observatory, Tucson

l-dependence. Hence it is sufficient to calculate averages of the a_i over certain ranges of l. The normalization of (5.96) is such that for uniform rotation a_1 is the rotational frequency, ν_{rot} (again up to a factor $1 - C$).

An example of the power distribution in the ν, m-plane is shown in Fig. 5.25. Although the signals from several adjacent degrees l are superposed, one can recognize the curvature in the power ridges that is due to the a_3 term of (5.96). For the differential rotation observed at the solar surface one finds $a_1 \approx 440\,\mathrm{nHz}$ and $a_3 \approx 21\,\mathrm{nHz}$ (sidereal rates). There is no significant variation of a_1 and a_3 for degrees l between 20 and 100. The conclusion is that in the outer part of the Sun (cf. Fig. 5.14) the latitudinal differential rotation is essentially the same as that of the surface. In Fig. 7.4 the angular velocity is shown as a function of radius for three heliographic latitudes.

Problem 5.19. Assume that the angular velocity $\Omega(r, \theta)$ is constant on cylinders, i.e., $\Omega = \Omega(s)$ where $s = r \sin \theta$. In particular, use

$$\Omega = \Omega_p + (\Omega_e - \Omega_p)(s/r_\odot)^2 \,, \tag{5.97}$$

where Ω_p and Ω_e are the polar and equatorial angular velocities at the surface (cf. Chap. 7). Substitute (5.97) into (5.88) and perform the angular integrations. Show that (5.96) has only terms with odd index i (cf. Hansen et al. 1977).

5.3.9 Travel Time and Acoustic Imaging

In addition to the amplitude, the *phase* of acoustic signals is measured at the solar surface. Following a first proposal of Roddier (1975), various techniques employing the phase information have been developed, named *tomography*, *holography*, or *acoustic imaging*.

The Time-Distance Method. Time-distance seismology has been used traditionally in geophysics. The application to the Sun was first proposed by Duvall et al. (1993 b); in the present section I follow the work of Kosovichev (1996) and Kosovichev et al. (2000).

The travel time of an acoustic wave on the Sun can be measured by means of the covariance function

$$\Psi(\tau, \Delta) = \int f(r_1, t) f^*(r_2, t + \tau)\, dt \,, \tag{5.98}$$

where Δ is the distance between two points r_1 and r_2 on the solar surface, $f(r_i, t)$ is an oscillation signal at point r_i, and τ is the time delay; the integral is taken over the time of observation. The signal f can be the velocity or the intensity variation. For a given Δ the covariance function attains maxima for certain delay times τ, namely for those that correspond to the travel times of the signal from r_1 to r_2 (or vice versa) via acoustic rays with 1, 2, ... reflections at the base of their respective cavities. Figure 5.15 shows examples of such rays, and Fig. 5.14 gives the depth of reflection as a function of degree l and frequency ν.

The travel time of an acoustic wave varies due to the variation of the sound speed $c(\boldsymbol{r})$ along the path, and due to local flows with velocity $\boldsymbol{v}(\boldsymbol{r})$:

$$\tau = \int \frac{ds}{c(\boldsymbol{r}) + \boldsymbol{v}(\boldsymbol{r}) \cdot \boldsymbol{n}(\boldsymbol{r})} , \tag{5.99}$$

where s is a coordinate along the path from $\boldsymbol{r}_1$ to $\boldsymbol{r}_2$, and $\boldsymbol{n}(\boldsymbol{r})$ is a unit vector tangent to the path. We write $c(\boldsymbol{r}) = c_0(\boldsymbol{r}) + \delta c(\boldsymbol{r})$ and consider the case where both $\delta c(\boldsymbol{r})$ and $\boldsymbol{v}(\boldsymbol{r}) \cdot \boldsymbol{n}(\boldsymbol{r})$ are small in comparison to $c_0(\boldsymbol{r})$, the sound speed of a horizontally uniform reference state. Then

$$\tau = \int \frac{ds}{c_0(\boldsymbol{r})} - \int \frac{\delta c(\boldsymbol{r}) + \boldsymbol{v}(\boldsymbol{r}) \cdot \boldsymbol{n}(\boldsymbol{r})}{c_0^2(\boldsymbol{r})} \, ds . \tag{5.100}$$

Here we can take the integral along the path that would be traveled in the reference state because, according to Fermat's principle, τ is an extremum, and first-order perturbations of the path do not contribute to the variation of τ. Equation (5.100) is an inversion problem: we want to know the functions $\delta c(\boldsymbol{r})$ and $\boldsymbol{v}(\boldsymbol{r})$ that appear in the integrand. This time there is no explicit parameter dependence. Nevertheless a solution is possible because a large number of travel times and wave paths are evaluated at the same time.

As first pointed out by Gough and Toomre (1983), the effects of the sound-speed and flow variations can be separated by measuring the travel times τ^+ and τ^- of signals traveling in opposite directions between two points. For, these two travel times respond in the same way to the variation δc, but in opposite ways to the local flow $\boldsymbol{v}$. Thus we have two inversion problems:

$$\frac{1}{2}(\tau^+ + \tau^-) = -\int \frac{\delta c(\boldsymbol{r})}{c_0^2(\boldsymbol{r})} \, ds \tag{5.101}$$

and

$$\frac{1}{2}(\tau^+ - \tau^-) = -\int \frac{\boldsymbol{v}(\boldsymbol{r}) \cdot \boldsymbol{n}(\boldsymbol{r})}{c_0^2(\boldsymbol{r})} \, ds . \tag{5.102}$$

The time-distance technique has been applied to the supergranulation, which is seen at the solar surface as a network of convective cells of 20 000–30 000 km extent. The result (Fig. 5.26) indicates that the flow structure continues several thousand km below the surface, and that associated temperature variations of up to 2%, positive in the upflow and negative in the downflow regions, confirm the convective nature of the supergranulation (cf. Sect. 6.5.2).

The temperature variation deduced for the supergranular pattern rests on the relation $\delta T/T \simeq 2\delta c/c$. In general, however, the sound speed may be perturbed additionally by a magnetic field. This complicates the inversion, but also opens the possibility of inferring the magnetic field underneath the Sun's surface.

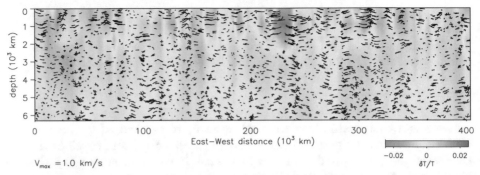

depth (10³ km)

0

100   200   300   400

East–West distance (10³ km)

V$_{max}$ =1.0 km/s

−0.02   0   0.02

δT/T

Fig. 5.26. A vertical cut showing the supergranular flow and the associated temperature variation below the Sun's surface (the vertical scale is stretched by a factor of ≈ 16). Time-distance inversion of data obtained with the Michelson Doppler Imager on SOHO. After Duvall et al. (1997)

Construction of an Acoustic Image. The acoustic signal at a point r_1 can be reconstructed from the signals at other points on the solar surface. The target point may be within the observed area, but also outside that area. Signals at points below the surface, and even on the far side of the Sun, can be reconstructed.

As an example we consider the case where measurements are made in a circular area around the target point (or around its epicenter if the target is below the surface); this case has been treated by Chou et al. (1999). We may then calculate

$$f(\boldsymbol{r}_1, t) = \sum w(\Delta)\langle f(\Delta, t + \tau)\rangle \,, \tag{5.103}$$

where the angular brackets denote an azimuthal average over an annulus with radius Δ, and where the sum is taken over all annuli within the observed area (the "aperture"), with weighting factors $w(\Delta)$ that express the change of amplitude during the transit of each contributing wave along its path. Each contribution is taken at time $t + \tau(\Delta)$, i.e., retarded by a travel time that can be determined by a correlation analysis if $\boldsymbol{r}_1$ is on the visible surface, but otherwise must be calculated from a solar model and acoustic ray theory. According to (5.40), modes with equal l/ν have approximately the same depth of internal reflection, and hence approximately the same travel time for a given distance Δ.

The acoustic signal can be reconstructed at all points within a given section of a plane or of a volume. Using data from the TON network, Chou et al. (1999) derived an acoustic image of an active region extending several ten thousand kilometers below the surface. With measurements of MDI on SOHO, Lindsey and Braun (2000) were able to image an active region on the far side of the Sun; for this purpose they used acoustic signals having one additional reflection from the solar surface between the points of measurement and the target.

5.4 Excitation and Damping

So far we have only considered eigenfrequencies and eigenfunctions of solar oscillations, but not the *amplitude* of the eigenfunctions. Also, nothing has yet been said about the *excitation* of solar oscillations. These subjects will be discussed briefly in the present section.

5.4.1 The κ Mechanism

We first mention the κ mechanism, which has been identified as the driving agent of stellar pulsations, although it is probably less important in the solar case. The κ mechanism acts like a valve: suppose that in a phase of compression the opacity increases. The compressed layer then absorbs energy out of the radiative flux toward the stellar surface, and thus will be heated in excess of the mere adiabatic heating. The subsequent expansion will be stronger than the preceding one was. Zhevakin (1953) demonstrated that this mechanism of *overstability* drives the pulsation of δ Cephei and related variable stars, where it is particularly effective in the layer of the second helium ionization.

The crucial parameter measuring the opacity variation is

$$\kappa_T \equiv \left(\frac{\partial \ln \kappa}{\partial \ln T} \right)_P \tag{5.104}$$

and is shown in Fig. 2.8. It has a high maximum in the layer of partial hydrogen ionization. In this layer there is strong driving, but we must include the contributions from *all* layers in order to see whether a particular mode is excited or damped. Indeed Ando and Osaki (1975) found positive growth rates for a wide range of solar oscillations with periods around 5 min. On the other hand there is not only excitation, but also *damping*, by radiative losses in the optically thin atmosphere and by interaction with the instationary convective motion. Hence the overstabilities could not be confirmed by subsequent studies of Goldreich and Keeley (1977 a) and Balmforth (1992).

There is another reason why the excitation of solar p modes by means of the κ mechanism appears unlikely. The excited (or damped) oscillator is symbolically represented by an equation of the form

$$\ddot{\xi} - 2\gamma\dot{\xi} + \omega_0^2 \xi = 0 , \tag{5.105}$$

where the net effect of all excitation and damping yields the coefficient γ. Now suppose excitation wins for a particular mode, that is, γ is positive. Then there is unlimited growth of this mode, because (5.105) is homogeneous and linear. It is only by means of the non-linear terms, which have been neglected in (5.105) as well as in our full system of oscillation equations, that the growth could be held. But before these terms can take effect the amplitude should be sizable, unlike the small amplitudes observed on the Sun. Cepheids are different: their pulsation amplitude, of order $\delta r/r \approx 0.1$ or larger, is indeed limited by non-linear effects.

5.4.2 Stochastic Excitation by Convection

In the preceding section we have mentioned the interaction with convection as a source of damping for the oscillatory motions. The idea was that parcels of gas moving back and forth during their convective motion provide a kind of friction, much like the atoms of a gas by their thermal motions and collisions.

But the convective motion that acts as friction at the same time provides excitation. Envisage the gas in the cavity between the two reflecting boundaries as a resonator which is continuously hit by convectively driven parcels of gas ("pistons"). A bell struck in a random way, or a trumpet excited with a random spectrum, and with random phase jumps, would be an analogue. Formally we may describe such an oscillator by an equation of type (5.105), except that the right-hand side is not zero, but a given (stochastic) function of frequency, the *forcing* function. Lighthill (1952) first treated the excitation of acoustic oscillations by a turbulent flow in this manner when he studied the generation of noise by jet aircraft. Kumar and Lu (1991) have advanced a simple model where the forcing is concentrated to a single surface, at $r = r_0$. Although the excitation is not stochastic in this model, the model illustrates some aspects of the solar oscillations, and we shall therefore discuss it here (with minor changes). We use the local analysis of Sect. 5.2.4 and set $l = 0$, although the case $l \neq 0$ can be treated as well. With $\xi_r = \psi \rho_0^{-1/2}$ the oscillation equations, supplemented by a forcing term, yield

$$c^2 \frac{\partial^2 \psi}{\partial r^2} - \frac{\partial^2 \psi}{\partial t^2} - V(r)\psi = f(r,t) \; , \tag{5.106}$$

where the "potential" V is given by the square, ω_A^2, of the acoustic cutoff frequency. We write V as a function of r because at $r = a$ the model is divided into two domains, 0 and 1, with potentials V_0 and V_1, as illustrated in Fig. 5.27. Domain 0 represents the hot solar interior, while domain 1 represents the relatively cool atmosphere which confines waves with frequencies below ω_A, but allows waves with higher frequencies to escape. In contrast to Eq. (5.105), there is no damping term in (5.106); the more essential difference, however, is the forcing term. Clearly, this forced, *inhomogeneous* oscillator is not subject to the difficulty of unlimited growth, encountered by the homogeneous oscillator driven by the κ mechanism. The spectrum, including the amplitudes, of the excited modes is entirely determined by the forcing function, which in the present model is given by a delta function,

$$f(r,t) = f_0 \delta(r - r_0) \exp(\mathrm{i}\omega t) \; . \tag{5.107}$$

Since the coefficients of Eq. (5.106) are constant in each domain, the solutions can be written in terms of trigonometric functions. We set $V = \infty$ for $r \leq 0$, so that $\psi(0) = 0$ is a boundary condition; we also require that there is no wave energy incoming from large r. In addition, ψ must be continuous at $r = a$ and $r = r_0$, while $\partial\psi/\partial r$ is continuous at $r = a$ and jumps by f_0/c_0^2 at $r = r_0$. With these conditions we obtain in domain 1 ($r \geq a$) the solution

$$\psi = - \left(\frac{f_0 \sin k_0 r_0}{c_0^2 (k_0 \cos k_0 a + i k_1 \sin k_0 a)} \right) \exp[i\omega t - i k_1 (r - a)] . \tag{5.108}$$

where $k_i = \sqrt{\omega^2 - V_i}/c_i$ for $i = 0, 1$. Figure 5.27 shows the power $\psi\psi^*$ as a function of cycle frequency ν, at $r = a$. The parameters are chosen so that the atmospheric cutoff is at $5\,\mathrm{mHz}$, as for the Sun, but otherwise the model is too crude for a detailed comparison. The essential result is the occurrence of resonances. Below the atmospheric cutoff ν_A, resonances occur at the eigenfrequencies of the interior (domain 0); they coincide with the zeros of the denominator of (5.108) and depend on the size a of the cavity. For frequencies larger than ν_A, k_1 is real; hence there are no further poles, although the denominator oscillates. In this case the dependence of the resonances on a decreases with increasing ν; at large ν they are determined by the source position r_0 alone. The resonances above ν_A have a finite amplitude. Of course, in a real system *all* resonances would be finite because of damping mechanisms such as non-adiabatic behavior.

The resonances above the atmospheric cutoff have their solar analogue in the high-frequency (above $\approx 5\,\mathrm{mHz}$) part of the ridges in the l, ν-diagram, as first noticed by Duvall et al. (1991), and seen in Figs. 5.6 and 5.7. Although the power rapidly decreases towards larger ν, those ridges apparently continue in a smooth fashion from low to high frequency. In the model this is achieved by choosing the source surface, $r = r_0$, very close to the surface of the cavity, $r = a$. Resonances below ν_A arise from the interference of waves reflected at $r = 0$ and $r = a$; those above ν_A are formed as waves excited at $r = r_0$ and reflected at $r = 0$ interfere. If $r_0 \simeq a$, the spacing of the peaks is about the same. Indeed the convective velocity is most vigorous close to the surface, exceeding $2\,\mathrm{km/s}$ in the mixing-length model of Table 6.1 at a depth of the order $500\,\mathrm{km}$. Comparing the results of the model to the observed spectrum

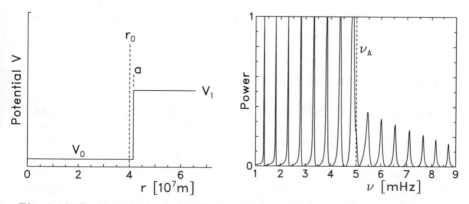

Fig. 5.27. Excitation of solar oscillations in a model similar to that of Kumar and Lu (1991). *Left*: Potential $V(r)$, with step at $r = a$, and point source at $r = r_0$; *right*: power at $r = a$ as a function of frequency, with resonances below and above the acoustic cutoff ν_A (the maxima below $\nu = \nu_A$ have been cut at finite hight)

Kumar (1994) concluded that the solar oscillations are excited in a layer only about 100–200 km below the surface.

A close inspection of Fig. 5.27 shows a slight asymmetry of the peaks in the calculated power spectrum. This asymmetry is caused by the diverse interferences of the excited and reflected waves; it was also noticed by Gabriel (1992), Nigam and Kosovichev (1998), and Rosenthal (1998), and contributes to the line asymmetry first discovered by Duvall et al. (1993 a). An additional contribution comes from the coupling of p modes with nearly equal frequencies by a large-scale convective flow (Roth and Stix 1999).

Although the average depth of the acoustic excitation may be several 100 km below the surface, individual sources of acoustic waves have been localized at the surface itself. These sources lie preferentially in the dark narrow downstream regions in between the bright granules (Brown 1991, Rimmele et al. 1995, Espagnet et al. 1996, Kiefer et al. 2000 b, Khomenko et al. 2001). In general the velocity in these downstream regions is higher than the velocity within the granules. It appears that the local sources, also called "acoustic events", feed sufficient energy into the solar oscillations as to compensate for the energy loss by interaction with small-scale motion or by leakage (Goode et al. 1998, Nigam and Kosovichev 1999, Strous et al. 2000).

We may ask why acoustic oscillations have not yet been observed below, say, 1 mHz, although p modes with still lower frequencies are possible according to theory. Here we must distinguish between the high-l and low-l cases. At high l, there are no eigenmodes at smaller frequencies, i.e., we see the smallest possible radial order n, including the fundamental, $n = 0$. At low l, the oscillations of low frequency have their upper reflection at a level that lies so deep below the photosphere that in the observable layer the amplitude has become too small to be detected with present methods (we recall that all observed oscillation signals emerge from the region of evanescent waves, i.e., from within the "tunnel"). The lowest mode of radial pulsation, with a period of slightly more than 1 hour, and a number of other low-l, low-n modes very probably are not yet detected for this reason.

5.5 Bibliographical Notes

Introductory texts to the field of solar and stellar oscillations are the books of Cox (1980) and Unno et al. (1989), or the early review of Ledoux and Walraven (1958). Basic concepts such as Eulerian and Lagrangian variables, linear perturbation theory, or the variational principle, are described in these publications. Detailed introductions to the subject, in particular to helioseismology, are also provided by Christensen-Dalsgaard (1998) and Christensen-Dalsgaard et al. (2000). Noyes and Rhodes (1984) and Stix (2000 b) give quick surveys.

The use of the diagnostic diagram (the k_{h}, ω-plane), the derivation of phase and group velocities for acoustic and internal gravity waves, and other results of the local analysis are described in the classic treatise of Eckart (1960) and by Moore and Spiegel (1964). The properties of internal gravity waves are also discussed in Landau and Lifschitz (1966 b, p. 49).

Much of the material on helioseismology is collected in conference proceedings, such as those edited by Hill and Dziembowski (1980), Gough (1983 a, 1986), Belvedere and Paternò (1984), Noels and Gabriel (1984), Ulrich (1984), and Christensen-Dalsgaard and Frandsen (1988). Some of these conferences are built around the results of major experiments like GONG and SOHO (Brown 1993, Ulrich et al. 1995, Hoeksema et al. 1995, Korzennik and Wilson 1998, Wilson 2001), others concentrate on special topics such as the convection-oscillation interaction (Pijpers et al. 1997) or the helioseismic diagnostics of convection and activity [Vols. 192 and 193 (2000) of *Solar Physics*].

Duvall et al. (1988) and Libbrecht and Kaufman (1988) present tables of observed oscillation frequencies which are more exhaustive than Table 5.1. Frequencies and mode spectra, as well as beautiful illustrations of various results, can also be found on the Internet pages of the mentioned projects.

Helioseismic results concerning the standard solar model – equation of state, convection-zone depth, diffusion of elements, etc. – have been obtained also by Kim et al. (1991), Christensen-Dalsgaard and Däppen (1992), Berthomieu et al. (1993), Christensen-Dalsgaard et al. (1993), Guzik and Cox (1993), and Guenther et al. (1996).

Lynden-Bell and Ostriker (1967) derived the variational principle for the case of rotation, a field treated also in the monograph of Tassoul (1978). Sobouti (1980) calculates eigenfrequencies of rotating stars by an expansion of the eigenfunctions in terms of those of the non-rotating star; Clement (1986) employs the variational principle to calculate the eigenfrequencies. A general approach, based on perturbation theory, to treat the effects of rotation and of a large-scale flow has been developed by Ritzwoller and Lavely (1991) and Lavely and Ritzwoller (1992, 1993). Schou et al. (1998) discuss the inversion of the rotational frequency splitting, and show results based on data from SOHO/MDI.

In addition to the permanent excitation by turbulent convection, flares and coronal mass ejections (Chap. 9) can lead to occasional seismic events on the Sun. Kosovichev and Zharkova (1998) describe a case where propagating circular wave packets were observed through their Doppler effect for 35 minutes after a flare.

Damping of solar oscillations by the turbulent velocity field of the solar convection zone has been discussed in terms of viscous damping (Goldreich and Keeley 1977 ab, Stix et al. 1993). Amplitudes and damping rates of stochastically excited solar and stellar oscillations have been estimated by Houdek et al. (1999). In addition to the damping effect, the turbulence also slightly shifts the p-mode frequencies as it changes the mean structure near the surface of the solar model (Christensen-Dalsgaard and Thompson

1997). Especially a contribution to the mean pressure caused by the turbulence induces a frequency shift (Rüdiger et al. 1997, Baturin and Mironova 1998, Rosenthal et al. 1999). Besides the structural effect, there are direct effects of the fluctuating convective velocity and sound speed on the p-mode frequencies (Brown 1984, Zhugzhda and Stix 1994, Stix and Zhugzhda 1998). In this case the frequency change occurs in combination with a modulation of the oscillation amplitude and phase (Stix 2000 a); such effects are visible when the oscillatory velocity field is studied simultaneously with the granular or mesogranular flow (Hoekzema et al. 1998, Hoekzema and Brandt 2000). If all the effects of solar convection are incorporated in a correct way, the small discrepancies between the observed and calculated oscillation frequencies, as mentioned in Sect. 5.3.5, probably will be diminished. Three-dimensional numerical simulation of convection confirms this expectation, and also provides hints on the spectrum of the excited modes (Stein and Nordlund 1998 b).

Besides the κ mechanism, another instability has been thought to drive solar oscillations: the "ε mechanism", consisting in an amplified energy production during the phase of maximum compression (an analogue to the diesel engine). The mechanism would operate in the region of ^{3}He accumulation (Fig. 2.11) and lead to growing perturbations because of the strong temperature sensitivity of the ^{3}He(^{3}He,2p)α reaction in the ppI chain. The g modes, having their peak amplitude in the deep core, would most likely be excited in this way. Dilke and Gough (1972), Christensen-Dalsgaard et al. (1974) and Shibahashi et al. (1975) found unstable g modes of low degree and argued that the instability, growing to finite amplitude, would smooth out the ^{3}He peak, so destroy its own basis, and therefore occur in an *intermittent* manner, with a time scale of $\approx 10^8$ years. It must however be said that Dziembowski and Sinkiewicz (1973) could *not* find unstable g modes, and Christensen-Dalsgaard and Gough (1975) pointed out that the damping in the outermost layers just beneath the photosphere strongly opposes, if not suppresses, the instability.

Several claims have been made at the identification of discrete g modes in solar Doppler or intensity signals (e.g., Delache and Scherrer 1983; Fröhlich 1988). In view of the upper limits determined by Appourchaux et al. (2000) such claims appear insubstantial. Since the spectrum of g modes is dense, a unique identification by means of the asymptotic formula (5.68) is difficult. Of course, it is an open question whether g modes have detectable amplitudes in the atmosphere: the convection zone constitutes a long tunnel, with much attenuation indeed! On the other hand, a number of studies confirm the existence of internal gravity waves in the solar atmosphere by means of their downward phase propagation (Deubner and Fleck 1989, Komm et al. 1991 b, Kneer and von Uexküll 1993).

The oscillations found in the limb-darkening function (Brown et al. 1978) are controversial. The periods of these oscillations are between 6.5 and 66 minutes, which is the range of the most interesting p modes of low degree and order! But the amplitude of the observed limb-darkening oscillations would correspond to a velocity of a few meters per second, which at disc center should be detected by the Doppler effect, but has not been found.

Also controversial is the 160-minute oscillation observed since 1974 at the Crimean Astrophysical Observatory and at the Wilcox Solar Observatory, Stanford. The originally reported regular phase shift of about 34 min/year (Fig. 5.23 in the first edition of this book) is absent in the later observations (Kotov et al. 1997), and the amplitude has decreased from 0.2–0.5 m/s to about one-half of this. – There is no theoretical identification of the 160-minute oscillation. Judging from the length of the period it could only be a g mode, but neither degree nor order are known. Yerle (1986) attributes the signal to effects of the terrestrial atmosphere.

6. Convection

The energy generated in the solar core must be carried to the surface. In Chap. 2 *radiation* was treated as one possible way of energy transport. Another is *convection*, where internal energy, sometimes including latent heat, is carried along with the motion of matter. We know that convection is quite efficient in distributing heat in rooms, and we shall see in the present chapter that convection is also an efficient carrier of solar energy, at least in the outer layer between a depth of $\approx 200\,000\,\mathrm{km}$ and the surface. The stratification of this outer layer is unstable, much like the stratification of the Earth's troposphere which becomes unstable when the ground is heated.

The mixing-length concept, where parcels of gas are envisaged to travel a certain distance and then to dissolve and to deposit their excess heat, will be used to describe the convective flux of energy.

As opposed to oscillatory motions, where parcels of gas move back and forth, the convective motion normally is *overturning*, although not necessarily stationary. We shall consider the overturning character as typical for convection. Thus, the cellular structure seen at the solar surface, from the small-scale granular velocity field to the pattern of giant cells, will be treated as a convective phenomenon in the present chapter. This widens the meaning of the word "convection", because most of the observable layer is stably stratified, and the energy transport is by radiation rather than by convection.

6.1 Stability

When a parcel of gas (also called a "bubble" or "blob", or an "eddy" or "fluid element") is adiabatically lifted from its equilibrium position, it will be either heavier or lighter than its new environment. In the latter case it continues to rise, i.e., the original equilibrium was *unstable*, in the former case it returns to its initial position, which means that the original equilibrium was *stable*. A criterion for such stability or instability was first derived by Schwarzschild (1906).

Let us consider a parcel of gas which has traveled a vertical distance δr, cf. Fig. 6.1. The density of the parcel is now ρ^*, and must be compared to the density ρ_0^* of its new environment. We assume that the motion is sufficiently fast so that the parcel behaves adiabatically, but still sufficiently slow so that

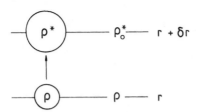

Fig. 6.1. Displacement of a convective "parcel"

at each point its internal pressure has adjusted itself to the local ambient pressure. These assumptions are reasonable whenever the time scale of energy exchange is long compared to the sound travel time across the parcel; in the optically thick solar interior this is always the case.

The density difference at the level $r + \delta r$ can be expressed in terms of the density gradient: expansion of ρ^* and ρ_0^* yields, to first order, the condition for instability,

$$\rho^* - \rho_0^* = \delta r \left[\left(\frac{d\rho}{dr} \right)_a - \frac{d\rho}{dr} \right] < 0 , \tag{6.1}$$

where $(d\rho/dr)_a$ is the density gradient under adiabatic conditions. We restrict the discussion now to the case where the pressure is given by the perfect gas equation, $P = \rho \mathcal{R} T / \mu$. But we shall allow for variations of the mean molecular weight μ arising from ionization and from the change of the chemical constitution due to hydrogen burning. Making use of the pressure equilibrium we thus obtain

$$\frac{dT}{dr} < \left(\frac{dT}{dr} \right)_a + \frac{T}{\mu} \frac{d\mu}{dr} - \frac{T}{\mu} \left(\frac{d\mu}{dr} \right)_a . \tag{6.2}$$

Inside the Sun $dT/dr < 0$. For instability to occur the temperature gradient must be steeper, i.e., more negative, than the adiabatic temperature gradient plus the corrections arising from the change of μ.

In the solar core the material is fully ionized, and we may assume that it remains so under an adiabatic perturbation. Then $(d\mu/dr)_a = 0$. The gradient $d\mu/dr$ is the result of helium accumulation towards the center and is negative. Therefore this gradient is *stabilizing*. This is of course expected since helium is heavier than hydrogen. However, the μ gradient is not really needed for stabilization of the Sun's core. The models actually calculated have a stable core even if the μ gradient in (6.2) is neglected. Of course, there might be other sources of instability: a magnetic field, or a large gradient in angular velocity, cf. Chap. 7. In a discussion of such phenomena it must not be forgotten that the molecular weight variation always acts against the instabilities.

In the outer layers of the Sun – the "envelope" – the mean molecular weight varies because the abundant elements hydrogen and helium are partially ionized. In this case μ is a function of P and T. Let us assume that within the upwards traveling parcel there is instantaneous adjustment of the ionization equilibrium. Then the molecular weight in the parcel is the same function of P and T as the exterior μ, that is,

$$\frac{d\mu}{dr} = \left(\frac{\partial\mu}{\partial P}\right)_T \frac{dP}{dr} + \left(\frac{\partial\mu}{\partial T}\right)_P \frac{dT}{dr} , \qquad (6.3)$$

and

$$\left(\frac{d\mu}{dr}\right)_a = \left(\frac{\partial\mu}{\partial P}\right)_T \left(\frac{dP}{dr}\right)_a + \left(\frac{\partial\mu}{\partial T}\right)_P \left(\frac{dT}{dr}\right)_a . \qquad (6.4)$$

We substitute these expressions into (6.2), recall that we have assumed pressure equilibrium, and find the Schwarzschild criterion for convective instability of the solar envelope,

$$\frac{dT}{dr} < \left(\frac{dT}{dr}\right)_a . \qquad (6.5)$$

The molecular weight does not enter explicitly because the factor $\delta \equiv 1 - (\partial\ln\mu/\ln T)_P$, cf. (2.51), appears on both sides and is positive. Instead of (6.5) one often writes

$$\nabla > \nabla_a , \qquad (6.6)$$

where $\nabla = d\ln T/d\ln P$, and ∇_a is the adiabatic value of ∇, defined in (2.49). Notice that these double-logarithmic temperature gradients must not be confused with the symbol ∇ that abbreviates the div, grad, or curl operators.

The criterion for instability is deduced here without recourse to the hydrodynamic equations governing the problem. A linear stability analysis of those equations leads however to the same result (Lebovitz 1966).

Problem 6.1. Show that the Schwarzschild criterion is equivalent to $N^2 < 0$, where N is the Brunt-Väisälä frequency, defined in (5.24). Which are the directions of the entropy gradient in stable and unstable layers, respectively?

6.2 Mixing-Length Theory

In a layer that is unstable according to the Schwarzschild criterion motions with growing amplitude occur. In a fluid of large viscosity these motions may develop into a regular cell pattern and become stationary, but in the almost inviscid stellar gas they generally will be *turbulent*. The ever changing velocity field observed at the solar surface supports this concept of a turbulent solar convection.

The quantitative description of turbulence is an essentially unsolved problem. Perhaps numerical simulation will be the answer; attempts in this direction will be discussed in Sects. 6.2.2 and 6.3.2 below. Here we shall outline a simplified concept, the *mixing-length theory*, which traditionally has been employed to describe convective energy transport in the interior of stars.

6.2.1 The Local Formalism

The mixing-length concept for turbulent flows was developed by G. I. Taylor, W. Schmidt, and L. Prandtl between 1915 and 1930. In analogy to the mean-free-path in the kinetic theory of gases it employs the *mixing length*, l, which is the distance over which a moving parcel of gas can be identified before it dissolves. The mixing length is usually taken as the distance to the nearest boundary of the moving fluid, or as some other typical length of the considered system.

Between 1930 and 1950 H. Siedentopf, L. Biermann, and E. J. Öpik introduced the concept into astrophysics, and various assumptions concerning the mixing length were made: the distance to the stellar center, the scale height of variables such as temperature or density, the mean observed diameter of granules, etc. The formalism presently used in most calculations is due to Vitense (1953) – see also Böhm-Vitense (1958) – and specifies

$$l = \alpha H_P , \tag{6.7}$$

where H_P is the pressure scale height and α is a constant, cf. (2.1). The following is a short outline of Vitense's formalism. It uses only mean quantities, such as the energy flux, F_C, the mean temperature excess ΔT of convecting gas parcels, and a typical velocity v of those parcels. Its aim is to calculate the mean temperature gradient, $(dT/dm)_C$, in the presence of convection (this quantity was needed for the solar models of Chap. 2).

Energy Flux. The basic equation to be used expresses the conservation of energy. In a layer such as the solar envelope, where energy sources are negligible, the total flux of energy, $L_\odot/4\pi r^2$, consists of radiative and convective contributions, F_R and F_C, so that

$$F_R + F_C = L_\odot/4\pi r^2 . \tag{6.8}$$

The radiative flux, F_R, has already been defined in Sect. 2.3.2:

$$F_R = -\frac{16\sigma T^3}{3\kappa\rho}\frac{dT}{dr} = \frac{16\sigma T^4}{3\kappa\rho H_P}\nabla . \tag{6.9}$$

For the convective flux we need the mean excess temperature ΔT of a gas parcel which has risen a distance δr. Expanding again up to first order we find

$$\Delta T = \left[\left(\frac{dT}{dr}\right)' - \frac{dT}{dr}\right]\delta r \equiv (\nabla - \nabla')\frac{T\delta r}{H_P} , \tag{6.10}$$

where $(dT/dr)'$ is a (mean) gradient corresponding to the actual temperature change in the parcel during its rise, and $\nabla' = -H_P(d\ln T/dr)'$.

Let us assume that the mean value (for many parcels) of δr is half the mixing length. Then, by (6.7),

$$\Delta T = (\nabla - \nabla')T\alpha/2 \ . \tag{6.11}$$

For an estimate of the convective energy flux F_{C} we must multiply ΔT by the specific heat capacity ρc_{P} and by an appropriate average v of the convection velocity:

$$F_{\mathrm{C}} = \alpha \rho c_{\mathrm{P}} v T(\nabla - \nabla')/2 \ . \tag{6.12}$$

Convection Velocity. Next we must determine v. To this end we consider the acceleration provided by the buoyancy of the parcel's density excess $\Delta \rho$, i.e.,

$$\partial^2 \delta r/\partial t^2 = -g\Delta\rho/\rho = g\delta \Delta T/T \ . \tag{6.13}$$

The second of these relations follows from the assumption of pressure equilibrium, $\Delta P = 0$, between the parcel and the ambient gas; the factor δ, given by (2.51), takes care of a possible variation of the mean molecular weight μ. We substitute (6.10) and assume that ∇, ∇', δ, g, and H_{P} all are constants over a mixing length. The integral of (6.13) then yields

$$(\partial \delta r/\partial t)^2 = \frac{g\delta}{H_{\mathrm{P}}}(\nabla - \nabla')(\delta r)^2 \ , \tag{6.14}$$

which means that the work done by the buoyancy force appears as kinetic energy of the parcel. It has been customary to multiply the right of (6.14) by $1/2$ in order to account for some expenditure of buoyancy work due to friction. Then, using again $l/2$ for the mean of δr, we obtain the mean convection velocity

$$v = l \left[\frac{g\delta}{8H_{\mathrm{P}}}(\nabla - \nabla') \right]^{1/2} \ . \tag{6.15}$$

Radiative Loss. During its rise the parcel radiates energy into its environment. For this reason the gradient ∇' differs from the adiabatic gradient ∇_{a}. In order to assess ∇' we write for the radiative flux across the surface of the parcel

$$F_{\mathrm{R}}' = -\frac{16\sigma T^3}{3\kappa\rho}\frac{\Delta T}{d} = \frac{8\alpha\sigma T^4}{3\kappa\rho d}(\nabla' - \nabla) \ , \tag{6.16}$$

where ΔT has been substituted by (6.11); d is the distance over which ΔT drops to zero (essentially the parcel's diameter).

The convective flux (6.12) can be written in the form

$$F_{\mathrm{C}} = \alpha \rho c_{\mathrm{P}} v T(\nabla - \nabla_{\mathrm{a}})/2 + \alpha \rho c_{\mathrm{P}} v T(\nabla_{\mathrm{a}} - \nabla')/2 \ . \tag{6.17}$$

The first term is the convective flux which we would have under ideal adiabatic conditions, the second (negative) term determines how much smaller than this the real convective flux is. We must multiply this second term by an effective cross section q of the parcel, and we must multiply F_R' by the parcel's surface S. Both results express the same quantity, namely the radiative loss per unit time. Hence

$$\frac{16\sigma T^3 S}{3\kappa\rho d}(\nabla' - \nabla) = \rho c_P v q(\nabla_a - \nabla') . \tag{6.18}$$

There is some arbitrariness in this derivation concerning the quantities S, d, and q. In order to remain consistent with Vitense's work, we choose $Sl/qd = 9/2$, corresponding, e.g., to spherical parcels and $d = (8/9)l$. Other choices could be made as well, but the result would be numerical factors of order 1 that occur in combination with the mixing length, much like the factor $1/2$ which was introduced in (6.15). In the end, the whole difference would be a different calibration of the dimensionless parameter α in (6.7).

We substitute (6.15), and obtain

$$\nabla' - \nabla_a = 2U(\nabla - \nabla')^{1/2} , \tag{6.19}$$

where

$$U = \frac{24\sqrt{2}\sigma T^3 P^{1/2}}{c_P\kappa g l^2 \delta^{1/2}\rho^{5/2}} . \tag{6.20}$$

The Cubic Equation. We return to our basic equation (6.8). We define a "radiative" gradient

$$\nabla_R = \frac{3\kappa\rho H_P L_\odot}{64\pi r^2 \sigma T^4} \tag{6.21}$$

which, in the absence of convection, is identical to the true gradient ∇. In a convection zone, ∇_R is the fictitious gradient that would be needed to transport all the energy by radiation; hence it simply measures the total energy flux.

Substitution of (6.9), (6.12), (6.15), and (6.21) into (6.8) yields

$$\nabla - \nabla_R + \frac{9}{8U}(\nabla - \nabla')^{3/2} = 0 . \tag{6.22}$$

This relation, in combination with (6.19), is a third-order algebraic equation determining ∇ in a convection zone. In the code describing the internal structure of the Sun (Chap. 2) this equation is numerically solved. It yields the convective temperature gradient which, by (2.8) and the definition of H_P, is related to ∇:

$$\left(\frac{dT}{dm}\right)_C = -\frac{T\nabla}{4\pi\rho r^2 H_P} . \tag{6.23}$$

Problem 6.2. Introduce $x = (\nabla - \nabla_a + U^2)^{1/2}$, and derive the cubic equation

$$\frac{9}{8U}(x - U)^3 + x^2 - U^2 - \nabla_R + \nabla_a = 0 . \tag{6.24}$$

Show that this equation has only one real root.

A Solar Model. For the standard solar model described in Chap. 2 some of the variables in the convection zone are listed in Table 6.1. In the outer part hydrogen is partially ionized. A large amount of latent heat (the ionization energy) is therefore available. This effectively lowers ∇_a, cf. Fig. 2.4, and so furthers the convective instability. It also raises c_P, see again Fig. 2.4, and hence raises the efficiency of the convective energy transport. For these reasons the convection zone of the Sun sometimes is called the "hydrogen convection zone". Nevertheless, the ionization of helium also contributes [η_i in Table 6.1 is defined as the number of atoms in the $(i + 1)$th state of ionization, divided by the total number of atoms of the element in question].

We also see from Table 6.1 that a very small temperature difference ΔT and a very small *superadiabaticity*, $\Delta\nabla \equiv \nabla - \nabla_a$, are sufficient for F_C to carry the entire solar luminosity over most of the convection zone. As a consequence, the stratification is essentially adiabatic, and the mixing-length formalism mainly serves to select the right adiabat for the stellar model.

The convection velocity v calculated according to (6.15) is about $2\,\mathrm{km/s}$ near the surface, but takes on much smaller values, $500\,\mathrm{m/s}$ and less, already at small depth. The numbers in the table must however not be understood too literally because v, unlike F_C, is subject to the arbitrary factor $1/2$ introduced above.

The total depth of a convection zone calculated with the recipe outlined here slightly depends on the physical input to the model, especially on the opacity which enters expression (6.21) for the radiative gradient. With the OPAL opacity table the depth is $\approx 200\,000\,\mathrm{km}$, consistent with the helioseismological result (5.80).

Problem 6.3. Show that the convective energy flows down the gradient dS/dr of the specific entropy. Give an interpretation in terms of a diffusion process, with a "turbulent diffusivity", $\kappa_t \simeq vl/2$.

Problem 6.4. Convince yourself that $\nabla_R > \nabla > \nabla' > \nabla_a$ in a convection zone. Show that the *efficiency of convection*

$$\Gamma \equiv \frac{\nabla - \nabla'}{\nabla' - \nabla_a} = \frac{c_P \rho^2 l \kappa v}{24 \sigma T^3} \tag{6.25}$$

is very large in the deeper part of the convection zone and that, therefore, $\nabla' \simeq \nabla_a$ is a very close approximation.

6.2.2 Numerical Test Calculations

The advent of large and fast computers made it possible to simulate numerically turbulent flows and their capability to transport heat. It is true, these three-dimensional and time-dependent hydrodynamical calculations are still too expensive (in terms of computer time and storage) to be incorporated

Table 6.1. Convection zone of a standard solar model ($Z + n$ means $Z \times 10^n$)

$r/r_\odot$	$P\,[\text{Pa}]$	$T\,[\text{K}]$	$\rho\left[\frac{\text{kg}}{\text{m}^3}\right]$	η_H	η_He	$\eta_\text{He+}$	$\Delta\nabla$	$\Delta T\,[\text{K}]$	$v\left[\frac{\text{m}}{\text{s}}\right]$	F_C/F
1.000	9.55+03	5.78+3	2.51−4	.00	.00	.00	−1.1−1	0.0+0	0	0.00
1.000	1.18+04	6.23+3	2.87−4	.00	.00	.00	9.7−2	7.7+0	152	0.00
1.000	1.34+04	6.74+3	3.01−4	.00	.00	.00	4.1−1	4.6+2	1181	0.05
1.000	1.45+04	7.21+3	3.05−4	.00	.00	.00	5.4−1	1.5+3	2172	0.32
1.000	2.19+04	9.26+3	3.52−4	.02	.00	.00	2.0−1	1.6+3	2427	0.96
1.000	3.33+04	1.05+4	4.57−4	.05	.00	.00	9.7−2	9.2+2	2042	0.99
.999	5.04+04	1.14+4	6.16−4	.09	.00	.00	5.7−2	5.9+2	1783	1.00
.999	7.64+04	1.22+4	8.44−4	.13	.00	.00	3.7−2	4.1+2	1581	1.00
.999	1.16+05	1.30+4	1.17−3	.17	.00	.00	2.5−2	2.9+2	1412	1.00
.999	1.76+05	1.37+4	1.63−3	.21	.00	.00	1.7−2	2.2+2	1266	1.00
.999	2.66+05	1.45+4	2.28−3	.25	.00	.00	1.3−2	1.7+2	1139	1.00
.998	4.04+05	1.53+4	3.18−3	.28	.00	.00	9.2−3	1.3+2	1026	1.00
.998	6.12+05	1.62+4	4.46−3	.32	.00	.00	6.9−3	1.0+2	925	1.00
.998	9.28+05	1.71+4	6.24−3	.36	.00	.00	5.2−3	8.0+1	837	1.00
.997	1.41+06	1.80+4	8.72−3	.40	.00	.00	3.9−3	6.4+1	758	1.00
.997	2.13+06	1.91+4	1.22−2	.44	.00	.00	3.0−3	5.3+1	688	1.00
.997	3.23+06	2.03+4	1.70−2	.48	.01	.00	2.4−3	4.3+1	626	1.00
.996	4.90+06	2.16+4	2.37−2	.52	.01	.00	1.9−3	3.6+1	571	1.00
.996	7.42+06	2.31+4	3.28−2	.56	.02	.00	1.5−3	3.1+1	524	1.00
.995	1.13+07	2.48+4	4.51−2	.61	.04	.00	1.2−3	2.7+1	481	1.00
.995	1.71+07	2.68+4	6.21−2	.65	.07	.00	9.3−4	2.3+1	441	1.00
.994	3.18+07	3.03+4	9.89−2	.72	.15	.00	6.8−4	1.9+1	393	1.00
.993	5.94+07	3.47+4	1.56−1	.78	.30	.00	4.9−4	1.5+1	350	1.00
.992	9.01+07	3.84+4	2.09−1	.81	.43	.00	3.9−4	1.4+1	326	1.00
.991	1.37+08	4.28+4	2.79−1	.84	.57	.00	3.2−4	1.2+1	305	1.00
.990	2.07+08	4.81+4	3.68−1	.87	.70	.00	2.6−4	1.1+1	286	1.00
.988	3.14+08	5.47+4	4.82−1	.89	.80	.00	2.1−4	1.0+1	269	1.00
.987	4.76+08	6.28+4	6.27−1	.91	.86	.01	1.6−4	9.3+0	251	1.00
.985	7.21+08	7.25+4	8.13−1	.93	.88	.03	1.2−4	8.1+0	232	1.00
.983	1.09+09	8.37+4	1.05+0	.94	.85	.09	9.0−5	6.8+0	212	1.00
.981	1.66+09	9.65+4	1.37+0	.94	.77	.19	6.5−5	5.6+0	194	1.00
.976	3.09+09	1.19+5	2.04+0	.95	.57	.40	4.1−5	4.4+0	171	1.00
.971	5.77+09	1.49+5	3.00+0	.96	.38	.61	2.6−5	3.5+0	153	1.00
.967	8.75+09	1.73+5	3.88+0	.96	.28	.72	2.0−5	3.1+0	142	1.00
.962	1.33+10	2.02+5	4.99+0	.96	.21	.79	1.5−5	2.7+0	132	1.00
.956	2.01+10	2.37+5	6.42+0	.96	.16	.84	1.1−5	2.3+0	123	1.00
.949	3.05+10	2.78+5	8.24+0	.97	.13	.87	8.1−6	2.0+0	114	1.00
.942	4.62+10	3.27+5	1.06+1	.97	.11	.89	5.9−6	1.8+0	105	1.00
.932	7.00+10	3.85+5	1.36+1	.97	.09	.91	4.3−6	1.5+0	98	1.00
.922	1.06+11	4.53+5	1.74+1	.97	.08	.92	3.2−6	1.3+0	90	.99
.910	1.61+11	5.34+5	2.24+1	.97	.07	.93	2.3−6	1.1+0	84	.99
.896	2.44+11	6.29+5	2.87+1	.97	.07	.93	1.7−6	9.7−1	78	.98
.880	3.70+11	7.42+5	3.68+1	.97	.06	.94	1.2−6	8.4−1	72	.97
.862	5.61+11	8.75+5	4.73+1	.98	.06	.94	9.1−7	7.2−1	67	.95
.841	8.50+11	1.03+6	6.07+1	.98	.05	.95	6.6−7	6.2−1	62	.91
.818	1.29+12	1.22+6	7.79+1	.98	.05	.95	4.7−7	5.2−1	57	.86
.778	2.41+12	1.56+6	1.14+2	.98	.04	.96	2.7−7	3.8−1	48	.71
.732	4.56+12	2.01+6	1.66+2	.98	.04	.96	1.1−7	1.9−1	34	.34
.717	5.44+12	2.15+6	1.85+2	.98	.04	.96	5.6−8	1.1−1	26	.15
.710	5.96+12	2.23+6	1.95+2	.98	.04	.96	1.9−8	3.8−2	15	.03

into a full stellar evolution code. They can nevertheless be used to test some of the assumptions made in the heuristic approach of the preceding section.

The calculations of Chan and Sofia (1987, 1989) and Kim et al. (1996) are of particular interest. These authors solve the Navier-Stokes equation together with the mass and energy equations for a compressible flow in an atmosphere having a prescribed energy flux density at its bottom. The calculated flow is turbulent, and three results are relevant to the mixing-length theory:

First, the correlation

$$A = \frac{\langle v_1 v_2 \rangle}{\langle v_1^2 \rangle^{1/2} \langle v_2^2 \rangle^{1/2}} \tag{6.26}$$

between the vertical velocity components at two levels, 1 and 2, in the atmosphere is calculated. The brackets denote an average over time and (or) the horizontal coordinates. The length over which A drops by a substantial amount (e.g., to 0.5) can be identified with the vertical distance after which a convective parcel looses its identity, i.e., with the mixing length. The calculations show that, at various levels in the atmosphere, and also for various density stratifications, the correlation length of A is the same *multiple of the pressure scale height*. This confirms the assumption (6.7). A similar proportionality to the density scale height has not been found.

The second result of relevance concerns the rms value of the temperature fluctuation ΔT and the mean square value v^2 of the vertical velocity component. Both of these quantities turn out to be proportional to $T\Delta\nabla$, with suitable constants of proportionality. Thus, we have a justification of equations (6.11) and (6.15), although it is necessary to set $\nabla' = \nabla_a$ and $\delta = 1$ for a detailed comparison.

Finally, Chan and Sofia calculated the convective energy flux from the exact formula

$$F_C = \rho c_P \langle v_r \delta T \rangle \ , \tag{6.27}$$

where δT is the temperature fluctuation. They found that the average $\langle v_r \delta T \rangle$ is proportional to the rms values of v_r and δT (which we identify with v and ΔT, respectively). In combination with the previous result this confirms the heuristic expression (6.12) of the convective flux.

The results of Chan and Sofia generally support the application of the mixing-length formalism, in particular in the deeper part of the convection zone. Only near the surface the mixing-length assumptions are less well justified; there the numerical simulations imply a steeper temperature gradient, and this result is supported by an improved agreement between predicted and observed p-mode frequencies (Kim et al. 1996, Schlattl et al. 1997).

6.2.3 Overshooting: A Non-local Formalism

Overshooting is the penetration of the convective motion into a stably stratified layer. The updraft in a strong thunderstorm may drive cumulus clouds up into the (stable) stratosphere; a convective current in a lake may pass across the $4°$ C level which, because of the density maximum of water at this temperature, may constitute the interface between a stable and an unstable layer. In the Sun, overshooting can occur both at the top and at the base of the convection zone. To both situations three-dimensional hydrodynamic calculations as well as modified versions of the mixing-length theory have been applied. In the present section we shall consider a version of the mixing-length formalism that allows for overshooting, with an application to the base of the convection zone.

The theory outlined in Sect. 6.2.1 is local, in the sense that ΔT, v, and the convective flux F_C are all completely determined by the local value of $\Delta \nabla$, the excess of the temperature gradient over the adiabatic gradient.

In the local theory ΔT, v, and F_C all become zero at the level $r_{\Delta \nabla}$ where $\nabla = \nabla_a$, i.e., beyond which the Schwarzschild criterion 6.6 for instability is no longer satisfied. On the other hand we know that convective parcels of gas acquire their temperature excess and their velocity over a distance of the order of the mixing length. A parcel coming from somewhere inside the region where $\nabla > \nabla_a$ will, therefore, generally have a non-zero temperature excess and velocity when it arrives at $r_{\Delta \nabla}$. That is, it will continue its journey, "overshooting" into the stably stratified region. There, the subadiabatic environment will tend to reverse the sign of the parcel's temperature excess, so that, after some distance, the buoyancy force becomes decelerating rather than accelerating. After a short further distance the parcel will be stopped.

It is clear that the treatment of convective overshooting requires a non-local formalism. One that has been proposed by Shaviv and Salpeter (1973) will be outlined in this section. It is an extension of the heuristic approach of Sect. 6.2.1, and has been adopted in most studies of overshooting in stellar convection zones.

We consider parcels of gas starting at level r_1. At level r_2 the mean temperature excess, ΔT, and the velocity, v, of these parcels are given by

$$\Delta T(r_2; r_1) = - \int_{r_1}^{r_2} \left[\frac{dT}{dr} - \left(\frac{dT}{dr} \right)_a \right] dr \tag{6.28}$$

and

$$\frac{1}{2} v^2(r_2; r_1) = \int_{r_1}^{r_2} \frac{g\delta}{T} \Delta T(r; r_1) dr . \tag{6.29}$$

Expression (6.28) replaces the temperature excess (6.10) of the local theory (if the radiative loss is neglected). Equally (6.29), which is the buoyancy work integral along the path from r_1 to r_2, replaces (6.15).

For rising parcels, we set $r_2 = r_1 + l/2$, for sinking parcels, $r_2 = r_1 - l/2$, where l is the mixing length. This is consistent with the local formalism where $\delta r = l/2$ was used. We should add a factor $1/2$ on the right of (6.29) on account of friction, as above. For the convective flux at level r we write

$$F_{\rm C} = f c_{\rm P} \rho v(r; r+l) \Delta T(r; r+l) + f' c_{\rm P} \rho v(r; r-l) \Delta T(r; r-l) \ . \quad (6.30)$$

The first term is the contribution of sinking parcels, the second that of rising parcels. The factors f and f' (with $f + f' \leq 1$) may be interpreted as the fractions of a given horizontal area (at level r) filled with sinking and rising parcels, respectively. These factors also may absorb the somewhat arbitrary scaling applied above to the velocity v.

We must substitute (6.30) into equation (6.8) that describes the total energy transport. For the mixing length l, which defines the range of integration in (6.28) and (6.29), we again use (6.7). As before, the dimensionless parameter α can be adapted – together with the initial helium content, Y_0 – to yield the correct radius and luminosity of the present Sun.

Overshooting at the Base of the Convection Zone. Let us apply the non-local treatment to the lower part of the solar convection zone. Here, in the optically thick regime, we can safely use the diffusion-like radiative part (6.9) of the energy flux. Deep in the convection zone we may also set $\nabla' = \nabla_{\rm a}$, cf. Problem 6.4.

We start the integration of the solar model at the surface with the usual local mixing-length formalism, and switch over to the non-local treatment at some level $r_{\rm c}$. As we are interested in downward overshooting, let us consider only sinking parcels. We then adjust f in (6.30) in such a way that $F_{\rm C}$ continuously matches to its local form that is used above $r_{\rm c}$ (Pidatella and Stix 1986, Skaley and Stix 1991).

With this prescription, but otherwise with the same program as described in Chap. 2, solar models have been calculated. Within the computational error the same values of Y_0 and α are found by calibrating the present luminosity and radius; the factor f in (6.30) is ≈ 0.2. The main characteristics of these models is a temperature gradient that continues to follow the adiabatic value, $\nabla_{\rm a} \simeq (\gamma-1)/\gamma \approx 0.4$, for some distance below the intersection with the radiative gradient, $\nabla_{\rm R}$, as illustrated in Fig. 6.2. The reason is that at that intersection the sinking parcels still have kinetic energy and so continue their descent into the stably stratified layer. Since radiative exchange is weak, the layer of convective overshooting becomes nearly isentropic. In fact the

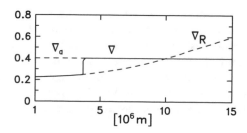

Fig. 6.2. Radiative and adiabatic temperature gradients (*dashed*), and real temperature gradient (*solid*) in a solar model with overshooting velocity at the base of the convection zone (schematic)

temperature gradient changes from slightly super-adiabatic to slightly sub-adiabatic. According to (6.28) the temperature excess of the sinking parcels becomes positive, thus reversing the sign of the buoyancy force. At the same time the convective energy flux becomes *negative*, so that within the layer of overshooting F_R must exceed the total flux.

The formalism described here breaks down when v^2, calculated on the basis of (6.29), turns negative. Clearly, we must set $v = 0$ downwards from the level r_v where this happens, because then the available kinetic energy has been used up. The level r_v marks the lower boundary of the convective overshooting; in the actual models the overshooting extends over a depth range of order 10^4 km, or 0.1–0.2 times the pressure scale height H_P at the base of the convection zone.

The abrupt transition at r_v between the convective layer and the radiative solar interior, with discontinuities in ∇, ΔT, and F_R, is a consequence of the non-local mixing-length formalism. The discontinuous temperature gradient gives rise to a characteristic signature in the p-mode frequencies: with increasing radial order n the spacing between these frequencies will vary in a periodic manner, instead of being equidistant. Such a signature has not yet been found, which places an upper limit of $\approx 0.1 H_P$ on the extent of the convective overshooting (Monteiro et al. 1994, Basu and Antia 1994, Christensen-Dalsgaard et al. 1995). It therefore appears that models with a more smoothed transition must be found. On the other hand, the idea of convective overshooting should not be abandoned entirely, mainly because a layer with a turbulent velocity field, but nevertheless stable mean stratification, offers a means to *store magnetic flux*, which may help to understand some characteristics of the solar magnetism (Sect. 8.2.4). Moreover, we recall that the destruction of lithium by nuclear reactions is very sensitive to the temperature and hence to the depth of mixing (Sect. 4.4.2). Therefore, convective overshooting may have been an important mechanism for the depletion of Li during the pre-main-sequence evolution of the Sun (Ahrens et al. 1992).

6.3 Granulation

6.3.1 The Observed Pattern

Under good conditions of atmospheric seeing, with a telescope of at least intermediate size ($D \geq 20$ cm, say), one sees a cellular pattern that covers the entire solar surface, except in sunspots. Bright isolated elements, the *granules*, appear on a dark background of multiply connected *intergranular lanes*; the total number of granules may reach 5 million. Figure 3.5 shows an example. The granulation was observed first by Sir W. Herschel, by J. Nasmyth, and by A. Secchi in the 19th century, and was first photographed by P. J. Janssen in 1877.

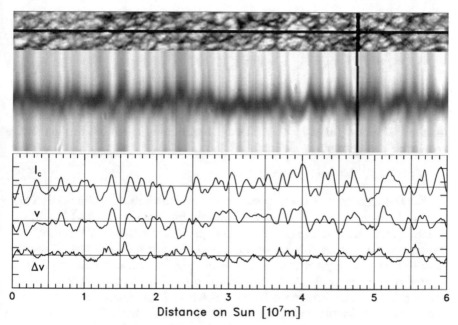

Fig. 6.3. Spectroscopy of solar granulation. *Upper panel*: a stripe of $60 \times 5.3 \, \mathrm{Mm}^2$ on the solar surface, with the spectrograph slit along its middle line, and a vertical black mark that serves to position the spectrum. *Middle*: the spectral band 491.186–491.219 nm (λ increasing upwards), with the Ni I line at 491.203 nm. *Lower panel*: intensity I_c of the continuum near the line, with variations up to $\pm 20\%$ around the mean; vertical velocity v, with variations up to $\pm 1.5 \, \mathrm{km/s}$; and full line width at half-maximum, corresponding to $\Delta v = 1.9$–$3.4 \, \mathrm{km/s}$ (combined thermal velocity and microturbulence ξ_t). The spectrum was taken on 30 July 1999 with the echelle spectrograph of the Vacuum Tower Telescope, Izaña, Tenerife (Nesis et al. 2001), exposure 0.6 s; a correlation tracker was used for stabilization. Courtesy H. Schleicher

The bright granules are upwards-moving, hot parcels of gas, while the dark intergranular lanes represent cooler, downwards-moving, material. This is clearly demonstrated in high-resolution spectra such as Fig. 6.3: these spectra show "line wiggles", i.e., the lines are blue-shifted in the brighter sections of the continuum, and red-shifted in the darker sections. Moreover, it appears that the width of the spectral line (lowest curve of Fig. 6.3) has maxima at those positions where the gradients of I and v are largest. The increased line width is ascribed to an increased non-thermal contribution ξ_t (Eq. 4.17), and has been interpreted as shear-generated turbulence (Nesis et al. 1993, 1996).

Intensity Contrast. The intensity contrast is important because it is a direct indication of the temperature fluctuations. For small fluctuations, and a Planck spectrum, we have $\delta I / I = 4 \delta T / T$. We may use the maximum and minimum observed intensities, $I_{\max}$ and $I_{\min}$, to define the maximum relative contrast

$$C = 2\frac{I_{\max} - I_{\min}}{I_{\max} + I_{\min}} . \tag{6.31}$$

More detailed information is provided by the power spectrum of the intensity fluctuation, $P_I(k_\mathrm{h})$, where $k_\mathrm{h} = (k_x^2 + k_y^2)^{1/2}$ is the horizontal wave number. The integral of P_I over all wave numbers k_h is related to the relative rms intensity fluctuation, defined as

$$(\delta I/I)_\mathrm{rms} = \langle (I - \langle I \rangle)^2 \rangle^{1/2} / \langle I \rangle . \tag{6.32}$$

For the best ground-based observations the rms intensity fluctuation may exceed 0.1, e.g., an 11-h sequence of granulation images obtained in a 10-nm spectral band around 468 nm on 5 June 1993 at the Swedish Vacuum Solar Telescope, La Palma (Simon et al. 1994). Figure 6.4 shows the intensity distribution of a typical image. Generally P_I, in particular at large k_h, is attenuated by atmospheric seeing, but also by the telescope. The telescopic attenuation can be overcome by a sufficiently large aperture, while seeing can be avoided by observing from above the atmosphere. For ground-based measurements a correction is necessary; the modulation transfer function, the MTF, should be known for this purpose.

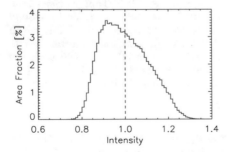

Fig. 6.4. Distribution of the granular intensity, relative to the mean (*dashed line*). A quiet-Sun area of $50'' \times 50''$ was observed with the Swedish Vacuum Solar Telescope, La Palma, in a 10-nm band at 468 nm. The rms contrast is 0.105. Courtesy P. N. Brandt

During a partial solar eclipse the intensity profile of the Moon's limb provides an opportunity to empirically determine the line spread function $\mathrm{LSF}(x)$ and, as its transform, the modulation transfer function MTF. An advantage of this method is that the same photographs contain both the lunar limb profile and the granulation to be investigated. One can therefore be sure to obtain the right MTF, for each observation. Also, in this way both the telescopic and the atmospheric effects are taken into account at once.

An intensity power spectrum, observed and corrected (i.e., divided by the MTF) by Deubner and Mattig (1975), is shown in Fig. 6.5. Except at very small k_h, around 1 Mm^{-1}, the correction is substantial. Moreover, again except for small k_h values, the correction strongly depends on the *wings* of the

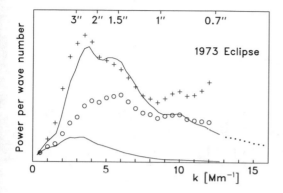

Fig. 6.5. Power spectrum of granular intensity fluctuations. Uncorrected (*lower*) and corrected (*upper continuous curve*). *Circles* and *crosses* mark the corrected spectrum if the spread function is approximated according to (6.33), with parameters as given in the text. Adapted from Deubner and Mattig (1975)

spread function. In order to show this, the spread function was approximated by a pair of Gauss functions,

$$\text{LSF}(x) \simeq \frac{1}{a}\exp(-(x/a)^2) + \frac{1}{b}\exp(-(x/b)^2) \ . \tag{6.33}$$

Two cases were considered. In both LSF(x) had a narrow core represented by $a = 180\,\text{km}$ and $a = 200\,\text{km}$, respectively. On the other hand, rather different wings were chosen, represented by $b = 540\,\text{km}$ and $b = 1500\,\text{km}$. The second case (crosses in Fig. 6.5) closely approximates the wings of the observed spread function, and indeed yields about the same power (up to $k_h \approx 8\,\text{Mm}^{-1}$) as the exact numerical calculation. The first case (circles) suppresses the extreme wings of LSF(x), with the result that almost half of the total power is lost.

The weight of the wings of the line spread function, and the great difficulty in measuring these wings correctly, has lead to a wide range of results concerning the solar intensity fluctuations. Also, there is a wavelength dependence of these fluctuations (Problem 6.5). For $(\delta I/I)_{\text{rms}}$, as defined in (6.32), Deubner and Mattig (1975) find the value 0.128 at $\lambda = 607\,\text{nm}$, Durrant et al. (1983) derive 0.113 at $\lambda = 556\,\text{nm}$, and Nordlund (1984 a), who used Deubner and Mattig's observed data, but a spread function with even more pronounced wings than in the second case above, obtained 0.20!

The intensity contrast C, defined in (6.31) above, of course exceeds $(\delta I/I)_{\text{rms}}$. The excess is roughly a factor 2, according to a large number of observations.

Problem 6.5. Approximate the emergent intensity by Wien's law, and show that

$$C \simeq (c_2/\lambda T^2)\Delta T \ , \tag{6.34}$$

where ΔT is the temperature difference between granules and intergranular space, and $c_2 \equiv hc/k = 1.438\,\text{cm}\,\text{K}$. Estimate ΔT from the numbers given in the text.

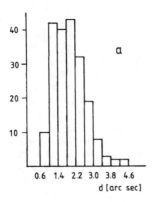 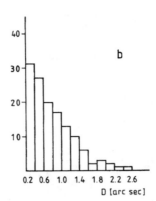

Fig. 6.6. Distribution of distances between centers of adjacent granules [**a**, after Bray and Loughhead (1977)], and of granular diameters [**b**, after Roudier and Muller (1986)]

Size. Two measures characterizing the spatial structure of the granulation are commonly in use. One is the distance d between the centers of adjacent granules; this distance is also called the "cell size". The other is the granular "diameter", defined as follows (Bray et al. 1984, p. 40). Let A be the area over which a granule's intensity is greater than a certain value, e.g., the mean intensity. This area generally is not circular; nevertheless, an effective diameter is

$$D = 2(A/\pi)^{1/2} \; . \tag{6.35}$$

The distances d between granular centers are distributed around a well-defined mean value, e.g., Fig. 6.6a. This "mean cell size" is $1.94''$ according to Bray and Loughhead (1977), and $1.76''$ according to Roudier and Muller (1986). The width of this distribution is about $1.5''$. The distribution of the "diameter" D is different: the number of granules appears to increase monotonically with decreasing D, see Fig. 6.6b.

The reason for the different distributions of d and D is the following (Namba and Diemel 1969): The diameter D of a granule may be arbitrarily small, but even for the smallest granule, with $D \to 0$, the distance to the center of a large neighbor remains finite.

Evolution. The granulation is a *non-stationary* phenomenon. Time sequences of high-quality pictures are necessary to study its evolution. Several balloon and space experiments have provided good material, but extended sequences of images with high resolution ($\approx 0.3''$) have been obtained from ground-based observatories as well (Simon et al. 1994, Hirzberger et al. 1997).

Some granules appear as brightenings of originally dark intergranular matter, but most are born as fragments of disintegrating older granules. The granules then grow in size, and typically develop a dark notch, or a dark spot

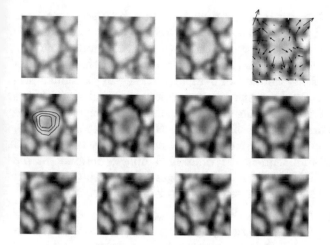

Fig. 6.7. Time sequence of an exploding granule. The width of each snapshot is 5.6″, the time step is 20 s. *Arrows* in the upper right image mark the horizontal velocity (derived from tracking small-scale features), *contours* in the middle left the divergence. From Rieutord et al. (2000)

that later connects to the intergranular lanes. Finally the granules split into two or more fragments, which fade away, or grow to become new granules.

When a large granule develops a dark spot in the center, a ring-shaped convection cell is formed: an *exploding granule*. The ring grows, and disintegrates into a number of segments (Fig. 6.7). The phenomenon is quite common: On a movie taken during a space experiment Title et al. (1986) recognized 44 examples within 1600 s in an area of 40″ × 40″. Because events near the edge of the frame, or at the beginning or end of the movie, may have escaped detection, the true density of exploding granules may be still larger.

The *lifetime* of solar granules can be measured by a correlation analysis. The time required for the auto-correlation of the photospheric intensity variation to decay to $1/e$ of an initial value is $\approx$ 6 minutes according to a number of studies, including the balloon experiment Stratoscope (Bahng and Schwarzschild 1961) and the space results of Title et al. (1986). Individual

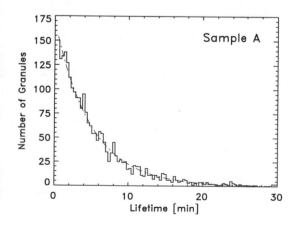

Fig. 6.8. Lifetime distribution for 2643 granules. From Hirzberger et al. (1999 a)

granules may however live much longer than this. In a time sequence from the balloon experiment Spektro-Stratoskop Mehltretter (1978) could identify single granules for over 8 minutes both backwards and forwards from any given instant of time; Dialetis et al. (1986) found a mean lifetime of 12 min for 200 individually tracked granules. Individual tracking, in particular performed manually, can account for the proper motion of granules and hence yields a longer lifetime. On the other hand, the result also depends on the criteria set for the birth and death of a granule. Hirzberger et al. (1999 a), using an automatic technique and more restrictive criteria, have tracked 2643 granules through an 80-min sequence of white-light images obtained at the Swedish Vacuum Solar Telescope, La Palma. Figure 6.8 shows their result: an exponential distribution of lifetimes; the mean is 6 minutes.

The Velocity Field. To measure the granular velocity (via the Doppler effect) a *spectrum* (Fig. 6.3) is necessary. Such is more difficult to obtain than granulation photographs, because a typical exposure time is 1 s, and good seeing conditions normally do not last that long. It is also necessary to recall that spectral lines originate higher in the solar atmosphere than the continuum. Therefore, the intensity and the velocity generally refer to different atmospheric levels.

Let us first concentrate on the rms value of the *vertical* velocity. This component is determined from spectra obtained at the center of the solar disc. The results of Durrant et al. (1979), listed in Table 6.2, illustrate three major problems.

First, a correction must be made for the telescopic and atmospheric image degradations. Durrant et al., who used observations made during the partial solar eclipse on 29 April 1976, achieve this correction by means of a modulation transfer function obtained from the profile of the Moon's limb, see above. The correction factor typically is between 3 and 4! It should be noted that here the *velocity* is corrected by means of an MTF obtained from *intensity* fluctuations. This is a plausible procedure since the spatial smearing (along the slit) "mixes" adjacent red- and blue-shifted parts of the line and so reduces the rms velocity signal; Mehltretter (1973) shows that the procedure is legitimate if the Doppler shift is small in comparison to the line width.

Table 6.2. Vertical rms velocity, in km/s, in the solar atmosphere. For the height of line formation see Fig. 6.9. From Durrant et al. (1979)

Line		v_{obs}	v_{corr}	$v_{k>2.3/\mathrm{Mm}}$	$v_{k<2.3/\mathrm{Mm}}$
Fe I	649.647 nm	0.32 ± 0.05	1.03 ± 0.16	0.89 ± 0.19	0.51 ± 0.05
Ca I	649.379 nm	0.24 ± 0.05	0.96 ± 0.18	0.81 ± 0.20	0.49 ± 0.03
Ba II	649.691 nm	0.22 ± 0.05	0.94 ± 0.22	0.82 ± 0.24	0.44 ± 0.04
Fe I	649.499 nm	0.30 ± 0.08	0.92 ± 0.20	0.75 ± 0.20	0.53 ± 0.07

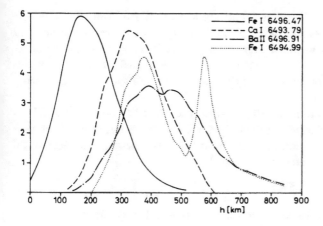

Fig. 6.9. Velocity contribution functions for the centers of the four lines listed in Table 6.2. From Durrant et al. (1979)

Second, the *granular* contribution to an observed Doppler shift must be separated from the *oscillatory* contribution. Strictly, this is possible only by means of simultaneous Fourier transforms in space *and* time, i.e., in the k_h, ω-plane. But a crude separation can be made for a single spectrum by a cut in the spatial power spectrum at some wave number k_0. Oscillations predominantly have smaller k_h, granules larger. The respective rms velocities are then obtained by integrations of the power spectrum from 0 to k_0, and from k_0 to ∞. For the results of Table 6.2 the cut was made at $k_0 = 2.3\,\mathrm{Mm}^{-1}$.

The third problem is the *height* determination. A rough measure of the height of formation of a line is its equivalent width. Better is the calculation of a velocity contribution function (or velocity weighting function) which, in general, differs from the intensity contribution function. Still, there are large uncertainties, as Fig. 6.9 demonstrates. The maxima of the functions are rather broad, and in the case of the Fe I 649.499 nm line there is even a double peak. Hence there is only little height resolution in the data of Table 6.2,

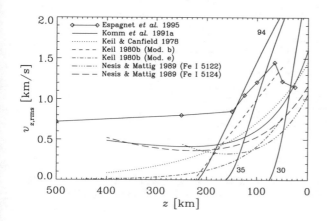

Fig. 6.10. Vertical rms velocity, as a function of height in the solar atmosphere, and results of three mixing-length models (*solid curves* with numbers) with convective overshooting. From Kiefer et al. (2000 a)

especially since the contribution functions are calculated only for the line *centers*. Nevertheless, several detailed studies suggest that both the vertical and the horizontal rms velocities decrease with height, so that the granular velocity field extends only over a few hundred km. Figure 6.10 shows some observations, and results of mixing-length models with convective overshooting of comparable magnitude.

The *horizontal* granular velocity is observed near the solar limb. Alternatively, we may measure the line-of-sight velocity as a function of angular distance, θ, from the disc center. Assuming that the vertical and horizontal velocity components are uncorrelated we have

$$v_{\mathrm{rms}}^2 = \mu^2 v_{r,\mathrm{rms}}^2 + (1 - \mu^2) v_{\mathrm{h,rms}}^2 , \tag{6.36}$$

where $\mu = \cos\theta$. Determinations of v_{rms} at several disc positions then yield both $v_{r,\mathrm{rms}}$ and $v_{\mathrm{h,rms}}$. The balloon experiment Spektro-Stratoskop made measurements at $\mu = 1$, $\mu = 0.6$, and $\mu = 0.2$, and showed that the two rms velocities are of the same magnitude (Mattig et al. 1981).

We may ask to what extent the granular velocity field contributes to the convective energy transport discussed in Sect. 6.2. A study of Nesis (1985), based on Spektro-Stratoskop, revealed that the vertical velocity v_r remains coherent with its value at the $\tau_{500} = 1$ level, and also with the intensity *at that level*, for the entire range of $\approx 500\,\mathrm{km}$ in height. On the other hand the coherences of the intensity fluctuations and of the horizontal velocity component decay rather rapidly with height. According to Krieg et al. (1999) the granular intensity fluctuation disappears at a height of $\approx 100\,\mathrm{km}$. The loss of intensity coherence indicates that upwards convective energy transport

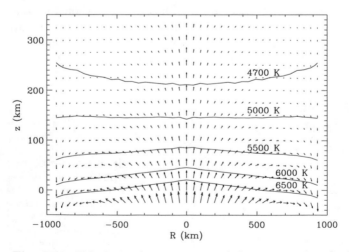

Fig. 6.11. Velocity and temperature of the average granule. The *arrow length* is proportional to $|v|$, with maximum 1.6 km/s. From Ruiz Cobo et al. (1996)

exists only in the deepest observable layers. Higher up, the (still coherent) vertical velocity pattern is interpreted as convective overshooting.

Ruiz Cobo et al. (1996) have used high-resolution spectra obtained at disc center to derive the line-of-sight (vertical) velocity component for a large number of granules. The horizontal component has been determined under the assumptions of mass conservation and azimuthal symmetry. Thus they derived the velocity distribution for an average granule shown in Fig. 6.11. In addition, the temperature field has been obtained from the spectral intensity; it shows a conspicuous reversal of the isotherms at a height of $\approx 140\,\mathrm{km}$.

6.3.2 Models

We have seen that the solar granulation is a non-stationary, overturning motion, i.e., "convection" in the loose definition above; moreover, in the deepest observable layers, there is a positive correlation between upward velocity and brightening, which means we have convection even in the narrower sense of convective upwards energy transport.

Unsöld (1930) pointed out that the Schwarzschild criterion for instability should be satisfied in the deep solar photosphere. In the atmospheric models described in Chap. 4 this happens first at some level between $\tau_{500} = 0.8$ and $\tau_{500} = 1$ (e.g., Gingerich et al. 1971). Below that level we must therefore replace the stable atmosphere (which is in radiative equilibrium) by a model which includes convection. This could be a mean model where the convection is described in terms of the mixing-length theory; Figure 6.10 shows results of such models that include overshooting into the stable atmosphere. The subject of the present section, however, is the simulation of the resolved granular pattern by means of a three-dimensional, time-dependent hydrodynamical calculation (Nordlund 1982, 1984b, 1985; Stein and Nordlund 1998 a).

Hydrodynamics. The conservation of mass and momentum is expressed by the equation of continuity, and the equation of motion, viz.

$$\frac{\partial \rho}{\partial t} + \nabla \cdot (\rho \boldsymbol{v}) = 0 \;, \tag{6.37}$$

and

$$\rho \frac{\partial \boldsymbol{v}}{\partial t} + \rho \boldsymbol{v} \cdot \nabla \boldsymbol{v} = -\nabla P + \rho \boldsymbol{g} + \rho \boldsymbol{f}_{\mathrm{visc}} \;, \tag{6.38}$$

where $\boldsymbol{g}$ is the gravitational acceleration, and $\rho \boldsymbol{f}_{\mathrm{visc}}$ is a viscous volume force.

For the granular motion, at any given point, there is little accumulation of mass, but rather a balance between inflow and outflow (into, and out of, a small volume which contains the point of reference). One therefore sometimes neglects the term $\partial \rho / \partial t$ in (6.37). This approximation, called *anelastic*, has been used in meteorology (e.g., Ogura and Phillips 1962). Its main consequence is that we cannot describe pressure waves, i.e., acoustic waves, by the truncated equations. To see this, take the divergence of (6.38) and obtain

$$\Delta P = \nabla \cdot [\rho(\boldsymbol{g} + \boldsymbol{f}_{\text{visc}} - \boldsymbol{v} \cdot \nabla \boldsymbol{v})] \ . \tag{6.39}$$

This equation states that, at any instant and everywhere, the pressure P is completely determined by the velocity field and the volume forces, and by the boundary conditions. In other words, a pressure perturbation does not propagate with the speed of sound, but rather with infinite speed. The more recent simulations do not rely on the anelastic approximation, because one of their aims is the interaction of the convective flow with acoustic waves.

According to the equation of motion, a vertical flow can be driven by the gravitational term (which contains the buoyancy), and (or) by the pressure gradient. On the other hand, since $\boldsymbol{g}$ is vertical, a horizontal motion can only be driven by a *horizontal pressure difference*. More specifically, there must be a pressure excess both in the granular center, to *accelerate* the horizontal outflow, and over the descending intergranular material, to *decelerate* the horizontal flow. This pressure excess *opposes* the buoyancy of the centrally ascending flow, an effect called *buoyancy breaking*, but *enforces* the acceleration of the sinking matter at the cell boundaries. As a consequence, the downdrafts are faster than the updrafts and, by the argument of mass continuity, the intergranular lanes are narrower than the granules (Hurlburt et al. 1984). The asymmetry of the forces driving updrafts and downdrafts may also contribute to the topological asymmetry, namely that the updrafts are isolated, while the downdrafts are connected.

Problem 6.6. Consider a mass flow $\rho \boldsymbol{v}$ that horizontally varies like $\sin k_{\text{h}} x$, and has a vertical scale height H_{m}. Show that the ratio of horizontal to vertical velocity components is

$$v_{\text{h}}/v_r \simeq 1/(k_{\text{h}} H_{\text{m}}) \ . \tag{6.40}$$

For given H_{m}, therefore, the ratio v_{h}/v_r increases with increasing cell size.

The Energy Balance. The hydrodynamic equations to be used in the present section differ in two important aspects from the equations we have used in Sect. 5.2.1 to describe the solar oscillations. First, they are *not* linearized, and second, as we must discuss now, we cannot use the adiabatic approximation. This is mainly because at the solar surface light is radiated into space, which is of course the very reason why we see the granulation.

The system (6.37) and (6.38) is complemented by an equation of state,

$$\rho = \rho(P, T) \ , \tag{6.41}$$

which contains the temperature; the latter is determined by the energy equation

$$\rho \frac{\partial H}{\partial t} + \rho \boldsymbol{v} \cdot \nabla H = Q + \frac{\partial P}{\partial t} + \boldsymbol{v} \cdot \nabla P \ . \tag{6.42}$$

Here, $H = U + P/\rho$ is the enthalpy per mass, and Q is the rate of energy gain (or loss) per volume. The specific internal energy U is assumed to be a known function of P and T.

If $Q = 0$, then (6.42) describes adiabatic changes. Let us restrict the attention to radiative exchange of heat. Then

$$Q = -\nabla \cdot \boldsymbol{F}_{\mathrm{R}} = \rho \int \kappa_\nu (I_\nu - S_\nu) d\Omega d\nu . \qquad (6.43)$$

Notice that $\boldsymbol{F}_{\mathrm{R}}$ is not simply proportional to the temperature gradient, as in the optically thick solar interior, but is derived from the intensity, which, at each location, depends on the direction. The intensity itself is a solution of the equation of transfer

$$\frac{dI_\nu}{ds} = -\rho\kappa_\nu (I_\nu - S_\nu) , \qquad (6.44)$$

where s is a coordinate in the direction considered, and S_ν is the source function. This completes the system. For its solution we require the knowledge of the functions $\rho(P,T)$, $U(P,T)$, of the absorption coefficient κ_ν, and of the source function S_ν; but for this moment we may consider all these functions as given, for example by the prescriptions outlined in Chaps. 2 and 4.

Inspection of (6.43) shows that heating, i.e., a positive contribution to Q, arises at frequencies and angles where the radiation field is more intense, or the "radiation temperature" is high. According to (6.42), such heating will compensate for the adiabatic cooling of a parcel of gas which expands as it ascends into the optically thin upper photosphere.

Deeper in the photosphere there is radiative *cooling* of the bright (hot) granule. This cooling weakens the buoyancy of the rising material. The effect of this, and of the pressure-induced buoyancy breaking (see above) is that the rise of new hot material to the surface is stopped. A dark spot develops in the granule's center, eventually leading to its decay.

Numerical Results. Solutions of equations (6.37), (6.38), and (6.41) to (6.44) can be obtained by numerical integration. To do this, even with the best available computers, it is necessary to content oneself with a volume that contains only a small number of granules, and with finite spatial resolution. The limited resolution raises the question of viscous dissipation. The viscosity of the solar gas is so small that the scale where viscous dissipation becomes important is never resolved. Yet it would be impossible to omit the viscous force in (6.38). The buildup of large gradients (due to the non-linear terms) would soon lead to a break-down of the whole calculation. An artificially high viscosity can be used as a remedy. The art is to choose $\boldsymbol{f}_{\mathrm{visc}}$ in (6.38) in such a way that the smoothing of large gradients is achieved at the limit of resolution without much influence on the spatially resolved scales.

The calculations of Stein and Nordlund (1998 a) allow a detailed comparison with the actual solar granulation. Figure 6.12 shows the flow and the

 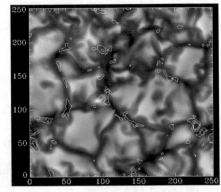

Fig. 6.12. Numerical simulation of solar granulation. *Left*: Velocity (*arrows*) and temperature (*color*, blue ≡ cool, red ≡ hot) of an individual granule. *Right*: Emergent intensity of a $6 \times 6 \,\mathrm{Mm}^2$ field of simulated granulation, with superimposed contours marking supersonic flow. From Stein and Nordlund (1998 a)

temperature field at one instant of time. The simulation resembles the observed pattern: bright material rising in isolated, and dark material sinking in connected, but otherwise irregular, areas. At the surface the horizontal often exceeds the vertical flow velocity, and occasionally is supersonic and forms shocks near the edges of granules (Fig. 6.12, right). There is some evidence for supersonic flows in spectroscopic data (Nesis et al. 1992); perhaps the bright edges of some granules shown in Figs. 3.5 and 3.11 are the intensity signature of shocks (de Boer et al. 1992).

The horizontal temperature variation of the simulation generally is much larger than observed. Nordlund suggests that most of this variation is masked by the temperature sensitivity of the opacity, which raises the visible surface over the granules and lowers it in the intergranular lanes. In the higher levels the temperature variation diminishes and even reverses its sign. This effect, which equally occurs in mean models based on mixing-length theory (Kiefer et al. 2000 a), is attributed to adiabatic cooling of granules that overshoot into the stable upper photosphere. It corresponds to the above-mentioned loss of coherence between intensity and upwards velocity.

The striking resemblance between the numerical simulation and images of the solar surface demonstrates that granulation is the visible manifestation of convection. What is the driver of the phenomenon? A numerical study of Rast et al. (1993) shows that about two-thirds of the enthalpy transported by granular convection is carried as latent heat of partially ionized hydrogen. The ionization energy is contained in the function $U(P, T)$, and the ensuing buoyancy force is contained in the term ρg. Thus, ionization plays an important role in the driving, in a similar way as in the heuristic theory of Sect. 6.2, except that single parcels, the granules, are now spatially resolved.

Rather narrow *downdrafts* are a main characteristic of the observed pattern as well as of the numerical simulation. The role of these downdrafts is greatly enhanced by the large temperature sensitivity of the opacity. As the gas reaching the surface cools by radiation, its opacity may decrease by two orders of magnitude; the increased transparency leads to further cooling and associated (downwards) buoyancy. If this thermal instability occurs at the center of a parcel arriving at the surface, it will enforce the mechanism of buoyancy breaking and thus lead to an exploding granule (Rast 1995). The intergranular downdrafts will be of particular interest in the context of the concentration of magnetic flux, cf. Sect. 8.2.1.

A further result of the numerical simulation is a reversal of the topology, from the connected intergranular lanes at the surface to isolated downflows at greater depth. Petrovay (1990) has suggested that such a reversal should occur at a depth of only a few hundred kilometers, as a consequence of the horizontal convergence and decreasing area filling fraction of the descending flow.

The depth where 50% of the hydrogen is ionized is about 2000 km, cf. Table 6.1. If the largest granular cells extend from this depth to the surface, and if the vertical and the horizontal extent of these cells are about the same, then we have an explanation for the observed diameter of the largest granules. Another limitation for the diameter of granules is buoyancy breaking: the numerical simulations show that only granules smaller than, say, 2000 km in diameter can survive for several minutes against the central disruptive force (Steffen et al. 1989). In general it now appears that the granulation structure seen on the Sun is governed by the details of the cooling process at the surface rather than by the heat flow from within the Sun.

6.3.3 Mean Line Profiles

Convective Blueshift. The lines in a solar spectrum that does *not* resolve the photospheric (granular) fine structure generally have wavelengths slightly different from the same lines measured in the laboratory. Such differences remain after corrections have been made for the Sun's motion relative to the observer, and for the gravitational redshift that corresponds to a velocity $Gm_\odot/r_\odot c$, or 636 m/s.

In most cases the residual is a *blueshift*. Its origin is the interplay between the granular intensity and the velocity field. The rising (blueshifted) granules are brighter and, although they may cover a slightly smaller fraction of the total area (cf. Fig. 6.4), they contribute more to the mean line profile than the darker, narrower, and descending (redshifted) intergranular lanes.

At disc center the convective blueshift typically is a few 100 meters per second, with a spread of about 1 km/s. Figure 6.13 shows the effect for a number of Fe I lines. There is a tendency, in particular among the weaker lines, to be more blueshifted if the lower excitation potential χ is larger. This is plausible since these lines are predominantly formed in the hot granules.

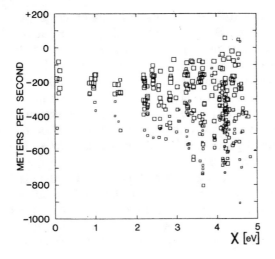

Fig. 6.13. Convective blueshift for Fe I lines, as a function of lower excitation potential. The area of each *square* is proportional to the line absorption depth. From Dravins (1982)

For strong lines the dependence on χ is less (except for the increasing scatter) because these lines are formed higher in the atmosphere, where, as we have seen, the correlation between intensity and upward velocity essentially is lost.

The Limb Effect. When the mean line position is measured as a function of position on the Sun it is found that the convective blueshift *decreases* from the center towards the limb of the solar disc. Of course the Doppler shift caused by the Sun's rotation must be eliminated in order to isolate the effect, for example by measuring along the central meridian. The decrease of the blueshift is called the *limb effect*. It is to be expected because the granular contrast decreases towards the limb and because the vertical velocity makes an increasing angle with the line of sight so that the Doppler shift decreases. The projection effect may even generate a redshift near the limb: the receding horizontal motion at the far side of granules is seen against bright granules behind.

Mean shifts of the Ti I line at 571.3 nm in *sunspots* have been measured by Beckers (1977). With respect to the laboratory wavelength there is only the predicted gravitational redshift, independent of the position on the solar disc. Since the convective energy transport is inhibited in sunspots (Sect. 8.3), this indirectly confirms that the blueshift with its center-to-limb variation is of convective origin.

Line Asymmetry. In addition to their Doppler shift, mean line profiles show an *asymmetry*. This asymmetry also has its origin in the granular convection, as demonstrated in Fig. 6.14. Schematically, the line is composed of granular and intergranular contributions. Both of these are already asymmetric by themselves because the center of the profile originates at greater height than the wings, and therefore has a different Doppler shift due to the dependence of the velocity field on height. Averaging over the blueshifted granular

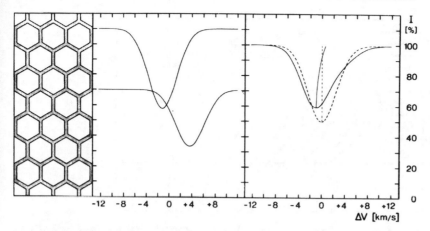

Fig. 6.14. Origin of the line asymmetry. Granular and intergranular regions (*left*) contribute different profiles (*center*). The average is an asymmetric line, with a C-shaped bisector (*right, solid*). The *dashed*, symmetric, line would result in the absence of convection. From Dravins et al. (1981)

and the redshifted intergranular contributions leads to the small net blueshift already described, but this blueshift is more pronounced in the line center than in the wings (the extreme wings may actually show a redshift), and thus a further asymmetry is introduced.

The line asymmetry is commonly represented by the *bisector*, which is the curve dividing any iso-intensity line across the profile into two halves. Often the bisector has the form of a "C". This *C shape* depends in a systematic manner on line strength and on the lower excitation potential (Fig. 6.15), as well as on other line parameters. Moreover, it varies with position on the

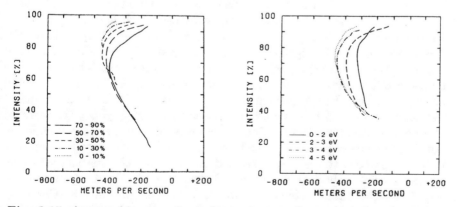

Fig. 6.15. Average bisectors, from observations at disc center, for solar lines of various strengths (*left*) and various lower excitation potentials (*right*). Adapted from Dravins (1982)

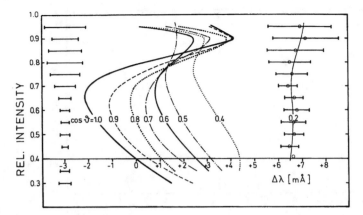

Fig. 6.16. Center-to-limb variation of the bisector for the Fe I line at $\lambda = 557.61$ nm. The bars show the rms variation at $\cos\theta = 1.0$ (*left*) and 0.2 (*right*). After Brandt and Schröter (1982)

solar disc: Figure 6.16 demonstrates this effect, but also shows the general limb effect as discussed above. Further, the oscillatory velocity field (Severino et al. 1986; Gomez et al. 1987) and the presence of a magnetic field in the solar atmosphere (Livingston, 1982) have an effect on the mean line profile, especially its asymmetry.

The main virtue of the analysis of mean line profiles is the possibility to investigate *stellar* granulation, which cannot be spatially resolved by direct observation. For such investigations the Sun plays a key role because only here are both spatially resolved *and* mean profiles available.

We have explained the convective blueshift, the limb effect, and the line asymmetry in a qualitative way by rising granules and sinking intergranular matter. Early quantitative calculations were based on compositions of normal one-dimensioned model atmospheres, each with is own temperature stratification, as described in Chap. 4, and with its own Doppler shift. The "3-stream" model of Voigt (1956) had redshifted and blueshifted fractions, and a fraction at rest; the "2-stream" model of Schröter (1957) had redshifted and blueshifted fractions only. More detailed calculations of mean line profiles are possible with the hydrodynamic model described in Sect. 6.3.2. In such a (3-dimensional) model the equation of radiative transfer, (6.44), is integrated along various selected rays to obtain theoretical line profiles as a function of position. The average, in space and (or) in time, is then the synthetic profile to be compared with the mean observed profile. The main characteristics described qualitatively in this section find their theoretical explanation in this way. Even the parameters of "micro" and "macro" turbulence (Chap. 4), which were introduced to explain non-thermal line broadening and line strengthening, appear to be unnecessary when the full 3-dimensional velocity field is taken into account.

6.4 Mesogranulation

At first sight the solar granulation displays the irregular, sometimes polygonal, pattern shown in Fig. 3.5, without any conspicuous organization at larger scales. A closer inspection however reveals that such organization might indeed exist.

An example is the distribution of "active" granules, i.e., granules which for some time continue the reproductive sequence: expansion, fragmentation, expansion of fragments, fragmentation of fragments ... . Using photographs taken at the Pic-du-Midi observatory, Oda (1984) found that these active granules form a network, with typical mesh size 10″, as shown in Fig. 6.17. The network is correlated with the photospheric brightness distribution. Hirzberger et al. (1999 b) found a related distribution for the exploding granules; their study was based on white-light images from the Swedish Vacuum Solar Telescope, La Palma.

The pattern shown in Fig. 6.17 is reminiscent of convection cells. A velocity field with the same cell size in fact had been discovered prior to Oda's work by November et al. (1981), and christened *mesogranulation*. Before the mesogranulation can be seen in the Doppler signal, some data reduction is necessary, as demonstrated in Fig. 6.18: the larger-scale (typically 40″) supergranulation must be subtracted, and the oscillatory velocity field must be removed by taking an average over time. For the Mg I line at $\lambda = 517.3\,$nm, which was used to measure the velocity shown in Fig. 6.18, the supergranular rms velocity is $\approx 40\,$m/s, comparable to the mesogranular flow, and the oscillatory signal is $\approx 500\,$m/s, i.e., larger by an order of magnitude!

As far as small scales, i.e., the granules, are concerned, the velocity signal is smoothed both by the seeing during the observation and by the time averaging. What is left of the granular contribution seems to have little effect on the mesogranulation: the signal is equally clear in the images with $1″ \times 1″$ and $3″ \times 3″$ resolution [(e) and (f) of Fig. 6.18; nevertheless, Wang (1989) and

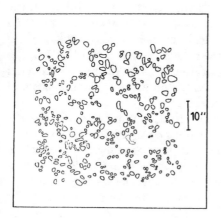

Fig. 6.17. Distribution of "active" granules. After Oda (1984)

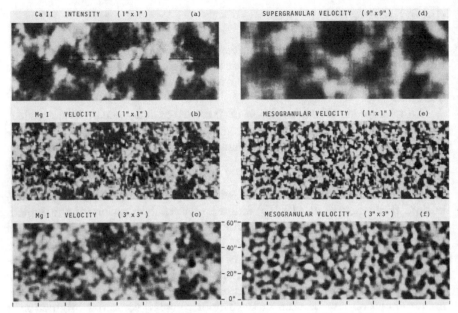

Fig. 6.18. Chromospheric network (**a**), and velocity maps (**b–f**; upflow *dark*, downflow *light*) showing super- and mesogranulation. The maps (**b**) to (**d**) contain the total velocity signal, at various resolutions as indicated. In (**e**) and (**f**) the supergranular velocity (**d**) is subtracted so that the mesogranulation is seen more clearly. From November et al. (1981)

Rieutord et al. (2000) attribute the phenomenon to residues of the averaging procedure].

The vertical component of the mesogranular velocity varies between $\pm 150\,\text{m/s}$, and the vertical rms velocity is $60\,\text{m/s}$. November et al. (1981) conjecture that the phenomenon is overshooting convection originating at greater depth than the ordinary granulation, and that the latent heat available in partially ionized He I provides the driving. According to Table 6.1, the first ionization of helium reaches 50% at a depth of $\approx 6000\,\text{km}$; if convection cells have comparable horizontal and vertical dimensions, this would explain the observed size of mesogranules. However, as is evident from Fig. 2.4, the ionization of He I does not result in a separate maximum of the specific heat; hence a distinct depth of enhanced buoyancy and ensuing distinct scale of convection cannot be expected on this ground.

Since the discovery by November et al. (1981) a number of studies (21 are listed by Rieutord et al. 2000) have confirmed dynamic phenomena at intermediate scales between granulation and supergranulation. Horizontal velocity components have been determined by the method of *local correlation tracking*, i.e., by using the (smaller-scale) granules as tracers (November 1986). Time sequences of high-quality granulation images are necessary for this purpose; moreover, one must be aware that granular evolution could be mis-interpreted as granular proper motion. Results from a space experiment (Title et al. 1987) as well as from the Swedish Solar Observatory at La Palma (Brandt et al. 1988) indeed show flow fields on the mesogranular scale. The horizontal velocity component varies around 500 m/s, but may reach 1 km/s; it generally exceeds the vertical component. Shine et al. (2000) present results of local correlation tracking, based on a 45-h sequence of MDI data in the high-resolution mode; in particular they confirm that mesogranules move toward the boundaries of the supergranulation cells.

Brandt et al. (1991) found that size, intensity, lifetime, and rate of expansion of granules vary with their location in the mesogranular pattern. An intensity modulation, with an amplitude of order 1 %, had been reported earlier by Koutchmy and Lebecq (1986), and a positive correlation with the vertical velocity indicates a convective character in the deep photosphere (Straus et al. 1992, Straus and Bonaccini 1997). On the other hand, these investigations could not establish the mesogranulation as convection cells with a distinct scale; it rather seems to be a large-scale extension of the granulation, best visible in regions of positive horizontal-flow divergence, which often coincides with exploding granules (e.g., Fig. 6.7).

6.5 Supergranulation

6.5.1 The Velocity Field and the Network

When Hart (1956) spectroscopically measured the solar rotation she found a velocity field that fluctuates across the visible disc. Moreover, the autocorrelation function of these fluctuations had a secondary maximum at a mean lag $d \approx 26000$ km between the two correlated points, in addition to the primary maximum at $d = 0$.

Subsequently Leighton et al. (1962), using their technique of photographic Doppler spectroheliograms (Sect. 3.4.7), found a cellular pattern, distributed uniformly over the Sun, with a typical cell diameter of 1.6×10^4 km and a mean spacing of $\approx 3 \times 10^4$ km between cell centers. They named this pattern the *supergranulation*. A study of Schrijver et al. (1997) shows that the distribution of supergranulation cells is in fact very similar to the distribution of granules, in spite of the large difference in length scale. As Fig. 6.19 demonstrates, the flow is almost invisible at the disc center by means of the Doppler effect; i.e., it is predominantly horizontal. Nevertheless it can be investigated at disc center by the method of local correlation tracking, where the granulation (and even the mesogranulation) that is carried along is used as a tracer. The measured horizontal velocity is most frequently in the range 300–500 m/s, directed radially from the center to the periphery of the cells; the distribution derived by Shine et al. (2000) extends to ≈ 1 km/s.

The vertical velocity of supergranulation cells is smaller. A photoelectric measurement, made by Küveler (1983) at the Locarno solar observa-

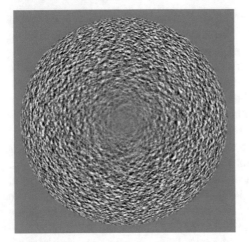

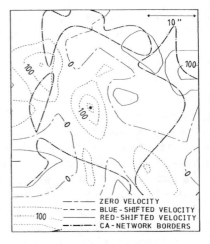

Fig. 6.19. Dopplergram of the entire Sun, showing the supergranulation (*dark*: approaching, *bright*: receding flow). From SOHO/MDI, courtesy G. W. Simon

Fig. 6.20. Velocity contours (labels in m/s) of a supergranulation cell, observed at disc center. Adapted from Küveler (1983)

tory of the University of Göttingen, is shown in Fig. 6.20. The Doppler shift of the Fe I line at 630.1517 nm was measured relative to the terrestrial O_2 line at 630.2005 nm. Both the central upflow (≈ 50 m/s) and the downflow (≈ 100 m/s) at the boundaries are clearly visible. However, unlike the granules, the supergranules seem to have *unconnected* areas of downflow distributed along their cell boundaries. Possibly this reflects the topological structure at deeper levels of the convection zone, which has isolated downdrafts according to numerical simulations. The downdrafts often coincide with the corners, or "vertices", where several supergranules meet (Frazier 1970).

Figure 6.20 also demonstrates the close coincidence of the supergranule boundaries with the *chromospheric emission network*, first proved by Simon and Leighton (1964). The connection is established by the magnetic field: we shall see in Chap. 8 that magnetic flux tends to be concentrated by the flow converging into the downdrafts, and we shall see in Chap. 9 that chromospheric heating predominantly occurs in regions of enhanced magnetic field strength.

The lifetime of supergranules is of the order of 1 day and therefore difficult to measure from the ground. Uninterrupted sequences of Doppler images such as Fig 6.19 have been obtained with the Michelson Doppler Imager on SOHO. A lifetime of 15–30 h has been derived from these sequences, although it had been known for some time (e.g., Kubičela 1976) that a few supergranules survive for 3 or 4 days.

The chromospheric network (Fig. 6.21) appears to have a similar mean life: Observations in Ca II, made on up to three consecutive days, yield a mean lifetime of 20 h (Simon and Leighton 1964); and 62 hours of continuous

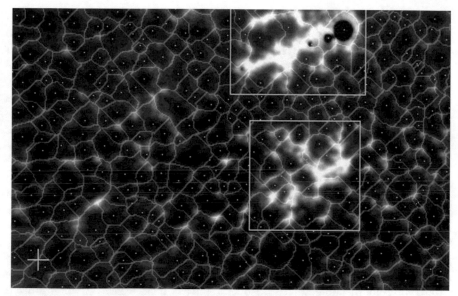

Fig. 6.21. Ca II K filtergram (1 nm bandpass), averaged over 8 h, with cell boundaries determined from the intensity gradient field. The white boxes enclose *plages*; the cross at the lower left is $31 \times 31 \, \mathrm{Mm}^2$. From Hagenaar et al. (1997)

observation during the arctic summer (in Thule) gave 25 h for the network observed in Hα (Rogers 1970). The material from Thule has also been analyzed in a morphological study by Janssens (1970). He found that, on the average, the network cells loose their identity after 21 hours.

6.5.2 Convective Nature

In order to understand the nature of the supergranulation it is necessary to observe *intensity* (i.e., temperature) fluctuations. Since the flow seen at the surface might be overshooting cells driven at greater depth, the deepest observable layer contains the most interesting information. For this reason Worden (1975) chose a wavelength at 1.64 μm, where the continuum opacity

Table 6.3. Intensity and temperature variation at supergranulation boundaries. From Worden (1975)

λ [μm]	Height [km] above $\tau_{500} = 1$	$\Delta I/I$	ΔT [K]	T_{B} [K]
1.71	200–500	0.055	230	4200
1.17	10	0.004	25	6200
1.64	−20	−0.007	−50	6800

has a minimum. The level observed at this wavelength is about 20 km below the level where $\tau_{500} = 1$. For comparison, intensity variations were also recorded at $\lambda = 1.17\,\mu\text{m}$ (corresponding to $\tau_{500} \approx 1$) and in a strong Mg I line at $1.71\,\mu\text{m}$ (corresponding to the temperature minimum). Table 6.3 summarizes the results for the intensity variation ΔI on the supergranular scale. This variation is large and (at the cell boundaries) *positive* at the highest level; in fact in the Mg I line we already see the chromospheric network. The intensity variation is still positive (but smaller) in the photosphere. Only at the deepest accessible level ΔI is *negative* at the cell boundaries, corresponding to a flow with upward convective energy transport.

The temperature variation listed in Table 6.3 is calculated according to $\Delta T = T_B \Delta I / I$, which follows from the Rayleigh–Jeans approximation of Planck's law; T_B is the brightness temperature (Sect. 1.5.3). Worden suggests that the ΔT values should be corrected by factors up to 10 because of the low spatial resolution of his observations.

Like the ordinary granulation (and the mesogranulation) supergranulation is generally interpreted as convection, with associated overshooting into the upper regions of the solar atmosphere. Latent heat, released when ionized He atoms recombine, has been proposed as the driving agent (Simon and Leighton 1964). We see from Table 6.1 that 50% of He II is ionized at a depth of about 20 000 km, and Fig. 2.4 shows that the specific heat has a small local maximum at that depth, which is of the same order as the horizontal extent of the supergranular cells. Although it has not yet been possible to prove the penetration of the supergranular flow to the depth range discussed here, the helioseismic time-distance method (Sect. 5.3.9) has shown that the flow as well as the associated temperature variation extend several thousand kilometers below the solar surface.

6.5.3 The Effect of Rotation

The larger the scale of a velocity field, the more sensitive is the flow to the action of the Coriolis force that arises in a rotating system. The rotational effect is estimated by the inverse of the Rossby number

$$Ro = u/(2\Omega l) \,, \tag{6.45}$$

which is essentially the ratio of the inertia and Coriolis forces; u and l are a typical velocity and a typical scale, and Ω is the average rate of rotation.

For supergranules $u \approx 500\,\text{m/s}$ and $l \approx 10^7\,\text{m}$; with $\Omega \approx 3 \times 10^{-6}\,\text{s}^{-1}$ we obtain $Ro \approx 10$. The effect of rotation, which is manifest in an azimuthal (vortex-type) velocity component within each cell, therefore is of order $1/10$: considering the dependence of the Coriolis force on latitude ψ we may expect an azimuthal motion of order $\sin\psi \times 50\,\text{m/s}$. This is small, but some spectroscopic evidence has been found by Kubičela (1973). For 32 supergranulation cells he found that the zero-velocity line across a cell generally is *not* perpendicular to the radius pointing from the disc center towards the cell. Instead, there is an average angle of $\approx 80°$. The sign of the deviation is such as expected from the Coriolis effect. An azimuthal velocity component was also found by Zhang et al. (1998) who traced the proper motion of magnetic flux concentrations within supergranulation cells; they found that the magnitude of this component increases from zero at the cell center to $\approx 400\,\text{m/s}$ near the cell boundaries, and that the direction is consistent with the action of the Coriolis force.

Another effect of solar rotation is the apparent decrease with latitude of the mean cell size of supergranules. High-latitude cells are smaller by $\approx 10\%$ than cells in the equatorial zone (Rimmele and Schröter 1989), an effect that also has been found for the cells of the chromospheric emission network by Brune and Wöhl (1982) and confirmed by Münzer at al. (1989). Presently there is no theoretical explanation for this dependence on latitude of the supergranular cell size.

6.6 Giant Cells

Patterns on the solar surface with typical length scales of 10^8 m and more are commonly called "giant cells". Other names are "giant granulation", indicating a physical analogy to the convective cells of smaller scale, and "global convection", which expresses the possible extent of the giant cells over the entire depth of the convection zone.

Giant cells are most elusive: although numerous observational studies have been made, no typical and persistent pattern has yet been established. Yet such giant cells are most interesting: with $l \approx 10^8$ m and $u \approx 100\,\text{m/s}$ (see below) their Rossby number (6.45) is much smaller than 1, and the effect of solar rotation upon giant cells should therefore be significant. Indeed, as we shall see in Chap. 7, giant cells possibly play a key role in the theory of the Sun's differential rotation.

6.6.1 Tracer Results

Giant convective cells have been inferred from the peculiar motions of smaller-scale features across the solar surface. Such tracers include the supergranulation and the related chromospheric network, inhomogeneities in the magnetic field distribution, and the visible consequences of such inhomogeneities, i.e., sunspots, filaments, etc. In fact, it was the appearance of complexes of (magnetically) active regions which, during the 1960s, first led to the concept of a convective pattern consisting of giant cells (Bumba 1967).

A particularly nice, although very rare, example of a regular distribution of filaments is shown in Fig. 6.22. As filaments are supported by the magnetic field, and the magnetic field is carried along and concentrated by the

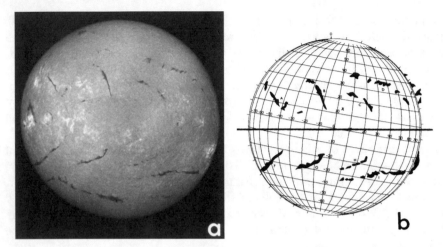

Fig. 6.22. Distribution of solar filaments on 11 June 1972 (Wagner and Gilliam 1976). Courtesy National Solar Observatory, AURA, Inc.

convective flow, a regular arrangement of elongated convective cells is very suggestive.

Because of their ubiquity the mottles forming the chromospheric network are particularly suited as tracers for the large-scale velocity field. Following an individual mottle for a period of at least several hours yields a velocity arrow; to improve the signal-to-noise ratio averages over a number of neighboring mottles may be taken, with still sufficient spatial resolution to detect the giant cells.

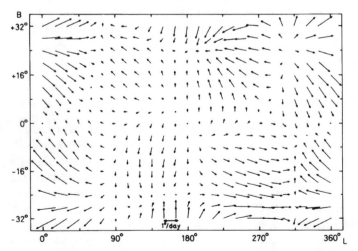

Fig. 6.23. Giant cell velocity field, derived during 1976 from the proper motion of Ca II mottles. From Schwan and Wöhl (1978)

For the period April to June 1975 Schröter and Wöhl (1976) in this way found a regular pattern of 4 cells distributed along the solar equator (i.e., longitudinal wave number $m = 2$). Only a narrow latitude zone ($\pm 12°$) was observed; therefore a divergence or a vorticity of the cell flow could not clearly be distinguished. But velocity differences of up to 80 m/s were measured, and flows *crossing the equator* have been identified. In another study, made during 1976 (Schröter et al. 1978), the latitude range was extended to $\pm 30°$. Again flows crossing the equator were found; in addition, areas of clearly divergent motion could be identified, cf. Fig. 6.23, which shows a spherical harmonic representation of the observed velocity vectors.

6.6.2 Spectroscopic Results

Just as the studies based on tracer methods, spectroscopic investigations have provided occasional evidence for the existence of giant cells, but not yet any permanent pattern, covering the entire solar surface.

An extensive data set has been obtained over many years with a Doppler compensator (Sect. 3.3.3) as a by-product of the solar magnetic charts recorded at the 150-foot tower telescope of Mt. Wilson Observatory. Unfortunately this material yields only upper bounds for the velocity of giant cells: these bounds (rms values) vary from 12 m/s, at longitudinal wave number $m = 1$, to 3 m/s for wave numbers $m \geq 20$ (LaBonte et al. 1981).

A positive spectroscopic result was reported by Pérez Garde et al. (1981). In September 1978 these authors found a sequence of cells each extending about 45° in longitude (corresponding to $m = 4$), with an amplitude of 40 m/s. Similar cell patterns were found by Scherrer et al. (1986) for several epochs during 10 years, cf. Fig. 6.24. Scherrer et al., who evaluated data

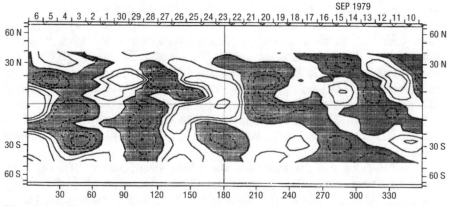

Fig. 6.24. East-west velocity on the Sun, derived from the Doppler shift of the Fe I line at 525.02 nm, during September 1979. Contour levels are ± 10, $\pm 20 \ldots$ m/s (*shaded* eastward). From Scherrer et al. (1986)

obtained at the Wilcox Solar Observatory at Stanford, also enumerate the great difficulties in detecting giant cells on the Sun:

The velocity field must be separated from several other signals, all of which have magnitudes comparable, or even exceeding the giant-cell velocity. On the large scale, the (non-uniform) solar rotation, and the limb effect (Sect. 6.3.3) must be subtracted. On the small scale, the supergranulation and the p-mode oscillations must be removed by proper averaging. Often the observing programs originally were not designed for measuring giant cells, and therefore the data are not well-suited for the various steps of reduction. Relatively easy is the elimination of the granular velocity: its scale is so much smaller than the scale of giant cells that observation at moderate spatial resolution sufficiently removes this signal.

Complications also arise from the solar magnetic field. Local flux concentrations, most prominent in sunspots, tend to reduce the convective motion by means of their Lorentz force. Even with all precautions in the data reduction, this effect could generate a spurious velocity signal on the scale of active regions, i.e., on the giant-cell scale. Another possible source of confusion is the use of magnetically sensitive spectral lines, such as the Fe I line at 525.02 nm, for velocity measurements.

Finally, we must keep in mind that spectroscopic measurements always yield the line-of-sight velocity only. Hence, assumptions concerning the space or time dependence must be made when individual components are inferred from measurements obtained at different times or locations on the disc.

6.7 Bibliographical Notes

The Solar Granulation of Bray et al. (1984) not only comprehensively treats the phenomenon mentioned in its title, but also constitutes a general guide to the field of solar convection. Accounts of the mixing-length theory, both historical and modern, can be found in the proceedings edited by Spiegel and Zahn (1977). The version used for the solar models of the present text is described by Kippenhahn et al. (1967), Baker and Temesváry (1963), or in the book of Cox and Giuli (1968). Simon and Weiss (1968) point out that heat transport of cells extending over *several* scale heights could be more efficient. In a mixing-length model they interpret the supergranulation and the giant cells in this way. The critical remarks on the mixing-length theory made by Renzini (1987) predominantly refer to the application of this formalism to convection in stellar cores.

Modifications and alternatives to the mixing-length theory have been proposed by Xiong (1985), Xiong and Chen (1992), Lumer et al. (1990), Forestini et al. (1991), Canuto and Mazzitelli (1991), and Canuto et al. (1996). These theories, like the original formalism, contain adjustable parameters and are calibrated with the help of the present Sun. Canuto (1993, 1996, 1999) presents a critical review and advocates the use of differential equations that govern the Reynolds stresses, the convective flux and related correlations (see also Grossman 1996, Canuto and Dubovikov 1998; the latter paper also comments on the asymmetry between up- and downdrafts). Chan and Sofia (1996) consider the problem of closure with respect to the higher-order moments. Lydon et al. (1992, 1993) have replaced the mixing-length equations altogether by relationships between the dynamic and thermodynamic quantities which they derived from numerical simulation; without adjustable parameters, they were able to predict the radius of the present Sun within an error of $\approx 1\,\%$. All these alternatives also allow for convective overshooting (it should be noted here that the proper fluid-dynamics term for this phenomenon is *penetrative convection* if, as in the case treated in Sect. 6.2.3, the mean stratification is substantially affected). Zahn (1991) gives a general discussion.

Spruit et al. (1990) and Spruit (1997) review stellar convection in comparison to laboratory experiments and numerical simulation. The main conclusion is that the observed pattern is essentially determined by the cooling at the solar surface. The downdrafts seen as intergranular lanes form isolated channels at greater depth, merging to successively larger scales. Such *plumes* may dive through the entire convection zone (Rieutord and Zahn 1995). At the surface their locations may appear as especially long-lived intergranular "holes" (Roudier et al. 1997). Numerical studies of convective overshooting show penetrative flows over a sizable fraction of a pressure scale height, both above and below the unstable region (Singh et al. 1994, 1995, 1998). Asplund et al. (2000 a) consider the effects of grid resolution in numerical simulations, while the anelastic approximation is discussed by Lantz and Fan (1999).

Balthasar (1984, 1988) has compiled parameter lists for 143 asymmetric spectral lines obtained with a Fourier transform spectrometer. Canfield and Beckers (1976) and Dravins (1982) describe the effects of unresolved motions on spectral lines. Kaisig and Durrant (1982) employ a perturbation analysis to model the line asymmetry, and Dravins et al. (1986) extend the work of Dravins et al. (1981) to Fe II lines. Márquez et al. (1996) found that internal gravity waves and a spatially variable microturbulence parameter yield a good match to observed asymmetric line profiles. Asplund et al. (2000 bc) find that the observed width and asymmetry of Fe lines can be matched in three-dimensional radiative-hydrodynamical simulations without recourse to micro- and macroturbulence concepts, and Gadun et al. (1999) reproduce a number of observed line-profile features even with two-dimensional models. *Stellar* line asymmetries and wavelength shifts, and implications on stellar granulation, were investigated by Dravins (1987 a), and first stellar bisectors, among others for α Cen A and Procyon, have been determined by Dravins (1987 b).

Reviews on resolved velocity patterns are those of Beckers (1981) and Beckers and Canfield (1976). Müller et al. (2001) study over 42 000 individual granules with an automatic tracking technique and find a mean lifetime of 7.5 min, slightly larger than the result of Hirzberger et al. (1999 a). The strong dependence of the granular intensity contrast on the form of the spread function has been confirmed by Collados and Vázquez (1987). A related discussion, on the restoration of the granular *velocity*, has been presented by Mattig (1980). Incidentally, there is not only atmospheric and instrumental smearing of granular information; through repeated scattering *within the solar atmosphere* horizontal fluctuations may be reduced in magnitude, i.e., the Sun itself sets limits to the resolution (Kneer 1979). Nevertheless, sharp edges caused, e.g., by the magnetic structure of the atmosphere may be recognized down to a scale of order 10 km (especially in polarized light) if a sufficiently large telescope is used (Bruls and von der Lühe 2001).

Meso- and supergranulation velocities, in particular their variation with height in the solar atmosphere, have been spectroscopically studied by November et al. (1979, 1982). The rotational effect on the supergranulation was measured by time-distance helioseismology (Duvall and Gizon 2000). Hill (1989) uses his method of ring diagrams to deduce flows of order 40 m/s on scales larger than the supergranulation. Volumes 192 and 193 (2000) of *Solar Physics* cover the helioseismic diagnostics of solar convection.

Bumba (1987) once more emphasizes the possible relationship between large-scale inhomogeneities of the solar magnetic field and giant-cell convection. A special spectroscopic search for giant cells at high solar latitude was made by Cram et al. (1983) and Durney et al. (1985). The strong rotational influence should render the giant convective cells elongated in the direction parallel to the axis of rotation (cf. Sect. 7.5.3) and thus, it was argued, their visibility should be best in the vicinity of the poles. But the evidence is marginal, with an upper bound of ≈ 5 m/s. In contrast to the theoretical prediction Ribes et al. (1985) report the discovery of an east-west oriented convective structure. Proper motions of sunspots, measured in spectroheliograms from the Paris observatory, are the basis of this finding.

The amplitude, or the upper bounds for the amplitude, of giant cells is about an order of magnitude smaller than predicted by the global circulation models described in Sect. 7.5.3. A possible answer to this discrepancy is a confinement of the giant cells to the lower part of the convection zone, and a screening by the above-lying, smaller-scale convection (Stix 1981 a; van Ballegooijen 1986).

7. Rotation

The discovery of solar rotation immediately followed the advent of the telescope and its use around 1610 by Galileo Galilei, Johannes Fabricius, Christoph Scheiner, and Thomas Harriot. These astronomers saw sunspots on their apparent move across the solar disc. Moreover, the early records such as Scheiner's *Rosa Ursina*, published in 1630, contain evidence that the sunspot passage took a little longer at higher solar latitude than near the equator, although the interpretation in terms of a *differential rotation* of the Sun came only much later. The subject has attracted solar physicists up to this day. New details are being observed, and old puzzles, such as the origin of the differential rotation, remain unsolved.

Observation and theory of the rotation seen on the solar surface constitute the major portion of this chapter. First, however, we shall address a few questions concerning the rotation of the Sun as a whole. The solar oblateness and the solar rotational history are two examples. We shall begin with a brief discussion of the direction of the Sun's axis of rotation.

7.1 Axis of Rotation

The direction of the solar axis of rotation in space is determined by two *rotation elements*. The first, i, is the angle of inclination between the ecliptic plane and the equatorial plane of the Sun. The second, Ω, is the ecliptic longitude of the ascending node of the Sun's equator, i.e., the angle, counted in the ecliptic, between the equinox direction and the direction where the solar equator cuts the ecliptic "from below" (in the sense of rotation). Due to the precession of the Earth's axis the equinox direction recedes by 0.01396 degrees per year. Therefore Ω increases by the same rate, and the epoch must be specified for any given value.

Most measurements of the rotation elements are derived from the apparent paths of sunspots. On the basis of his own measurements 1853–1861, and earlier results of other observers, Carrington (1863) recommended the values $i = 7.25°$ and $\Omega(1850) = 73.67°$ (cf. Stark and Wöhl 1981). Since 1928 these values have been used exclusively in calculations of heliographic coordinates. Nevertheless, recent investigations, based on much larger data sets than used by Carrington, suggest that i is slightly smaller: Balthasar et al. (1986 a), using sunspot group observations made between 1874 and 1976 at Greenwich, and between 1947 and 1984 at Kanzelhöhe, Austria, found $i = 7.137° \pm 0.017°$. For recurrent spots alone (those which are seen a second time after passing the Sun's far side) the Greenwich data yield $i = 7.12° \pm 0.05°$ (Balthasar et al. 1987), which is also smaller than Carrington's result. Carrington's Ω, on the other hand, lies within the range of modern determinations.

Heliographic Coordinates. Once the axis of solar rotation is known the solar latitude is easily defined as the angular distance ψ from the equator. The distance to the North Pole is $\theta = \pi/2 - \psi$; this quantity, also called the colatitude, is often used in theoretical work.

Longitude is more difficult since there are no fixed points on the Sun. Following R. C. Carrington, we divide the *time* into intervals of 27.2753 days, called *Carrington rotations* because this period was determined by Carrington as the mean synodic rotation period of sunspots. These intervals have consecutive numbers: rotation number 1 commenced on 9 November 1853, number 1978 was completed on 30 June 2001. At the date of commencement of a new rotation the center of the solar disc has longitude $\phi = 0$.

The coordinates ψ and ϕ are measured in a system envisaged to rotate with the Sun. In addition to these we may define coordinates β and λ in an *ecliptic* system. These angles are related to ψ and ϕ and to the rotation elements trough (cf. Waldmeier 1955, p. 42)

$$\sin \psi = \cos i \sin \beta - \sin i \cos \beta \sin(\lambda - \Omega) , \tag{7.1}$$

$$\tan \phi = \cos i \tan(\lambda - \Omega) + \sin i \tan \beta \sec(\lambda - \Omega) . \tag{7.2}$$

Problem 7.1. Yearbooks such as *The Astronomical Almanac* contain for each date the values of B_0 and L_0, the heliographic coordinates of the disc center, and the value of P, the position angle of the northern extremity of the axis of rotation, measured eastwards from the north point of the disc. Let ρ be the observed heliocentric angular distance of a sunspot from disc center, and θ its position angle on the disc. Show that the spot has heliographic coordinates ψ and ϕ given by

$$\sin \psi = \sin B_0 \cos \rho + \cos B_0 \sin \rho \cos(P - \theta) , \tag{7.3}$$

$$\cos \psi \sin(\phi - L_0) = \sin \rho \sin(P - \theta) , \tag{7.4}$$

$$\cos \psi \cos(\phi - L_0) = \cos \rho \cos B_0 - \sin B_0 \sin \rho \cos(P - \theta) . \tag{7.5}$$

Convince yourself that the solar North Pole is best visible in September, and the South Pole is best visible in March.

7.2 Oblateness

7.2.1 Origin

As a rotating, non-rigid body the Sun must have an oblateness. At the surface, where the oblateness can be observed, there are two sources for such an effect: one is the rotation of the surface layer itself, the other a possible internal quadrupole moment. We shall now derive expressions for these two contributions.

For simplicity, consider an inviscid star in the state of pure steady rigid rotation, i.e., (in spherical polar coordinates)

$$\boldsymbol{v} = (0, 0, r\Omega \sin\theta) , \tag{7.6}$$

and

$$\rho \boldsymbol{v} \cdot \nabla \boldsymbol{v} = -\nabla P - \rho \nabla \Phi . \tag{7.7}$$

The equilibrium condition (7.7) is written in a frame at rest, and Φ is the gravitational potential. The surface of our rotating star is characterized by a constant value of the pressure P. Substitution of (7.6) into (7.7) then implies that, at the surface, the expression

$$\Phi - \frac{1}{2}(r\Omega \sin\theta)^2 \tag{7.8}$$

is a constant. In addition, there must be, at the surface, a continuous transition of Φ to the outer gravitational potential, which consists of the familiar monopole and a small quadrupole:

$$\Phi = \Phi_{\text{ext}} \equiv -\frac{Gm_\odot}{r}\left[1 - J_2\left(\frac{r_\odot}{r}\right)^2 P_2(\theta)\right] . \tag{7.9}$$

Here J_2 is the quadrupole moment, and $P_2(\theta)$ is the second Legendre polynomial.

We approximate the oblate surface by

$$r(\theta) = r_\odot(1 - cP_2(\theta)) , \tag{7.10}$$

where $r_\odot$ is now the mean solar radius and c is a small quantity related to the oblateness:

$$\Delta r/r_\odot \equiv (r_{\text{equator}} - r_{\text{pole}})/r_\odot = 3c/2 . \tag{7.11}$$

Only in combination with (7.10) is expression (7.8), with (7.9) substituted, a constant. Using $\sin^2\theta = 2(1 - P_2)/3$ we finally obtain, to leading order,

$$\frac{1}{3}(\Omega r_\odot)^2(1 - P_2) + \frac{Gm_\odot}{r_\odot}(1 + cP_2) - \frac{Gm_\odot J_2 P_2}{r_\odot} = \text{const. .} \tag{7.12}$$

We collect the terms proportional to P_2, eliminate c by means of (7.11), and use $g_\odot = Gm_\odot/r_\odot^2$:

$$\frac{\Delta r}{r_\odot} = \frac{1}{2}\frac{\Omega^2 r_\odot}{g_\odot} + \frac{3}{2}J_2 \ . \tag{7.13}$$

In the derivation of (7.13) we have ignored that the Sun rotates differentially. On the other hand, we have admitted a quadrupole moment, which possibly arises from a core rotating more rapidly than the surface. This apparent inconsistency can however be removed in a more general treatment which allows for differential rotation (cf. Problem 7.2). Then the value of Ω to be used in (7.13) is a mean angular velocity at the surface.

If we take Carrington's synodic rotation period (Sect. 7.1), we obtain the synodic angular velocity $\Omega_{\mathrm{syn}} = 2.67 \times 10^{-6} s^{-1}$, and $\Omega_{\mathrm{sid}} = 2.87 \times 10^{-6} s^{-1}$ as the sidereal rate. The contribution of the solar surface rotation to the oblateness then is 1.04×10^{-5}.

Problem 7.2. For a rotating star described by (7.6) and (7.7) introduce an effective gravitational acceleration, defined by $\rho g_{\mathrm{eff}} = \nabla P$. Show that the following three statements are equivalent: (a) g_{eff} can be derived from a potential, Ψ; (b) Ω is constant on cylinders, i.e., $\Omega = \Omega(s)$, where $s = r \sin\theta$; (c) the surfaces $P = \mathrm{const.}$ and $\rho = \mathrm{const.}$ coincide with each other (and with the surfaces $\Psi = \mathrm{const.}$).

7.2.2 Measurements

Observed values of $\Delta r/r_\odot$ are listed in Table 7.1. The first value given is considerably in excess of the surface contribution. It would mean that $J_2 \approx 2.5 \times 10^{-5}$; with this quadrupole moment the gravitational potential of the Sun would deviate from the spherically symmetric form by an amount that would cause the perihelion of Mercury's orbit to rotate by about $3.5''$ per century. This in turn would require a modification of Einstein's general theory of relativity which already accounts for the full $43''$ per century not explained by classical celestial mechanics and special relativity. Moreover, the larger quadrupole moment, if due to a rapidly spinning core, would imply an internal angular velocity about 20 times larger than Ω at the surface, in contradiction to the results obtained from helioseismology. On the other hand, the more recent oblateness measurements are quite consistent with the notion that the Sun's visible oblateness is essentially caused by the surface rotation alone. In fact the internal rotation derived from helioseismology (Sect. 7.4.1) has been used by Pijpers (1998) to estimate $J_2 = (2.18 \pm 0.06) \times 10^{-7}$.

The measurement of $\Delta r/r_\odot$ is difficult. Dicke et al. project the Sun onto an occulting disc which is slightly smaller than the solar image. The light passing the edge is scanned by two diametrically opposed rotating windows, and the signal measured in this way is converted into information concerning

Table 7.1. Measurements of the solar oblateness $\Delta r/r_\odot$

Dicke and Goldenberg (1967)	$(5.0\ \pm 0.7)\ \times 10^{-5}$
Hill and Stebbins (1975)	$(0.96 \pm 0.65) \times 10^{-5}$
Dicke et al. (1985)	$(2.0\ \pm 0.15) \times 10^{-5}$
Dicke et al. (1986)	$(0.58 \pm 0.14) \times 10^{-5}$
Sofia et al. (1994)*	$(0.92 \pm 0.13) \times 10^{-5}$
Lydon and Sofia (1996)	$(0.88 \pm 0.1)\ \times 10^{-5}$
Rozelot and Rösch (1997)	$(0.93 \pm 0.22) \times 10^{-5}$
Kuhn et al. (1998)	$(0.83 \pm 0.06) \times 10^{-5}$

* as quoted by Lydon and Sofia (1996)

the Sun's shape. Hill and Stebbins also use two windows at diametrically opposed edges of a solar image. At a fixed position angle, these windows are scanned across the solar limb; the scan signal is fed into a servo system to adjust the distance between the two windows. The adjustment itself is measured interferometrically. The whole apparatus can be turned so that equatorial and polar solar diameters can be measured. The further results listed in Table 7.1 have been obtained with the balloon-borne *Solar Disk Sextant* of Maier et al. (1992), with a rotatable scanning heliometer at the Observatoire de Pic du Midi, and with the Michelson Doppler Imager during roll maneuvers of the SOHO satellite.

An oblateness of 10^{-5} corresponds to a difference of $14\,\mathrm{km}$ between the equatorial and polar solar diameters, or to $0.02''$. This is an order of magnitude less than can be resolved with the best solar telescopes during excellent observing conditions! Clearly, a large number of measurements must be made before a statistically meaningful result is obtained. Even more difficult is the assessment of all possible *systematic* errors. The latter may arise both in the instrumentation (e.g., inaccurate centering of the solar image) and on the Sun itself (e.g., latitude-dependent brightness of the limb).

7.3 Rotational History

7.3.1 The Initial State

We have seen in Chap. 2 that the total mass of a star, its initial chemical constitution, and its age essentially determine the present state. There, by "initial" we meant the beginning of the hydrogen-burning phase, i.e., of the star's main-sequence life. For the evolution of a *rotating* star we should now, additionally, know the initial angular velocity.

There are several observational and theoretical aspects indicating that the initial Sun had a more rapid rotation than the present Sun. The first is that the specific angular momentum of the material that eventually formed the

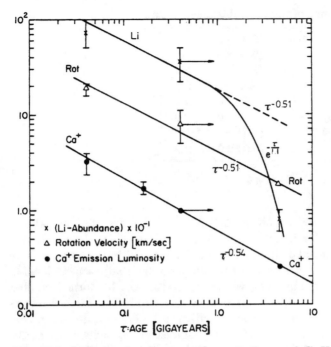

Fig. 7.1. Lithium abundance, surface rotation, and Ca II emission versus stellar age. From Skumanich (1972)

Sun was much larger than that of the whole present solar system. Although much of this angular momentum must have been lost by magnetic braking during an early phase, this loss probably was not complete. The T Tauri stars, which populate a region of the Hertzsprung–Russell diagram passed by the Sun before it arrived on the main sequence, rotate with surface velocities around 15 km/s (the group with $m/m_\odot < 1.25$; see Bouvier et al. 1986), as compared to the Sun's 2 km/s. During their final approach to the main sequence these stars contract and therefore even accelerate their rotation.

The second piece of evidence is that for otherwise similar main-sequence stars the rate of rotation decreases with increasing age. This is demonstrated in Fig. 7.1, together with two other age-dependent stellar properties: the lithium abundance, and the emission in the core of the Ca II lines H and K.

A third argument is indirect. Early main-sequence stars, spectral types O to F, rotate more rapidly, by factors up to 100, than later main-sequence stars, types F to M (Fig. 7.2). The reason is that magnetic braking, first proposed by Schatzman (1959), is only effective among the latter. Stars earlier than, say, F5, have no deep outer convection zone; hence they cannot generate a magnetic field that would provide the lever arm of the braking mechanism. By implication, the initial Sun was a fast rotator, magnetically braked down to the present state.

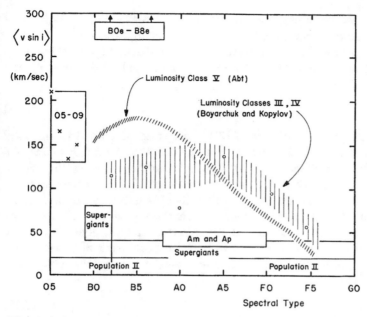

Fig. 7.2. Rotational velocity of main-sequence stars. From Slettebak (1970)

For an evolutionary calculation of the rotating Sun it is appropriate to start a little earlier than at the beginning of the hydrogen burning. Pre-main-sequence solar models (those representing the Sun as it evolved down the Hayashi line, Fig. 2.9) are fully convective. The convection is turbulent, and the associated rapid exchange of momentum soon evens out any gradient in angular velocity. This process, called turbulent friction or turbulent diffusion, allows us to start with a pre-main-sequence star of one solar mass and *uniform* angular velocity Ω. At this time the total angular momentum, J_0, may be obtained by an extrapolation from the earlier spectral types. Pinsonneault et al. (1989) take values of J_0 between 5×10^{42} and $5 \times 10^{43}\,\mathrm{kg\,m^2/s}$, which is up to 260 times the present Sun's angular momentum (assuming the present Sun rotates with $\Omega = \mathrm{const.}$; cf. also Problem 2.4); similarly, Rüdiger and Kitchatinov (1996) start their calculation with a rotation rate that is several tens of times larger than the present.

7.3.2 Torques

Magnetic Braking. Any matter that escapes from the surface of the rotating Sun takes with it some angular momentum. The essence of *magnetic braking* is that this angular momentum does not correspond to the solar surface but to a distance, called the *Alfvén radius* r_A, far above this surface. In a simplified view we may consider the magnetic lines of force as a lever arm that out to r_A forces the escaping material to rotate rigidly with the Sun.

Therefore, the density of angular momentum *increases* for $r < r_A$. Beyond r_A the field becomes too weak to enforce rigid rotation, and the angular momentum of the escaping matter is conserved. The total loss of angular momentum is then the linear (azimuthal) velocity at r_A, multiplied by the lever r_A and by the rate $\dot{m}$ of the mass loss:

$$\dot{J} \simeq \Omega r_A^2 \dot{m} \ . \tag{7.14}$$

Of course the right-hand side of (7.14) may also be viewed as the total torque exerted upon the Sun by the magnetic brake. Although (7.14) is a simple formula, it is difficult to specify this torque for an evolutionary calculation. The Alfvén radius depends on the magnetic field, whose strength in the past, $\boldsymbol{B}(t)$, is not known. The field itself is generated by a dynamo (Sect. 8.4), with Ω and its derivatives as essential ingredients. Therefore r_A depends on Ω. In a similar chain of arguments the mass loss depends on coronal heating and thus on $\boldsymbol{B}$ and finally on Ω. To circumvent all these poorly known processes it is common to parameterize the torque by a power law, and replace (7.14) by

$$\dot{J} = K\Omega^\alpha \ . \tag{7.15}$$

An estimate of α must follow from the physics just discussed; the constant K is fixed by the requirement that the present Sun rotates at the observed rate.

Angular Momentum Transport in the Interior. Magnetic braking directly affects only the *surface* of the Sun. Still, we assume that a state of uniform rotation was maintained as long as the Sun was fully convective. The internal torque required for the outward transport of angular momentum was provided by turbulent friction (strictly, a gradient of Ω must be present for the transport to occur; but since the process of turbulent friction is so efficient a rather small gradient suffices, and we may consider Ω as a constant). Thus, during this early phase the Sun was essentially slowed down as a whole.

But then, during the final contraction to the main sequence, a radiative core developed, leaving only an outer convective shell. Only this shell continued to rotate uniformly (and to loose angular momentum by magnetic braking). At the same time the core, because of its contraction, began to rotate more rapidly. In the absence of other processes the result would have been a large negative gradient $\partial\Omega/\partial r$ in the core of the present Sun.

Two questions arise immediately: How large a gradient of Ω is compatible with the observed splitting of p-mode frequencies, and with the observed oblateness of the solar surface? And how large a gradient of Ω can be maintained against the diverse instabilities?

The p-mode frequency splitting so far yields no evidence of a fast rotation in the core. Down to $0.05 r_\odot$ the rate of rotation appears to be rather constant at $\approx 435\,\text{nHz}$, cf. Fig. 7.5. Hence a rapidly rotating core, if present, must be very small indeed, although the oblateness measurements would still permit a central rotation rate several times the surface value.

The question of stability is more difficult. Large $|\partial\Omega/\partial r|$ means a strong *shearing* motion which, at the low viscosity of the solar gas, must be unstable. Another instability arises from the accumulation of ^{3}He and the ensuing (perhaps intermittent) growth of *internal gravity waves* in the core. A third is of *baroclinic* nature: a finite angle between the surfaces of constant temperature and constant pressure makes potential energy available which may drive a flow, comparable to the driving of winds in the terrestrial atmosphere. The effect of these and many other instabilities is that the generated flows transport angular momentum and hence tend to smooth out gradients of Ω.

Internal Magnetic Torque. But the instabilities also *mix* matter, which brings about two complications. First, as model calculations show, the mixing associated with sufficient transport of angular momentum would cause a much stronger than observed lithium depletion. Second, because of the gradient in the mean molecular weight, the mixing would require work, so much work in fact that it possibly does not occur after all. For these reasons Spruit (1987) favors the torque exerted by an *internal magnetic field*. It is readily seen that a very weak field is sufficient to slow down the Sun's core:

Let B_p and B_t be the poloidal and toroidal components of an axisymmetric internal field (i.e., with lines of force in meridional planes and along parallel circles, respectively). To B_t belongs a poloidal current density of order $B_t/(r\mu)$ which, together with B_p, forms an azimuthal Lorentz force of order $B_pB_t/(r\mu)$ per unit volume. Multiplication by r (the lever arm) yields the torque density, and again by r^3 (the volume) yields the total torque. To be effective within a time t this total torque must be of order $\Theta\Omega/t$, i.e.,

$$\Theta\Omega/t \simeq B_pB_tr^3/\mu , \tag{7.16}$$

where Θ is the moment of inertia; $\Theta \simeq \rho r^5$, with an appropriate mean density ρ. Now if there is shear inside the Sun, then B_t is generated from B_p, as described by the induction equation (Sect. 8.1.1). By order of magnitude this means, after a time t,

$$B_t \simeq B_p\Omega t , \tag{7.17}$$

and therefore,

$$B_p \simeq (\mu\rho)^{1/2}r/t . \tag{7.18}$$

As typical values for the solar interior take $\rho = 10^3 \, \text{kg/m}^3$ and $r = 3 \times 10^8$ m. For t to be less than the solar age we then need

$$B_p \geq 10^{-10} \, \text{T} . \tag{7.19}$$

If the field is stronger then the braking of the core proceeds more quickly. Incidentally, the magnetic torque also acts as a restoring force of a torsional oscillation; the oscillation is superimposed on the general core braking, and its period decreases with increasing poloidal field strength [if (7.19) is satisfied, the period is shorter than the Sun's age].

Problem 7.3. Derive the magnetic torque per volume from the general expression $r \times (j \times B)$, and write down the correct differential equations approximated by (7.16) and (7.17).

7.3.3 Evolution of the Solar Rotation

The evolution of a rotating star is described by a sequence of equilibrium models, just as outlined in Chap. 2 for the non-rotating star. Of course the star is no longer spherically symmetric, and all variables depend on latitude, in addition to the dependence on radius. The case where an effective (i.e., gravitational plus centrifugal, cf. Problem 7.2) potential Ψ exists is the easiest, and has been considered in most calculations so far. The recipe to be used has been described by Kippenhahn and Thomas (1970). It replaces spheres by equipotential surfaces, and $m(r)$ by m_Ψ, the mass contained within the surface corresponding to the value of Ψ. In the stellar structure equations m_Ψ functions as the independent coordinate.

The change of the total angular momentum is modeled by (7.15). Within the star, the transport of angular momentum can be described as a diffusion process, supplemented by the action of the magnetic torque. If s is the distance from the axis of rotation, then $\rho s^2 \Omega$ is the density of angular momentum, and

$$\frac{\partial}{\partial t}(\rho s^2 \Omega) = \nabla \cdot (\nu \rho s^2 \nabla \Omega) + [r \times (j \times B)]_z \ . \tag{7.20}$$

The the first term on the right-hand side is the divergence of the (negative) diffusive flux of angular momentum down the gradient of Ω. The diffusion constant, ν, may contain information about the instabilities mentioned in the previous section. For turbulent diffusion in a convection zone ν is of order $10^9 \, \mathrm{m}^2/\mathrm{s}$, cf. (7.32) below; for the various instabilities which possibly occur in the radiative core it is much smaller. The second term on the right of (7.20) is the component parallel to the axis of rotation of the magnetic torque.

From the remarks in this and the two preceeding sections it is clear that all conclusions concerning the Sun's rotational history must be very uncertain at the present time. One result is that the present state is almost independent of the initial angular momentum, see Fig. 7.3. The reason is that, by (7.15), the larger the original angular velocity, the stronger the braking.

Figure 7.3 shows the evolution of the angular velocity over the Sun's main-sequence life. The earlier calculations illustrate that the final angular velocity is virtually independent of the total initial angular momentum, and that, due to a large diffusion coefficient, the outer convection zone rotates rigidly. On the other hand, because no internal magnetic torque had been

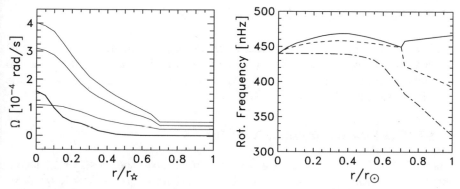

Fig. 7.3. *Left*: angular velocity $\Omega(r)$ at an age of 3×10^7 yr, from calculations without magnetic torque, for a total initial angular momentum of 50, 16.3, and 5×10^{42} kg m^2/s (*thin curves*, from above), and the final angular velocity of the present-day Sun in all three cases (*thick curve*). *Right*: the final rotational frequency $\Omega(r)/2\pi$ at the equator (*solid*), 45° latitude (*dashed*), and at the poles (*dash-dotted*) for a model that includes the magnetic torque in the interior; this model also includes the Λ effect which leads to differential rotation in the convection zone. After Pinsonneault et al. (1989) and Rüdiger and Kitchatinov (1996)

incorporated, a rapidly rotating central region is preserved up to the present. With the magnetic torque included, the final function $\Omega(r)$ is much flatter, and so reproduces the helioseismological result. Within the convection zone, the model of Rüdiger and Kitchatinov (1996) includes the non-diffusive part of the Reynolds stress tensor (the Λ effect described in Sect. 7.5.1); as a consequence, a non-uniform angular velocity $\Omega(r, \theta)$ is maintained. However, before we discuss these models in more detail, we shall, in the two subsequent sections, review the observational evidence concerning the solar rotation.

Problem 7.4. Assume that the magnetic torque in Eq. (7.20) is negligible and show that a substantial change in Ω occurs in a time r^2/ν, the time scale of diffusion.

Problem 7.5. Rayleigh's dynamical instability. In a rotating star, exchange two rings of matter with different radii. Show that this exchange liberates energy if the angular momentum per unit mass *decreases* with distance s from the axis of rotation, i.e., if

$$\frac{\partial}{\partial s}(s^2\Omega)^2 < 0 . \tag{7.21}$$

[In the solar core this instability is inhibited by the entropy gradient, cf. Chap. 6; in addition, the radial change of Ω appears to be too small to satisfy (7.21). For a stability analysis of the internal rotation inferred from helioseismic data see Charbonneau et al. (1999)].

7.4 The Angular Velocity of the Sun

In this section we shall collect results concerning the angular velocity $\Omega(r, \theta)$ of the Sun. In view of the theory to be presented in Sect. 7.5, we shall also include some data for meridional circulation, as well as for the large-scale velocity field already treated in Sect. 6.6.

7.4.1 The Internal Angular Velocity

As explained in Sect. 5.3.8, the solar rotation removes the $(2l+1)$-fold degeneracy of the frequencies of a non-radial p mode of degree l. The frequencies are split into $(2l + 1)$-fold multiplets. Since different modes of oscillation penetrate into different depths of the Sun, the frequency splitting can be inverted; the angular velocity as a function of depth can thus be obtained. Moreover, multiplets with non-equidistant frequency spacing contain information on the dependence of Ω on latitude. The reason for this dependence is that the spherical harmonics with $|m| \approx l$ (the outer multiplet components) are rather concentrated towards the equator, while harmonics with $|m| \ll l$ (the inner components) cover the entire sphere, cf. Problem 5.15.

The results (Fig. 7.4) show small changes of the angular velocity between the surface and $\approx 0.95 r_\odot$. Except for this near-surface region, the angular velocity within the convection zone has approximately the same latitude dependence as at the surface. At the base of the convection zone, around

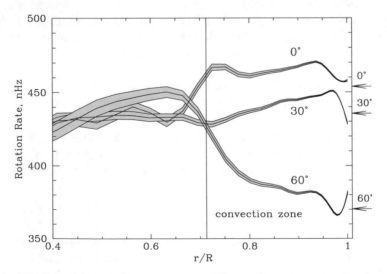

Fig. 7.4. Solar rotation rate $\nu_{\rm rot} = \Omega/2\pi$, as a function of r, for three heliographic latitudes. An inversion of data obtained with the Michelson Doppler Imager on SOHO. The *vertical line* marks the base of the convection zone; *arrows* indicate the rate measured spectroscopically at the surface. From Kosovichev et al. (1997)

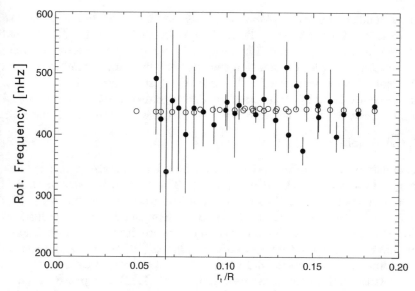

Fig. 7.5. Solar rotation rate $\nu_{\mathrm{rot}} = \Omega/2\pi$, as a function of r, derived from the experiments GOLF (*filled circles*) and MDI (*open circles*) on the SOHO satellite. After Bertello et al. (2000)

$r = 0.7r_\odot$, there is a transition region of thickness $\Delta r \approx 0.1r_\odot$ to a core with approximate solid-body rotation; this transition region has been named the *solar tachocline* by Spiegel and Zahn (1992). In the core the angular velocity is of the same order as around $\pm 30°$ latitude at the surface. Figure 7.4 reaches only down to $0.4r_\odot$, but other helioseismic investigations (Tomczyk et al. 1995 b; Chaplin et al. 1996 b, 1999) have demonstrated that the rotation rate of the inner core does not substantially deviate from the value in the outer parts of the core. Figure 7.5 illustrates that the rotation frequency essentially stays at $\approx 435\,$nHz, although for $r < 0.1r_\odot$ the error bars grow considerably.

The marked change of the angular velocity near $r = 0.7r_\odot$ has important consequences for the generation of the solar magnetic field and its reversal during the 22-year magnetic cycle. The shear arising from the non-uniform rotation deforms magnetic lines of force originally lying in meridional planes and thus builds up the strong toroidal field that is necessary for the formation of sunspots. This process is one essential ingredient to the $\alpha\Omega$ dynamo treated in Sect. 8.4.3, and we shall see there that the positive derivative $\partial\Omega/\partial r$ at low latitude (the upper curve in Fig. 7.4) probably plays a crucial role.

7.4.2 The Angular Velocity at the Surface

When we discussed the measurement of giant cells in Sect. 6.6 we mentioned solar rotation as one of the complications. Of course the converse is also true:

the measurement of solar rotation is difficult, among other reasons, because of the presence of other velocity fields. Proper averaging must again be used to reduce the statistical error. The elimination of systematic errors is more difficult: as an example, we mention the spurious line shift introduced when the Doppler compensator (Sect. 3.3.3) is used for an asymmetric line; and of course the center-to-limb variation of the line asymmetry makes things worse!

In most investigations the observational result is represented in the form

$$\Omega = A + B \sin^2 \psi + C \sin^4 \psi \,. \tag{7.22}$$

The coefficients A, B, and C are obtained by a least-square fit to the measurements. Examples are listed in Table 7.2. The standard deviation of A is of order 0.01 degrees/day, that of B and C of order 0.1 degrees/day. The functions $\sin^{2n} \psi$ are not orthogonal. As a consequence there is a (negative) correlation between B and C if these coefficients are determined individually for a number of data sets. For this reason Snodgrass (1984) has fitted the spectroscopic line-of-sight velocity data from Mt. Wilson with a set of functions that are orthogonal on the solar disc. Instead of (7.22), a representation in Gegenbauer polynomials is thus obtained (the expansion coefficients of these polynomials are subsequently converted into the A, B, C's listed in the table).

Problem 7.6. Relate the coefficients A, B, C to the first three coefficients in the expansions in terms of Gegenbauer and Legendre polynomials.

Table 7.2. Coefficients for the Sun's differential rotation (sidereal, in degrees/day)

		A	B	C
Newton and Nunn (1951)	Greenwich; recurrent spots, 1878–1944	14.368	−2.69	
Howard et al. (1984)	Mt. Wilson; all spots, 1921–1982	14.522	−2.84	
Balthasar et al. (1986 b)	Greenwich; all spots, 1874–1976	14.551	−2.87	
Snodgrass (1984)	Mt. Wilson; Doppler shifts, 1967–1984	14.050	−1.492	−2.606
Snodgrass and Ulrich (1990)	Mt. Wilson; Doppler signal correlations, 1967–1987	14.71	−2.39	−1.78
Komm et al. (1993 a)	Kitt Peak; magnetogram correlations, 1975–1991	14.42	−2.00	−2.09
Brajša et al. (2001)	SOHO/EIT; bright points, 1998–1999	14.6	−3.0	
Timothy et al. (1975)	Skylab; coronal holes, 1973	14.23	−0.4	

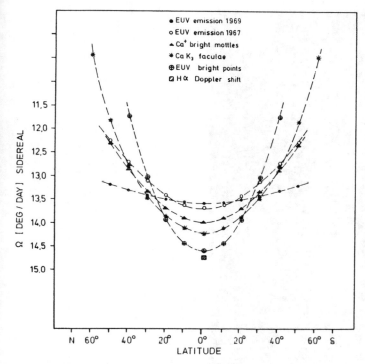

Fig. 7.6. Rotation of various chromospheric features. After Schröter (1985)

The coefficient A represents the equatorial rate of rotation. The results listed in Table 7.2 show remarkable differences, above all a rather small spectroscopic value. It is true that light scattered into the spectrograph, either by the atmosphere or by the telescope, can lead to an apparently reduced rotational Doppler shift, if the width of the spread function is of the order of the solar radius (Schröter 1985). On the other hand the lower rate A found for recurrent (i.e., old) spots indicates that these have been braked by the slower rotation of the plasma; such braking has been confirmed by Balthasar et al. (1982).

The difference in the equatorial rotation rate between recurrent sunspots and the mean of all spots has been confirmed by many investigations. Thus, it appears that sunspots slowly plough through the solar gas and so counterfeit a faster rotation. For the latitude range where sunspots occur helioseismological studies indeed indicate a rate of rotation that increases with depth between the surface and $\approx 0.95 r_\odot$ (Fig. 7.4). If *all* spots are taken into account, the data from Mt. Wilson and Greenwich yield consistent results, for both A and B.

To determine C would be meaningless for sunspots as tracers, because spots normally do not occur at latitudes higher than $\approx 40°$. Spectroscopic data can be obtained over a wider range than sunspot positions. Similarly, the

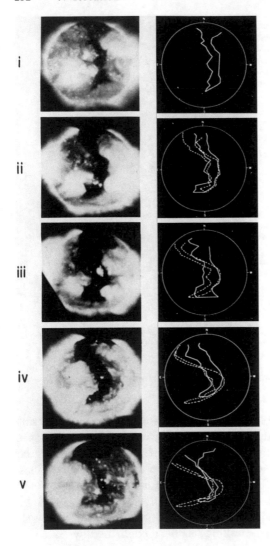

Fig. 7.7. The "Italian Boot" coronal hole, seen 1973 during five successive solar rotations: (**i**) June 1, (**ii**) June 28, (**iii**) July 25, (**iv**) August 21, (**v**) September 28. Soft x-ray photographs (*left*); outlines of real position (*right, solid*), and of an imaginary position (*right, dotted*) that would have followed from the sunspot rotation law. From Timothy et al. (1975)

correlation between the magnetic and velocity structure of consecutive solar maps can be calculated up to high heliographic latitude. The spectroscopic method and the method of local correlation tracking therefore yield all three coefficients A, B, and C, although the results show noticeable differences.

Besides sunspots, other tracers, e.g., prominences or emission inhomogeneities seen in the chromosphere and corona, have been used to determine the profile of Ω. Figure 7.6 shows some results for the chromosphere. The bright points observed with the Extreme ultraviolet Imaging Telescope (EIT) on SOHO refer to coronal emission. A rule of thumb is that the larger the scale and the longer the lifetime of a certain feature is, the more uniform is its rotation. An extreme case is the almost rigid rotation of the coronal holes observed during the Skylab mission in 1973; it is evident in comparison with the sunspot rotation, Fig. 7.7, but also in the rather small value of B, Table 7.2. It is however necessary to add that coronal holes studied at other occasions show more differential rotation than the example shown here (Shelke and Pande 1985).

Variability. The results communicated so far are either valid for short periods of time (such as the rotation of coronal holes), or averages over long periods (the sunspot results of Table 7.2). Many observers have looked for time variations of the Sun's angular velocity. If there are any, they must be small. The Greenwich data show a slight (≈ 0.1 degree/day) decline of the equatorial rate of rotation during the first decade of the 20th century (e.g., Balthasar et al. 1986 b). Of similar magnitude are the periodic changes seen in both the Greenwich and Mt. Wilson data. For example, if in each 10-degree latitude interval the average over the *whole* period of observation is subtracted from the *yearly* averages, and if the residuals thus calculated are then averaged over latitude, the variation shown in Fig. 7.8 is obtained. Most conspicuous (and significant) is the maximum near the minimum of activity; a secondary, smaller, maximum occurs near the maximum of activity. The result holds if individual latitude ranges are considered separately, and implies that the whole solar surface slowly speeds up and slows down in unison. Of course the conservation of angular momentum requires a compensating variation somewhere below the surface.

In the spectroscopic data, the net change with the solar cycle of the surface angular velocity is only marginally visible. But an interesting result, somewhat misleadingly named the "torsional oscillator", has been isolated by Howard and LaBonte (1980). They subtracted daily rotation fits of the form (7.22) from daily angular velocity values measured in 2-degree latitude zones. On the one hand large day-to-day variations in the measured data are

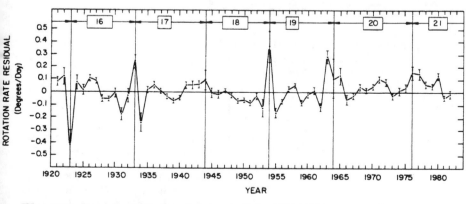

Fig. 7.8. Annual residual sunspot rotation, 1921–1982, averaged over latitudes 30° S to 30° N. *Vertical* lines mark the minima of the solar activity cycle (cycle *numbers* in boxes). From Gilman and Howard (1984 a)

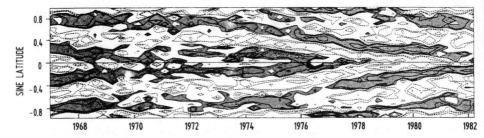

Fig. 7.9. Variation of the solar differential rotation, derived from Doppler shifts measured at Mt. Wilson Observatory. Velocity contours are ±1.5, ±3, ±6 m/s; areas with velocity $\geq +1.5$ m/s are *shaded*. After Howard and LaBonte (1983)

removed by this procedure, on the other hand spatial variations of small latitudinal extent [which are not represented by (7.22)] are made visible. Among these variations is a steady narrow equatorial retardation (the equator rotates slower than the adjacent low latitude zones by about 0.04 degrees/day), and a wave-like periodic variation of the differential rotation: alternating faster and slower latitude zones, migrating towards the equator in each of the two hemispheres. Moreover, helioseismic studies (Howe et al. 2000 b) show that this variation is not confined to the solar surface, but extends downward at

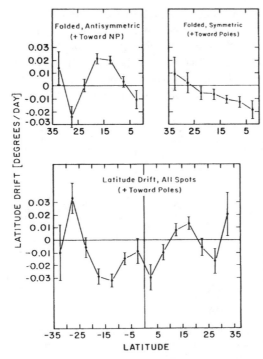

Fig. 7.10. Average latitudinal motion of sunspots observed at Mt. Wilson, 1917 − 1983. *Lower panel*: full latitudinal dependence; *upper panels*: symmetric and antisymmetric parts. *Error bars* represent the standard deviation. From Howard and Gilman (1986)

least 6×10^7 m, about one-third of the convection zone. The phenomenon, shown in Fig. 7.9, is clearly related to the solar magnetic cycle: at any fixed point it takes 11 years to repeat. Although the details of the interaction of the magnetic and velocity fields are far from being understood, it is clear that the mean magnetic force density, $\langle \boldsymbol{j} \times \boldsymbol{B} \rangle$, must have a period of 11 years rather than 22 years, because it is of second order in the field strength.

7.4.3 Meridional Circulation

By *meridional circulation* we mean the longitudinally averaged north-south and radial velocity components; per definitionem these are axisymmetric.

In principle, the line-of-sight component of the meridional circulation can be measured spectroscopically. However, no unique result has been obtained so far. The error bars are of the same magnitude as the measurements themselves (≈ 10 m/s), and it is not even clear whether the observed signals hint toward an equatorward or poleward flow. The main reason of the uncertainty is again the presence of Doppler shifts arising from other sources. Most detrimental is a *latitude dependence* of the line asymmetry (Schröter 1985).

The situation is somewhat better when sunspots are used as tracers of the north-south component. Here long-time averages yield a marginally significant signal. Figure 7.10 shows a typical result. Its symmetric part (with respect to the equator) confirms the circulation pattern first discovered by Tuominen (1941): an equatorward flow at latitudes lower than $\approx 25^\circ$, and

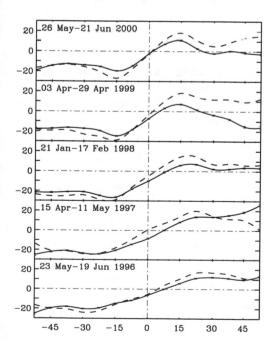

Fig. 7.11. Meridional flow (in m/s, positive towards north) at depths 0.9 Mm (*dashed*) and 7.1 Mm (*solid*), as a function of latitude. MDI data for 1996–2000 are evaluated in a ring-diagram analysis, as described by Haber et al. (2000). Courtesy D. A. Haber

a poleward flow at higher latitudes. But there seems to be an antisymmetric part, too: a northward flow around 15° in *both* hemispheres, and a flow apparently crossing the equator from north to south. Neither the symmetric nor the antisymmetric part presently has a theoretical explanation. Perhaps it is relevant that both symmetries are also present in the giant-cell pattern described in Sect. 6.6, as well as in the large-scale solar magnetic field. In fact, the dividing line between equatorward and poleward flow apparently coincides with the central latitude of the sunspot zone, as was first noticed by Becker (1954). This line therefore participates in the latitudinal migration of the spot zones, in a similar manner as the zonal-flow maxima and minima shown in Fig. 7.9. A helioseismic ring-diagram analysis of Haber et al. (2000) shows that this migrating meridional-flow pattern extends over a depth of order 10^7 m at least. Figure 7.11 shows a result for the period 1996–2000.

Problem 7.7. Estimate the spurious meridional circulation arising if the rotation element i is wrong by 0.1 degree. What is its time dependence?

7.4.4 Correlation of Flow Components

Before we turn to the theory we must mention an observational result which could have been discussed already in the context of giant cells, Sect. 6.6. This is the correlation between the longitudinal and latitudinal components of the surface velocity field. It has been found from the proper motion of various tracers. Let us use spherical polar coordinates (r, θ, ϕ). The quantity in question then is

$$Q_{\theta\phi}(r_\odot, \theta) = \langle v_\theta v_\phi \rangle \,, \tag{7.23}$$

where the brackets mean an average over longitude ϕ. The observations indicate that $Q_{\theta\phi}$ is positive in the northern hemisphere, and negative in the southern, with a nearly linear variation in the latitude range covered by sunspots (e.g., Pulkkinen and Tuominen 1998). It will soon be clear – cf.

Table 7.3. Correlation $|Q_{\theta\phi}|$, in $\mathrm{m^2/s^2}$, between latitudinal and longitudinal velocity components

Ward (1965)	sunspot groups	1.5	$\times 10^3$
Belvedere et al. (1976)	Ca II faculae	4	$\times 10^3$
Schröter and Wöhl (1976)	Ca II network mottles	4	$\times 10^3$
Gilman and Howard (1984 b)	sunspot groups	2	$\times 10^3$
Gilman and Howard (1984 b)	single spots	0.6	$\times 10^3$

(7.27) below – that this sign means transport of angular momentum from both sides towards the equator. Averages over latitude of $|Q_{\theta\phi}|$ are listed in Table 7.3. Although there are differences between the various tracer results we conclude that there exists a real effect of order $10^3 \, \mathrm{m}^2/\mathrm{s}^2$.

7.5 Models of a Rotating Convection Zone

7.5.1 Conservation of Angular Momentum

As a model of the solar convection zone let us consider a spherical shell of compressible gas, with inner and outer radii r_v and $r_\odot$. Within this shell we have turbulent convection, described by a velocity field $\boldsymbol{u} = (u_r, u_\theta, u_\phi)$, superimposed on the general (non-uniform) rotation, $\langle v_\phi \rangle$, and (possibly) a meridional circulation, $\boldsymbol{v}_m = (\langle v_r \rangle, \langle v_\theta \rangle)$; the brackets again mean an average over ϕ. The total velocity is, therefore,

$$\boldsymbol{v} = \langle \boldsymbol{v} \rangle + \boldsymbol{u} \ . \tag{7.24}$$

The viscosity of the solar gas is very small, so that we may begin with Euler's equation in order to describe the conservation of momentum:

$$\frac{\partial \boldsymbol{v}}{\partial t} + \boldsymbol{v} \cdot \nabla \boldsymbol{v} = -\frac{1}{\rho} \nabla P - \nabla \Phi \ , \tag{7.25}$$

which we have already considered in the special case of pure rotation, expressed by (7.6). Let us, for the moment, neglect fluctuations of the density ρ, as in the anelastic approximation (strictly, this is not allowed, especially since we know that the convective motion is *driven* by density fluctuations). The conservation of mass is then expressed by the equation of continuity in the steady form

$$\nabla \cdot (\rho \boldsymbol{v}) = 0 \ . \tag{7.26}$$

We now consider the mean longitudinal motion, i.e., the average over the ϕ component of (7.25). The gradients do not contribute to this average (Problem 7.8). We multiply by the lever arm $s = r \sin \theta$, substitute (7.24), make use of (7.26), and write $s\Omega$ in place of $\langle v_\phi \rangle$; the result is

$$\frac{\partial}{\partial t}(\rho s^2 \Omega) + \nabla \cdot (\rho s^2 \Omega \boldsymbol{v}_\mathrm{m} + \rho s \langle u_\phi \boldsymbol{u} \rangle) = 0 \ , \tag{7.27}$$

and expresses the conservation of angular momentum. At any given place, the density $\rho s^2 \Omega$ of angular momentum changes either because angular momentum is carried along by meridional circulation $\boldsymbol{v}_\mathrm{m}$, or because its transport is mediated by the *Reynolds stresses* (Reynolds 1895),

$$Q_{ij} = \langle u_i u_j \rangle \ . \tag{7.28}$$

These two means of transporting angular momentum play the key roles in most theories of the Sun's differential rotation, and we see now why it is important to gather as much observational information as possible about them.

The observed variation in time of the Sun's angular velocity is so much smaller than the spatial variation (in latitude) that it is reasonable to restrict the attention to *steady* models. Then the diverse transport processes must compensate each other. Moreover, we consider the convection zone in isolation, with no flux (of matter or angular momentum) across its surfaces at r_v and $r_\odot$. That is, at these boundaries,

$$\langle v_r \rangle = 0 \ , \qquad Q_{r\phi} = 0 \ . \tag{7.29}$$

Theories of the differential rotation can be classified according to their treatment of the Reynolds stresses Q_{ij}. In principle these quantities must be calculated from the velocity components u_i, which means that the fluctuating part of (7.25) should be solved. Such is not feasible at present. In fact one may ask whether it is necessary to know all details of the turbulent flow in order to explain a large-scale phenomenon such as the Sun's differential rotation.

In the following two sections we shall deal with two kinds of models: those which employ the Reynolds stresses as mean quantities only, and those which explicitly attempt to calculate u_i, and therefrom Q_{ij}.

Problem 7.8. Confirm (7.27); show that the gradients occuring in (7.25) make no contribution.

Problem 7.9. Show that $\langle \boldsymbol{u} \cdot \nabla \boldsymbol{u} \rangle$ may be written in terms of a divergence of the Reynolds-stress tensor.

7.5.2 Mean-Field Models

Let us now follow the work of Rüdiger (1980) and divide the Reynolds stresses into their "diffusive" and "non-diffusive" parts. The diffusive part was already mentioned in the context of the internal solar rotation: it models the smoothing effect of turbulence upon large-scale gradients. In the present case this means that the transport of angular momentum occurs downwards along the gradient of Ω, i.e.,

$$(Q_{r\phi}, Q_{\theta\phi}) = -\nu_t s \nabla \Omega \tag{7.30}$$

(the factor $s = r \sin \theta$ ensures that $Q_{r\phi}$ and $Q_{\theta\phi}$ vanish at the poles, as they must according to their definition, and that the parameter ν_t has the dimension of a diffusivity). It is easy to see, cf. Problem 7.4, that there will be uniform rotation if there is no meridional circulation and no contribution other than (7.30) to the Reynolds stresses. For $\nu_t = 10^9 \, \mathrm{m}^2/\mathrm{s}$, the state of uniform rotation would be reached within a few years.

Latitude-dependent Heat Transport. Equation (7.27) suggests that one way to sustain non-uniform rotation against diffusive decay is meridional circulation. Such circulation may arise because, as the Sun rotates, its convective flow is distorted by the Coriolis force, and hence the convective energy transport will depend on latitude (Weiss 1965; Durney and Roxburgh 1971). The circulation restores the energy balance, and at the same time provides the required transport of angular momentum.

Following this concept a large number of models have been calculated. A typical result is shown in Fig. 7.12. In this example the circulation is in two layers, with the main equatorward flow at an intermediate depth, and a small poleward flow both at the surface and at the bottom of the shell. Many of these models succeed in maintaining a differential rotation of the observed size with rather small circulation velocities at the surface, and so are consistent with observations. But some have a large (100 m/s, say) circulation speed at $r = r_\odot$ and must be sorted out; others have a too large difference between the effective temperature values at the pole and the equator, in comparison to the ≈ 1 K allowed by observations (e.g., Kuhn et al. 1998); these models must be sorted out, too.

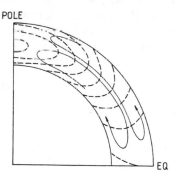

POLE

EQ

Fig. 7.12. Surfaces of constant angular velocity (*dashed*), and streamlines of meridional circulation for a model with latitude-dependent heat transport. From Stix (1987 a)

Anisotropic Viscosity and Λ Effect. Another class of models also uses meridional circulation to balance the diffusive decay of differential rotation. But this time the circulation is not due to a distorted energy balance, but to a special non-diffusive part of Q_{ij}. As a generalization of (7.30) we write

$$Q_{r\phi} = -\nu_t r \sin\theta \frac{\partial\Omega}{\partial r} + \Lambda_r \sin\theta\Omega ,$$

$$Q_{\theta\phi} = -\hat{s}\nu_t \sin\theta \frac{\partial\Omega}{\partial\theta} + \Lambda_h \cos\theta\Omega \tag{7.31}$$

(Rüdiger 1980). The additional, non-diffusive, transport is the Λ *effect*, described by the parameters Λ_r and Λ_h for the radial and the horizontal directions, respectively. The new terms are proportional to Ω because they owe their existence to the rotation itself. Recalling the definition (7.28) of the Q_{ij} we see that Λ_r could be a function of r alone, but that Λ_h should have at least a $\sin^2 \theta$-dependence (the factor $\cos \theta$ takes care of the antisymmetry of the Coriolis effect upon $\boldsymbol{u}$, and the dimensionless parameter $\hat{s}$ is included in order to allow for a difference in the vertical and horizontal smoothing effect of the turbulence).

Historically the development was slightly different. In view of their smoothing effect the Reynolds stresses were written as a linear function of $\langle v_i \rangle$ and the derivatives of $\langle v_i \rangle$. Upon substitution into the averaged Euler equation a term resembling the viscous force in the Navier-Stokes equation was then obtained. In its simplest form the diffusivity thus defined is a scalar quantity, and is identical to the coefficient ν_t introduced above. Going back to the definition (7.28) of Q_{ij} an order-of-magnitude estimate immediately shows that

$$\nu_t \simeq lu \, , \tag{7.32}$$

where l and u are the typical scale and velocity of the field $\boldsymbol{u}$ (in the convection zone, $\nu_t \approx 10^9 \, \mathrm{m^2/s}$).

It was however noticed by Biermann (1951 a) that in a stellar convection zone the transport should generally be different in the radial and horizontal directions. Therefore, ν_t was replaced by a tensor ν_{ij} which, in spherical polar coordinates, was given the form (Kippenhahn 1963)

$$\nu_{ij} = \nu_t \begin{pmatrix} 1 & 0 & 0 \\ 0 & \hat{s} & 0 \\ 0 & 0 & \hat{s} \end{pmatrix} \, , \tag{7.33}$$

which is known under the name "anisotropic viscosity".

A detailed calculation shows that $\hat{s}$ is the parameter already introduced above (Kippenhahn writes s for $\hat{s}$) and that (7.33) is identical to (7.31) provided

$$\Lambda_r = 2\nu_t(\hat{s} - 1) \, , \quad \Lambda_h = 0 \, . \tag{7.34}$$

We shall now consider this special case. The necessity of a meridional circulation is seen in the following way. Assume first that $\boldsymbol{v}_m = 0$ and substitute (7.31) into the (steady) angular momentum balance. A little calculation shows that the angular velocity Ω depends on r. The centrifugal force arising from such Ω generally is not conservative, and hence cannot be balanced by a pressure gradient. Turning around the argument of Problem 7.2, we conclude that the assumption of pure rotation, i.e., $\boldsymbol{v}_m = 0$, was false. Thus we do have meridional circulation, and may use it to sustain the observed $\partial \Omega / \partial \theta$. This is the essence of the anisotropic viscosity models (Kippenhahn 1963; Köhler 1970).

A variety of such models has been calculated; their circulation and angular velocity pattern often look similar to the case shown in Fig. 7.12. The models are superior to the latitude-dependent heat transport models in that they all have a very small ($\approx 1\,\mathrm{K}$) pole-equator temperature difference. Their difficulty is that the choice $\hat{s} > 1$ yields the observed equatorial acceleration; this choice is in contrast to the intuitive notion that convection would provide more transport in the radial than in the horizontal directions (a notion that is confirmed by attempts to calculate the Reynolds stresses ab initio in a theory of rotating turbulence, e.g., Rüdiger 1980, 1983).

Models without Meridional Circulation. Let us finally include the horizontal non-diffusive part of the Reynolds stress $Q_{\theta\phi}$. In fact this *must* be done because in any model which yields the correct $\partial\Omega/\partial\theta$ a merely diffusive $Q_{\theta\phi}$ would have the opposite sign of what is observed, cf. (7.31) and Sect. 7.4.4 (this argument is due to Rüdiger, see e.g., Rüdiger and Tuominen 1987). Thus, we now admit $\Lambda_\mathrm{h} \neq 0$. In this case the assumption of pure rotation (i.e., $\boldsymbol{v}_\mathrm{m} = 0$) would generally again lead to a contradiction, because there would again be a non-conservative centrifugal force. But now the ensuing circulation is not the only agent to maintain a finite $\partial\Omega/\partial\theta$. The Λ_h-term directly fulfills the same purpose. Indeed, a numerical experiment of Schmidt and Stix (1983) shows that the model produces nearly the same angular velocity $\Omega(r, \theta)$, regardless whether the circulation is – at the surface – poleward or equatorward (depending on the total shell depth, both kinds of flows did occur). Perhaps this is an indication that meridional circulation is rather accidental to the problem of the Sun's differential rotation. Indeed the model of Rüdiger and Kitchatinov (1996), which ignores meridional circulation, yields a rotation law that is consistent with the helioseismic result, as a comparison of Figs. 7.3 and 7.4 shows.

7.5.3 Explicit Models

There is no hope (and, as already said, perhaps no need) to simulate all details of turbulent convection in a numerical calculation. But perhaps we are able to simulate the *interesting part*. This part consists of the largest cells in the convection zone. For these the rotational influence is strong, i.e., the Rossby number (6.45) is small; we may expect that the ensuing distortion of the cells produces the Reynolds stresses that we need to maintain the right $\Omega(r, \theta)$.

Strictly speaking, the velocity field is now divided into three parts. The axisymmetric part is the same as in (7.24) above. But the part $\boldsymbol{u}$ is subdivided: the larger-scale structure is treated explicitly, which is possible with a reasonable number of grid points or harmonics; the smaller-scale structure is parameterized as in the mean-field models, but in the crudest possible way, namely by a scalar turbulent diffusivity ν_t.

Calculations of this kind have been performed especially by P. A. Gilman and G. A. Glatzmaier since about 1975. They use the anelastic approximation already mentioned in Sect. 6.3.2 in the context of granulation. Further, they define a reference state of the convection zone. This state is characterized by

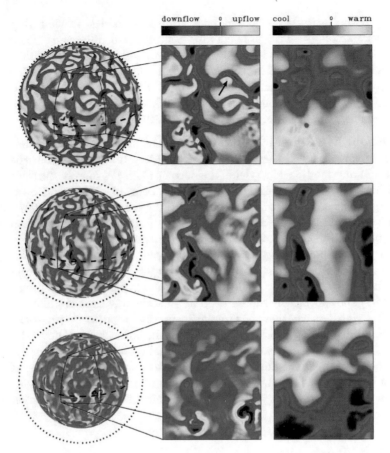

Fig. 7.13. "Global convection". *Left*: snapshots of the radial velocity at three levels in the convection zone (*dotted*: solar surface, *dashed*: equator). *Right*: enlarged segments of $50° × 66°$, showing radial velocity and temperature. The *arrow* points to an area of high vorticity. From Miesch et al. (2000)

purely diffusive energy transport, cf. Problem 6.3, and by a pure rotation with a mean angular velocity Ω_0. Depending on the rates κ_t and ν_t of turbulent diffusion, and on Ω_0, the reference state itself becomes unstable, and global convection sets in. In the numerical models this global convection is simulated by solving the time-dependent equations of motion.

The results of such a calculation depend on the model parameters, in particular on the turbulent diffusivity ν_t. Miesch et al. (2000) treat two cases, with $\nu_t = 1.9 \times 10^9 \,\mathrm{m}^2/\mathrm{s}$ and $\nu_t = 3 \times 10^8 \,\mathrm{m}^2/\mathrm{s}$, respectively, for a spherical shell confined between $r = 0.62 r_\odot$ and $0.96 r_\odot$. Their simulation covers the main part of the convection zone, including a layer of convective overshooting at its base, but excluding a layer immediately below the photosphere where the scales are too small to be resolved. In the case with the larger diffusivity the global convection is essentially laminar, although slowly variable. It has the form of rolls, oriented in the north-south direction. The reason for this orientation is that rotation removes the degeneracy present in a plane fluid layer heated from below, or in a (non-rotating) spherical shell heated from within. In such systems the rolls orient themselves according to accidental initial conditions. In contrast, the *rotating* shell has a distinguished direction; here north-south rolls are the most unstable mode of convection. For slow rotation the rolls are bent according to the shell curvature ("bananas").

The simulation with the smaller value of ν_t yields a more turbulent flow. Figure 7.13 illustrates the complex morphology of this case. In the upper part of the convection zone there is a network of narrow downflow lanes enclosing broad regions of slow upflow. In the deeper part the downflow lanes first become disconnected and then acquire a plumlike character. The plumes are especially conspicuous in the region of convective overshooting; the two dark patches in the lower middle panel of Fig. 7.13 are examples. The right panels of the figure show the temperature. The downflows generally are cooler.

Presently the explicit models do not yet correctly describe the solar convection zone. In both simulations the velocity field is influenced by the Coriolis force in such a way that angular momentum is transported towards lower latitudes. An equatorial acceleration of the observed magnitude is easily obtained. However, in the interior of the convection zone the angular velocity tends to be constant on cylinders, which is in contrast to the result derived from the rotational splitting of the p-mode frequencies. In addition to this discrepancy, both the poloidal velocity components and the latitudinal variation of the heat flow exceed the limits set by observation; at low latitude, the meridional surface flow converges toward the equator, opposite to some of the observations (e.g., Fig. 7.11).

The cylindrical shape of the constant-angular-velocity contours is a consequence of the rotational constraint treated in Problem 7.10. The constancy of $\Omega(r, \theta)$ on cylinders can be broken by a baroclinic stratification, where the surfaces of constant pressure and constant density make a finite angle (e.g., Stix 1989, Miesch 2000). One way to achieve this is imposing the condition of constant heat flow at the outer boundary of the shell (Elliott et al. 2000).

Problem 7.10. Assume that the balance of forces is dominated by the equilibrium between pressure gradient and Coriolis force (known in meteorology as the *geostrophic* balance), and assume $P = P(\rho)$. Prove the Taylor–Proudman theorem, namely that, in the rotating frame, $\partial \boldsymbol{v}/\partial z = 0$, where z is the coordinate parallel to the axis of rotation.

7.6 Bibliographical Notes

The subject of internal solar rotation is covered in the proceedings edited by Durney and Sofia (1987). An earlier volume, edited by Belvedere and Paternò

(1978), and the articles of Gilman (1974) and Stix (1989) also give general reviews; the books of Tassoul (1978, 2000) are comprehensive introductions.

The rotation elements, i and Ω, have also been measured spectroscopically (Wöhl 1978), but with lesser precision. The measurements of the Sun's oblateness and their implications are discussed by Dicke (1974), Godier and Rozelot (2000) include the differential rotation into their theoretical study. Ulrich (1986) gives an account of the stresses for the internal transport of angular momentum.

Schröter (1985) reviews the surface measurements, and especially the sources of errors, which are plentiful. Beck (2000) compares the results obtained in a large number of studies based on spectroscopic, tracer, and helioseismic techniques. As a new helioseismic result, variations of the angular velocity at the base of the convection zone, with an amplitude of order 1 % and a time scale of 1.3 years, have been reported by Howe et al. (2000 a). A critical discussion of the wave-like variation of the angular velocity at the surface (Fig. 7.9) is due to Snodgrass (1985). Hathaway (1987) proposes a general representation of the surface velocity field in terms of spherical harmonics, and discusses how to separate the various contributions, i.e., rotation, circulation, giant cells, etc.

Many studies based on sunspots or other tracers have confirmed that the meridional circulation diverges from the zones of sunspot occurrence, and its concomitant dependence on the cycle phase (e.g. Tuominen and Kyröläinen 1982, Komm et al. 1993 b, Nesme-Ribes et al. 1993, Pulkkinen and Tuominen 1998); Wöhl and Brajša (2001) show that the magnitude of the velocity increases with the distance from the spot zone. The spectroscopic determinations of the circulation velocity have been reviewed by Cavallini et al. (1992) and interpreted as a real temporal variation.

Rüdiger (1989) treats the foundations for the mean-field theory of the solar differential rotation. The Λ effect is quenched by rapid rotation (Kitchatinov and Rüdiger 1993), which has implications for the prediction of stellar differential rotation. In addition to the Λ effect, the anisotropy of the convective heat transport and the meridional circulation have been included in models of rotating convection zones by Kitchatinov and Rüdiger (1995, 1999) and Rüdiger et al. (1998). Mean-field models of the solar differential rotation have also been calculated by Pidatella et al. (1986). Durney (1987) has outlined a model in which the rotation makes the excess, $\nabla - \nabla_a$, of the temperature gradient over the adiabatic gradient dependent on latitude. Thus he generalizes the mixing-length theory to the case of a rotating convection zone. Not only the mixing length, but also the lateral dimension of the convecting parcels must be specified in such a model; if these quantities are known, the Reynolds stresses can be estimated. A rigorous discussion of the Reynolds stress tensor can also be found in Gough (1978). Miesch (2000) reviews the diverse models that describe the convection-rotation coupling.

8. Magnetism

For at least 2000 years astronomers have been attracted by solar activity: At first mainly the appearance and the variation of sunspots, later sunspot structure, prominences, coronal variations, eruptions with all their terrestrial consequences from aurorae to blackouts in radio transmission, – the literature on these phenomena is enormous.

It is now known that all solar activity is an immediate consequence of the existence of a magnetic field on the Sun. In the present chapter we shall discuss this field. The main emphasis will be on the photosphere and the convection zone. The chromosphere and the corona, where the magnetic field often is the dominant agent, will be treated in the subsequent chapter.

Magnetism on the Sun occurs on all scales. The diameters of the smallest "tubes" of magnetic flux are at or below the present limit of spatial resolution, while the largest manifestations of the field, the *mean-field* components, may cover an entire hemisphere. In three sections we shall go from the smallest to the largest scales. But first let us recollect some fundamental facts about the interaction of electrically conducting matter with a magnetic field.

8.1 Fields and Conducting Matter

8.1.1 The Induction Equation

We begin with Maxwell's equations for the magnetic field $\boldsymbol{B}$, the electric field $\boldsymbol{E}$, and the electric current density $\boldsymbol{j}$:

$$\operatorname{div} \boldsymbol{B} = 0 , \tag{8.1}$$

$$\operatorname{curl} \boldsymbol{B} = \mu \boldsymbol{j} , \tag{8.2}$$

$$\operatorname{curl} \boldsymbol{E} = -\dot{\boldsymbol{B}} , \tag{8.3}$$

where the dot denotes the time derivative, and μ is the magnetic permeability (which will always be taken as that of free space, i.e., $\mu = 4\pi \times 10^{-7}$ Vs/Am). In (8.2) we have neglected the displacement current, which is an excellent approximation for non-relativistic, or "slow" phenomena (Problem 8.1).

If the magnetic and electric fields are embedded in a material with electric conductivity σ, then the current density is σ times the electric field strength.

This proportionality is known as Ohm's law; if the material is in motion, we must take into account that the law is valid in the co-moving frame of reference:

$$\tilde{\boldsymbol{j}} = \sigma\tilde{\boldsymbol{E}} \ . \tag{8.4}$$

If the motion, say $\boldsymbol{v}$, is non-relativistic ($|\boldsymbol{v}| \ll c$) the contribution of the convected volume charge to the electric current density is negligible, and the transformation to the frame at rest is $\tilde{\boldsymbol{j}} = \boldsymbol{j}$ and $\tilde{\boldsymbol{E}} = \boldsymbol{E} + \boldsymbol{v} \times \boldsymbol{B}$. Hence

$$\boldsymbol{j} = \sigma(\boldsymbol{E} + \boldsymbol{v} \times \boldsymbol{B}) \tag{8.5}$$

is the form of Ohm's law which we shall use in the present chapter.

It is an easy matter to eliminate $\boldsymbol{E}$ and $\boldsymbol{j}$ from (8.2), (8.3), and (8.5). We obtain the *induction equation*

$$\dot{\boldsymbol{B}} = \operatorname{curl}(\boldsymbol{v} \times \boldsymbol{B}) - \operatorname{curl}(\eta \operatorname{curl} \boldsymbol{B}) \ , \tag{8.6}$$

where

$$\eta = \frac{1}{\mu\sigma} \tag{8.7}$$

is called the magnetic diffusivity [because, for $\boldsymbol{v} = 0$, (8.6) is a diffusion equation].

The first term on the right of (8.6) describes the inducing effect of the material motion upon the magnetic field, while the second term manifests the "ohmic" decay of the field due to the finite electrical resistance. By order of magnitude we may compare the two terms if we replace the diverse vectors by their absolute magnitudes, and the curl operator by $1/l$, where l is the scale of field variation in space. We obtain

$$R_{\mathrm{m}} = vl/\eta \tag{8.8}$$

as the ratio of the induction term over the decay term; R_{m} is called the *magnetic Reynolds number*; it can also be understood as a ratio of two time scales, namely the time scale of ohmic decay

$$\tau_{\mathrm{D}} = l^2/\eta \tag{8.9}$$

and the advection time scale, l/v. In solar physics one often speaks of "high conductivity"; the precise meaning of this is $R_{\mathrm{m}} \gg 1$, or that τ_{D} is very much longer than l/v or than any other time scale of interest.

Problem 8.1. Give an estimate of the displacement current which was neglected in (8.2).

8.1.2 Electrical Conductivity on the Sun

Everywhere on the Sun a sufficient number of free electrons is available so that an electric current can flow: a current that consists in a drift of the negative electrons relative to the positive ions. The ease with which such a drift is possible determines the electrical conductivity of the ionized gas, or *plasma*. It depends on the frequency and nature of collisions between the diverse charged and neutral particles.

For a *fully ionized gas* an important result has been obtained by L. Spitzer and others (Spitzer 1962):

$$\sigma = \frac{32\,\varepsilon_0^2\sqrt{\pi}(2kT)^{3/2}}{\sqrt{m_e}e^2 Z \ln \Lambda}\,\gamma_E \; . \tag{8.10}$$

In this expression Z is the charge number of the ions, and $\Lambda = r_D/p_0$; r_D is the Debye radius (2.41), and p_0 is the impact parameter for a 90-degree collision of an electron with an ion. The first factor of (8.10) is obtained under the assumption that the ions are at rest and the electrons do not interact with each other (the *Lorentz gas*), and the proportionality to $(2kT)^{3/2}$ reflects the electron velocity distribution. The second factor, γ_E, is a correction arising from collisions among electrons. When $Z = 1$, which is the most interesting case for the Sun, Spitzer gives $\gamma_E = 0.582$.

The quantity $\ln \Lambda$ weakly depends on density and temperature. From a table presented by Spitzer (1962) we may take typical values: $\ln \Lambda \approx 5$ for the solar interior, $\ln \Lambda \approx 10$ for the chromosphere, and $\ln \Lambda \approx 20$ for the

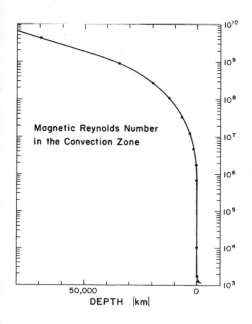

Magnetic Reynolds Number
in the Convection Zone

DEPTH |km|

Fig. 8.1. Magnetic Reynolds number in the outer part of the solar convection zone. From Stix (1976 b)

corona. For a crude estimate of σ in the solar convection zone let us assume complete ionization and apply (8.10). With all constants substituted, we have $\sigma \simeq 0.003T^{3/2}$ A/Vm. We may use the results of Chap. 6 (Table 6.1) to see that this is a very large conductivity indeed. Figure 8.1 shows the magnetic Reynolds number (8.8) calculated with convection velocities and scales obtained in the mixing-length theory.

Problem 8.2. Show that $\Lambda \simeq (kT/E_{\mathrm{ES}})^{3/2}$ and use the results of Fig. 2.2 in order to assess the value of $\ln \Lambda$ in the Sun's interior.

In the solar photosphere and lower chromosphere the temperature is so low that the main constituents, hydrogen and helium, are almost entirely neutral. Heavier elements with low ionization potentials still donate enough electrons for an electric current to flow, but – in comparison to (8.10) – the conductivity is greatly reduced due to the encounters of electrons with neutrals. Nagasawa (1955) has calculated the velocity distribution of the electrons in this case; his result was used by Kopecký (1957) to derive the following formula, valid for weak ionization (electron density n_{e} small compared to the density n_{n} of the neutrals):

$$\sigma = \frac{3\,e^2}{8S(2\pi m_{\mathrm{e}}kT)^{1/2}} \frac{n_{\mathrm{e}}}{n_{\mathrm{n}}} \, . \tag{8.11}$$

The main uncertainty of this result (a factor 2, say) lies in the cross section S for collisions of electrons with neutral particles. Values around $10^{-19}\,\mathrm{m}^2$ have been measured, and this value has been used for the results presented in Fig. 8.2.

Several remarks must be made here. First, there might be large *horizontal gradients* of the conductivity. In sunspots, in particular, the temperature inhomogeneity can cause a conductivity variation of an order of magnitude.

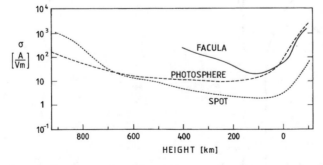

Fig. 8.2. Electrical conductivity σ in the solar atmosphere; photosphere and facula after Kopecký and Soytürk (1971); sunspot after a formula of Kopecký (1966), and data from Avrett (1981 a)

Second, when collisions are rare and the magnetic field is strong, the conductivity becomes *anisotropic*. Expressions (8.10) and (8.11) above would then give the conductivity for a current along the field, while the "transversal" conductivity would be substantially reduced (because of the guiding of electrons around the magnetic lines of force). The anisotropy of σ becomes important in the chromosphere and the corona. Third, the effect of *plasma turbulence* (i.e., a variety of waves depending on the electromagnetic restoring force which acts on charged particles) is to reduce the conductivity, mostly in combination with an anisotropy. Plasma turbulence must not be confused with the *hydrodynamic turbulence* of conducting matter which increases the diffusivity of the *average* magnetic field. This important aspect will be treated in Sect. 8.4.2 below.

8.1.3 Frozen Magnetic Field

The magnetic Reynolds number R_m shown in Fig. 8.1 is large, and a large R_m is also obtained even if we take the smallest possible σ in a sunspot, say $1\,\mathrm{A/Vm}$ (Fig. 8.2), together with $l = 100\,\mathrm{km}$ (the order of the scale height) and $v = 1\,\mathrm{km/s}$ (a typical measured velocity). Thus we see that the solar plasma generally has high conductivity, and in this limit the important concept of a *frozen magnetic field* has wide applications.

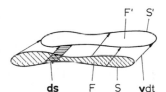

Fig. 8.3. Motion of a closed curve S with a flow $\boldsymbol{v}$

The meaning of a frozen field is that the magnetic flux is transported along with the material motion. To see this take the total magnetic flux

$$\varPhi = \int_F \boldsymbol{B} \cdot \boldsymbol{df} \tag{8.12}$$

across an area F, and consider the circumference S of F at two instants separated by an infinitesimal interval dt of time (Fig. 8.3). At the later instant the circumference is made up by the same material "particles" (in the sense of fluid mechanics) but it may be deformed or stretched according to the various paths $\boldsymbol{v}dt$ followed by these particles. The magnetic flux $\varPhi'$ across the new area F' may differ from $\varPhi$ either because the field $\boldsymbol{B}$ itself has undergone a change in time, or because some of the flux has left the volume between F and F' through the side walls. The (outwards directed) area element of the latter is $-\boldsymbol{v}dt \times \boldsymbol{ds}$; hence

$$\Phi' - \Phi = dt \left[\int_F \dot{\boldsymbol{B}} \cdot \boldsymbol{df} + \int_S \boldsymbol{B} \cdot (\boldsymbol{v} \times \boldsymbol{ds}) \right] , \tag{8.13}$$

where the second integral is taken along the closed curve S. We use $\boldsymbol{B} \cdot (\boldsymbol{v} \times \boldsymbol{ds}) = (\boldsymbol{B} \times \boldsymbol{v}) \cdot \boldsymbol{ds}$, apply Stokes' theorem to convert the line integral into an integral over F, take the limit $dt \to 0$, and obtain

$$\frac{d\Phi}{dt} = \int_F \left[\dot{\boldsymbol{B}} - \mathrm{curl}\,(\boldsymbol{v} \times \boldsymbol{B}) \right] \cdot \boldsymbol{df} . \tag{8.14}$$

By virtue of the induction equation (8.6), the right-hand side is zero for $\sigma \to \infty$. That is, in the limit of infinite conductivity the total magnetic flux enclosed by the curve S is conserved. Since the choice of S is arbitrary, and since the constant flux can be represented by a fixed number of field lines, we may say that the field lines behave as if they were firmly attached, or "frozen", to the moving fluid. The concept of a frozen field will help to comprehend many of the solar phenomena treated in the present chapter, as well as in the subsequent one.

Problem 8.3. Use the equation of continuity to show that for a frozen field the equation

$$\frac{d}{dt} \left(\frac{\boldsymbol{B}}{\rho} \right) = \left(\frac{\boldsymbol{B}}{\rho} \cdot \nabla \right) \boldsymbol{v} \tag{8.15}$$

holds, a result first derived in 1946 by C. Walén (cf. Cowling 1976). Give an interpretation.

Problem 8.4. A velocity field $\boldsymbol{v} = (-\alpha x, -\alpha y, 2\alpha z)$ with $\alpha > 0$ is given. Find the steady solution of the induction equation (with constant η) under the assumption that $\boldsymbol{B}$ points into the z-direction. What is the central field strength of the flux tube generated by the converging flow if the total flux is given? Calculate the radius of the circle that encloses 90 % of the flux (Moffatt, 1978).

8.1.4 The Magnetic Force

A single particle which carries a charge e and moves with velocity $\boldsymbol{v}$ across a magnetic field $\boldsymbol{B}$ experiences a force

$$e\boldsymbol{v} \times \boldsymbol{B} . \tag{8.16}$$

This force is perpendicular to both $\boldsymbol{v}$ and $\boldsymbol{B}$. It causes the well-known gyration of the particle around the magnetic lines of force.

In the present chapter we shall not study individual particle motions, nor shall we separately consider individual species with their diverse charges, i.e., ions, electrons, and neutrals. Instead we shall treat the solar plasma as a *single fluid*. In combination with the neglect of the displacement current in (8.2), and the non-relativistic form (8.5) of Ohm's law, this constitutes the *magnetohydrodynamic* approximation.

The volume force exerted by a magnetic field on a conducting fluid is the *Lorentz* force,

$$\boldsymbol{j} \times \boldsymbol{B} \; . \tag{8.17}$$

We do not rigorously derive this expression here. But it is clear that the product $e\boldsymbol{v}$ in (8.16) describes the transport of electric charge, i.e., the moving particle contributes to the electric current. The sum over those products for all charged particles contained in a small volume δV, divided by δV, yields the net current density $\boldsymbol{j}$. Since in the one-fluid model the forces that are felt by the individual charged particles are communicated to the fluid as a whole by means of collisions, we obtain a net force that is proportional to $\boldsymbol{j}$, namely (8.17).

The volume force (8.17) can be divided into a magnetic pressure gradient and a magnetic tension: using (8.2), and B for the absolute magnitude of $\boldsymbol{B}$, we find

$$\boldsymbol{j} \times \boldsymbol{B} = -\mathrm{grad}\,\frac{B^2}{2\mu} + (\boldsymbol{B} \cdot \mathrm{grad})\,\frac{\boldsymbol{B}}{\mu} \; . \tag{8.18}$$

Instead of pressure and tension we also speak of *Maxwell stresses*. With the definition

$$B_{ik} = \frac{1}{\mu}\left(\frac{1}{2}\delta_{ik}B^2 - B_i B_k\right) \tag{8.19}$$

the ith cartesian component of the volume force is

$$-\frac{\partial}{\partial x_k}B_{ik} \; . \tag{8.20}$$

As described by the first term of (8.18), a bundle ("rope", "tube") of magnetic flux applies a lateral pressure to the gas into which it is embedded. To obtain equilibrium, this magnetic pressure must be balanced by the gas pressure. A pressure of 10^4 Pa is typical for the solar atmosphere; such pressure may balance a field of up to ≈ 0.15 T, which indeed is the strength of the field concentrations seen on the solar surface. In sunspots, where we see into the somewhat deeper level corresponding to $\approx 2 \times 10^4$ Pa, the field strength may reach a maximum value about twice as high.

The effect of magnetic tension is that the lines of force have a desire to shorten themselves. This effect (but also the pressure effect) provides a restoring force to perturbations. The fluid is thus able to support particular wave motions, in addition to those which depend on the compressional and gravitational restoring forces.

Problem 8.5. Confirm (8.18) to (8.20). Convince yourself that the two terms of (8.18) have opposite components in the direction of $\boldsymbol{B}$. Write the volume force in cylindrical coordinates and calculate the magnetic force for the field of Problem 8.4.

8.2 Flux Tubes

8.2.1 Concentration of Magnetic Flux

We have already seen in Problem 8.4 how a convergent motion is capable of magnetic flux concentration. However, the flow considered there was rather idealized, and in particular unrealistic at large distance from the axis of symmetry. For the Sun, we are more interested in *cellular* flows, possibly resembling the granules or supergranules (cf. Chap. 6). Flux concentration by cellular flows was first investigated by Parker (1963 a) and Weiss (1964), and since then, mainly by means of numerical integration, by Weiss and his collaborators.

Let us concentrate here on solutions of the induction equation (8.6) alone, with *given* flow $\boldsymbol{v}$. In this approximation, called *kinematic*, the accelerating (or braking) effect of the Lorentz force (8.17) upon the motion is neglected. Because the force $\boldsymbol{j} \times \boldsymbol{B}$ is of second order, the kinematic approximation is justified at small magnetic field strength.

A time-dependent kinematic field solution is illustrated in Fig. 8.4. In this case the flow is stationary and two-dimensional, i.e., consists of rolls described by

$$
\begin{aligned}
v_x &= -u \sin(\pi x/l) \cos(\pi z/l) , \\
v_y &= 0 , \\
v_z &= u \cos(\pi x/l) \sin(\pi z/l) .
\end{aligned}
\tag{8.21}
$$

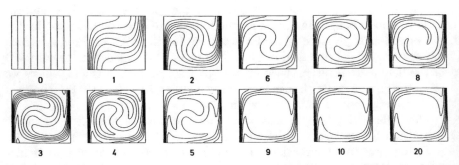

Fig. 8.4. Concentration of magnetic flux at the edges of convection cells. Each box is a vertical cross section; the flow is clockwise, the numbers give the time in units of $5l/8u$. From Galloway and Weiss (1981)

The magnetic Reynolds number, here ul/η, is 250; the boundary conditions are such that the field is vertical at all times at all boundaries. Initially the field is homogeneous and of strength B_0. The clockwise flow first takes this field along as if it was frozen into the fluid. But later on the lines of force are more and more deformed and finally sharply bent. This means that the diffusion term of the induction equation (8.6), which contains the now growing second derivative, is no longer negligible. Dissipation, in the figure visible in the form of *field-line reconnection*, sets in and finally dominates almost the whole volume. The magnetic flux is thus *expelled* from the cell interior, and accumulated in sheets near the cell edges.

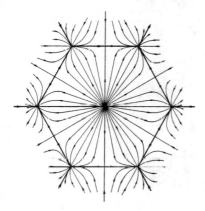

Fig. 8.5. Streamlines at the top of hexagonal convection cells. After Clark and Johnson (1967)

A second example is the hexagonal flow illustrated in Fig. 8.5. At the bottom, $z = 0$, of each hexagon the flow converges at the center; at the top, $z = 1$, convergence occurs at the six vertices. As shown in Fig. 8.6, each point of convergence marks a column of concentrated vertical magnetic flux. The larger the magnetic Reynolds number, the narrower becomes the column (or *flux tube*). In the example shown R_m is 400.

For the final steady state we could calculate the width d of the flux sheet of Fig. 8.4, or of the flux tubes of Fig. 8.6, by a local application of the analysis of Problem 8.4. For the moment, however, it may suffice to present a crude argument, which also yields the correct result. In the steady state the time scale d^2/η of field decay (in the column) must be equal to the time scale l/u of field advection (into the column). Therefore $d^2 \simeq \eta l/u$, or

$$d \simeq l/R_m^{1/2} . \tag{8.22}$$

We may also estimate the strength B of the concentrated field. To this end let us assume that the *whole* flux across the surface of a cell is accumulated in the sheets or tubes, respectively. In the first example we must consider the flux per unit length along the convection rolls (8.21); this is $\simeq B_0 l$ initially, and $\simeq Bd$ in the final steady state. In the second example the flux across

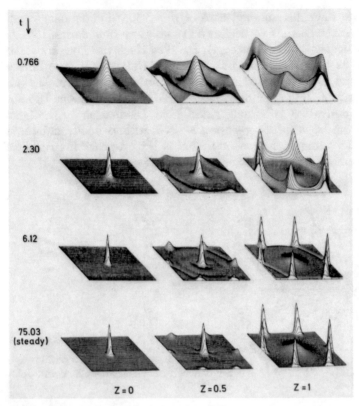

Fig. 8.6. Concentration of magnetic flux by hexagonal convection, at three levels z in the convection cell; time in units of (cell height)/(max. vertical velocity). After Galloway and Proctor (1983)

one of the hexagonal cells is $\simeq B_0 l^2$ initially, and $\simeq B d^2$ in each of the final tubes. We equate the initial and final fluxes, use (8.22), and obtain

$$B \simeq R_{\mathrm{m}}^{1/2} B_0 \tag{8.23}$$

in a flux sheet, and

$$B \simeq R_{\mathrm{m}} B_0 \tag{8.24}$$

in a flux tube. The numerical results confirm these estimates, although a non-negligible fraction of the total flux appears at the top center of a hexagonal cell (see the lower right of Fig. 8.6).

The process of field amplification is rapid: it takes only a time of order l/u, the advection or "turnover" time, to produce the sheets or tubes from the initial homogeneous field. The subsequent expulsion of flux from the cell interior is somewhat slower and depends on R_{m}. In the examples shown in Figs. 8.4 and 8.6 the final state is essentially reached after about $5\,l/u$.

In the following, we shall only discuss the flux *tubes*, although sheets, which are produced in a two-dimensional flow geometry, may as well exist on the Sun, cf. the chain-like crinkles described in Sect. 8.2.2 below or the elongated intergranular feature reported by von der Lühe (1987).

The main question now is the final strength up to which the field is amplified. Expressions (8.23) and (8.24) only give the final field in terms of the initial field B_0. However, as the magnetic force at some stage must become strong enough to counteract the motion, there should be a limit which is independent of B_0. Several estimates have been made at this dynamical field limitation. Here I mention only the simplest of them: namely that there is *equipartition* between the kinetic and magnetic energy densities when the Lorentz force (8.17) begins to matter (the *dynamic* regime). In this case

$$B = B_e \equiv (\rho\mu)^{1/2}u \; . \tag{8.25}$$

The numerical calculations confirm that in the dynamic regime regions of motion and regions of field mutually tend to exclude each other. Using (8.24) and (8.25) we can calculate the total flux across a cell which is necessary to reach the dynamic regime. This critical flux is

$$\Phi_c = l\eta(\rho\mu)^{1/2} \; . \tag{8.26}$$

The equipartition field (8.25) and the critical flux (8.26) are listed in Table 8.1 for the three prominent cell types that we have treated in Chap. 6. It is interesting to compare the flux values given there to the fluxes of observed magnetic features. The latter range from $\approx 10^9$ Wb for the smallest flux tubes to $\approx 10^{14}$ Wb for an active region.

For comparison, the field B_P which corresponds to equilibrium between magnetic and gas pressure (for the deep part of the respective convection cells) is also listed in Table 8.1. Generally B_P is much larger than the equipartition field (8.25).

Table 8.1. Critical flux and magnetic field strength in the Sun's convection zone, after Galloway and Weiss (1981). The magnetic Reynolds number R_m is calculated for $\eta_t = 2 \times 10^7$ m^2/s

	Granules	Supergranules	Giant cells
Depth [10^6 m]	0.5	10	150
l [10^6 m]	1	15	100
ρ [kg/m^3]	10^{-3}	1	10
u [m/s]	2000	400	70
Φ_c [Wb]	7×10^8	3×10^{11}	2×10^{13}
B_e [T]	0.07	0.45	0.9
B_P [T]	0.5	50	2500
R_m	100	300	350

For the evaluation of Φ_c in Table 8.1 a diffusivity has been adopted which is orders of magnitude larger than the value corresponding to the conductivity given in Sect. 8.1.2. The argument used by Galloway and Weiss (1981) is that there is still a turbulent motion *within* the flux tube that tends to diffuse the field (cf. the result of Sect. 8.4.2 below). The actual value taken, $\eta_t = 2 \times 10^7\,\mathrm{m^2/s}$, is however not derived from a rigorous theory; it corresponds to the observed decay rate of sunspots (Sect. 8.3.2). It must be said that the whole problem of diffusion in flux tubes is not really understood: On the one hand, we argue that the field B is strong enough to exclude a flow of order u from the tube, on the other hand we need motions inside the tube. And what is worst, those internal motions should be larger than u in order to provide the desired diffusive effect!

Problem 8.6. Consider magnetic sheets and tubes with average distance l from each other. How does the rms field strength depend on R_m in the two cases?

Convective Collapse. At the solar surface the density is $3 \times 10^{-4}\,\mathrm{kg/m^3}$. That is, for granules with $u = 2\,\mathrm{km/s}$, the equipartition field is $0.04\,\mathrm{T}$; for the larger cells it is still smaller. This is at least a factor 3 short of the field strength found observationally. Fortunately, further amplification is possible (e.g., Parker 1978) by a process known as *convective collapse*. Essentially this process is the convective instability of Sect. 6.1, but greatly altered by the presence of the magnetic field. The present treatment follows the analysis of Spruit and Zweibel (1979; but note the different orientation of the z coordinate).

We consider a *thin* vertical flux tube in the limit of perfect conductivity. Let z be the height (along the tube) and s the distance from the tube axis. The thin-tube approximation, which we shall treat in more detail in Sect. 8.2.3 below, assumes $B_s \ll B_z$ and $v_s \ll v_z$. To leading order, then, the equations of motion are

$$\rho \frac{dv}{dt} = -\frac{\partial P}{\partial z} - \rho g \tag{8.27}$$

and

$$P + \frac{B^2}{2\mu} = P_e , \tag{8.28}$$

where $v = v_z$, $B = B_z$; P is the pressure inside the tube, and P_e is the external pressure. Let F be the cross section of the tube, so that ρF is the mass per unit length. Conservation of mass then requires $\partial(\rho F)/\partial t + \partial(v\rho F)/\partial z = 0$. Since the flux is conserved, we have $BF = \mathrm{const.}$; hence

$$\frac{\partial}{\partial t}\left(\frac{\rho}{B}\right) + \frac{\partial}{\partial z}\left(v\frac{\rho}{B}\right) = 0 . \tag{8.29}$$

Finally, the assumption of *adiabatic motion,* which was already used in Chap. 5, yields

$$\frac{dP}{d\rho} = \left(\frac{dP}{d\rho}\right)_S = c^2 . \tag{8.30}$$

The four equations for v, ρ, P, and B are now perturbed around the static equilibrium ($v = 0$). The perturbation has a time dependence of form $\exp(\mathrm{i}\omega t)$. The equations are linearized, and the perturbations of ρ, P, and B are eliminated. A simplification consists in the assumption that P_e is not affected by the perturbation inside the flux tube. Further, we assume that in the static state $\beta = \mathrm{const.}$, where

$$\beta = \frac{2\mu P}{B^2} . \tag{8.31}$$

Because of (8.28) this means that the pressure inside the tube is lower than P_e by the constant factor $1/(1 + 1/\beta)$.

We thus obtain a second order equation for v:

$$v'' + C(z)v' + D(\omega^2, z)v = 0 , \tag{8.32}$$

where $' = \partial/\partial z$, and

$$C(z) = -\frac{1}{2H_P} + \frac{\Gamma_1' c_T^2}{\Gamma_1 c^2} , \tag{8.33}$$

$$D(\omega^2, z) = \frac{\omega^2}{c_T^2} + \frac{1+\beta}{2H_\Gamma^2}\left[\delta(\nabla - \nabla_a) + \frac{\Gamma_1' H_P c_T^2}{\Gamma_1 c^2}\right] . \tag{8.34}$$

In these expressions H_P is the pressure scale height; the quantities $\Gamma_1 = \rho c^2/P \equiv (\partial \ln P/\partial \ln \rho)_S$, $\delta = 1 - (\partial \ln \mu/\partial \ln T)_P$, $\nabla = \partial \ln T/\partial \ln P$, and $\nabla_a = (\partial \ln T/\partial \ln P)_S$ have all been introduced earlier (Chaps. 2, 5, 6). The *tube speed* or *cusp speed* c_T will be defined in Sect. 8.2.3 below, and obeys

$$c_T^2 = c^2/(1 + \beta\Gamma_1/2) . \tag{8.35}$$

For a given static state (the mixing-length model of Chap. 6) equation (8.32) has been solved numerically by Spruit and Zweibel, with the boundary condition that $v = 0$ at two levels, z_0 and z_1. These levels were chosen sufficiently far away from the layer in which $\nabla - \nabla_a$ is large. The instability found in the analysis then depends only weakly on the exact values of z_0 and z_1.

The problem thus formulated is a homogeneous boundary value problem, and as such has non-trivial solutions only for certain values of ω^2, the eigenvalues. In general the eigenvalues depend on the parameters of the problem, here in particular on β. The important result is that for large β there always exist negative eigenvalues ω^2, which means that for a weak magnetic field the static state is unstable. Only if β is smaller than a critical value β_c are *all* eigenvalues positive, which is necessary for stability. For the lower boundary at a depth $z_1 = 5000$ km the result is $\beta_c = 1.83$, for $z_1 = 195\,000$ km (about the total depth of the convection zone) $\beta_c = 1.51$. From these values of β_c and from (8.28) and (8.31) it is straight forward to see that stable static flux tubes are possible for a minimum field strength of order 0.1 T. At such strength the field is capable of suppressing the convective instability. Numerical work confirms the transition into the strong-field state (Grossmann-Doerth et al 1998), although this final state appears to be time-dependent rather than stationary or static. An example is shown in Fig. 8.7.

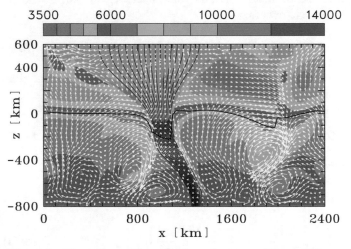

Fig. 8.7. Vertical cut through the solar atmosphere, with gas flow (*arrows*) and concentrated magnetic field (*thin solid curves*) in a two-dimensional numerical simulation. The *thick curve* near $z = 0$ marks optical depth $\tau = 1$, the temperature scale (in K) is given at the top. Courtesy O. Steiner

The general picture of the photospheric magnetic field can now be summarized. The field is very weak in the major fraction of the solar surface, but has values around or above 0.1 T in local spots, the flux tubes. We may say that the field is *intermittent*. The first cause for this field structure is concentration of flux by converging motions in the highly conducting solar plasma. The second cause is the convective collapse, which transforms the weak concentrations into strong and narrow tubes. We shall see in the following section that observations confirm this picture (in fact the observations preceeded the theoretical interpretation outlined here!).

8.2.2 Observational Evidence for Flux Tubes

Pores and Magnetic Knots. Of course, sunspots have been known as large tubes of magnetic flux for a long time. However, we postpone the treatment of sunspots to Sect. 8.3, and concentrate first on smaller features.

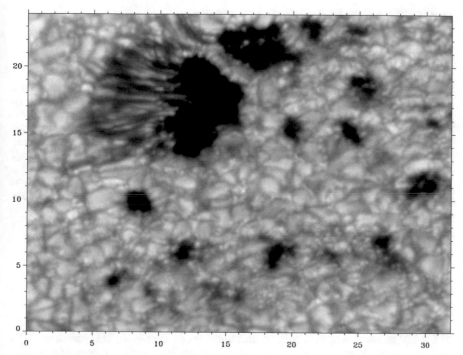

Fig. 8.8. Pores and a small sunspot with a one-sided penumbra, in a $32'' \times 24''$ area. Notice the regions showing the fuzzy "abnormal" granulation. From the Dutch Open Telescope, La Palma

The smallest magnetic phenomena on the Sun which can be distinguished in white light are the *pores*, cf. Fig. 8.8. Pores can be almost as dark as sunspots, but have no penumbra and are generally much smaller. Their size is that of one or several granules. A lifetime of ≈ 1 day has been found for pores, which is comparable to the lifetime of supergranulation cells. The magnetic field strength in pores can be deduced from Zeeman splitting, and is 0.15 T or more.

Almost invisible in white light, but very similar in their magnetic structure, are the *magnetic knots*. These magnetic flux concentrations are seen in the spectrum in form of "line gaps", which are local sections (1–2″ long) of magnetically sensitive spectral lines where the absorption profile is shallower and broader (Fig. 8.9). These gaps are caused by Zeeman broadening, and by the higher temperature in the region where the line is formed. Beckers and Schröter (1968) have observed the intensity and circular polarization in such line gaps, and have applied Unno's equations (Sect. 3.5.3) to obtain the magnetic field that best fits the observed profiles. The result is $B \approx 0.1$ to 0.2 T.

The field strength in magnetic knots has been measured also by polarimetry in the infrared. In Fig. 8.10 the circular polarization $V(\lambda)$ of an infrared

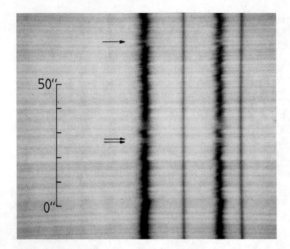

Fig. 8.9. Solar spectrum with "line gaps" (*arrows*) in the 630.15-nm and 630.25-nm lines of Fe I. Photograph W. Mattig, Domeless Coudé Telescope, Capri, exposure 0.2 s with RCA image intensifier (line curvature and black dots are imperfections of the intensifier). Notice that the gaps are absent in the narrow terrestrial lines

line clearly shows the Zeeman splitting. Because, according to (3.55), $\Delta\lambda_B/\lambda$ is proportional to λ, the splitting is easily detectable. A field strength of 0.17 T has been measured in this example.

Magnetic knots are abundant in the vicinity of sunspots. Up to a distance of 8×10^7 m from a spot's center, there are ≈ 10 knots per 100 granules. The polarity of the knot field predominantly is opposite to the polarity of the spot to which they "belong". The combined net magnetic flux of all knots is of the same order as the flux in the spot. This suggests that the lines of force which leave a sunspot essentially return to the solar surface and re-enter in the surrounding knots.

The lifetime of magnetic knots is at least 1 hour. Since the knots mostly are associated with downward motion, and are situated in the dark intergranular lanes, it is suggestive to assume a close relationship between their existence and the converging granular and supergranular velocity field.

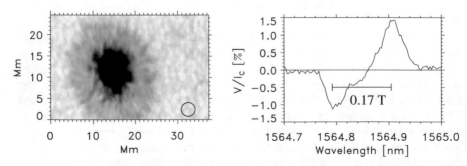

Fig. 8.10. A magnetic knot in the vicinity of a sunspot near disc center. The profile $V(\lambda)$ of the Fe I line at 1564.8 nm has been measured at the center of the small circle. From the Tenerife Infrared Polarimeter at the German Vacuum Tower Telescope. Courtesy R. Schlichenmaier

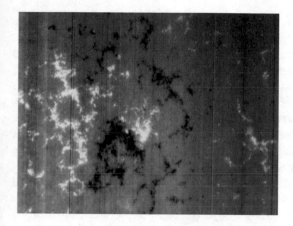

Fig. 8.11. Magnetogram of an active region. The field points toward the observer in the bright areas, and into the Sun in the dark areas. The network of 20 000–30 000 km mesh size is clearly discernable. Courtesy San Fernando Observatory, The Aerospace Corporation (Vrabec, 1971)

The connection of magnetic flux concentrations in particular to the supergranular flow has been confirmed by magnetograms (Fig. 8.11) that show the typical pattern of the 20 000–30 000-km cells. As with the magnetic knots, the strength of this "network field" has been convincingly shown by the Zeeman splitting in the infrared.

Unresolved Field Structure. A magnetic map of the solar surface is obtained by a measurement of the V profile of a magnetically sensitive spectral line as a function of position on the disc. An example with a resolution of $\approx 1''$ is shown in Fig. 8.12; this magnetogram is taken near disc center, in the "quiet" Sun. The field structure is rather irregular, with intermittent flux concentrations above a nearly field-free background.

The field strength cannot directly be read off a magnetogram. One reason is that the magnetograph may saturate: only for a weak field (small Zeeman displacement) is the V signal proportional to B. At larger field strength the maxima of $V(\lambda)$ do no longer grow with B; they also move out of the two

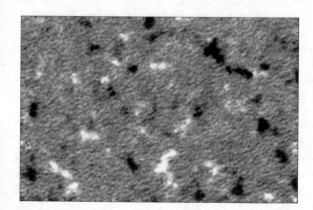

Fig. 8.12. "Quiet-Sun" magnetogram, observed with SOHO/MDI. *White* represents positive, *black* negative polarity. The displayed area is approximately $180'' \times 115''$

exit slits of the magnetograph (Fig. 3.40), and thus the measured V no longer increases.

The second reason why the field strength is not immediately measured in the magnetograph is the uncertainty about spatial resolution. In Fig. 8.12, for example, features down to about $1''$ are discernable; but we do not know whether we measure a field of uniform strength (over the resolved area) or a field that occupies only a fraction of the resolved area (but is much stronger there), or even a field that has sign reversals within the resolved area: What is measured is the average field strength B over the resolved area or, which is equivalent, the total magnetic flux enclosed by that area.

In order to obtain information about the true field structure and field strength, the above-mentioned saturation of the polarimeter signal can be exploited. It is necessary for this purpose to measure simultaneously in at least two lines having different Landé factors g^*. Except for the g^* values, the two lines should be formed in the same layer of the solar atmosphere, and should have the same excitation energy of the lower level so that their temperature dependence is the same.

Let us concentrate on the longitudinal Zeeman effect. As long as the field is weak we may expand (3.78):

$$V(\lambda) = \frac{dI_0}{d\lambda} \Delta\lambda_{\mathrm{B}} , \tag{8.36}$$

which means that $V(\lambda)$ is proportional to the field strength B and has its maximum value V_{max} at the wavelength where the line profile is steepest. In addition, (8.36) says that $V(\lambda)$ is proportional to g^*, cf. (3.55). The two iron lines at 524.706 nm and 525.022 nm shown in Fig. 3.41 satisfy the conditions mentioned, and have g^* values of 2 and 3 respectively, but obviously their circular polarization is *not* in the ratio 2 : 3. Instead, the two V_{max} values are almost equal. Hence (8.36) is not applicable; the V profile saturates because $\Delta\lambda_{\mathrm{B}}$ becomes comparable to the line width $\Delta\lambda_{\mathrm{D}}$. The example of Fig. 3.41 is of particular beauty because the measurement has been made with the Fourier transform spectrometer and the saturation therefore has nothing to do with the position of a magnetograph exit slit.

If saturation can be recognized, a crude estimate at B is obtained by setting $\Delta\lambda_{\mathrm{B}} = \Delta\lambda_{\mathrm{D}}$, and by using $g^* = 3$, $\Delta\lambda_{\mathrm{D}} = 42\,\mathrm{m\AA}$ (Sect. 3.5.4), and (3.55). The result is $B = 0.11\,\mathrm{T}$. The comparison of the V profiles of two lines thus leads to the conclusion that strong unresolved field concentrations must exist on the Sun. Since the average field $\langle B \rangle$ generally is smaller by orders of magnitude, we must conclude further that the field strength is high only in a small fraction of the observed area. Numerous studies confirm this result.

The next question is the *structure* of the strong unresolved field. Is there a single large flux concentration or a larger number of small flux tubes? A possible answer comes from counts of bright points (see below) or spicules (Chap. 9), which are seen in filtergrams and are related to the magnetic field.

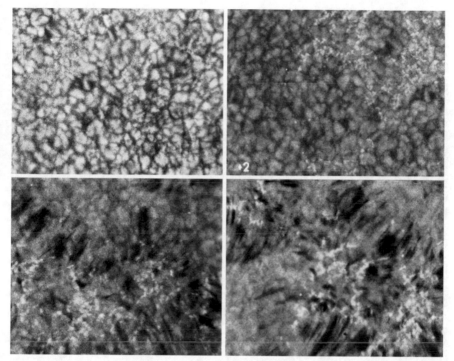

Fig. 8.13. The solar filigree at distances $+7/8$ Å (*lower left*), $-7/8$ Å (*lower right*), and $+2$ Å (*upper right*) from the center of Hα. The continuum (*upper left*) shows "abnormal granulation" (Dunn and Zirker, 1973). Courtesy National Solar Observatory, AURA, Inc.

Bright Points and the Filigree. If we tune a narrow-band filter away from the core of a strong absorption line towards the line wings we see into increasingly deeper layers. Half an ångström away from the core of Hα, for example, rosettes of dark mottles are most prominent (Fig. 9.1). Dunn and Zirker (1973), using moments of excellent seeing conditions, took a number of filtergrams at various distances $\Delta\lambda$ from the line center, cf. Fig. 8.13. At $\Delta\lambda = \pm7/8$ Å the mottles are still clearly seen, but further out in the wings smaller features appear. At $\Delta\lambda = 2$ Å a pattern of fine grains, $\approx 0.25''$ in size, is visible. Often these grains are connected to chains or crinkles. The name *filigree* has been given to this pattern by Dunn and Zirker.

Even in the continuum the filigree can be seen: the granulation shows a peculiar fine structure and a changed contrast. Such "abnormal granulation" can be recognized in Fig. 8.8 and in the upper left portion of Fig. 8.13. The grains and crinkles of the filigree seem to sit preferentially in the dark intergranular lanes.

Another manifestation of sub-arcsecond fine structure are the *facular points* discovered by Mehltretter (1974). Faculae are bright features seen in

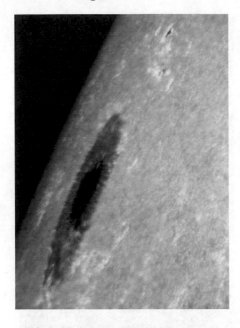

Fig. 8.14. Sunspot and faculae near the solar limb. Photograph D. Soltau, Newton Telescope, Izaña, Tenerife

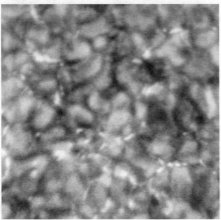

Fig. 8.15. Facular points in a photograph taken in 1973 by J. P. Mehltretter with the R. B. Dunn Solar Telescope at Sacramento Peak Observatory; wavelength range 393.4 ± 0.8 nm (Ca II K). Displayed area is approximately $20'' \times 20''$; the scale is the same as in Fig. 8.16

white light near the solar limb (Fig. 8.14), and with the help of filters also elsewhere on the disc. Observing at disc center during extremely good seeing, Mehltretter was able to resolve these features into a number of small bright points. The facular points also have sizes of about $0.25''$ or less, and clearly sit within the dark intergranular lanes (Fig. 8.15). Simultaneous photographs in Ca II and in the wing of Hα suggest a close correspondence to the grains and crinkles. We may assume that the two phenomena represent the same solar feature, the filigree.

Fig. 8.16. G-band bright points. CCD image obtained at the R. B. Dunn Solar Telescope at Sacramento Peak Observatory, with a filter of 1 nm width centered at 430.5 nm; the exposure time was 6 s, and an adaptive optics system, locked on the small pore in the center, was used. The contrast and resolution is best in the central part, but gradually deteriorates as one moves towards the edges, away from the isoplanatic patch. The displayed area is approximately $40'' \times 40''$. Courtesy T. R. Rimmele

The sharpest images of bright points have been obtained in the G band (Fig. 8.16). The G band is a band head at $\approx 430.5 \pm 1$ nm which is dominated by absorption lines due to electronic transitions of the CH molecule, along with changes of the rotational and vibrational energy. Due to the enhanced temperature in the visible layer of a small magnetic element (Fig. 8.26) there is enhanced dissociation of CH, and therefore less absorption, as illustrated in Fig. 8.17. The high contrast of the G-band images is due to the large sensitivity to temperature variations of the CH dissociation; the bright-point intensity may exceed the average quiet-Sun intensity by up to 30 %.

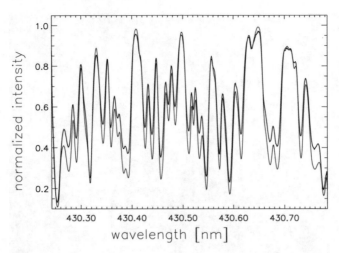

Fig. 8.17. Mean spectrum of several G-band bright points (*thick curve*), and mean spectrum of the quiet Sun (*thin curve*). The CH lines are weaker in the bright-point spectrum, while the Ca I line at 430.254 nm and the Fe I line at 430.318 nm have about the same strength in both spectra. After Langhans et al. (2001)

Magnetograms such as Fig. 8.12 do not resolve a flux tube of the size of a crinkle or bright point. But counts made by Mehltretter have shown that the number density of facular points is roughly proportional to the mean magnetic field strength, see Fig. 8.18. If we divide this mean field strength by the number density of facular points, we obtain the average magnetic flux per point. Mehltretter's result is 4.4×10^9 Wb, which, with a field of 0.1 T as deduced above, leads to $d \approx 200$ km for the diameter of the flux tubes. The fact that the visible features become larger at greater height (i.e., towards the core of Hα) can be explained by the spatial divergence of the magnetic lines of force.

Finally, a direct proof that bright points are magnetic features was presented by Keller and von der Lühe (1992). Using speckle interferometry they were able to reconstruct images in polarized light with a diffraction limited

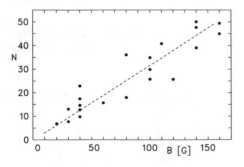

Fig. 8.18. Number of facular points in $5'' \times 5''$ squares, plotted against the average field strength B in the same areas. After Mehltretter (1974)

resolution (at the Swedish 50-cm telescope in La Palma). On these images magnetic flux concentrations with diameters of $\approx 200\,\mathrm{km}$ were found that could be identified with the bright points seen in the filtergrams.

8.2.3 Vertical Thin Flux Tubes

In this and the following section we shall describe the magnetic field and the motion in an *isolated* tube of magnetic flux, i.e., within a bundle of lines of force that is surrounded by a plasma with zero magnetic field. The complete treatment of such a configuration would consist of a solution of the appropriate equations both in the tube interior and in the exterior space, and of combining these two solutions by a fit across the tube boundary. In the special cases where such treatment has been applied it confirms the results obtained by the method of approximation used in the following.

The approximation of *thin* or *slender* flux tubes is valid as long as all length scales along the tube are large in comparison to the tube diameter. In the present case, where we consider an axisymmetric vertical tube, with its axis in the z-direction (upwards, cf. Fig. 8.19), and with a cross section of radius $R_\mathrm{T}(z)$, this means that

$$R_\mathrm{T}/H \ll 1\,, \qquad kR_\mathrm{T} \ll 1 \tag{8.37}$$

for all z; H is the scale height of any quantity, e.g., pressure, in the tube, and k is the vertical wave number of any perturbation propagating along the tube.

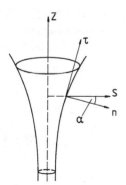

Fig. 8.19. Thin vertical flux tube

We shall here neglect all dissipative processes. That is, we assume that the plasma is inviscid and of perfect electrical conductivity, and that all changes of state occur adiabatically. The set of equations which must be solved is then: the equation of motion,

$$\rho(\dot{\boldsymbol{v}} + \boldsymbol{v} \cdot \nabla \boldsymbol{v}) = -\nabla P + \rho \boldsymbol{g} + \boldsymbol{j} \times \boldsymbol{B}\,, \tag{8.38}$$

the equation of continuity,

$$\dot{\rho} + \nabla \cdot (\rho \boldsymbol{v}) = 0 \ , \tag{8.39}$$

the induction equation, (8.6) with $\eta = 0$,

$$\dot{\boldsymbol{B}} = \mathrm{curl}\,(\boldsymbol{v} \times \boldsymbol{B}) \ , \tag{8.40}$$

and the condition of adiabatic change, cf. (8.30), (8.35),

$$\dot{P} + \boldsymbol{v} \cdot \nabla P = (P\Gamma_1/\rho)(\dot{\rho} + \boldsymbol{v} \cdot \nabla \rho) \ . \tag{8.41}$$

The dot denotes $\partial/\partial t$; in a system of cylindrical coordinates (s, ϕ, z) around the tube axis the gravitational acceleration is $\boldsymbol{g} = (0, 0, -g)$.

We now expand all variables in powers of s, the distance from the axis of symmetry. If we restrict our attention to flux tubes with small radial variation within $s < R_\mathrm{T}$, we may hope to obtain satisfactory results even when the expansions are truncated after only one or two terms.

We require that our variables are *regular* on the axis, and can therefore be expanded in a Taylor series around $s = 0$. As a consequence of the axial symmetry then P, ρ, v_z, and B_z have only *even* terms in their expansions, viz.

$$P = P_0(z, t) + P_2(z, t)s^2 + \dots \tag{8.42}$$

etc., while the s and ϕ components of $\boldsymbol{v}$ and $\boldsymbol{B}$ have only odd terms, e.g.,

$$v_s = v_{s1}(z, t)s + v_{s3}(z, t)s^3 + \dots \ . \tag{8.43}$$

Let us make two modifications of the scheme just outlined. The first is to replace v_ϕ by the angular velocity $\Omega = v_\phi/s$; Ω has an even expansion such as (8.42). The second is the introduction of a vector potential $\boldsymbol{A}$ for the poloidal field components B_s and B_z. Thus

$$\boldsymbol{B} = \boldsymbol{B}_\mathrm{p} + \boldsymbol{B}_\mathrm{t} \equiv \mathrm{curl}\,\boldsymbol{A} + (0, B_\phi, 0) \ , \tag{8.44}$$

where

$$\boldsymbol{A} = (0, A, 0) \ , \tag{8.45}$$

and

$$B_s = -\frac{\partial A}{\partial z} \ , \qquad B_z = \frac{1}{s}\frac{\partial}{\partial s}(sA) \ . \tag{8.46}$$

In this way we reduce the number of variables by one, and at the same time ensure that always div $\boldsymbol{B} = 0$. From (8.46) it is clear that A has an odd expansion of type (8.43).

We should stress here that the expansions (8.42), (8.43) must not be confused with a linearization. Even in the lowest order our problem will keep the essential non-linearities of the full magnetohydrodynamic equations (8.38) to (8.41). Of course we shall employ linearized forms later, in particular when we discuss waves of small amplitude.

As an illustrative example, let us carry out the expansion and substitution procedure for the induction equation (8.40). To this end we divide the velocity into its poloidal and toroidal parts:

$$v = v_{\mathrm{p}} + v_{\mathrm{t}} , \tag{8.47}$$

where

$$v_{\mathrm{p}} = (v_s, 0, v_z) , \qquad v_{\mathrm{t}} = (0, s\Omega, 0) . \tag{8.48}$$

Then, by (8.44),

$$\dot{B} = \mathrm{curl}\,(v_{\mathrm{p}} \times B_{\mathrm{p}}) + \mathrm{curl}\,(v_{\mathrm{p}} \times B_{\mathrm{t}} + v_{\mathrm{t}} \times B_{\mathrm{p}}) . \tag{8.49}$$

The first term on the right-hand side is poloidal, i.e., has no ϕ component; the second is toroidal, i.e., has no s and z components. Hence (8.49) can immediately be separated. The poloidal part is

$$\dot{A} = (v_{\mathrm{p}} \times B_{\mathrm{p}})_\phi = -v_z \Lambda' - \frac{v_s}{s} \frac{\partial}{\partial s} (s\Lambda) , \tag{8.50}$$

where the dash denotes $\partial/\partial z$. In (8.50) we have omitted the curl operator; that is, we have integrated, and a constant of integration should appear. However, since $\mathrm{curl}\,\mathrm{grad} = 0$ this constant would be a gradient. As (8.50) is a ϕ component the integration constant would be of form $\partial C/\partial \phi$, which is zero in the present, axisymmetric, case.

The toroidal part of (8.49) yields, after a small calculation,

$$\dot{B}_\phi = -(B_\phi v_z)' - \frac{\partial}{\partial s}(B_\phi v_s) + \Omega' \frac{\partial}{\partial s}(sA) - sA' \frac{\partial \Omega}{\partial s} . \tag{8.51}$$

Now insert the expansions into (8.50) and compare the coefficients of the diverse powers of s. The result is, to orders s and s^3,

$$\dot{A}_1 = -v_{z0} A_1' - 2v_{s1} A_1 , \tag{8.52}$$

$$\dot{A}_3 = -v_{z0} A_3' - 4v_{s1} A_3 - v_{z2} A_1' - 2v_{s3} A_1 . \tag{8.53}$$

In a similar way, (8.51) gives, to order s

$$\dot{B}_{\phi 1} = -(B_{\phi 1} v_{z0})' - 2B_{\phi 1} v_{s1} + 2\Omega_0' A_1 . \tag{8.54}$$

To third order, we obtain an equation for $B_{\phi 3}$ which we shall not consider further [we have written down (8.53) because A_3 corresponds to B_{z2}, i.e., to a *second-order* correction of the main field component, B_z, in the tube].

The other equations are treated in complete analogy. We shall take the resulting relations as given whenever we need them further below.

We must, however, consider the transition at $s = R_{\mathrm{T}}$, which will yield an additional equation. The tube boundary constitutes a *tangential discontinuity*

(Landau and Lifschitz 1967, p. 265). Let $\hat{n}$ be the local unit vector normal to the boundary, and n and τ coordinates in the normal and tangential directions (Fig. 8.19). Because of the perfect conductivity there is no flow across the magnetic field, i.e., $v_n = 0$ at $s = R_T$, and $\boldsymbol{v} = (0, \boldsymbol{v}_\tau)$ in the n, τ-system. Hence

$$v_s = v_\tau \sin \alpha \, , \tag{8.55}$$

$$v_z = v_\tau \cos \alpha \, . \tag{8.56}$$

Since $\boldsymbol{B}$ is tangential to the boundary, we have

$$\tan \alpha = B_s / B_z \, . \tag{8.57}$$

Using (8.55) to (8.57), and expression (8.18) for the magnetic force, we may calculate the projection of the equation of motion, (8.38), onto the n-direction. We obtain

$$\frac{\partial}{\partial n} \left(P + \frac{B^2}{2\mu} \right) = C \, , \tag{8.58}$$

where C is a combination of tangential derivatives, time derivatives, and other *bounded* terms. Integration of (8.58) across the boundary from $n = -\varepsilon$ to $n = +\varepsilon$ yields (in the limit $\varepsilon \to 0$) the condition that $P + B^2/2\mu$ must be continuous, or that, at $s = R_T$,

$$P + \frac{B^2}{2\mu} = P_{\mathrm{e}} \, . \tag{8.59}$$

The external magnetic field is zero; P_{e} is the external pressure. We again use our expansion, this time at $s = R_T$, and the transformation (8.55), (8.56) to obtain

$$P_0 + \frac{2}{\mu} A_1^2 + R_T^2 \left[P_2 + \frac{1}{2\mu} \left(16 A_1 A_3 + A_1'^2 + B_{\phi 1}^2 \right) \right] = P_{\mathrm{e}} \, . \tag{8.60}$$

The radius R_T may be eliminated in terms of the total flux

$$\Phi = 2\pi \int\limits_0^{R_T} B_z s \, ds = 2\pi R_T(z) A(z, R_T(z)) \simeq 2\pi R_T^2 A_1 \, . \tag{8.61}$$

The last expression on the right is of sufficient accuracy because R_T^2 itself is a second-order term.

If we go to the second order in the expansion we must use (8.60) as it stands; for problems which we treat only in the lowest order we may drop the term proportional to R_T^2. Notice that in the latter case the total flux does not matter: That is, to leading order, tubes with various total fluxes behave in the same way on their axis, $s = 0$.

Leading Order. We apply the recipe outlined above and obtain

$$\rho_0(\dot{v}_{z0} + v_{z0}v'_{z0}) = -P'_0 - \rho_0 g \; , \tag{8.62}$$

$$\dot{\rho}_0 + (\rho_0 v_{z0})' + 2\rho_0 v_{s1} = 0 \; , \tag{8.63}$$

$$\dot{A}_1 = -v_{z0}A'_1 - 2v_{s1}A_1 \; , \tag{8.52}$$

$$\dot{P}_0 + v_{z0}P'_0 = (P_0\gamma/\rho_0)(\dot{\rho}_0 + v_{z0}\rho'_0) \; , \tag{8.64}$$

$$P_0 + \frac{2}{\mu}A_1^2 = P_e \; . \tag{8.65}$$

It is straight forward to eliminate v_{s1} from this set of equations, to replace A_1 by $B_{z0}/2$ and so to recognize that equations (8.27) to (8.30) are the same set. Hence, we have confirmed the equations used to describe the convective collapse. The occurrence of v_{s1} and (via A_1) of the horizontal field component B_{s1} illustrates that the collapse does not merely involve a vertical field and a vertical motion, as it had appeared from the more heuristic derivation of (8.29).

At static equilibrium, $v = 0$ and $\partial/\partial t = 0$, we have only (8.65) and $P'_0 = -\rho_0 g$. Let us assume that the temperature, and hence the pressure scale height H_P, is the same inside and outside the tube. Differentiation of (8.65) then yields $B'_{z0} = -B_{z0}/2H_P$, or

$$B_{z0} \propto P_e^{1/2} \; . \tag{8.66}$$

This result, together with (8.61), gives

$$R_T \propto P_e^{-1/4} \; , \tag{8.67}$$

which means that it takes four pressure scale heights for the tube radius to expand by a factor e. In the solar atmosphere this is about 500 km. Thus, a tube which is thin at optical depth $\tau_{500} = 1$ may no longer be thin [in the sense of (8.37)] at the level of the temperature minimum. On the other hand, (8.67) ensures that even for tubes that violate (8.37) in the photosphere the present approximation will be quite satisfactory in the underlying convection zone.

Problem 8.7. For given $T(z)$ and given total flux Φ calculate explicitly the functions $B_{z0}(z)$, $P_0(z)$, and $R_T(z)$ for a static thin vertical tube.

Longitudinal Tube Waves. Now as we have justified the equations governing the convective collapse we may as well exploit them further. We are interested in small perturbations of the static equilibrium and hence may use the linearized equation (8.32). For simplicity, let us neglect the variation of Γ_1 and of the mean molecular weight, so that $\Gamma_1 \equiv \gamma = $ const., $\delta = 1$, and $\nabla_{\mathrm{a}} = (\gamma - 1)/\gamma$. In addition, let us consider the special case of an isothermal atmosphere where $\nabla = 0$, and where H_{P} is constant. Because β is also a constant (Problem 8.7) we then have constant coefficients C and D [expressions (8.33) and (8.34)], and the solutions are waves of the form

$$v_{z0} \propto \exp\left(\mathrm{i}\omega t + \mathrm{i}kz + z/4H_{\mathrm{P}}\right) . \tag{8.68}$$

The dispersion relation is

$$\omega^2 = c_{\mathrm{T}}^2 \left[k^2 + \frac{1}{2H_{\mathrm{P}}^2} \left(\frac{1}{8} + (1+\beta)(\gamma-1)/\gamma \right) \right] . \tag{8.69}$$

As in the case of the acoustic waves treated in Chap. 5 there is a cutoff frequency below which the wave cannot propagate. The cutoff frequency is given by

$$\omega_1^2 = \frac{c_{\mathrm{T}}^2}{2H_{\mathrm{P}}^2} \left[\frac{1}{8} + (1+\beta)(\gamma-1)/\gamma \right] . \tag{8.70}$$

Problem 8.8. Confirm (8.69). For very large scale height the phase velocity ω/k approaches the tube speed c_{T}. Show that

$$c_{\mathrm{T}}^2 = \frac{c^2 c_{\mathrm{A}}^2}{c^2 + c_{\mathrm{A}}^2} , \qquad \text{where} \tag{8.71}$$

$$c_{\mathrm{A}} = B_{z0}/(\mu\rho_0)^{1/2} \tag{8.72}$$

is the *Alfvén velocity*.

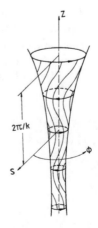

Fig. 8.20. Longitudinal tube wave   **Fig. 8.21.** Torsional tube wave

Although the wave (8.68) is mainly longitudinal, it does comprise (to first order in s) a periodic lateral expansion and contraction of the tube – cf. (8.63), which gives v_{s1} in terms of the density and longitudinal velocity. Because of this the longitudinal tube wave is also called the "sausage mode". Figure 8.20 is an illustration.

Torsional Alfvén Waves. As another possible wave motion let us shortly discuss the torsional Alfvén wave. Alfvén waves are well known in magnetohydrodynamics as periodic motions perpendicular to the magnetic lines of force. These lines are distorted and their tension, cf. (8.18), provides the restoring force. In a homogeneous equilibrium field Alfvén waves exist as solutions even of the non-linear magnetohydrodynamic equations, and we shall see presently that our thin flux tube also is capable of supporting such a wave.

For simplicity let us consider a tube without stratification, i.e., $g = 0$. Since the scale height is infinite such a tube does not fan out, cf. (8.67). The transverse direction, in which we seek our wave motion, is then horizontal, and in order to avoid compression (so that the magnetic force is the only restoring force) we choose $v_s = 0$. The main tube field B_z is distorted in the azimuthal direction only, and B_s remains zero. It is not difficult to see that in the lowest order of the thin-tube expansion $\dot{\rho}_0$, $\dot{P}_0$, P_0', $\dot{A}_1$, A_1' all must vanish, and that the ϕ components of the equations of motion and induction yield, to order s

$$\rho_0 \dot{\Omega}_0 = \frac{2}{\mu} A_1 B_{\phi 1}' , \tag{8.73}$$

$$\dot{B}_{\phi 1} = 2\Omega_0' A_1 . \tag{8.74}$$

Elimination of $B_{\phi 1}$ gives

$$\ddot{\Omega}_0 = c_{\mathrm{A}}^2 \Omega_0'' , \tag{8.75}$$

where c_{A} is the Alfvén velocity introduced in (8.72). The last equation describes an oscillation of Ω_0 (and of $B_{\phi 1}$) that propagates in the z-direction, i.e., along the tube. This is the torsional Alfvén wave. It is remarkable that no linearization was required to derive (8.75).

Torsional Alfvén waves are modified by stratification, in which case the tube widens in the z-direction, cf. (8.67). Such (more general) situations have been investigated numerically. The distortion of the lines of force is illustrated in Fig. 8.21.

Problem 8.9. Calculate the second-order (in s) terms of the torsional Alfvén wave in a thin flux tube without stratification.

8.2.4 Curved Thin Flux Tubes

Like any magnetic line of force a magnetic flux tube may be bent in an arbitrary way. As long as the curvature of such bending is small, i.e., as long as the radius of curvature is large compared to the tube diameter, we may again apply the approximation of *thin* flux tubes.

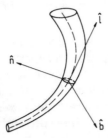

Fig. 8.22. Curved thin flux tube

The axis of the flux tube constitutes a curve in 3-dimensional space. Let l be the length measured along this curve. At each point we may define a set of three orthonormal vectors, namely the tangent vector $\hat{\boldsymbol{l}}$, the principal normal $\hat{\boldsymbol{n}}$, and the binormal $\hat{\boldsymbol{b}} = \hat{\boldsymbol{l}} \times \hat{\boldsymbol{n}}$ (Fig. 8.22). The differential geometry of space curves tells us that, going along the curve, these three unit vectors change according to Frenet's equations:

$$\partial \hat{\boldsymbol{l}}/\partial l \;=\; \kappa \hat{\boldsymbol{n}} \quad , \tag{8.76}$$

$$\partial \hat{\boldsymbol{n}}/\partial l \;=\; -\kappa \hat{\boldsymbol{l}} \quad + \tau \hat{\boldsymbol{b}} \;, \tag{8.77}$$

$$\partial \hat{\boldsymbol{b}}/\partial l \;=\; -\tau \hat{\boldsymbol{n}} \quad , \tag{8.78}$$

where κ is the *curvature* (the inverse of the *radius of curvature*), and τ the *torsion* of the curve.

The motion of the flux tube is governed by (8.38). We decompose the volume force, $\boldsymbol{F}$, on the right of that equation into its components parallel to the three unit vectors, thereby essentially following Spruit (1981a). We do this in the lowest order of the thin-tube approximation, in which the field is in the direction of $\hat{\boldsymbol{l}}$, and the Lorentz force (8.17) is perpendicular to $\hat{\boldsymbol{l}}$. In the direction of $\hat{\boldsymbol{l}}$ we therefore have

$$F_l = -\partial P/\partial l + \rho \hat{\boldsymbol{l}} \cdot \boldsymbol{g} \;. \tag{8.79}$$

For the transverse components we use the form (8.18) of the Lorentz force. To lowest order $\boldsymbol{B} = B\hat{\boldsymbol{l}}$ and

$$\boldsymbol{B} \cdot \nabla \boldsymbol{B} = B \frac{\partial}{\partial l}(B\hat{\boldsymbol{l}}) = \frac{1}{2} \hat{\boldsymbol{l}} \frac{\partial B^2}{\partial l} + \kappa B^2 \hat{\boldsymbol{n}} \;, \tag{8.80}$$

where (8.76) has been used. Let n and b be coordinates in the directions of $\hat{\boldsymbol{n}}$ and $\hat{\boldsymbol{b}}$. Then

$$F_n = -\partial P/\partial n + \rho \hat{\boldsymbol{n}} \cdot \boldsymbol{g} - \frac{1}{2\mu} \frac{\partial B^2}{\partial n} + \frac{\kappa}{\mu} B^2 \;, \tag{8.81}$$

and

$$F_b = -\partial P/\partial b + \rho\,\hat{\boldsymbol{b}}\cdot\boldsymbol{g} - \frac{1}{2\mu}\frac{\partial B^2}{\partial b}\ . \tag{8.82}$$

Now we have already seen in the preceding section that, to lowest order, the internal pressure plus the magnetic pressure are laterally balanced by the external pressure P_e. This result holds for curved tubes as long as the curvature is weak (here expressed through $\kappa R_T \ll 1$). We may therefore replace $P + B^2/2\mu$ by P_e. Using $\nabla P_e = \rho_e\boldsymbol{g}$ we thus obtain

$$F_n = (\rho - \rho_e)\,\hat{\boldsymbol{n}}\cdot\boldsymbol{g} + \kappa\,B^2/\mu\ , \tag{8.83}$$

$$F_b = (\rho - \rho_e)\,\hat{\boldsymbol{b}}\cdot\boldsymbol{g}\ . \tag{8.84}$$

Static Tube. Since $\rho \neq \rho_e$ in general, an immediate consequence of (8.84) is that a static curved tube, i.e., one with $\boldsymbol{F} = 0$, must lie in a single *vertical plane*. For, differentiation of $\hat{\boldsymbol{b}}\cdot\boldsymbol{g} = 0$ and use of (8.78) yields $\tau\hat{\boldsymbol{n}}\cdot\boldsymbol{g} = 0$. The case $\hat{\boldsymbol{n}}\cdot\boldsymbol{g} = 0$, i.e., the vertical tube, is uninteresting, hence $\tau = 0$: the tube axis has no torsion.

Let θ be the angle of the axis with the horizontal direction x, and let z be vertically upwards. In the x, z-plane, then $\hat{\boldsymbol{l}} = (\cos\theta, \sin\theta)$, $\hat{\boldsymbol{n}} = (-\sin\theta, \cos\theta)$, and, by (8.77),

$$\kappa = -\hat{\boldsymbol{l}}\cdot d\hat{\boldsymbol{n}}/dl = d\theta/dl = \sin\theta\, d\theta/dz\ . \tag{8.85}$$

Substitution into (8.83) gives, with $\hat{\boldsymbol{n}}\cdot\boldsymbol{g} = -g\cos\theta$,

$$\frac{d\theta}{dz} = \frac{\rho - \rho_e}{B^2}\,\mu g\cot\theta\ . \tag{8.86}$$

This equation describes the shape of a curved thin flux tube in static equilibrium.

Magnetic Buoyancy. A special case is the straight horizontal flux tube. In this case $\hat{\boldsymbol{l}}\cdot\boldsymbol{g} = 0$; as $\hat{\boldsymbol{n}}$ and $\hat{\boldsymbol{b}}$ are not distinguished we may arbitrarily choose $\hat{\boldsymbol{n}}$ parallel and $\hat{\boldsymbol{b}}$ perpendicular to $\boldsymbol{g}$. Then the only component of $\boldsymbol{F}$ which does not vanish is

$$F_n = (\rho - \rho_e)\,\hat{\boldsymbol{n}}\cdot\boldsymbol{g}\ . \tag{8.87}$$

We have $P < P_e$ and, assuming that the temperature is the same inside and outside the tube, $\rho < \rho_e$. Hence $\boldsymbol{F}$ points into the opposite direction of $\boldsymbol{g}$, i.e., upwards. This upwards directed force is called *magnetic buoyancy*. Its effect is a rise of the horizontal flux tube as a whole (Parker 1955 a, Jensen 1955).

Transverse Tube Waves. We now study the dynamic response of a flux tube to the force F in a special case: a tube which is straight and vertical in its undisturbed state, but which has curvature as it is locally elongated in a transverse direction, say x, by a small amount $\xi(z,t)$ (Fig. 8.23). The tube axis always remains in the x, z-plane, i.e., $\hat{b} \cdot g = 0$, and the only interesting transverse force component is F_n.

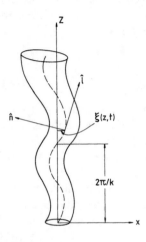

Fig. 8.23. Transverse tube wave

Let α be the angle between the tube axis and the vertical, so that $\tan \alpha = \partial \xi / \partial z \equiv \xi'$. As ξ is small, α is also small, and $\hat{l} = (-\xi', 1)$, $\hat{n} = (-1, -\xi')$ to first order in α. Using again (8.77) we find $\kappa = \xi''$. Hence

$$F_n = (\rho - \rho_e)\, g\xi' + \frac{B^2}{\mu}\, \xi'' \ . \tag{8.88}$$

Before we substitute this into the equation of motion we must find a way to incorporate the perturbation of the *external* medium. This perturbation was neglected in our discussion of the convective collapse in Sect. 8.2.1 above and also in the derivation of (8.83), but this time the flux tube will move fully into the external medium and this certainly will have some effect. An approximate way is to neglect both compression and curvature at this point. Then we may use the result valid for a circular cylinder in a liquid, namely that the inertia of the cylinder is *apparently* increased by the mass of the displaced liquid (Basset 1961, Vol. I, p. 186). This result is plausible: obviously some of the surrounding material must be accelerated to give room for the transverse motion of the tube. Following this reasoning, and making use of the small amplitude of the perturbation, we obtain $(\rho + \rho_e)\, \partial v_n / \partial t = F_n$, or, with $v_n = \dot{\xi}$ and (8.72),

$$\ddot{\xi} = \frac{\rho - \rho_e}{\rho + \rho_e}\, g\xi' + \frac{\rho}{\rho + \rho_e}\, c_A^2 \xi'' \ . \tag{8.89}$$

This is the equation of the *transverse tube wave*.

In the case of no stratification, $g = 0$, the wave propagates with velocity $c_A/(1 + \rho_e/\rho)^{1/2}$, i.e., with the velocity of an Alfvén wave slowed down by the additional inertia of the exterior mass.

We now include stratification, but do this in the special case of equal internal and external (constant) temperature, where $\rho_e/\rho = P_e/P = 1+1/\beta$, with β given by (8.31). Using $\beta c_A^2 = 2gH_P$ we find

$$\ddot{\xi} = g(-\xi' + 2H_P\xi'')/(2\beta + 1) , \tag{8.90}$$

which is solved by

$$\xi \propto \exp(i\omega t + ikz + z/4H_P) . \tag{8.91}$$

The dispersion relation

$$\omega^2 = \frac{gH_P}{\beta + 1/2} \left(k^2 + \frac{1}{16H_P^2} \right) \tag{8.92}$$

shows that transverse tube waves cannot propagate if ω is smaller than the cutoff frequency ω_2 given by

$$\omega_2^2 = \frac{g}{8H_P} \frac{1}{2\beta + 1} . \tag{8.93}$$

Problem 8.10. Calculate the cutoff frequency (8.93) of transverse tube waves for parameter values corresponding to the solar atmosphere and compare it to the acoustic cutoff frequency (Chap. 5).

Problem 8.11. Use Frenet's equations, (8.76) to (8.78), to bring Walén's frozen field relation, (8.15), into the form

$$\frac{d}{dt} \left(\frac{B}{\rho} \right) = \frac{B}{\rho} \left(\frac{\partial v_l}{\partial l} - \kappa v_n \right) . \tag{8.94}$$

Instability of Horizontal Tubes. Above we have seen that horizontal tubes filled with gas of density ρ that is lower than the exterior density ρ_e are buoyant and therefore rise towards the solar surface. We shall now show that a horizontal tube with $\rho = \rho_e$, i.e., a tube with no buoyancy, nevertheless cannot reside in the convection zone because it is *unstable*. Spruit and van Ballegooijen (1982) have described this instability.

We consider small perturbations of a thin flux tube which, in equilibrium, is horizontal and has $\rho_0 = \rho_{e0}$. Let ξ be the displacement of a tube parcel in the longitudinal ($\simeq$ horizontal, x) direction, and ζ the displacement in the transverse ($\simeq$ vertical, z) direction. The concomitant perturbations of pressure, density, and field strength are denoted by P_1, ρ_1, B_1 respectively – not to be confused with the first-order terms of the expansions of Sect. 8.2.3. In the x, z-plane we immediately find $\hat{l} = (1, \partial\zeta/\partial x)$ and $\hat{n} = (\partial\zeta/\partial x, -1)$;

equation (8.77) then gives $\kappa = -\partial^2 \zeta / \partial x^2$. Hence, (8.79) yields the following equation for the longitudinal perturbation:

$$\ddot{\xi} = -\frac{1}{\rho_0} \frac{\partial P_1}{\partial x} - g \frac{\partial \zeta}{\partial x} \ . \tag{8.95}$$

The transverse equation follows from the perturbation of (8.83):

$$\ddot{\zeta} = -\frac{\rho_1}{2\rho_0} g - \frac{\zeta g (1 - \nabla)}{2 H_P (1 + 1/\beta)} + \frac{\partial^2 \zeta}{\partial x^2} \frac{B_0^2}{2\mu\rho_0} \ , \tag{8.96}$$

where, in the same way as explained above in the context of transverse tube waves, the external medium has been included by dividing the force F_n by $(\rho + \rho_e)$ instead by ρ only. The external equilibrium density has been replaced by ρ_0, and $\rho_{e1} = -\rho_0 \zeta / H_{\rho e} = -\rho_0 \zeta (1 - \nabla)/[H_P(1 + 1/\beta)]$ has been substituted.

The remaining equations are the linearized forms of the lateral equilibrium (8.28), the condition of adiabatic variation (8.41), and the frozen-field condition, which we take in the form (8.94):

$$\frac{P_1}{P_0} + \frac{2 B_1}{\beta B_0} = -\frac{\zeta}{H_P} \ , \tag{8.97}$$

$$\frac{P_1}{P_0} = \gamma \frac{\rho_1}{\rho_0} \ , \tag{8.98}$$

$$\frac{\dot{B}_1}{B_0} - \frac{\dot{\rho}_1}{\rho_0} = \frac{\partial \dot{\xi}}{\partial x} \ . \tag{8.99}$$

Following Spruit and van Ballegooijen (1982), we seek solutions of form $\exp(i\omega t + ikx)$, introduce a dimensionless frequency $\tilde{\omega}$ by $\tilde{\omega}^2 = \omega^2 \beta H_P / g$ and a dimensionless wave number $\tilde{\alpha} = k H_P$, and obtain for the system (8.95) to (8.99) the dispersion relation

$$\tilde{\omega}^4 + A \tilde{\omega}^2 + C = 0 \ , \tag{8.100}$$

where [with $\nabla_a = (\gamma - 1)/\gamma$]

$$A = -\tilde{\alpha}^2 - \frac{\gamma \beta \tilde{\alpha}^2}{1 + \beta\gamma/2} + \left(\frac{1}{2} - \frac{1}{\gamma} \right) \frac{\beta^2}{(1+\beta)(2+\beta\gamma)} + \beta^2 \frac{\nabla - \nabla_a}{2(1+\beta)} \ , \tag{8.101}$$

and

$$C = \gamma \tilde{\alpha}^2 \frac{\beta}{1 + \beta\gamma/2} \left\{ \tilde{\alpha}^2 - \frac{\beta}{2(1+\beta)} \left[\frac{1}{\gamma} + \beta(\nabla - \nabla_a) \right] \right\} \ . \tag{8.102}$$

Problem 8.12. Confirm (8.100) to (8.102) and show that the condition for real $\tilde{\omega}^2$, namely $A^2 > 4C$, is always satisfied.

The horizontal flux tube is stable against the perturbations admitted in the present analysis if always $\omega^2 > 0$; the tube is unstable if at least one ω with $\omega^2 < 0$ can be found, because then the corresponding perturbation varies exponentially with time. Since C is the product of the two roots of (8.100), a *sufficient* condition for instability is $C < 0$. Such C can certainly be found (by the choice $\tilde{\alpha} = 0$) if

$$\beta(\nabla - \nabla_a) > -1/\gamma \ . \tag{8.103}$$

Vanishing $\tilde{\alpha}$ means that the horizontal tube moves as a whole, and condition (8.103) says that such a motion is accelerated exponentially within the convection zone where $\nabla > \nabla_a$. Of course we have $C < 0$ even for non-vanishing $\tilde{\alpha}$ whenever

$$\tilde{\alpha}^2 < \frac{\beta}{2(1+\beta)} \left[\frac{1}{\gamma} + \beta(\nabla - \nabla_a) \right] \ . \tag{8.104}$$

We must conclude that *horizontal flux tubes in the convection zone are unstable*. Moreover, even in a stably stratified layer, where $\nabla < \nabla_a$, criterion (8.103) predicts instability if the magnetic field is sufficiently strong. Using (8.31), the critical field strength beyond which the tube is unstable is

$$B_c = [2\mu P \gamma (\nabla_a - \nabla)]^{1/2} \ . \tag{8.105}$$

A layer of special interest here is the region of overshooting convection at the bottom of the solar convection zone. At a depth of 2×10^8 m the pressure is

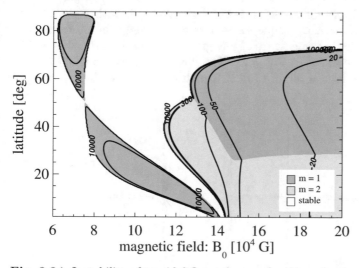

Fig. 8.24. Instability of toroidal flux tubes, as function of magnetic field strength and heliographic latitude (Ferriz Mas and Schüssler 1995), for $\nabla - \nabla_{ad} = -2.6 \cdot 10^{-6}$. The labels on the contours are growth times in days, m is the azimuthal wave number of the unstable modes. Courtesy P. Caligari

of the order 10^{13} Pa (Table 2.4 or 6.1); and $|\nabla_a - \nabla| \approx 10^{-6} \ldots 10^{-5}$. Using (8.105) and models of overshooting convection as described in Sect. 6.2.3, Pidatella and Stix (1986) showed that stable flux tubes with a field strength of several tesla are permitted in such a layer.

A general stability analysis must include the effects of rotation. In this case the critical field strength depends on latitude. Instead of (8.100) one obtains a dispersion relation of sixth order for a generally complex frequency (Ferriz-Mas and Schüssler 1995), which must be solved numerically. Non-axisymmetric perturbations now become especially important; Figure 8.24 shows a typical result with unstable modes of azimuthal wave number $m = 1$ and $m = 2$. In a non-axisymmetrically perturbed tube the gas flows from the outward displaced crests along the tube toward the troughs. Hence the crests become lighter and continue to rise. Parker (1966) has considered this process first in the context of the galactic magnetic field. Therefore the instability discussed here has been named the *Parker instability*.

8.2.5 Thermal Structure of Photospheric Tubes

If a facular or G-band bright point represents a thin flux tube, and a pore represents a thick flux tube, then we have yet to explain why the former appears as a *bright* feature, while the latter is *dark* (and why magnetic knots are entirely invisible in the continuum).

A strong magnetic field inhibits the convective transport of energy, as was first recognized by Biermann (1941) in the context of sunspots. The reason is that the field is frozen to the plasma, but too strong to suffer the kinematic effects discussed in Sect. 8.2.1 above; it rather *prevents* the overturning motion. A consequence is a reduced energy input from below and hence a reduced temperature at any given level, both in the thin and the thick tube, as illustrated in Fig. 8.25. The difference between thin and thick tubes is that the lateral influx of heat, which is only by radiative transport

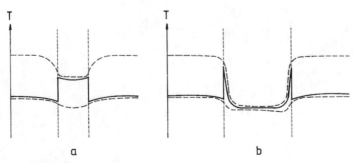

Fig. 8.25. Temperature profiles across a very thin (**a**) and a moderately thin (**b**) flux tube. The *lower dashed curve* is for the geometric level where $\tau = 1$ in the photosphere, the *upper* for the (deeper) level where $\tau = 1$ in the tube. The *solid curve* indicates the temperature actually observed. Adapted from Spruit (1981 b)

and therefore inefficient, cannot heat the thick tube (except a thin outer part), but does suffice to heat the thin tube up to a temperature that is above the photospheric temperature (at a geometrically higher level). Now, because the opacity decreases with decreasing temperature and decreasing density (the density is reduced in the tube) we see to deeper geometrical levels in the tube than in the surrounding photosphere. The result is that in the thin tube we see material which is hotter, and thus brighter, than the photospheric material, while the opposite is true for the thick tube. Spruit (1976), who first presented this argument, found that the transition from thin bright tubes to thick dark tubes should occur around 600 km for the tube diameter. In this intermediate region the magnetic knots should then be found.

The influx of heat from the surrounding photosphere into the tube produces a "hot wall", which becomes visible when the tube is inclined. As a consequence, tubes that are invisible or dark near the disc center may appear as bright features, e.g., faculae, towards the limb.

More detailed calculations confirm the qualitative reasoning of Fig. 8.25. Semi-empirical models of thin flux tubes, of the kind explained in Chap. 4, have been constructed by Solanki (1986) and have a higher temperature than the surrounding photosphere at the same optical depth, as anticipated in Fig. 8.25a. In order to obtain an intensity for the tube models Solanki took the profile $V(\lambda)$ of the Stokes parameter describing the circular polarization, measured with a Fourier transform spectrometer, and inverted (8.36):

$$I_V(\lambda) = I_C + \frac{1}{\Delta\lambda_B} \int_{\lambda_1}^{\lambda} V(\lambda')\, d\lambda' \ . \tag{8.106}$$

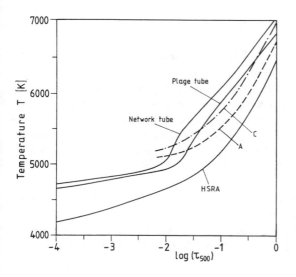

Fig. 8.26. Temperature in thin magnetic flux tubes, as a function of optical depth. *Solid*: semi-empirical models for tubes in the network and in plages (with small and moderate magnetic filling factors, respectively). *Broken*: theoretical atmospheres; A and C denote tubes where the gas pressure is reduced by factors 0.5 and 0.3 respectively, relative to the exterior pressure. The Harvard–Smithsonian Reference Atmosphere is shown for comparison. From Schüssler (1987 b)

Here I_C is the continuum intensity, and λ_1 is a wavelength lying far in the wing, where $V = 0$. Since the validity of (8.36) is limited due to the saturation of the V profile with increasing field strength, I_V is only an approximation to the tube intensity. However, other means to separate the radiation emerging from the interior of spatially unresolved tubes do not exist at present.

Using the intensity profiles (8.106) of many spectral lines one can construct semi-empirical mean models of the atmosphere within the flux tubes, as described in Sect. 4.3. Such mean models depend on the magnetic filling factor of the considered region on the Sun; tubes in plage regions appear to be slightly cooler than tubes in the chromospheric network. Fig. 8.26 shows results of Solanki, together with theoretical results obtained by Deinzer et al. (1984), who integrated the magnetohydrodynamic equations, including an energy equation. In the latter, the convective transport was reduced at large field strength, according to the above-mentioned idea of Biermann.

8.3 Sunspots

It has been known since the work of G. E. Hale in the early 20th century that sunspots possess a magnetic field. Indeed, ever since the field has been considered as the primary cause of the spots' darkness.

Because the magnetic field is source-free, the total magnetic flux through the solar surface is always zero. What we can expect, then, is eruption of flux in form of loops. These loops lead to the characteristic bipolar structure of spots. Nevertheless, such structure is not a necessity. Often the flux of one polarity is more concentrated than that of the other, so that only a single, "unipolar" visible spot is formed.

8.3.1 Evolution and Classification

Let us discuss the various forms of sunspots and sunspot groups in terms of the *Zürich classification* (Waldmeier 1955). This classification is largely a time sequence which, if all stages come about, may take several months. It distinguishes unipolar and bipolar configurations, the size and complexity of the spot (group), and the presence or absence of a *penumbra*, i.e., the ray-structured halo of intermediate mean intensity that in general surrounds the dark central part, the *umbra*, of large spots. The sunspot darkness depends on wavelength, cf. the intensity scans shown in Fig. 8.27.

A: *A single spot, or a group of spots, appears. There is no penumbra, and no obvious bipolar configuration.*

B: *A group of spots without penumbrae. The group is dominated by two spots that mark the two magnetic polarities; that is, the bipolar character is clearly noticeable.*

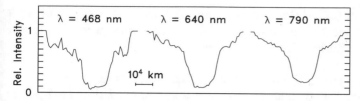

Fig. 8.27. Intensity profiles across a sunspot, at three wavelengths. Adapted from Wittmann and Schröter (1969)

The orientation of bipolar configurations is roughly east-west on the Sun, but with the leader spot (in the sense of rotation) slightly closer to the equator than the follower spot. The inclination of the p,f-axis (for "preceding" and "following", respectively) decreases during the early part of the group evolution. Typically it may start at 10°–20°, and end up at 5°, largely as a consequence of the group's growth in east-west direction. The inclination also increases with increasing heliographic latitude: mean values vary between 3.6° for latitudes 0°–4° and 10.8° for latitudes 30°–34°, according to a study of A. H. Joy (cf. Hale et al. 1919).

Even without proper magnetic field measurements the bipolar character becomes clear from Hα observations. The *arch-filament systems* (Fig. 8.28) studied by Waldmeier (1937) and Bruzek (1967) clearly mark the loops of the emerging field; they generally occur in the emerging flux regions recognized by Zirin (1971); see also Liggett and Zirin (1985).

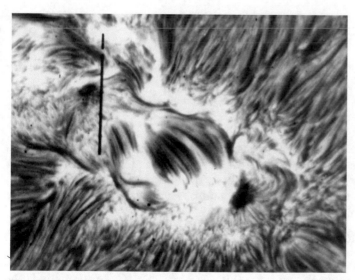

Fig. 8.28. Arch-filament system (*center*) in a small sunspot group, seen in Hα. Slit-jaw photograph (the *black line* is the entrance slit to the spectrograph), A. Bruzek, Domeless Coudé Telescope, Capri

For more than half of the sunspot groups the evolution terminates, after a day or a few days, with state A or B. They never develop a penumbra. For those that do the Zürich classification continues.

C: *Bipolar group; one of the principal spots has a penumbra.*
D: *Bipolar group; both principal spots have a penumbra, but at least one of them still has a simple structure. The group extends over less than $10°$ on the Sun.*

Mostly, but not always, the larger spots have a penumbra, while the smaller ones have none. Experienced observers have seen spots without a penumbra as large as 11 000 km in diameter, and spots with penumbra as small as 2500 km in (umbral) diameter (McIntosh 1981). The typical radial extent of the penumbra is 5000 km. It appears as if the penumbra forms out of radially aligned granules surrounding the umbra, a process that may be completed within one hour.

Again, the growth of a sunspot group may end with state C or D, and be followed immediately by the decay states explained below. On the other hand there are further growing groups, classified as E and F.

E: *Large bipolar group, extending over more than $10°$ on the Sun. Both principal spots have a penumbra and often a complex structure. Numerous small spots are present.*
F: *Very large, and very complex, bipolar group, extending over $\geq 15°$.*

The following three classes describe the decay of the group, and of the last remaining single spot.

G: *Large bipolar group, of extent $\geq 10°$. The small spots in between the principal ones have disappeared.*
H: *Unipolar spot with penumbra, diameter (including penumbra) $\geq 2.5°$.*
I: *As H, but diameter $< 2.5°$.*

The H spot often is rather symmetric, and stays without change for several weeks, except for a very slow decrease in size. It is the "theoreticians' spot". Figure 8.29 shows examples. It has been demonstrated (e.g., Bumba 1963) that the slow decay of the area, A, of many H spots proceeds at a rate that is nearly constant:

$$\dot{A} \approx -1.5 \times 10^8 \, \text{m}^2/\text{s} \; ; \tag{8.107}$$

(sunspot areas are often measured in millionths of the visible solar hemisphere, in which units this rate is -4.2 per day). The H spot normally is the principal leader spot of an originally bipolar group, and may survive long after the follower spot has been dissolved.

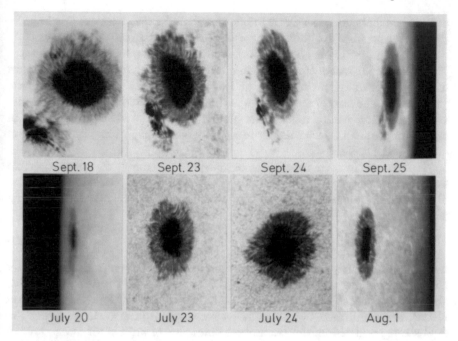

Fig. 8.29. Sunspots on their path across the solar disc. From Wittmann and Schröter (1969)

More recent studies have included other spot types, in addition to the long-lived H spots. Moreno-Insertis and Vázquez (1988) and Petrovay and van Driel-Gesztelyi (1997) found that a decay rate proportional to the instantaneous spot radius s generally is a better approximation than (8.107):

$$\dot{A} \simeq -C_D \, s/s_0 \, , \tag{8.108}$$

where s_0 is the maximum radius, and $C_D \approx 32$ millionths of the visible hemisphere per day, i.e., $C_D \approx 1.1 \times 10^9 \, \mathrm{m^2/s}$. The rate (8.108) yields a concave (parabolic) decay law for the sunspot area, and it includes the fast initial decay phase. As the decay proceeds, the ratio of the umbral to total radius approximately remains constant at a value somewhat less than $1/2$ (e.g., Martínez Pillet et al. 1993).

McIntosh (1981, 1990) has added two more parameters to the Zürich classification. These parameters have to do with the shape and complexity of the largest spot within the group, and whether the group is compact or open, i.e., densely or sparsely filled with spots (like star clusters). The additional classification has proved valuable in the prediction of flares (Sect. 9.5): the more complex the configuration, the more likely is the magnetic instability that gives rise to the flare.

Sunspots and sunspot groups are only the most conspicuous part of what is called an *active region*. The active region comprises field concentrations outside sunspots, e.g., magnetic knots, as well as faculae, chromospheric emission, etc., and may live much longer than the sunspots that it contains.

McIntosh (1981) also describes the obvious relation of sunspot configurations to the four scales of convection discussed in Chap. 6. The fine structure of spots, especially during birth and decay, is related to the photospheric granulation; the size of small spots is that of mesogranulation cells; the distribution within groups reflects the supergranulation; and the distribution of groups on the solar surface sometimes suggests a relationship with convection on the global scale.

The magnetic field in a sunspot reflects the complexity of the visible structure. The maximum field strength is of order 0.3 T. In circular symmetric H spots the absolute magnitude of the field varies approximately according to

$$B(s) = B(0)/(1 + s^2/s_*^2) \,, \tag{8.109}$$

where s is the distance to the spot's center, and s_* is the spot radius, including the penumbra (Beckers and Schröter 1969). The angle of the field vector with the local vertical is $\simeq (\pi/2)(s/s_*)$, as already found by Hale and Nicholson (1938), and confirmed by later observations. According to these results the field strength at the outer penumbral boundary of an H spot is one-half of the central field strength, and the field is horizontal at that boundary.

We conclude this section by defining, as a measure of solar activity, the *sunspot relative number* introduced by R. Wolf in 1848:

$$R = k\,(10\,g + f) \,, \tag{8.110}$$

where g is the number of spot groups and f is the total number of spots (an isolated spot is also a group). The calibration constant k accounts for the instrumentation and seeing conditions at the individual observatory. By coincidence, R is nearly proportional to the total area on the Sun covered by spots. Since the field strength in sunspots is always of the same order ($\approx 0.2 - 0.3$ T), R is also a rough measure of the total absolute magnetic flux penetrating the visible hemisphere within sunspots.

8.3.2 Sunspot Models

The solar astronomer distinguishes two types of "sunspot models". The first essentially consists in the physical quantities, notably temperature, as functions of height in the sunspot atmosphere, and is determined empirically by the methods outlined in Chap. 4. The second one is the magnetohydrostatic model, which is derived theoretically and aims to explain the balance of forces and energy.

The Atmospheric Model. As explained in Chap. 4, it is the center-to-limb variation, and the variation with wavelength, of the emitted intensity which allows us to determine the diverse atmospheric parameters as a function of height. Both line and continuum intensities contribute to the model.

The continuum intensity, relative to the quiet solar atmosphere, decreases with decreasing wavelength. This is already discernible in the scans shown in Fig. 8.27, but becomes more pronounced as a wider wavelength range is considered, cf. Table 8.2. Converted to absolute intensities, the values listed in this table correspond to brightness temperature values between 3400 K and 5000 K. Thus, the sunspot atmosphere may roughly be compared to the atmosphere of a star of spectral type K or M. The characteristics of late stars are also found in the *line* spectrum of sunspots which includes strong lines of CH, CN, and many other molecular lines (Wöhl 1971).

Table 8.2. Broad-band umbra/quiet-photosphere intensity ratio. From Albregtsen and Maltby (1981)

λ [μm]	Bandwidth [nm]	Number of observed spots	Intensity ratio		
			min.	max.	aver.
0.378	2	4	0.008	0.014	0.010
0.579	6	7	0.031	0.079	0.058
0.669	10	10	0.062	0.141	0.087
0.876	20	10	0.168	0.249	0.207
1.215	40	6	0.319	0.375	0.340
1.54	50	6	0.451	0.535	0.480
1.67	80	15	0.455	0.608	0.525
1.73	50	6	0.481	0.611	0.558
2.09	50	5	0.530	0.627	0.581
2.35	50	6	0.544	0.613	0.581

A sunspot umbra is greatly inhomogeneous. Therefore, the atmospheric sunspot models generally are calculated as *mean* models, much like the atmospheric models of the quiet photosphere. The temperature stratification shown in Fig. 8.30 represents such a mean umbral model. Its lower portion is considerably cooler than the quiet Sun (at the same optical depth), but this trend is reversed in the chromosphere, and the transition to the corona occurs at a deeper level than outside the spot.

Geometrically, the spot temperature profile given in Fig. 8.30 must be shifted by several hundred km downwards relative to the photospheric profile. The reason is the decrease of the opacity (with decreasing temperature and density), which was already discussed in Sect. 8.2.5. Thus, the sunspot represents a *dip* in the solar surface. This is the Wilson effect, named after the Scottish astronomer A. Wilson who first observed it in 1769. For spots near the solar limb, the effect renders the disc-side penumbra apparently narrower

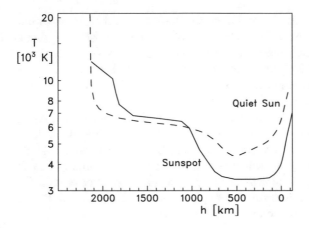

Fig. 8.30. Temperature as a function of height in sunspot and quiet-Sun models. For both curves h is measured from the level where $\tau_{500} = 1$. Data from Maltby et al. (1986)

than the limb-side penumbra; this is clearly seen in the examples shown in Fig. 8.29.

Tables of empirical sunspot models have been published, e.g., by Avrett (1981 a), and by Maltby et al. (1986). In these tables the pressure at $\tau_{500} = 1$ exceeds the normal photospheric pressure (Table 4.1). If there was no Wilson effect this would be in contrast to the magnetohydrostatic model treated below. The geometrical depression, however, does permit the desired pressure deficit of the spot at any given (geometrical) height.

Problem 8.13. Use the absolute photospheric intensities given in Chap. 1 in order to calculate the brightness temperature T_B corresponding to the sunspot intensities of Tab. 8.2. Why has T_B a maximum near $\lambda = 1.6\,\mu\text{m}$?

The Magnetohydrostatic Model. A sunspot consists of a tube of magnetic flux embedded in the solar plasma. Following Meyer et al. (1974) and Meyer et al. (1979), we assume that the flux has been brought to the surface, perhaps in small portions, by the action of magnetic and convective forces, and that it has been swept together by the supergranular flow. We further assume that the magnetic field has throttled the flux of convective energy (Biermann 1941). Since heat cannot be accumulated underneath the visible spot it must somehow be removed, and Meyer et al. (1974) suggest that this leads to a reversal of the originally convergent supergranular motion into an annular cell whose flow direction (at the surface) is *divergent*, i.e., away from the spot. Around stable H spots such an annular cell, the "moat", has indeed been observed. An indication of it can also been recognized in the magnetogram shown in Fig. 8.11.

We neglect both the dynamic effect of the (slow) moat flow and the (also slow) decay of the spot. The equilibrium is then magnetohydrostatic, i.e.,

$$-\nabla P + \rho \boldsymbol{g} + \boldsymbol{j} \times \boldsymbol{B} = 0 \; . \tag{8.111}$$

We assume an axisymmetric configuration, and use cylindrical polar coordinates (s, ϕ, z), with z upwards, so that $\boldsymbol{g} = (0, 0, -g)$. The electric current, with density $\boldsymbol{j}$, may be distributed over the entire volume of the spot. In addition, or as an alternative, there might be currents that are concentrated into thin current sheets.

Let us first consider a model with a volume current. In order to isolate one essential point, namely the lateral equilibrium, we adopt the similarity assumption of Schlüter and Temesváry (1958):

$$B_z(s, z) = D(a)\zeta^2 , \tag{8.112}$$

where $a(s, z) = s\,\zeta(z)$. Up to the normalization constant $D(0)$, the function $\zeta(z)$ is the square root of the field strength on the spot axis, and the function $D(a)$ defines the distribution of B_z over any horizontal plane. The dimensionless coordinate a is chosen so that $a = $ const. is the equation of a field line, cf. Problem 8.14. Therefore the distribution of B_z is transformed from one level z to another in a self-similar fashion along the field lines.

To make the field solenoidal, the horizontal component must be

$$B_s(s, z) = -aD(a)d\zeta/dz . \tag{8.113}$$

Substitution into the horizontal (s) component of (8.111), and integration over a yields an ordinary differential equation for $y(z) \equiv (B_z(0, z))^{1/2}$:

$$fyy'' - y^4 + 2\mu\Delta P = 0 , \tag{8.114}$$

where f is a constant which depends on the spot's total magnetic flux, and $\Delta P(z)$ is the pressure difference between the spot's exterior and its axis.

If $\Delta P(z)$ were known we could integrate (8.114). In a consistent model, however, $\Delta P(z)$ must be determined by the magnetic field itself: the reduced convective energy flux leads to a decreased thermal pressure in the spot. It is this latter part of the model which is difficult to formulate. Deinzer (1965) found a solution in terms of the mixing-length theory outlined in Chap. 6. The mixing length served as a free parameter and was reduced inside the spot in comparison to the exterior (where he took $l/H_P = 1$). Alternatively – and the results shown in Tab. 8.3 are calculated in this way – the spot's effective temperature is chosen freely, whereupon the mixing length is obtained as an eigenvalue. Each of the models also predicts a Wilson depression, z_D, of the

Table 8.3. Effective temperature, mixing length/scale height, central field strength, and Wilson depression for model sunspots of total flux 5×10^{13} Wb (Deinzer 1965)

T_{eff} [K]	l/H_P	B [T]	z_D [km]	T_{eff} [K]	l/H_P	B [T]	z_D [km]
3800	0.185	0.38	815	4200	0.277	0.31	590
4000	0.226	0.34	700	4400	0.345	0.27	480

right order of magnitude. At large depth, $> 10^7$ m say, the field in Deinzer's model becomes independent of depth.

Problem 8.14. Show that (8.112) and (8.113) satisfy div $\boldsymbol{B} = 0$, and that $a = $ const. describes a line of force. Which geometrical meaning has the function $\zeta(z)$? Confirm (8.114) and, for given total flux, determine the constant f. Calculate f for a field distributed according to (8.109).

While the self-similar sunspot model demonstrates that the convective heat transport in sunspots should be substantially reduced (but not entirely suppressed), it does not account for the discontinuous transitions between the umbra and the penumbra, and between the penumbra and the surrounding field-free gas. Such can be achieved in a model containing current sheets, as first proposed by Gokhale and Zwaan (1972).

Figure 8.31 shows a more recent example, described by Schmidt (1991) and Jahn and Schmidt (1994). There are two interfaces, tangential to the magnetic field, separating three domains of different convective energy transport, the umbra, penumbra, and the exterior convection zone. No energy flows between umbra and penumbra, but some heat is allowed to enter the penumbra from the exterior. The electric currents that flow in the azimuthal direction along the two interfaces are related to discontinuities of the magnetic field strength. An additional volume current is allowed in the penumbra of the earlier model of Jahn (1989).

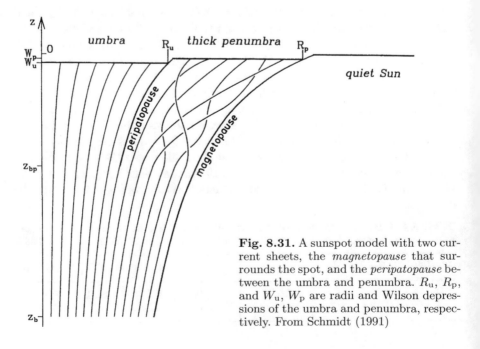

Fig. 8.31. A sunspot model with two current sheets, the *magnetopause* that surrounds the spot, and the *peripatopause* between the umbra and penumbra. R_u, R_p, and W_u, W_p are radii and Wilson depressions of the umbra and penumbra, respectively. From Schmidt (1991)

The model equations have been solved as a free-surface problem: the geometrical position of the two interfaces, the magnetic field, and the thermodynamic properties of the three domains were determined in such a way that
the magnetohydrostatic balance was satisfied; in addition, the observed magnetic field configuration (8.109) and the observed mean heat fluxes in the
three domains were matched. Jahn and Schmidt have calculated their models for diverse values of the total magnetic flux, total depth, and lateral inflow
of heat. The models require a reduced convective heat transport in the umbra and penumbra; they predict separate values of the Wilson depression in
these two domains. Another result is that the penumbra has a depth that is
comparable to its horizontal extent and contains more than 50 % of the total
magnetic flux, in consistence with (8.109). No penumbra was found if the
total magnetic flux was below $\approx 3 \times 10^{13}$ Wb, because in such small spots the
heat flux available to the penumbra was insufficient. If a penumbra existed,
the magnetopause had a minimum inclination of $\approx 25°$ from the vertical.

Stability. A perturbation of a magnetohydrodynamic system typically propagates with the Alfvén velocity (8.72). The time a perturbation needs to tra-
verse a sunspot is of the order of 1 hour, or even smaller. Thus, the spot must
be stable against perturbations; i.e., the perturbation must decay, because
if it was growing it would disrupt the whole configuration in a time short
compared to the spot's lifetime.

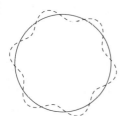

Fig. 8.32. Interchange instability. The originally circular
flux tube (*solid* cross section) develops flutes (*dashed*)

A special instability which has been discussed in the sunspot context
is the *interchange* or *flute instability* (Fig. 8.32). In the absence of other
forces such as gravity a flux tube embedded in a field-free plasma is unstable against fluting in the tube-plasma interface if the magnetic pressure
increases as one approaches this interface from within the tube. In this case
the magnetic pressure is enhanced in the crests between the flutes and decreased in the flutes themselves. Hence the flutes tend to grow, or magnetic
and non-magnetic columns try to "change places". A field that is concave (in
a vertical cross section, as seen from the exterior) is prone to have the lines
of force crowded near the interface; this is the unstable situation. And the
funnel-shaped sunspot field *is* concave, cf. Fig. 8.31.

Nevertheless, Meyer et al. (1977) were able to show that the spot is stabilized by *gravity* if the radial component B_s of the field decreases upwards

along the interface. This is because it takes gravitational energy to form a flute by the exchange of heavy exterior plasma with the lighter, magnetized, interior material. The stabilizing was found to be effective in the upper part of the spot. Deep down, where the field presumably is more vertical (at least in the magnetohydrostatic models), the question of sunspot stability remains unresolved.

Decay of sunspots. We have seen that the observed decay of the sunspot area can be approximated in two ways: either with a constant rate, or with a rate proportional to the radius. It will be clear presently that the constant rate (8.107) means that the destruction mechanism works over the entire cross section of the spot, while a rate proportional to s, such as (8.108), indicates that the spot is destructed mainly along its periphery.

In both cases the essential assumption is that the spot decays in a *diffusive* way, and that a turbulent diffusivity η_t, due to irregular motions in the deeper part (several 1000 km), describes the process in an adequate manner. A general derivation of η_t will be described in Sect. 8.4.2 below. Here we simply use the induction equation (8.6), with $v = 0$ and η replaced by η_t. For a vertical field B_z that is independent of depth this means

$$\dot{B}_z = \frac{1}{s} \frac{\partial}{\partial s} \left(s \eta_t \frac{\partial B_z}{\partial s} \right) . \tag{8.115}$$

We first assume that η_t is constant. In this case a solution to (8.115) is (Meyer et al. 1974, Krause and Rüdiger 1975)

$$B_z = \frac{\Phi}{4\pi \eta_t t} \exp \left(-\frac{s^2}{4 \eta_t t} \right) , \tag{8.116}$$

where Φ is the *total* flux. According to this solution, the central field strength B_m is very large for small t, whereas the field strength in real sunspots is limited to $\approx 0.3\,\mathrm{T}$ by the lateral magnetohydrostatic equilibrium. For this reason Meyer at al. (1974) used an empirical expression for the flux Φ_* through the visible spot, namely

$$\Phi_* \simeq 0.4\,A\,B_m \tag{8.117}$$

(Bray and Loughhead 1964, p. 216), and a constant B_m. Since the visible spot is a consequence of reduced energy transport, it is reasonable to assume that $\Phi_*(t)$ is that part of the total original flux Φ which at time t is within a circle where B_z exceeds a value B_c, the minimum field strength to affect the convective energy transport. If s_c is the radius of that circle then

$$\Phi_* = 2\pi \int_0^{s_c} B_z s \, ds = \Phi - 4\pi \eta_t B_c t , \tag{8.118}$$

which, by (8.117), yields the desired linear decrease of the spot area:

$$\dot{A} \simeq -10\pi\eta_{\mathrm{t}}B_{\mathrm{c}}/B_{\mathrm{m}} \ . \tag{8.119}$$

The critical field strength B_{c} is not precisely known; Meyer et al. (1974) take $B_{\mathrm{c}}/B_{\mathrm{m}} = 1/2$. Comparing (8.119) with (8.107) we then find the required value of the diffusion constant in the spot,

$$\eta_{\mathrm{t}} \approx 10^{7}\,\mathrm{m^2/s} \ . \tag{8.120}$$

We may also identify B_{c} with B_{e}, the equipartition field strength defined in (8.25); then the required η_{t} would be somewhat larger. In any case we find values that are smaller than the turbulent diffusivities ν_{t} and κ_{t} encountered earlier (Chaps. 6, 7). But, then, convection is (partially) suppressed by the magnetic field of the sunspot.

The quenching of η_{t} by the electromagnetic force has been modeled in in the form

$$\eta_{\mathrm{t}}(B) = \frac{\eta_{\mathrm{t}0}}{1 + |B/B_{\mathrm{e}}|^\alpha} \ , \tag{8.121}$$

where B_{e} is again the equipartition field strength, $\eta_{\mathrm{t}0} = \eta_{\mathrm{t}}(0)$, and α is a free parameter. With such a field-dependent diffusivity Petrovay and Moreno-Insertis (1997) have solved (8.115) numerically for a spot of original radius s_0 and field strength B_{m}. They found the remarkable result that, if the quenching is sufficiently steep ($\alpha > 2$, say), an azimuthal current sheet around the spot develops within one diffusion time $s_0^2/\eta_{\mathrm{t}0}$. The current sheet slowly moves inward with approximately constant velocity w. Thus the visible spot gradually shrinks by erosion at its boundary, caused be the turbulent motion (mainly of the surroundings). As the eroding front moves inwards, magnetic flux is transported outwards. This transport is wB_{m} on the inner side of the spot boundary, and $-\eta_{\mathrm{t}0}\,\partial B/\partial s$ on the outer side. Since the spot boundary is at the radius where $B \simeq B_{\mathrm{e}}$, we have $\partial B/\partial s \simeq -B_{\mathrm{e}}/s_0$ and, therefore,

$$w \simeq \frac{\eta_{\mathrm{t}0}B_{\mathrm{e}}}{s_0 B_{\mathrm{m}}} \ . \tag{8.122}$$

In this model the lifetime of the sunspot is s_0/w, the time after which the current sheet reaches the axis $s = 0$. The decay of the spot area is parabolic:

$$A(t) = \pi s^2 = \pi(s_0 - wt)^2 = A(0) - 2\sqrt{\pi A(0)}\,wt + \pi(wt)^2 \ , \tag{8.123}$$

and the decay rate is $\dot{A} = -2\pi sw$. A comparison with (8.108) yields $C_{\mathrm{D}} = 2\pi\eta_{\mathrm{t}0}B_{\mathrm{e}}/B_{\mathrm{m}}$. We take $B_{\mathrm{e}} = 0.04\,\mathrm{T}$, $B_{\mathrm{m}} = 0.3\,\mathrm{T}$, and for $\eta_{\mathrm{t}0}$ the diffusivity arising from the non-magnetic turbulent convection of the spot surroundings, $\eta_{\mathrm{t}0} \approx 10^{9}\,\mathrm{m^2/s}$. Thus we find $C_{\mathrm{D}} \approx 10^{9}\,\mathrm{m^2/s}$, which is of the same order as the value deduced by Petrovay and van Driel-Gesztelyi (1997).

8.3.3 Sunspots and the "Solar Constant"

What happens to the energy flux that is blocked by a sunspot? Solar observers often have searched for bright rings, i.e., areas of enhanced energy flux around spots. The evidence is casual, and entirely insufficient to account for the full spot deficit (Bray and Loughhead 1964).

Measurements made with the Active Cavity Radiometer (Sect. 3.6.1) show that we must not expect bright rings around spots. The two largest dips in the curve of Fig. 8.33, which occurred in early April and late May 1980, both coincide with the appearance of sunspot groups. Moreover, size and intensity of the spots approximately explain the magnitude of these dips. We must conclude that the sunspot flux is truly missing. It reappears neither as a bright ring nor as small excess flux distributed over a large fraction of the solar disc (the latter possibility had occasionally been proposed as a replacement of the evasive bright ring). This result has been confirmed by numerous further coincidences of spots with luminosity dips.

Problem 8.15. Calculate the expected drop of the solar energy flux at 1 astronomical unit for a sunspot of area A and intensity contrast C, at angular distance θ from the disc center. Assume that the limb darkening is the same for spot and undisturbed photosphere and use the result (4.53) of the Eddington approximation (Foukal, 1981).

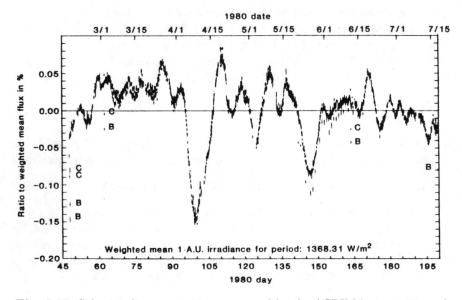

Fig. 8.33. Solar irradiance variation, measured by the ACRIM instrument on the Solar Maximum Mission (Willson et al. 1981). Courtesy Jet Propulsion Laboratory, California Institute of Technology, Pasadena, California

An interesting implication for stars other than the Sun is that transient drops of their luminosity may be interpreted in terms of stellar spots.

For the Sun, let us apply a model of time-dependent diffusive energy transport (Foukal 1981; Spruit 1981 c). We may thereby understand that the missing spot energy is *stored* in the convection zone.

For the present purpose we neglect the small difference between ∇' and ∇_a discussed in Sect. 6.2.1. The result of Problem 6.3 is then, in vectorial form

$$\boldsymbol{F}_C = -\frac{1}{2} \rho l v \, T \nabla S . \tag{8.124}$$

We need this vectorial form of the convective transport because with the sunspot as obstacle the situation is of course no longer spherically symmetric. We also need time-dependence, because at the time when the spot is formed the obstacle is "switched on", and we want to see how the energy begins to flow around it. Hence

$$\rho T \frac{\partial S}{\partial t} = -\text{div} \, \boldsymbol{F}_C \tag{8.125}$$

is the energy equation to be solved (for the sake of simplicity, we neglect the radiative energy flux). As boundary condition we require that, at any time, the flux is radiated into space according to the surface temperature. For the perturbation caused by the obstacle this means

$$\delta F_r = 4 \sigma T^3 \delta T . \tag{8.126}$$

Foukal and Spruit have solved the problem posed by (8.124) to (8.126), and have calculated the total energy released by the model surface. The main result is that as soon as the obstacle is introduced this energy drops by an amount corresponding exactly to the surface fraction of the obstacle. To return to the original luminosity, it would be necessary to heat the surface slightly in order to compensate for the reduction of its radiating part. However, because of the large thermal inertia (or heat capacity) of the layers to be heated this recovery takes very long.

A measure for the time required to reach thermal equilibrium is the internal energy U divided by the luminosity $L_\odot$:

$$t_{\text{th}} \simeq U/L_\odot , \tag{8.127}$$

For the whole star this thermal time scale is equivalent (Problem 2.10) to the Kelvin–Helmholtz time originally introduced in terms of the gravitational energy.

Using (8.127) we can define the thermal time scale in dependence of the total depth D of the layer considered. Foukal (1981) takes $D = 7500\,\text{km}$ for his model and finds $t_{\text{th}} \approx 1$ year. Spruit (1981 c) uses the entire convection zone and obtains $t_{\text{th}} \approx 2 \times 10^5$ years (which demonstrates the large heat capacity of the deeper layers). Both values exceed the lifetime of sunspots.

Therefore, the short-term drops of the "solar constant" essentially must be explained by the transient partial occultation of the solar surface by sunspots.

The *long-term* effects of sunspots on the "solar constant" are not precisely known. If, as Foukal's model suggests, the thermal adjustment occurs within a year, then no changes on the 11-year time scale of the sunspot cycle can be expected. If the adjustment takes much longer than 11 years then the "solar constant" should vary in anti-phase with the sunspot frequency. However, the evidence obtained from a number of space missions is opposite (Fig. 1.3): the solar irradiance varies *in phase* with the sunspot cycle, as first noticed during the Solar Maximum Mission (Willson et al. 1986). *Faculae* have been made responsible for this in-phase variation, as well as for some of the short-term variations seen in records like Fig. 8.33, but not explained by spots (Chapman et al. 1984; Foukal and Lean 1986, 1988). The area on the Sun covered by faculae typically is 15–20 times larger than the area covered by sunspots (Chapman et al. 1997). Hence a mean facular excess of a few percent is sufficient to over-compensate the sunspot deficit and thus cause the observed irradiance variation.

8.3.4 Dots and Grains

In order to assess the energy transport in the magnetohydrostatic spot model above, we have merely *reduced* the mixing length rather than suppressing convection altogether. We have, in addition, used a *turbulent* diffusivity, η_t, to describe the decay of sunspots. Both these approaches are justified only if there exists a small-scale, irregular velocity field within sunspots.

Some evidence for such a velocity field lies in the non-thermal broadening of spectral lines. Using two lines with Landé factor $g = 0$, observing at various disc positions, and assuming that the non-thermal broadening is due to unresolved waves, Beckers (1976) found 2.4 km/s and 1.5 km/s as most likely rms amplitudes in the vertical and horizontal directions, respectively.

Time-dependent and small-scale structure has been observed in sunspot umbrae as well as in penumbrae. Examples have been shown in Figs. 3.11 and 3.12; further detail can be seen in Fig. 8.34. Sufficiently long exposed umbral photographs show the "umbral granulation" reminiscent of the normal photospheric granulation (e.g., Bumba and Suda 1980). Because single bright points rather than an uniform granular pattern are typical, the name *umbral dots* is now common. Umbral dots were first investigated on photographs taken with the Stratoscope balloon experiment (Danielson 1964). In the penumbra, the radially aligned penumbral filaments have been known for a long time; A. Secchi made beautiful hand-drawings around 1870. Modern images resolve the filaments into elongated units called *penumbral grains* (Fig. 8.34). Except for their shape the grains are similar to umbral dots and, near the outer boundary of the penumbra, to photospheric granules. Some properties of dots and grains are summarized in Table 8.4, and Fig. 8.35 shows the distributions in size and brightness of umbral dots. The double-peaked

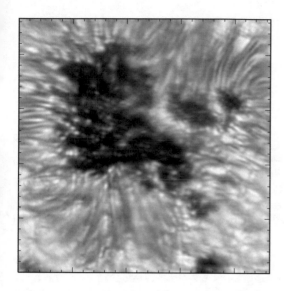

Fig. 8.34. A sunspot image, reconstructed by the phase-diversity method, showing light bridges, penumbral grains, and umbral dots. The spectral band is $569.5 \pm 1\,\mathrm{nm}$; tick marks are $1''$ apart. German Vacuum Tower Telescope, Izaña, Tenerife. From Tritschler and Schmidt (2002)

brightness distribution corresponds to a spatial separation: the brighter dots preferentially sit near the umbral border or near a light bridge, the darker ones are more frequent well inside the umbra.

As in the convection zone surrounding the sunspot, the temperature stratification in the deeper part of the spot is superadiabatic. However, the insta-

Table 8.4. Properties of umbral dots and penumbral grains. After Muller (1973), Moore (1981), and Sobotka (1999)

	Umbral dots	Penumbral grains
Size [km]: width	200 – 500	≤ 300
length	–	≤ 2000
Filling factor	≤ 0.1	≤ 0.5
Relative intensity	0.3 – 0.9	≤ 1.2 average: 0.95, on background of ≈ 0.6
Magnetic field [a]	weaker than in dark surroundings	less inclined than in dark surroundings
Lifetime [h]	0.1 – 2	0.5 – 4
Motion: Doppler velocity	$\leq 25\,\mathrm{m/s}$ with respect to dark surroundings	upward with respect to dark surroundings
proper motion	100 – 400 m/s	$\leq 500\,\mathrm{m/s}$, mainly toward umbra

[a] There is no unequivocal result concerning the field strength in penumbral grains; see the discussion of Wiehr (1999)

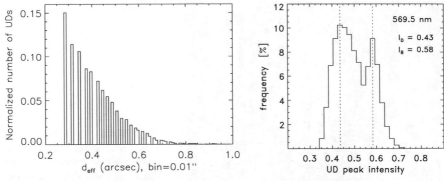

Fig. 8.35. *Left*: Number of umbral dots as a function of the diameter, from Sobotka et al. (1997 a). *Right*: Brightness distribution of umbral dots. Courtesy A. Tritschler

bility treated in Sect. 6.1 cannot develop in the same manner, because the magnetic field provides a restoring force which is absent outside the spot. Instead of the overturning convective motion a growing oscillatory motion is possible under certain conditions. Such *oscillatory convection* has been proposed as an explanation for umbral dots. Once the instability is fully developed (which we must assume in view of its short time scale) the phenomenon needs not be periodic. Thus, the idea is that parcels of hot gas squeeze through the magnetic lines of force from beneath and so cause the visible structure of the spot.

The inhomogeneous sunspot model of Parker (1979 b) is an extreme, but consistent view of the situation just described. In this model the spot is a cluster of flux tubes, intermixed with columns of field-free gas. The dots and grains would be the uppermost tips of those columns. Support for this model comes from the difficulty to stabilize the deeper, more vertical, part of the sunspot field against the flute instability, cf. Sect. 8.3.2 above. If the field-free columns are isolated, then the oscillatory convection is the only means of transporting energy in the spot; if they are interconnected (while the magnetic tubes are isolated), then the convection could take on a more normal form, including some overturning flow. The true subsurface structure of sunpots is not known at present; the approach with a reduced mixing length is a substitute for a comprehensive magnetohydrodynamic description.

For the penumbra, a more detailed picture has been proposed by Schmidt (1991): Convective energy transport by the interchange of entire magnetic flux tubes over a large depth range. Such flux tubes would move back and forth from the outer edge of the penumbra (the *magnetopause* of Fig. 8.31) to the penumbral photosphere. The part of the tube that just arrives at the surface has the highest temperature and moves towards the umbra, as the observed penumbral grains do. Below we shall further discuss this model in the context of the Evershed effect.

8.3.5 Oscillations in Sunspots

The sunspot atmosphere is compressible, stably stratified, and has a magnetic field. For each of these circumstances it is able to support oscillations. Periodic variations of the Doppler shift and intensity of spectral lines provide evidence for such oscillations. The measurements are made in the same manner as those in the undisturbed photosphere, although the resolution in frequency and wave number generally is lower. Also, care must be taken to avoid a signal introduced from outside the spot by scattered light. For example, one may use a spectral line which exists only in the spot, such as the 649.62 nm line of the TiO molecule (Soltau et al. 1976); Bogdan (2000) gives a list of lines that have been used to measure oscillations in sunspots.

The power of sunspot oscillations lies predominantly at periods around 3 minutes and around 5 minutes; the latter are prominent when a photospheric line is used, the former dominate in chromospheric lines. Individual power peaks such as shown in Fig. 8.36 have contributions from diverse modes of oscillation and may therefore have different positions and heights, depending on the observed sunspot.

In addition to the velocity and intensity, oscillations of the magnetic field in sunspots have been reported, from ground-based measurements (Horn et al. 1997, Balthasar 1999, Kupke et al. 2000; but see also Lites et al. 1998) as well as from space observations (Rüedi et al. 1998, Norton et al. 1999). Oscillation periods are around 3 min and 5 min, the amplitudes are of order 10 G; the occurrence is intermittent and unevenly distributed within the spots.

Five-Minute Oscillations. The amplitude of the five-minute oscillations seen in sunspots is markedly smaller than the amplitude of such oscillations elsewhere on the Sun. Thomas (1981) has proposed that the five-minute oscillations in sunspots are forced upon the spot by the p-mode oscillations of the surrounding convection zone. If, instead of the Fourier transform or spherical harmonic analysis, a Hankel transform (in a cylindrical polar-coordinate system centered at the spot) is used, the waves traveling inward and outward can be separated (Braun et al. 1987). It has been shown in this way that the

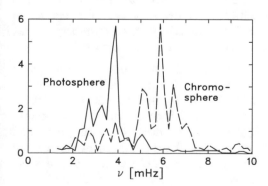

Fig. 8.36. Power spectrum of sunspot oscillations. *Solid*: photospheric (Fe I, 630.25 nm), in units of 10^6 (m/s)^2/Hz. *Dashed*: chromospheric (Ca II H), in 10^7 (m/s)^2/Hz. Composite after Thomas et al. (1984)

outgoing power is less than the incoming power, sometimes reduced to one-half. Apparently there is wave absorption in sunspots, but actually it may be conversion of acoustic into magnetohydrodynamic modes of oscillation (e.g., Bogdan 2000).

Three-Minute Oscillations and Umbral Flashes. The three-minute oscillations in sunspots are distinct in that the spot has *resonances* in this range of periods. As explained in the context of the Sun's global oscillations, resonances (or *modes*) are the consequence of reflecting boundaries. And as there, the reflecting boundaries are provided by the steep gradient of the atmospheric parameters that are important for wave propagation.

The speed of sound is proportional to $T^{1/2}$ and therefore increases both towards the interior and towards the corona. The Alfvén velocity, proportional to $\rho^{-1/2}$, increases outwards. As model calculations show, a waveguide is formed between the inwards increasing sound speed and the outwards increasing Alfvén velocity. In a quantitative treatment this yields the three-minute resonance. Geometrically this resonance occurs rather deep, in the sunspot's *photosphere*. This is not necessarily in contrast to the observation that the three-minute oscillation has the largest amplitude in the spot's chromosphere: because of its density dependence the *kinetic energy* is at least five times larger in the photosphere (Abdelatif et al. 1984; Lites and Thomas 1985).

The three-minute oscillations also have a signature in the *intensity* of spectral lines. Most conspicuous are the *umbral flashes*, discovered even prior to the velocity signal (Beckers and Tallant 1969). The flashes are seen in the sunspot chromosphere as periodic brightenings in the emission cores of the Ca II lines (Fig. 9.2). In a sequence of K-line filtergrams one can see them arising in the central part of the umbra and then moving towards the umbra/penumbra boundary (Moore 1981). The umbral flashes seem to occur whenever the velocity oscillation reaches a large amplitude, of order 5 km/s.

Excitation of the resonant umbral oscillations is possible by means of the short-period tail of the global solar oscillations (the p modes), i.e., from outside the spot. The exciting amplitude is, however, small; moreover, within the umbra the flashes move *outwards*. Therefore a more plausible hypothesis is excitation from within the sunspot itself. The above-mentioned irregular velocity field beneath the visible layer offers a broad spectrum of perturbations. From these, the spot atmosphere can select its own resonant modes.

Running Penumbral Waves. In stable, circular sunspots the phenomenon of *running penumbral waves* can be observed. These waves were discovered by Zirin and Stein (1972) in the intensity of Hα, and by Giovanelli (1972) in the Doppler shift of the same line. In a filtergram taken in off-center Hα, for example, the periodic Doppler shift produces alternating dark and bright bands. The bands emerge from the umbra/penumbra boundary and propagate radially outwards across the penumbra. The period is in the range 3 to 5 minutes; the horizontal phase velocity is typically 10 to 20 km/s.

The running penumbral waves could be excited by umbral oscillations. Lites et al. (1982) found indications for such a connection, but in general this is not well documented. The propagation speed is of the order of the Alfvén speed in the penumbral *photosphere*. The photospheric wave guide described above for the umbral oscillations may therefore channel the penumbral waves as well (actually it was in this context that this wave guide was first proposed by Nye and Thomas 1974).

8.3.6 The Evershed Effect

At the Kodaikanal Observatory in India, Evershed (1909) first studied the horizontal outflow in the penumbrae of sunspots, which since bears his name. The flow is observed in the penumbral photosphere and rather abruptly ceases at the boundary to the undisturbed photosphere. High-resolution spectra show that the phenomenon is related to the variable inclination of the magnetic field (Degenhardt and Wiehr 1991, Schmidt et al. 1992, Title et al. 1993, Rimmele 1995): the outflow is largest in channels of nearly horizontal magnetic field, where it may reach several kilometers per second. In between those channels the inclination from the vertical is less and more closely follows the mean inclination, $\gamma \simeq (\pi/2)(s/s_*)$, cf. Sect. 8.3.1. In addition to the horizontal outflow, vertical flow components have been observed: patches of upflow in the inner penumbra, and patches of downflow in the outer penumbra (Stanchfield et al. 1997, Westendorp et al. 1997). The vertical flow components are correlated with the brightness. A study of Schlichenmaier and Schmidt (1999, 2000) leads to the picture of narrow upstream channels in the inner penumbra that turn into the horizontal direction and finally bend downwards near the outer edge of the sunspot.

The Evershed effect has been interpreted as a *siphon flow* along magnetic flux tubes (Meyer and Schmidt 1968, Thomas 1988). If a tube forms an arch (Fig. 8.37) between two foot points with different values of the gas pressure (at the same geometrical level, i.e., the same gravitational potential), a flow is driven from the high-pressure end to the low-pressure end. The flow velocity increases along the ascending part of the arch and reaches its sonic point at the summit. On the downstream side it first accelerates further, but then

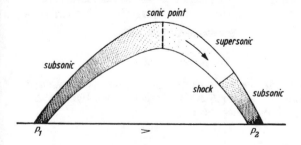

Fig. 8.37. Siphon flow along a flux tube. Adapted from Meyer and Schmidt (1968)

undergoes a shock, thereby adjusting its pressure to the given end pressure. Since the flux tubes laterally are in magnetohydrostatic equilibrium, where a balance of type (8.59) holds, the pressure at each foot point may be derived from the measured strength of the magnetic field. If a flux tube connects the inner penumbra, with $B_1 \approx 0.15\,\mathrm{T}$, to a magnetic knot or a pore outside the sunspot with, say, $B_2 \approx 0.2\,\mathrm{T}$, then we have $P_1 > P_2$, which drives the desired flow.

The *moving-tube* model of Schlichenmaier et al. (1998 ab) is more general than the siphon model, as it includes time dependence and does not rely on a given pressure difference. Magnetic flux tubes emerging from the deep penumbra are able to transport heat to the penumbral photosphere. Figure 8.31 schematically shows this convection by the interchange of entire flux tubes. The path of a single tube starting at the magnetopause has been modeled in the thin-tube approximation. The tube first rises adiabatically. At the point where it meets the photosphere it sharply bends into the horizontal. At this point a high temperature is sustained by the upflow of hot gas within the tube; this has been interpreted as a penumbral grain (Fig. 8.38). The model also yields a horizontal pressure gradient along the tube which drives an outward flow, as in the siphon model. As the gas in the tube reaches the optically thin photospheric environment, it cools down, and the tube becomes

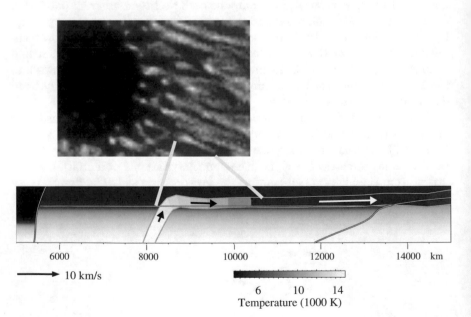

Fig. 8.38. The moving-tube model for the Evershed flow in a sunspot penumbra (*lower panel*). Penumbral grains (*upper panel*) are interpreted as hot upflows within emerging flux tubes. Courtesy R. Schlichenmaier

transparent itself. Thus the bright grain tails off, and the horizontal outflow is accelerated to its largest velocity in the outer, dark part of the tube.

The observation that the Evershed flow is so sharply confined to the penumbra may find an explanation such that the flow channels become invisible at the outer penumbral edge. The downflow patches found near that edge support this view. But even a horizontal channel may cut across the $\tau = 1$ surface, because near the magnetopause this surface is inclined upwards.

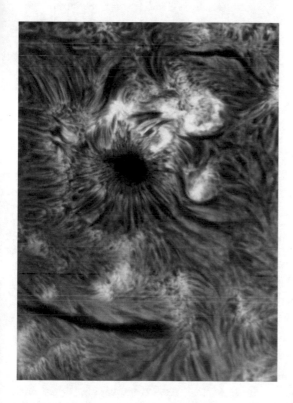

Fig. 8.39. Hα filtergram showing the superpenumbra of a sunspot. Photograph A. Bruzek, Domeless Coudé Telescope, Capri

Large round sunspots often have a *superpenumbra*. The superpenumbra is seen in the chromosphere, notably in the center of Hα (Fig. 8.39), consists of dark fibrils more or less aligned radially, and extends far beyond the penumbra. Along the dark fibrils there is an *inward* horizontal flow. The velocity of this "inverse Evershed flow" is of order 20 km/s. The siphon model can be applied to this situation as well: Let a set of flux tubes, arching into the chromosphere, connect magnetic knots or pores, with $B \approx 0.2$ T as before, to the sunspot umbra, with a field strength $B \approx 0.25$ T. The associated pressure difference between the outer and inner foot points then drives the inverse chromospheric flow. In many cases this superpenumbral flow has a vortex structure; this is a consequence of the Coriolis force, but also depends on the strength of the magnetic field (Peter 1996).

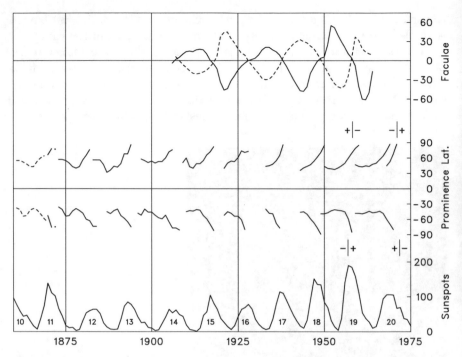

Fig. 8.40. Solar-cycle variation of sunspots, latitude of polar prominences, and faculae. The numbers of northern (*solid*) and southern (*dashed*) polar faculae are drawn with a sign in order to indicate the alternating magnetic polarity (Sheeley 1964). The ± signs mark magnetic reversals at the poles. After Stix (1974)

8.4 The Solar Cycle

8.4.1 Global Magnetism

We shall now describe the solar magnetic field on the largest possible scale, that is the field of the Sun as a whole. At the same time, we shall deal with very long time scales, notably the 11-year cycle of the Sun's magnetic activity.

The best-known indicator of the solar cycle is the sunspot relative number, (8.110), yearly means of which are shown by the lower curve of Fig. 8.40. The cycle was first discovered by Schwabe (1844). Each cycle, lasting from one minimum to the following, is given a number, beginning with the minimum around 1755 (Wolf evaluated the notes of many earlier solar observers and so was able to establish the cycle backwards in time). But not only the spots appear in cycles. Figure 8.40, for example, shows the latitude where the *polar crown* of prominences appears, and the frequency of polar faculae.

G. E. Hale had measured the magnetic field in sunspots for the first time in 1908. By 1923 numerous spots, of three consecutive cycles, had been observed, confirming the polarity rules formulated by Hale et al. (1919):

The magnetic orientation of leader and follower spots in bipolar groups remains the same in each hemisphere over each 11-year cycle.

The bipolar groups in the two hemispheres have opposite magnetic orientation.

The magnetic orientation of bipolar groups reverses from one cycle to the next.

These rules are illustrated in Fig. 8.41. They mean that the magnetic cycle does not last 11 years, but 22 years. Only very few exceptions to the polarity rules are known.

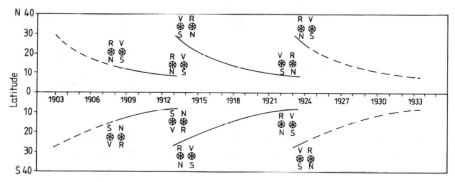

Fig. 8.41. Polarity rules for sunspots: N, north, S, south. R and V indicate the red and violet σ-component of the Zeeman triplet; the preceding spot (in the sense of rotation) is at the right of each pair. The curves indicate the migration in latitude of the spot zones. After Hale (1924)

Another important and well-known result is the latitude migration of sunspot zones. Already noticed by C. Scheiner in the early 17th century, it was studied by G. Spörer since 1860 and hence is often called *Spörer's law*: the first spots of each cycle appear at about 30°– 35° latitude, in both hemispheres. As the cycle advances, the zones of sunspot occurrence migrate towards lower latitude, and the last spots of a cycle are normally within ±10° of the equator. The difference in latitude between the beginning and the end of a cycle allows a clear distinction as to which of two overlapping cycles a certain spot (pair) belongs. The latitude migration is also indicated in Fig. 8.41, but is better documented in the famous *butterfly diagram* (Maunder 1922); a more recent example is shown in Fig. 8.42.

The systematic behavior of bipolar sunspot groups is readily understood in terms of a subsurface, mean toroidal magnetic field. A *toroidal* field in this context is a field whose lines of force are circles around the solar axis. Locally the field may be more concentrated than elsewhere and, driven to the surface by the convective or magnetic buoyancy, may form bipolar spot groups.

As an estimate of the mean toroidal field B_t we may divide the total flux observed in spot groups by the total available area. According to obser-

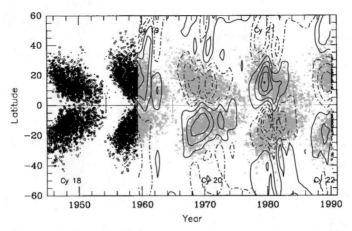

Fig. 8.42. Butterfly diagram of sunspots, according to Harvey (1992 b), and contours of radial mean field (from Mt. Wilson and Stanford magnetograms). The contours are for $\pm 20\,\mu$T, $\pm 60\,\mu$T, $\pm 100\,\mu$T, ..., *solid* positive, *dashed* negative. From Schlichenmaier and Stix (1995)

vations of Howard and LaBonte (1981) the total flux at any one time is of order 10^{15} Wb. For the horizontal extent of the available area we take the latitude range where sunspots occur, corresponding to, say, 2.5×10^8 m. If we take the depth of the convection zone, 2×10^8 m, for the vertical extent, then $B_{\mathrm{t}} \approx 0.02$ T. We should not take more than the depth of the convection zone because the field is oscillatory and therefore has a very small skin depth toward the core (Problem 8.16). On the other hand, the toroidal flux may well occupy only a small fraction of the convection zone, notably the layer of convective overshooting at its bottom, with a thickness of order 5×10^6 m. This yields the estimate $B_{\mathrm{t}} \approx 1$ T. Depending on whether the latitudinal flux distribution is smooth or in the form of concentrated flux tubes, the actual field strength may be still larger, possibly reaching 10 T. As we have seen, tubes of this field strength become unstable even in the stably stratified layer of overshooting convection (Fig. 8.24). After the onset of the instability the tube rises in loops toward the surface, as illustrated in Fig. 8.43.

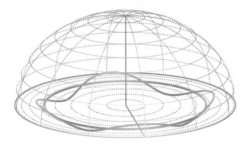

Fig. 8.43. The northern hemisphere of the Sun, with a toroidal magnetic flux tube. The original circular tube becomes warped after the onset of the instability. After Ferriz-Mas and Schüssler (1994); courtesy P. Caligari

Problem 8.16. Calculate the skin depth for a field which alternates with a period of 22 years when the conductivity σ is given by (8.10), and for an "effective" σ derived from (8.145).

In addition to the mean toroidal field there is a mean poloidal magnetic field, $\boldsymbol{B}_{\mathrm{p}}$. We shall use the term *poloidal* for a field having its lines of force in meridional planes, i.e., planes containing the solar axis. In spherical polar coordinates a poloidal field has components in the r and θ directions.

Magnetograms of the entire solar surface are composed in a "synoptic" way: the line-of-sight component of $\boldsymbol{B}$ is measured in a strip along the central meridian of the visible disc; as the Sun rotates, a full magnetogram is obtained in ≈ 27 days. Therefore the mean poloidal field, which is an average over longitude and hence axisymmetric by definition, may be calculated as an average over *time*. Strictly these two averages are not the same because $\boldsymbol{B}$ itself depends on time. Still, a meaningful mean field can be obtained by taking an average over more than one rotation. The short-term fluctuations may thus be eliminated, while the long-term variation is retained, in particular the 11-year variation in which we are interested presently.

A second difficulty with the measurement of the mean poloidal field is that the line-of-sight component contains not enough information to derive the *vector* $\boldsymbol{B}_{\mathrm{p}}$. A least-square fit to a potential field, as outlined in Sect. 9.3.2, is a possible answer. Hence we may obtain the mean poloidal magnetic field simply as the axisymmetric $(m = 0)$ contribution of the multipole expansion (9.30), that is

$$B_{\mathrm{p}}(r_{\odot}, \theta) \simeq \sum_{l=1}^{N} \langle g_{l}^{0} \rangle \left((l+1) P_{l}(\theta), -\frac{dP_{l}}{d\theta} \right). \tag{8.128}$$

This is an approximation to the field at the solar surface, $r = r_{\odot}$; the contribution from the source surface at $r = r_{\mathrm{w}}$ (Sect. 9.3.2) has been discarded. Contours of such a measured mean field (radial component) have been superimposed on the butterfly diagram of Fig. 8.42.

The most important property of the Sun's mean poloidal field is its cyclic reversal. As direct measurements show, this reversal takes place at the poles around the maximum of sunspot activity, as indicated in Fig. 8.40; more recent reversals have been documented as well, e.g., Snodgrass et al. (2000). It is also significant that, like the mean toroidal field, the mean poloidal field is *antisymmetric* with respect to the equatorial plane.

The obvious relationship between the toroidal and poloidal mean-field components leads to the more general question of the origin of solar magnetism. It is well-known that a moving conductor may generate a magnetic field by induction: this is the principle of the self-excited dynamo. We shall assume in the following sections that the cyclic mean magnetic field of the Sun is the direct result of such a dynamo, operating in an alternating-current mode.

Industrial dynamos have multiply connected conductors. It is therefore not trivial that a dynamo would work in a star, which is simply connected and permits arbitrary short-circuits. In order to demonstrate the possibility, a solution, $\boldsymbol{B}$, of the induction equation (8.6) must be found, and this solution must not be generated by sources exterior to the Sun. Two theorems are relevant in this context. The first, originally due to Cowling (1934), states that the solution $\boldsymbol{B}$ must not be axisymmetric; the second, proved by Bullard and Gellman (1954), says that a field cannot be maintained if the velocity that appears in (8.6) is a pure rotation, no matter how non-uniform.

In spite of these two anti-dynamo theorems it is possible to generate an axisymmetric *mean* solar field in a dynamo, and even to employ the Sun's non-uniform rotation as an essential ingredient. This will be demonstrated in the two following sections.

8.4.2 Mean-Field Electrodynamics

We seek a solution to the dynamo problem in terms of a *mean* magnetic field. Hence we write

$$\boldsymbol{B} = \langle \boldsymbol{B} \rangle + \boldsymbol{b} \ , \tag{8.129}$$

where $\langle \boldsymbol{B} \rangle$ may be understood as an average over longitude or, more generally, as an ensemble average (i.e., an average over a large number of realizations of a stochastic variable). In any case, we shall assume that the operation of averaging commutes with the operations of differentiation and integration. The field $\boldsymbol{b}$ is the fluctuating part of $\boldsymbol{B}$, and obeys $\langle \boldsymbol{b} \rangle = 0$. Likewise we write

$$\boldsymbol{v} = \langle \boldsymbol{v} \rangle + \boldsymbol{u} \ . \tag{8.130}$$

By $\langle \boldsymbol{v} \rangle$ we shall mean a motion on the Sun which is of global scale, notably the solar differential rotation. The field $\boldsymbol{u}$ describes the irregular, or turbulent, convective motion treated in Chap. 6.

In this section the dynamo problem will be treated in its *kinematic* form, where $\boldsymbol{v}$ is assumed to be given, and independent of $\boldsymbol{B}$. It is however not necessary to know all details of the velocity field. We shall be content to have the mean $\langle \boldsymbol{v} \rangle$, as well as the *statistical properties* of $\boldsymbol{u}$, i.e., mean values, correlations, etc. The kinematic mean-field dynamo was first treated by Parker (1955 b), Steenbeck et al. (1966), and Steenbeck and Krause (1969); the following outline rests on their work, and on the treatise of Moffatt (1978).

Substitution of (8.129) and (8.130) into the induction equation (8.6), and separation of the mean and fluctuating parts yields

$$\frac{\partial}{\partial t} \langle \boldsymbol{B} \rangle = \mathrm{curl}\, (\langle \boldsymbol{v} \rangle \times \langle \boldsymbol{B} \rangle + \boldsymbol{\mathcal{E}} - \eta \, \mathrm{curl}\, \langle \boldsymbol{B} \rangle) \ , \tag{8.131}$$

and

$$\frac{\partial \boldsymbol{b}}{\partial t} = \mathrm{curl}\,(\langle \boldsymbol{v} \rangle \times \boldsymbol{b} + \boldsymbol{u} \times \langle \boldsymbol{B} \rangle + \boldsymbol{G} - \eta\,\mathrm{curl}\,\boldsymbol{b})\,, \qquad (8.132)$$

where

$$\boldsymbol{\mathcal{E}} = \langle \boldsymbol{u} \times \boldsymbol{b} \rangle\,, \qquad (8.133)$$

$$\boldsymbol{G} = \boldsymbol{u} \times \boldsymbol{b} - \langle \boldsymbol{u} \times \boldsymbol{b} \rangle\,. \qquad (8.134)$$

The mean electric field $\boldsymbol{\mathcal{E}}$ is the crucial quantity. If it was known, we could solve (8.131) for $\langle \boldsymbol{B} \rangle$. In principle $\boldsymbol{\mathcal{E}}$ must be calculated in terms of $\langle \boldsymbol{B} \rangle$ by substituting into (8.133) a solution $\boldsymbol{b}$ of (8.132). That is too difficult in general, but inspection of (8.132) and (8.134) immediately shows that there must be a *linear* relationship between $\boldsymbol{b}$ and $\langle \boldsymbol{B} \rangle$, and hence also between $\boldsymbol{\mathcal{E}}$ and $\langle \boldsymbol{B} \rangle$. We write the latter relationship as an expansion:

$$\mathcal{E}_i = \alpha_{ij}\langle B_j \rangle + \beta_{ijk}\partial\langle B_j \rangle/\partial x_k + \dots\,. \qquad (8.135)$$

The coefficients α_{ij} and β_{ijk} are *pseudo-tensors*, because they relate an *axial* vector $\langle \boldsymbol{B} \rangle$ to a *polar vector* $\boldsymbol{\mathcal{E}}$. In the kinematic approach they are statistical properties of the velocity field $\boldsymbol{u}$, and are independent of $\boldsymbol{B}$.

First-order Smoothing. The mean electric field $\boldsymbol{\mathcal{E}}$ has been calculated explicitly in the special case where the term $\boldsymbol{G}$ in (8.132) is negligible. This neglect of second-order terms (in the fluctuating quantities) has been called the *first-order-smoothing approximation* or, since no higher than second-order terms appear in (8.133), the *second-order-correlation approximation*. This approximation is justified if either of the following conditions is met:

$$ul/\eta \ll 1 \qquad \text{or} \qquad (8.136)$$

$$u\tau/l \ll 1 \qquad (8.137)$$

(Steenbeck and Krause 1969). Here u is a typical magnitude of $\boldsymbol{u}$, and l and τ are typical scales of the variation of $\boldsymbol{u}$ (and $\boldsymbol{b}$) in space and time, respectively. Condition (8.136) ensures that $\boldsymbol{G}$ is much smaller than the last term on the right of (8.132); this is of course the case of small magnetic Reynolds number. If (8.137) holds, then $\boldsymbol{G}$ is negligible in comparison to $\partial \boldsymbol{b}/\partial t$.

Unfortunately neither of the two conditions is satisfied on the Sun. The magnetic Reynolds number is large, and observations suggest that $u\tau/l \approx 1$ rather than $u\tau/l \ll 1$. It is nevertheless instructive to proceed. In order to obtain qualitative results let us, in addition to $\boldsymbol{G}$, omit the term containing $\langle \boldsymbol{v} \rangle$ in (8.132). Moreover, due to the high conductivity we may omit the diffusion term in that equation. A solution to (8.132) is then

$$\boldsymbol{b} = \int_{-\infty}^{t} \mathrm{curl}\,(\boldsymbol{u} \times \langle \boldsymbol{B} \rangle)\,dt'\,. \qquad (8.138)$$

The integrand is to be taken at time t'; the initial field, $\boldsymbol{b}(-\infty)$, need not be written down because certainly it is uncorrelated to $\boldsymbol{u}(t)$ and hence does not contribute to the subsequent calculation of $\boldsymbol{\mathcal{E}}$.

Problem 8.17. Use (8.138) to calculate $\boldsymbol{\mathcal{E}}$ in the special case where $\boldsymbol{u}$ represents "weakly" isotropic turbulence; that is the case where the statistical properties of $\boldsymbol{u}$ are invariant under rotation but not generally under reflection of the frame of reference. Show that (8.135) takes the form

$$\boldsymbol{\mathcal{E}} = \alpha\langle\boldsymbol{B}\rangle - \beta\,\mathrm{curl}\,\langle\boldsymbol{B}\rangle + \dots \;, \tag{8.139}$$

where

$$\alpha = -\frac{1}{3}\int\limits_0^\infty \langle\boldsymbol{u}(t)\cdot\mathrm{curl}\,\boldsymbol{u}(t-t')\rangle\,dt' \;, \tag{8.140}$$

and

$$\beta = \frac{1}{3}\int\limits_0^\infty \langle\boldsymbol{u}(t)\cdot\boldsymbol{u}(t-t')\rangle\,dt' \;. \tag{8.141}$$

Substitution of (8.139) into (8.131) yields the mean-field induction equation

$$\frac{\partial}{\partial t}\langle\boldsymbol{B}\rangle = \mathrm{curl}\left(\langle\boldsymbol{v}\rangle \times \langle\boldsymbol{B}\rangle + \alpha\langle\boldsymbol{B}\rangle - \eta_{\mathrm{t}}\,\mathrm{curl}\,\langle\boldsymbol{B}\rangle\right) \;, \tag{8.142}$$

where

$$\eta_{\mathrm{t}} = \eta + \beta \;. \tag{8.143}$$

The mean-field induction equation, (8.142), is distinguished from the original induction equation (8.6) in two ways. The first is a substantial increase of the diffusivity. By order of magnitude, (8.141) gives

$$\beta \simeq \frac{1}{3}u^2\tau \simeq \frac{1}{3}ul \gg \eta \;. \tag{8.144}$$

In the solar convection zone we thus have, again by order of magnitude,

$$\eta_{\mathrm{t}} \simeq \beta \approx 10^8 \dots 10^9\,\mathrm{m}^2/\mathrm{s} \;. \tag{8.145}$$

As a consequence the global time scale of diffusive decay, $r_\odot^2/\eta_{\mathrm{t}}$, is of order 10 to 100 years. We may expect that this time scale plays an important role in any solution of (8.142).

The second, and novel, feature of the mean-field induction equation is the term involving α, also called the α *effect*. This term ensures that the solution $\langle\boldsymbol{B}\rangle$ is not subject to the anti-dynamo theorems mentioned in the preceding section. In fact quite simple dynamos can be conceived with the help of the α effect (Problems 8.18 and 8.19).

A quantitative estimate of α in the solar convection zone is difficult. Expression (8.140) indicates that for non-vanishing α the flow $\boldsymbol{u}$ should be *helical*; that is, $\boldsymbol{u}$ should be correlated to its own vorticity, curl $\boldsymbol{u}$. Krause (1967) has shown that in the convection zone the Coriolis force produces such helicity, in combination with either the density stratification or the radial gradient of the rms velocity of convection (or both). For the effect of density stratification consider a rising parcel of gas which expands and, due to the Coriolis force, acquires a left-handed (in the northern hemisphere) helical component; and consider a sinking parcel which (again in the northern hemisphere) also acquires a left-handed helical component. Left-handedness therefore dominates, and the flow has (in this case negative) helicity. The effect of the velocity gradient is illustrated by considering the base of the convection zone where the flow of descending parcels must diverge, while the flow of ascending parcels must converge. In both cases the Coriolis force causes a right-handed helicity (again, in the northern hemisphere).

The global convection models of Sect. 7.5.3 confirm the different signs of the helicity in the bulk of the convection zone and near its lower boundary. For the α effect, the result obtained in the calculations of Krause and of other authors is, by order of magnitude,

$$\alpha \simeq \pm l\Omega \, , \tag{8.146}$$

where Ω is the mean angular velocity of the Sun. The sign of α is opposite to the sign of the helicity. Its magnitude is less than the estimate (8.146) if $\boldsymbol{u}$ and curl $\boldsymbol{u}$ are not perfectly correlated. Moreover, α depends on the scale l, which widely varies in the convection zone. Values of α between a few cm/s and $\approx 100 \, \mathrm{m/s}$ have been derived by various authors.

Numerical Determination of α. As an alternative to first-order smoothing, numerical simulation of magneto-convection has been used to calculate the tensors α and β which appear in (8.135). In this way it is possible to go beyond the limitations set by (8.136) and (8.137), although the magnetic Reynolds number R_{m} of the solar convection zone is still much larger than the values used in actual calculations. Moreover, a large R_{m} implies small scales of the magnetic field, which cannot be resolved in a global numerical model, such as shown in Fig. 7.13. Therefore the calculations have been made in a rectangular box, intended to simulate the situation in a particular limited section of the solar convection zone.

The result shown in Fig. 8.44 is obtained for the simplest case where the vectors of angular velocity and gravity are parallel to each other; the depth range covers several density scale heights and includes a stably stratified region in the lower part. Thus, the location of the computational box would be on the southern axis of the Sun, at the base of the convection zone. An external mean horizontal field $\boldsymbol{B}_{\mathrm{H}}$ is given. After the magnetohydrodynamic simulation has settled to a statistical equilibrium, the component of $\boldsymbol{\mathcal{E}}$ in the direction of $\boldsymbol{B}_{\mathrm{H}}$ is evaluated and yields

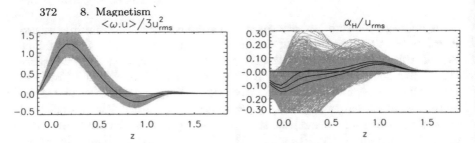

Fig. 8.44. Numerical simulation of magneto-convection near the base of the solar convection zone. *Left*: normalized helicity. *Right*: component α_H of the α tensor. The abscissa z points downward; sign reversals occur at $z \approx 0.7$, close to the level $z = 1$ where the stratification becomes stable. *Shading* indicates the variation, the curves give averages and error estimates for a calculation lasting ≈ 20 coherence times of the turbulent flow. From Ossendrijver et al. (2001)

$$\alpha_H = \frac{\langle u \times b \rangle \cdot B_H}{|B_H|^2} . \tag{8.147}$$

By applying different external mean fields, other components of the α tensor have been determined. The present example confirms the close relationship of the α effect with the helicity of the convective flow: In the southern hemisphere there is positive helicity within the convection zone, and accordingly negative α_H; both quantities reverse their signs near the transition to the stable stratification, where sinking parcels are converted into horizontally diverging flows.

The mean-field induction equation has been the basis for much work on the solar dynamo, as well as for planetary, stellar, and galactic dynamos. But although some beautiful results have been obtained, the approximations made must not be forgotten. On the Sun, first-order smoothing is marginally valid at best, and the capability of numerical simulation is limited. We should also be aware that, whenever a turbulent diffusivity η_t is used (e.g., Sects. 8.2.1 and 8.3.2), we deal with the *mean* field $\langle B \rangle$, rather than with B itself.

Problem 8.18. Interpret the α effect in terms of a mean volume current parallel to the mean magnetic field. Conceive a dynamo with two interconnected, ring-shaped, current/field systems.

Problem 8.19. Set $\langle v \rangle = 0$, and find solutions $\langle B \rangle$ of (8.142) which satisfy $\langle B \rangle \times \mathrm{curl}\,\langle B \rangle = 0$. What is the condition for exponential growth of $\langle B \rangle$?

8.4.3 The Kinematic $\alpha\Omega$ Dynamo

The α effect is, by itself, capable of generating a mean magnetic field. But in the solar case the more interesting solutions are obtained in combination with the differential rotation. As before, we shall take the kinematic point of view. That is, we shall assume that $\alpha(r,\theta)$ and $\Omega(r,\theta)$ are given functions (in spherical polar coordinates r, θ, ϕ). In accord with what was said in the preceding section, let α be *antisymmetric* with respect to the equatorial plane, i.e.,

$$\alpha(r, \pi - \theta) = -\alpha(r, \theta) . \tag{8.148}$$

The angular velocity will be taken as a *symmetric* function,

$$\Omega(r, \pi - \theta) = \Omega(r, \theta) , \tag{8.149}$$

and, besides rotation, no mean motion shall exist:

$$\langle v \rangle = (0, 0, \Omega r \sin \theta) . \tag{8.150}$$

We separate the mean field into its poloidal and toroidal parts

$$\langle B \rangle = B_{\mathrm{p}} + B_{\mathrm{t}} , \tag{8.151}$$

where

$$B_{\mathrm{p}} = \mathrm{curl} \ (0, 0, A(r, \theta, t)) , \tag{8.152}$$

$$B_{\mathrm{t}} = (0, 0, B(r, \theta, t)) . \tag{8.153}$$

The existence of a vector potential of B_{p} follows from $\mathrm{div} \langle B \rangle = 0$ and from the axisymmetry; the function B had already been introduced as B_{t} in Sect. 8.4.1.

The mean-field induction equation can now also be separated into its poloidal and toroidal parts. In the simplest case, where η_{t} is a constant, one obtains

$$\dot{A} = \alpha B + \eta_{\mathrm{t}} \Delta_1 A , \tag{8.154}$$

$$\dot{B} = \frac{\partial \Omega}{\partial r} \frac{\partial}{\partial \theta} (A \sin \theta) - \frac{1}{r} \frac{\partial \Omega}{\partial \theta} \frac{\partial}{\partial r} (r A \sin \theta)$$

$$- \frac{1}{r} \frac{\partial}{\partial r} \left(\alpha \frac{\partial}{\partial r} (r A) \right) - \frac{1}{r^2} \frac{\partial}{\partial \theta} \left(\frac{\alpha}{\sin \theta} \frac{\partial}{\partial \theta} (A \sin \theta) \right) + \eta_{\mathrm{t}} \Delta_1 B , \tag{8.155}$$

where

$$\Delta_1 = \Delta - (r \sin \theta)^{-2} . \tag{8.156}$$

Equations (8.154) and (8.155) demonstrate the crucial role of the α effect. With $\alpha = 0$, the poloidal field would decay exponentially, because its equation is of diffusive nature. And with A gone, B would decay as well, for the same reason.

By (8.155), the toroidal field is generated from a parent poloidal field by means of differential rotation *and* α effect. In solar models the latter effect is commonly neglected in this equation. The condition for this is

$$|\alpha| \ll r_\odot^2 |\nabla \Omega| . \tag{8.157}$$

Although the magnitude of α is not precisely known, condition (8.157) is probably satisfied. The whole regenerative cycle is then

$$\boldsymbol{B}_{\rm t} \;\; \overset{\alpha}{\to} \;\; \boldsymbol{B}_{\rm p} \;\; \overset{\nabla\Omega}{\longrightarrow} \;\; \boldsymbol{B}_{\rm t} \; . \tag{8.158}$$

This is the $\alpha\Omega$ dynamo [more precisely, it should be called $\alpha\nabla\Omega$ dynamo; as (8.155) shows, only the *derivatives* of Ω matter]. Now let α_0 and $\Delta\Omega_0$ be typical values of α and of internal differences of Ω; scale α and Ω in terms of these two constants, and scale r and t in terms of $r_\odot$ and $r_\odot{}^2/\eta_{\rm t}$, respectively. It is then readily seen that the essential parameter of the $\alpha\Omega$ dynamo is the *dynamo number*

$$D = \alpha_0 \Delta\Omega_0 r_\odot^3 / \eta_{\rm t}^2 \; . \tag{8.159}$$

The dynamo number measures the strength of the two induction effects, α_0 and $\Delta\Omega_0$, relative to the diffusivity. If either α_0 or $\Delta\Omega_0$ is zero the solution decays exponentially. Hence it is clear, and confirmed by numerical calculations, that D must exceed a certain critical value in order to prevent the decay. Before we discuss these calculations let us consider an analytical model which was first presented by Parker (1955 b).

Problem 8.20. Confirm (8.154) and (8.155). Transform these equations so that the role of the dynamo number becomes clear.

Problem 8.21. Use (8.148) and (8.149) to show that separate solutions $\langle \boldsymbol{B} \rangle$ of either symmetry with respect to the equatorial plane exist.

Parker's Migratory Dynamo. In place of the spherical polar coordinates we now use a cartesian system (x, y, z). We may imagine that this system sits somewhere in the northern hemisphere, with x pointing southwards, y eastwards, and z upwards. We set $\alpha = \text{const.}$, $\langle \boldsymbol{v} \rangle = (0, \Omega_0 z, 0)$ with $\Omega_0 = \text{const.}$, and seek solutions $A(x, t)$ and $B(x, t)$. In this case

$$\boldsymbol{B}_{\rm p} = \left(0, 0, \frac{\partial A}{\partial x} \right) \; , \tag{8.160}$$

$$\boldsymbol{B}_{\rm t} = (0, B, 0) \; , \tag{8.161}$$

and equations (8.154) and (8.155) are replaced by

$$\dot{A} = \alpha B + \eta_{\rm t} \partial^2 A / \partial x^2 \; , \tag{8.162}$$

$$\dot{B} = \Omega_0 \partial A / \partial x + \eta_{\rm t} \partial^2 B / \partial x^2 \; . \tag{8.163}$$

The solution is

$$(A, B) = (A_0, B_0) \exp\left[{\rm i}\,(\omega t + kx) \right] \; , \tag{8.164}$$

with the dispersion relation

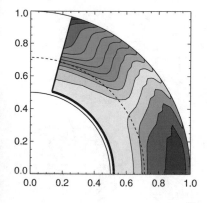

Fig. 8.45. Contours of constant angular velocity in a meridional section of the Sun, derived by the Global Oscillation Network Group, GONG. The labels are $r/r_\odot$, *red* indicates fast rotation at low latitude, *blue* slow rotation at high latitude. The *dashed* curve marks the base of the convection zone. Courtesy M. Roth

$$(i\omega + \eta_t k^2)^2 = ik\Omega_0\alpha . \tag{8.165}$$

Let us consider the special case $\alpha\Omega_0 < 0$. This would be the situation in the transition region from the convection zone to the radiative core: the tachocline found by helioseismology (Figs. 7.4 and 8.45) is a layer with $\partial\Omega/\partial r > 0$ (at low latitude, where sunspots occur), and – on the northern solar hemisphere – we have $\alpha < 0$ at this depth, see above. Without loss of generality we may assume $k > 0$. The frequency ω, which is generally complex, is then

$$\omega = i\eta_t k^2 \pm (1+i)\,|k\alpha\Omega_0/2|^{1/2} . \tag{8.166}$$

For a successful dynamo the growth rate,

$$-\mathrm{Im}(\omega) = -\eta_t k^2 \mp |k\alpha\Omega_0/2|^{1/2} , \tag{8.167}$$

must not be negative. Hence the solution with the upper sign in (8.166) and (8.167) can be discarded. The other solution is marginally stable if

$$|k\alpha\Omega_0/2|^{1/2} = \eta_t k^2 , \tag{8.168}$$

and grows exponentially for larger $|k\alpha\Omega_0/2|$. As in the more general case treated above, condition (8.168) states that the product of α and Ω_0 must exceed a critical magnitude, i.e., that the dynamo number must exceed a critical value.

For the marginal (or unstable) solution the frequency of oscillation is

$$\mathrm{Re}(\omega) = -|k\alpha\Omega_0/2|^{1/2} , \tag{8.169}$$

which is *negative*. Therefore, the mean field (8.164) is a wave propagating in the positive x direction. If we had started with $k < 0$, the result would have been the same, if only $\alpha\Omega_0 < 0$. In the opposite case, $\alpha\Omega_0 > 0$, the propagation would be in the negative x direction. By an appropriate placing of the local cartesian frame of reference one can show that generally the dynamo

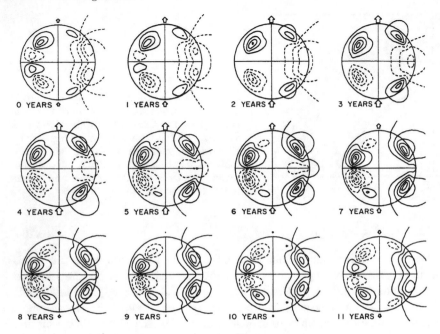

Fig. 8.46. Oscillatory kinematic $\alpha\Omega$ dynamo. The meridional cross sections show contours of constant toroidal field strength on the *left*, and poloidal lines of force on the *right*. *Arrows* indicate strength and sign of the polar field, the time scale is adjusted to 11 years for each half-cycle. From Stix (1976 b)

wave migrates along the surfaces of constant angular velocity (Yoshimura 1975 b). The direction of migration therefore is given by the vector

$$\alpha\nabla\Omega \times \boldsymbol{e}_\phi \,, \tag{8.170}$$

where $\boldsymbol{e}_\phi$ is the unit vector in the azimuthal (ϕ) direction.

Expression (8.169) shows that the period of the oscillatory dynamo is given by the geometric mean of the time scales on which the two induction effects operate. In the marginal case, (8.168) implies that this geometric mean, and hence the dynamo period, is equal to the time scale of diffusion.

Problem 8.22. Discuss the phase of the poloidal relative to the toroidal field in the two cases $\alpha > 0$, $\Omega_0 < 0$ and $\alpha < 0$, $\Omega_0 > 0$. Derive the phase relation for the solar field from Figs. 8.41 and 8.42 (Stix 1976 a; Yoshimura 1976).

Numerical results. In spherical geometry, solutions to (8.154) and (8.155) have been obtained by various numerical techniques. In most cases these solutions have been subject to the boundary condition that the mean field smoothly joins an external potential field. Since the equations as well as the boundary condition are homogeneous, one must solve an *eigenvalue* problem. That is, for a solution of the form

$$\langle \boldsymbol{B} \rangle \propto \exp\left(i\omega t\right) \tag{8.171}$$

the complex frequency, ω, is an eigenvalue once the functions $\alpha(r,\theta)$ and $\Omega(r,\theta)$ and the constant η_{t} are specified. In order to isolate the marginally stable solutions one may keep the *form* of $\alpha(r,\theta)$ and $\Omega(r,\theta)$ fixed and vary their *amplitude*, i.e., vary the dynamo number D. The critical dynamo number D_{c} is then defined as the value of D for which $\mathrm{Im}(\omega) = 0$.

Because the mean field migrates along the surfaces of constant Ω, and because it is the aim to simulate the latitude migration of the sunspot zones, most models have concentrated on the *radial shear* $\partial\Omega/\partial r$.

The numerical calculations essentially confirm Parker's analytical migratory dynamo. There exist oscillatory solutions and, with α and Ω specified such that $\alpha_{\mathrm{north}}\partial\Omega/\partial r < 0$, the mean field in both hemispheres migrates from higher heliographic latitude towards the equator. The period of the oscillatory solutions is essentially determined by two time scales: the diffusion time $r_{\odot}{}^{2}/\eta_{\mathrm{t}}$ and the combined induction time of α effect and differential rotation [the inverse of (8.169)]. By order of magnitude this result agrees with the period of the solar cycle. A more precise prediction is not possible, mainly because $\alpha(r,\theta)$ in the convection zone is not sufficiently known. The evolution of calculated mean fields is illustrated in Figs. 8.46 and 8.47.

The solution to (8.154) and (8.155) may be symmetric (even), or antisymmetric (odd) with respect to the equatorial plane (Problem 8.20). Of these

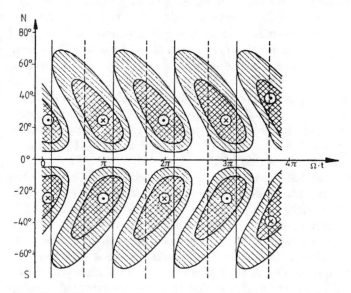

Fig. 8.47. Theoretical butterfly diagram (contours of constant toroidal field) in an oscillatory kinematic $\alpha\Omega$ dynamo. After Steenbeck and Krause (1969)

two parities the latter corresponds to the main characteristics of the observed mean field. It obeys

$$B(r, \pi - \theta) = -B(r, \theta) , \tag{8.172}$$

$$A(r, \pi - \theta) = A(r, -\theta) . \tag{8.173}$$

For $\alpha_{\text{north}} \partial \Omega / \partial r < 0$ this antisymmetric field is the preferred solution of the numerical models, in the sense that it becomes marginal at a smaller critical dynamo number than the symmetric field. This lends further support to the above choice of $\alpha(r, \theta)$ and $\Omega(r, \theta)$.

8.4.4 The Magnetohydrodynamic Solar Dynamo

For a supercritical dynamo number the solution of the kinematic $\alpha\Omega$ dynamo grows without limit. In reality this is impossible; it is prevented by the growing magnetic force density, the Lorentz force $j \times B$, which modifies the velocity field in such a way that the dynamo efficacy is reduced (Lenz's rule). We may expect that in an equilibrium state the dynamo operates at a finite amplitude of the magnetic field. In contrast to the kinematic dynamo, such an equilibrium is called a *magnetohydrodynamic* dynamo.

For an $\alpha\Omega$ dynamo the effect of the Lorentz force will be a dependence of α and Ω on the mean field $\langle B \rangle$. In order to establish this dependence it is necessary to generalize the theory of mean-field electrodynamics to the full magnetohydrodynamic case, including an equation of motion and the Lorentz volume force. This problem has not been solved generally, but below we shall mention a few simplified models that have been discussed in the solar context.

An alternative is the generalization of the global convection models of Sect. 7.5.3. Again not all the details of the magnetic and velocity fields need to be simulated. The smallest scales are parameterized by a parameter η_{t}, which is of the same order as the ν_{t} and κ_{t} used in the earlier models. The larger and intermediate scales are treated explicitly. Thereby we may expect to capture the interesting aspects of the dynamo action because these aspects result from the rotational influence upon convection, and this influence is largest for the largest convection cells. The proportionality (8.146) of α to the scale l is a manifestation of this fact.

The point of the explicit models is that the induction equation does *not* contain an α term. Since the models are non-axisymmetric, and include convective motions, no restrictions in form of anti-dynamo theorems exist.

Numerical integrations indeed demonstrate working dynamos. Even oscillatory and migratory magnetic fields have been obtained (Gilman 1983; Glatzmaier 1985). Unfortunately these fields are at variance with the solar mean field. The main difference is their poleward, rather than equatorward, migration. Such a discrepancy might be due to the differences between the angular velocity $\Omega(r, \theta)$ derived from global convection models and the angular velocity of the Sun, as inferred by helioseismology, but also to differences of the helicity between the earlier models and the Sun.

The continuous loss of unstable and buoyant magnetic flux from a dynamo within the convection zone has led to the idea that the dynamo mainly operates in a layer of overshooting turbulence at the base of the convection zone, where the situation is more stable (Sect. 8.2.4). But the details of such a deep-seated dynamo are unknown, although a number of models have been offered.

Dynamic Systems. One of the important dynamic aspects of solar magnetism is the instability and buoyancy of horizontal flux tubes in the convection zone. As the flux tubes drift upwards they acquire a tilt that is observed in the emergent bipolar regions; this tilt yields a contribution to the mean-field equation that is equivalent to the α effect. Adding a non-linear term to equation (8.155), Leighton (1969) was able to construct a dynamo that works at finite field amplitude. The resulting field resembles the solar field, but, since α and Ω are freely given, the model is essentially still kinematic. The same is true for models that simulate the dynamic effects by a quenching or cutoff at large $\langle B \rangle$ of the α effect and (or) the shear, but otherwise use the equations of the $\alpha\Omega$ dynamo (Stix 1972; Yoshimura 1975 a).

As a dynamic system the magnetohydrodynamic dynamo is capable of *chaotic behavior*. Such can be seen from the numerical integrations mentioned above, but still more clearly by considering the following system of coupled ordinary differential equations (Weiss et al. 1984):

$$\dot{A} = 2DB - A , \tag{8.174}$$

$$\dot{B} = iA - \frac{1}{2}i\Omega A^* - B , \tag{8.175}$$

$$\dot{\Omega} = -iAB - \nu\Omega . \tag{8.176}$$

The first two of these equations are scaled versions of (8.162) and (8.163); D is the dynamo number. The spatial dependence has been eliminated by a Fourier expansion, of which only the leading terms are retained; A^* is the complex conjugate of A. Equation (8.176) is the ϕ component of the equation of motion, truncated in an analogous manner, and $\nu = \nu_t/\eta_t$. This equation describes the effect of the Lorentz force, here $-iAB$, upon the angular velocity, $\Omega(t)$.

System (8.174)–(8.176) is a complex generalization of a system of three ordinary differential equations first studied by Lorenz (1963) as a model of turbulent convection. Like the Lorenz system it has *chaotic* solutions but, because it is complex, it also has solutions that are periodic in time. We may identify the latter with the dynamo wave of Sect. 8.4.3.

The critical dynamo number for the complex Lorenz system is $D = 1$. Above $D = 1$ there exists a periodic solution, for which the period of Ω

Fig. 8.48. *Left*: toroidal magnetic field in a dynamic $\alpha\Omega$ dynamo, as described by (8.174)–(8.176). Adapted from Weiss et al. 1984. *Right*: butterfly diagram from a two-dimensional generalization of the model. After Beer et al. (1998)

is $1/2$ times the period of A and B. In this respect the periodic solution resembles the cyclic variation of Ω described in Sect. 7.4.2. If D is sufficiently large and $\nu < 1$, the periodic solution becomes unstable; it is replaced first by multiply periodic and finally by chaotic solutions. Figure 8.48 (left panel) shows an example where we find "normal" cycles with variable amplitude, but also periods of low activity, much as the *Maunder Minimum* in the sunspot record of the 17th century (Fig. 8.49, lower panel).

During the period from about 1645 to 1715 very few spots were seen on the Sun. Eddy (1976) has named this prolonged sunspot minimum after E. W. Maunder who had considered the subject in the 1890's. Recent studies (Ribes and Nesme-Ribes 1993, Hoyt and Schatten 1996) confirm the nearly complete absence of sunspots. On the other hand, records of aurorae (Schröder 1988) and the concentration of the ^{10}Be isotope in ice cores (Beer et al. 1998) indicate that the activity cycle actually might have persisted with low amplitude during the whole period, although the cycle maxima and minima deduced from these two sources differ greatly.

The complex Lorenz system has been generalized by including the spatial variation in two dimensions, depth and latitude (Tobias 1997, Knobloch et al. 1998). Again an irregular modulation of the field amplitude is obtained; in addition, the butterfly diagram appears as a characteristic of the $\alpha\Omega$ dynamo, as shown in Fig. 8.48 (right panel). As a further interesting feature the model exhibits the non-linear interaction of the magnetic field modes with different parities with respect to the equator. Such interaction is absent in linear models with symmetric Ω and antisymmetric α. A close inspection of the butterfly diagram of Fig. 8.48 reveals a small contribution of even parity. Indeed, the solar mean magnetic field does not strictly follow the antisymmetric parity defined by (8.172) and (8.173), and the few spots observed during the Maunder Minimum predominantly occured on the southern hemisphere!

Problem 8.23. Find the analytic periodic solution of the complex Lorenz system (8.174)–(8.176).

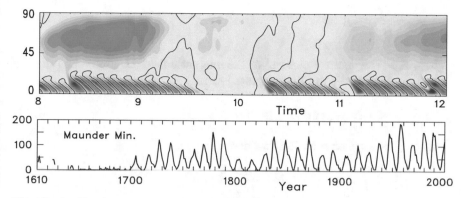

Fig. 8.49. *Upper panel*: butterfly diagram derived from an $\alpha\Omega$ dynamo with a stochastic contribution to the α effect; time in diffusion time units. After Ossendrijver (2000). *Lower panel*: sunspot relative number, adapted from Eddy (1977)

Stochastic α Effect. A second class of models that account for an irregular modulation of the solar cycle relies on the fact that the coefficient α is itself a fluctuating quantity.

The α effect results from the influence of the Coriolis force that renders convective eddies helical. The Coriolis influence is largest for the largest eddies. Of these there is only a finite number within the volume of averaging, or along a circle of constant latitude if one considers averages over longitude. Therefore α will consist of a time-independent term and a fluctuation,

$$\alpha = \alpha_0(r,\theta) + \alpha_1(r,\theta,t) \ . \tag{8.177}$$

The large variation of the α coefficient obtained from numerical simulations (Fig. 8.44) supports this concept of α fluctuations.

The fluctuations of α are of particular relevance if one considers the instability of the toroidal magnetic field itself, again under the Coriolis influence, as the origin of the α effect. This is a *dynamic* α effect; such has been envisaged originally by Babcock (1961) and Leighton (1969), and worked out by Ferriz-Mas et al. (1994) for the case of thin flux tubes, and by Brandenburg and Schmitt (1998) in a numerical simulation. In this situation the α effect depends on a finite strength of the magnetic field, and the fluctuation of α can lead to a turn-off of the dynamo. Of course, some kinematic dynamo action must be there in addition to the dynamic α; otherwise the cycle could not resume after an extended pause. Figure 8.49 (upper panel) illustrates the interruption of the cycle that can occur in an $\alpha\Omega$ dynamo of this type.

Magnetic Helicity and Current Helicity. In the preceeding sections we have seen that the helicity of turbulent convection, i.e., a correlation between the flow and its vorticity, plays a crucial role for the dynamo. We shall call this the *kinetic helicity*, as opposed to *magnetic helicity* (Moffatt 1978), defined for a volume V,

$$H_{\mathrm{m}} = \int\limits_V \boldsymbol{A} \cdot \boldsymbol{B} \; dV \; , \tag{8.178}$$

where $\boldsymbol{A}$ is the vector potential of the magnetic field $\boldsymbol{B}$. From the induction equation (8.6) and from its integrated form

$$\frac{\partial \boldsymbol{A}}{\partial t} = \boldsymbol{v} \times \boldsymbol{B} - \eta \, \mathrm{curl} \, \boldsymbol{B} - \mathrm{grad} \, \psi \tag{8.179}$$

(where ψ is a scalar function) we derive

$$\begin{aligned}
\frac{dH_{\mathrm{m}}}{dt} &= \int\limits_V \left(\boldsymbol{A} \cdot \frac{\partial \boldsymbol{B}}{\partial t} + \boldsymbol{B} \cdot \frac{\partial \boldsymbol{A}}{\partial t} \right) \; dV \\
&= -2 \int\limits_V \eta \, \boldsymbol{B} \cdot \mathrm{curl} \, \boldsymbol{B} \; dV + \; \text{surface terms} \; .
\end{aligned} \tag{8.180}$$

This equation means that in ideal magnetohydrodynamics, where $\eta = 0$, and under boundary conditions such that the surface terms vanish, the magnetic helicity is conserved. If $\eta \neq 0$, but small as in the solar convection zone, then H_{m} is still nearly constant (since the η considered here is not the turbulent diffusivity η_{t}). The conservation of H_{m} is independent of the function ψ, i.e., invariant under a gauge transformation of the vector potential $\boldsymbol{A}$. The integrand $\boldsymbol{A} \cdot \boldsymbol{B}$ is not gauge invariant, in contrast to the integrand on the right of (8.180), which is proportional to the *current helicity*

$$h_{\mathrm{j}} = \boldsymbol{j} \cdot \boldsymbol{B} \; . \tag{8.181}$$

Seehafer (1996) has pointed out that the constraint of magnetic helicity conservation, even as it is not exact for $\eta \neq 0$, has an impact on mean-field dynamos; it may lead to magnetic helicity of opposite sign at large and small scales, respectively. It appears necessary to dispose the large-scale part by allowing for a flow of magnetic helicity across the boundaries (Brandenburg and Dobler 2001).

Problem 8.24. Calculate the surface terms that appear in Eq. (8.180). Under which boundary conditions do these terms vanish?

8.5 Bibliographical Notes

Comprehensive introductions to magnetohydrodynamics are provided by the books of Roberts (1967), Biskamp (1993), and Choudhuri (1998). The monographs of Parker (1979 a), Priest (1982) and Zeldovich et al. (1983), and the volumes edited by Rutten and Schrijver (1994), Schüssler and Schmidt (1994), and Tsinganos (1996) cover the subject of solar magnetism in general, the books of Moffatt (1978), Krause and Rädler (1980), and Childress and

Gilbert (1995) specialize on the generation of fields by fluid motions. Zwaan (1987) generally discusses the diverse patterns of the solar magnetic field, the book of Schrijver and Zwaan (2000) puts the solar case into the general context of stellar magnetic activity. For the definition of the numerous special terms used in solar physics the glossary of Bruzek and Durrant (1977) should be consulted.

The electrical conductivity of the solar atmosphere has been recalculated by Kovita and Cram (1983), with results similar to those presented in Fig. 8.2.

The small-scale magnetic field, and the physics of flux tubes in particular, are treated in the proceedings edited by Schröter et al. (1987), Rimmele et al. (1999), Schmieder et al. (1999), and Sigwarth (2001 a). Solanki (1993) gives an overview to observation and modeling of the solar magnetic field on small scales, and Frutiger and Solanki (2001) construct models of flux tubes by inversion of spectral line profiles in polarized light. Observational indications for the convective collapse of flux tubes have been found by Lin and Rimmele (1999) and by Bellot Rubio et al. (2001). The G-band bright points have been studied in depth by Berger et al. (1995, 1998), Steiner et al. (2001) explain why these points are bright. Leka and Steiner (2001) compare numerical simulations to observed magnetic features such as pores and the *azimuth centers* that have been defined first by Keppens and Martínez Pillet (1996); the azimuth centers are probably related to the magnetic knots. Flux concentration by cellular motions is comprehensively reviewed by Proctor and Weiss (1982), Weiss et al. (1996), Rucklidge et al. (2000), and Weiss (2001).

Caligari et al. (1995) theoretically describe the rise of thin magnetic flux tubes to the solar surface, and the agreement that is obtained with the observed latitude of emergence and tilt of bipolar groups if that rise starts at a field strength of 10 T at the base of the convection zone. The fragmentation of a rising tube of finite diameter into two vortex rolls, which was first observed by Schüssler (1979) in a numerical study, can be suppressed by twisting the tube field (Moreno-Insertis and Emonet 1996, Emonet and Moreno-Insertis 1998), although twisted tubes may be susceptible to the kink instability (Wissink et al. 2000 b). Longcope et al. (1998) suggest that the twist originates from the helicity of the turbulent convection.

The thin-tube approximation for magnetic flux concentrations, and the possibility to describe the diverse wave modes supported by them, has been elaborated by Ferriz-Mas and Schüssler (1989), Ferriz-Mas et al. (1989), Zhugzhda (1996), Schmitt (1998), and Bennett et al. (1999). Thin tubes expanding into the upper atmosphere have been treated by Pneuman et al. (1986).

The volumes edited by Cram and Thomas (1981), by Thomas and Weiss (1992), and by Schmieder et al. (1997) are modern introductions to sunspots; a classical treatise is the book of Bray and Loughhead (1964). Pizzo (1986, 1990) and Pizzo et al. (1993 ab) calculate two-dimensional magnetohydrostatic models of flux tubes, while Parker (1979 c) and Choudhuri (1986) discuss details of the cluster model. Moreno-Insertis and Spruit (1989) investigate the stability of sunspots to convective motion; the interchange instability of flux tubes is examined by Bünte (1993) and Bünte et al. (1993). Oscillations in sunspots have been reviewed by Staude (1999) and Bogdan (2000). Montesinos and Thomas (1997, with further references therein) have elaborated the siphon-flow model of penumbral grains, and Degenhardt et al. (1993) find some evidence for the shocks that occur in the flow according to that model.

Solar variability and its influence on Earth has been reviewed by Lean (1997). Fligge et al. (2000) relate the irradiance variation to the variable magnetic features on the Sun's surface. As for the luminosity dips caused by sunspots, it appears that they are caused predominantly by *young* spots, during the phase of rapid development (Pap 1985). Section 1.6 gives further references on this subject.

The solar cycle and the solar dynamo are subjects of the proceedings edited by Bumba and Kleczek (1976), Harvey (1992 a), Krause et al. (1993), Proctor et al. (1993), and Núñez and Ferriz-Mas (1999), of the reviews of Cowling (1981), Rädler (1990), or Stix (1981, 1987 b, 1991, 2001). Mean-field electrodynamics is the subject of Roberts and Soward (1975 a,b), and Knobloch (1977, 1978). The layer of overshooting turbulence at the base of the convection zone has been treated by van Ballegooijen (1982), Schmitt et al. (1984), and Pidatella and Stix (1986). Schüssler (1983, 1987 a) gives arguments why the dynamo should operate in this layer.

Since helioseismology has revealed the variation of Ω near the base of the convection zone, $\alpha\Omega$ dynamo models employ this tachocline. With an analytical model Parker (1993) illustrates that the dynamo takes on the character of a surface wave; Charbonneau and McGregor (1997) extend this interpretation to the spherical geometry. Dikpati and Charbonneau (1999) apply the observed rotation profile $\Omega(r, \theta)$ to the earlier model of Leighton (1969), and Belvedere et al. (2000) consider the asymptotic case of large dynamo number (Ruzmaikin et al. 1988) in the light of the helioseismic result. It is another problem whether the toroidal field generated by the differential rotation has the form of isolated flux tubes, or a more even distribution. Rempel et al. (2000) and Wissink et al. (2000 a) have considered the stability of a magnetic layer and the breakup of such a layer into individual tubes.

There are further observations that may provide keys for a better understanding of the Sun's dynamo. One is the variation of sunspot darkness over the cycle (Albregtsen and Maltby 1978). A second is the apparent change of the number density of granules with the solar cycle (Macris and Rösch 1983) and, probably related, a variation of the atmospheric temperature gradient (Holweger et al. 1983). A third is the meridional circulation (Sect. 7.4.3), which may severely alter the propagation of the dynamo wave, as a calculation of Choudhuri et al. (1995) shows. Finally, the p-mode eigenfrequencies vary with the phase of the solar cycle, as first noticed by Woodard and Noyes (1985) and Fossat et al. (1987), and confirmed later (e.g., Howe et al. 1999); at 3 mHz, for example, the frequencies increase by $\approx 0.4\,\mu$Hz from activity minimum to maximum. Dziembowski et al. (2000) discuss possible reasons for this variation, one being changes in the structure of the turbulent convection and the ensuing variation of the convection-oscillation interaction (Zhugzhda and Stix (1994).

Apart from questions concerned with the 11-year solar cycle, there are other, more fundamental, problems in dynamo theory. One is related to the origin of the small-scale magnetic flux that is found outside active regions, notably in the *ephemeral active regions* (Martin and Harvey 1979). It was suggested earlier that this kind of magnetism is "waste" from the global dynamo (Golub et al. 1981). More recently, numerical simulations led Cattaneo (1999; see also Cattaneo and Hughes 2001) to advance the hypothesis of local field amplification by turbulent convection. Other questions arise in the context of magnetic helicity conservation. Brandenburg et al. (2002) review this subject, Berger and Ruzmaikin (2000) especially discuss the solar case.

9. Chromosphere, Corona, and Solar Wind

In the preceding chapters we have learned about convection and rotation, the main causes of solar magnetism. And we have learned about this magnetism itself. We shall now recognize that the Sun's magnetic field is the main source of almost all structure and variability which we find in the outermost layers: the chromosphere, the corona, and the solar wind. Moreover, we shall see that the physical state and the extent of these outer layers are most certainly determined by the solar magnetic field.

Thus, inhomogeneity and time-dependence are characteristics of the topics to be treated in this chapter. Nevertheless, we shall often retreat to the conveniences of spherically symmetric and stationary models. Although such models are idealized and far from reality, they may serve to illustrate basic concepts such as the expanding hot corona, or may yield order-of-magnitude estimates which otherwise would be difficult to obtain.

To begin with, we shall review the most important observations.

9.1 Empirical Facts

9.1.1 The Chromosphere

For a few seconds, just after the beginning and before the end of a total eclipse, the solar limb presents a most colorful view (*chromosphere*: the "colored sphere"). The spectrograph reveals the *flash spectrum* which shows – in emission – a large number of the lines that are dark in the normal solar spectrum, in particular the strongest ones. A prominent example is the red Hα line at 656.3 nm. These lines are seen in emission because at their frequencies the solar atmosphere is still opaque at a level where, even for a tangential line of sight, it is transparent in the continuum (the visible limb lies at $\tau_{500} \approx 0.004$, cf. Sect. 1.3). That is, in the lines we see the hot solar atmosphere, while in the continuum we look into cool and dark space.

Mottles and Spicules. The large optical thickness of the chromospheric lines also allows us to observe the chromosphere on the disc, by means of a narrow-band filter. The Hα filtergram of Fig. 6.22 is an example. Another is Fig. 9.1, taken in the wing of the Hα line profile. Even more than the limb observations, such disc filtergrams demonstrate the extreme inhomogeneity of the solar chromosphere. The Sun is covered by numerous dark *mottles*. Certainly these mottles are related to the *spicules* seen at the limb (e.g., Fig. 9.7), although a unique identification with these spicules is difficult. As Fig. 9.1 shows, the dark mottles more or less outline a network; detailed studies show that this network follows the supergranulation pattern. Since magnetic flux is continuously convected towards the supergranular boundaries, it appears that the mottles owe their existence to local enhancements of the solar magnetic field.

An average spicule seen at the limb has a height of 5000 km, a width of no more than 500 km, a temperature of 10^4 K, and a density of 3×10^{-10} kg/m^3. Upward velocities of, typically, 25 km/s have been measured. These numbers are puzzling for several reasons. First, one has to find a mechanism that accelerates the gas; this mechanism should work along the spicule channel because in a ballistic model the observed height is not reached even with an initial velocity of 25 km/s. This problem appears particularly severe for the *macrospicules* which are concentrated in the polar regions and in Hα images extend up to 20 000 km beyond the solar limb. Second, since the total mass flux in spicules exceeds the solar-wind mass flux by two orders of magnitude, most of the mass moving upwards in spicules must return to the solar surface. At present little is known about this return-flow, although many spicules seem to fall back to the solar surface after a lifetime of 5–10 minutes.

Fig. 9.1. Hα filtergram from Sacramento Peak Observatory, showing dark mottles on the solar disc. The filter is tuned to 7/8 Å from the line center, the bandwidth is 0.22 Å. Courtesy National Solar Observatory, AURA, Inc.

Increasing Temperature. But there is more to the chromospheric phenomenon: the outwards *increasing* temperature. We have already seen that such an outwards rising temperature is obtained in the mean atmospheric models of Chap. 4, in particular when the departures from LTE become substantial. We shall now describe an effect which is typical for the chromosphere, namely the *emission reversal* which occurs in the cores of strong resonance lines. The most important examples are the Ca II lines H and K at 396.85 nm and 393.37 nm, and the Mg II lines h and k at 280.27 nm and 279.55 nm. A spectrum of Ca II K and its vicinity is shown in Fig. 9.2, and intensity scans across such a spectrum are presented in Fig. 9.3.

In a qualitative way we may explain the emission reversal as follows. Imagine that we observe the Sun through a narrow-band filter, which we tune from the continuum towards the center of Ca II K. Apart from the weak lines visible in both Figs. 9.2 and 9.3, we find monotonically increasing absorption. Hence, we see into increasingly higher layers of the atmosphere. As long as the temperature decreases with height the line becomes darker, which explains the wings of the line profile. However, at a wavelength about 0.25 Å from the line center, the absorption has become so large that we begin to see the layers above the temperature minimum. Although the absorption still increases, the source function essentially follows the Planck function and, therefore, also increases; this yields the emission reversal of the line profile. As detailed calculations show, the emission is also obtained in LTE models.

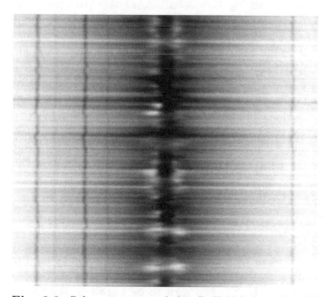

Fig. 9.2. Solar spectrum of the Ca II K line region. The spectral band is from 393.17 nm to 393.56 nm, the vertical extent is 160 Mm on the Sun. Photograph Schauinsland Observatory, Kiepenheuer-Institut, Freiburg (Grossmann-Doerth et al. 1974)

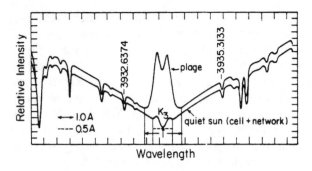

Fig. 9.3. Ca II K line profiles of quiet-Sun and plage regions. The wavelength range is 3930.1 to 3937.4 Å. From Skumanich et al. (1984)

In fact it is much stronger in the LTE models than in the models based on statistical equilibrium only.

Finally, as we tune the filter towards the line core, we arrive so high in the atmosphere that the assumption of LTE completely breaks down. The source function is smaller than the Planck function; as a result we have the intensity minimum at the line center.

The intensity minima in the red and violet wings of the K line are usually designated by K_{1r} and K_{1v}, the emission peaks by K_{2r} and K_{2v}, and the central minimum by K_3.

It is evident from the spectrum shown in Fig. 9.2 that the Ca II emission strongly varies with position on the solar surface. Filtergrams, or spectroheliograms such as the one shown in Fig. 9.4, reveal that there are mainly two sources of Ca II emission: larger areas, the *plages*, which spatially coincide with the active regions, and the finer *chromospheric network*, which we have already identified with the cell boundaries of the supergranular velocity field.

The close relationship between plages and active regions, and between the network and the supergranulation (which, as we have seen, accumulates magnetic flux, cf. Fig. 8.11), suggests that the magnetic field plays a key role for the chromospheric emission. The chromospheric heating must be greatly enhanced in the areas where the magnetic field strength is large. We postpone the discussion of chromospheric heating but note that a study of Skumanich et al. (1975) shows that the total Ca II K emission in the network is roughly proportional to the local strength of the magnetic field.

The connection between Ca II emission and magnetism allows us to study magnetic activity of stars other than the Sun. Indeed, from measurements initiated in 1966 by O. C. Wilson (Wilson 1978), and extended by Baliunas et al. (1995), it is possible to identify activity cycles for about 50 late-type stars. Moreover, as the sources of Ca II emission are unevenly distributed on the stellar surface, it is also possible to measure the *rotational modulation* of the emission. Stellar rotation that is too slow to be visible in the typical rotational line-broadening could still be detected in this way (Vaughan et al. 1981). Rapidly rotating young stars usually have high average levels of calcium emission, but no smooth cyclic variation; slow rotators like the Sun

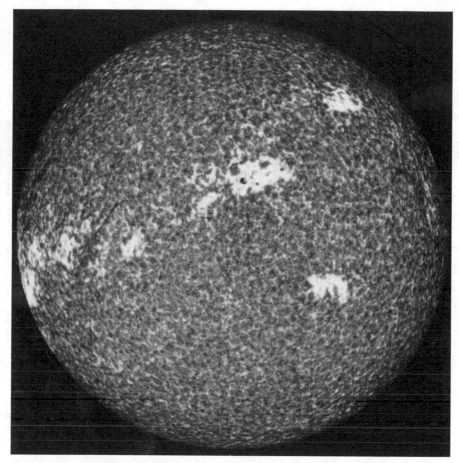

Fig. 9.4. Ca II K spectroheliogram of 31 May 1984, from Sacramento Peak Observatory, showing plages and the chromospheric network. The bandwidth is 0.5 Å (*dashed horizontal line in Fig. 9.3*). Courtesy National Solar Observatory, AURA, Inc.

clearly exhibit cycles, while their average emission is lower. When the Sun is observed as a star, i.e., in the light integrated over the disc, the cyclic as well as the rotational variation of the Ca II emission is clearly discernable.

Because of the importance of the outward temperature increase, especially for the Ca II emission, a more precise definition of the chromosphere is used frequently: the layer between the temperature minimum and the level where $T = 25\,000\,\mathrm{K}$. In the one-dimensional mean model, e.g., Fig. 4.9, this layer comprises some 2000 km; Fontenla et al. (1990, 1993) give tables of such models. On the other hand the spicules, which also have a chromospheric temperature, cover a range of $\approx 5000\,\mathrm{km}$ when observed at the limb. This apparent discrepancy once more illustrates how much care must be taken with the application of a mean model!

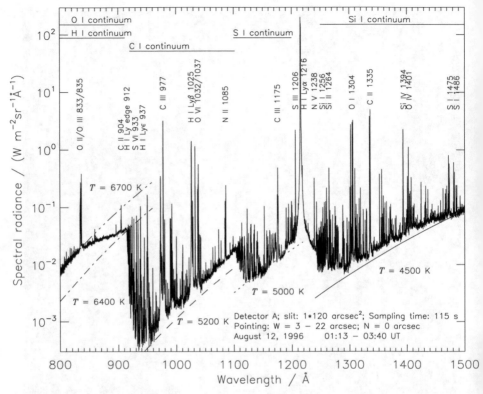

Fig. 9.5. Solar spectrum in the extreme UV, obtained from a quiet-Sun region with the spectrograph SUMER on the SOHO satellite. Prominent lines, and continua with black-body fits, are identified. The relative uncertainty of the spectral flux is ≈ 0.2. Courtesy K. Wilhelm

9.1.2 The Transition Region

The "layer" between the relatively cool ($\approx 10^4$ K) chromosphere and the hot ($\approx 10^6$ K) corona is called the *transition region*. It is appropriate to think of a temperature regime rather than a geometric layer. This is not only because of the extreme spatial inhomogeneity of this region, but also because the transition is so sharp that there is virtually a discontinuity in T (and hence in ρ, because the pressure, $P \propto \rho T$, must remain continuous).

The ultraviolet spectrum from 50 nm to 160 nm is most appropriate to investigate the transition region. It contains numerous emission lines, and emission continua, originating from ions existing at various temperatures in the above-mentioned range (Fig. 9.6). Observation from space is necessary to acquire an ultraviolet spectrum. Examples of successful instruments are the Harvard College Observatory spectrometer on Skylab (Reeves et al. 1977), and the High Resolution Telescope and Spectrograph (HRTS) of the Naval Research Laboratory (Brueckner 1981) flown on rockets and on Spacelab 2.

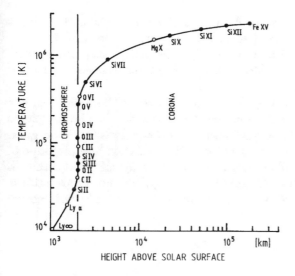

Fig. 9.6. Temperature as a function of height in a mean solar model atmosphere. *Dots* indicate the temperatures of formation for some atomic species (*open circles*: observed with the Harvard College Observatory spectrometer). After Reeves et al. (1977)

Figure 9.5 shows an EUV spectrum of the wavelength range 800–1500 nm obtained with the spectrograph SUMER (*Solar Ultraviolet Measurements of Emitted Radiation*, Wilhelm et al. 1995) on board of the SOHO satellite.

The concept of a rather thin transition region follows from the measurement of the absolute intensity of EUV lines, in combination with the one-dimensional models described in Chap. 4. The result of such models is the continuation of the steep temperature rise already seen in the upper chromosphere (Table 4.1). Figure 9.6 is a schematic illustration; the rather narrow ranges of temperature where each of the various ions can exist are also marked.

As with the chromosphere, the mean model of the transition region must be understood as a starting point of discussion rather than as a re-

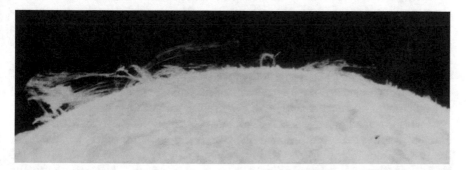

Fig. 9.7. Lyman α filtergram (Bonnet et al. 1980), showing the structure of the transition region with a resolution of $1''$. This photograph (exposure 1 s) was obtained during a rocket flight by the Laboratoire de Physique Stellaire et Planétaire, Paris, in cooperation with Lockheed Corporation and NASA

alistic description. Again we know from limb observations that the sources of transition-zone radiation are distributed over several thousand km. Pictures such as the Lyman α filtergram shown in Fig. 9.7 directly demonstrate the substantial inhomogeneity. Moreover, there are material flows (Doppler shifts), and there is variation in time.

The spatial inhomogeneity becomes less pronounced in the higher-temperature part of the transition-zone radiation. In particular, the chromospheric network, which is clearly visible in the lines of the lower transition zone, gradually disappears as we go towards the corona. The common interpretation of this effect is that the governing agent, namely the magnetic field, becomes more homogeneous as it fans out from the photospheric flux tubes into the corona; as the coronal gas pressure is small the confinement into such narrow tubes is no longer possible.

9.1.3 The Corona

In the upper part of the transition region the mean temperature profile flattens (Fig. 9.6), and we come to an extended region with $T \approx 10^6$ K, the solar *corona*. Strictly we should now distinguish between the ion and electron temperatures, T_{ion} and T_e, but presently we discuss the mere fact that the coronal temperature is so much higher than the temperature of the photosphere, which was recognized around 1940, earlier than the corresponding information about the chromosphere and transition region was available. Still, the discovery came rather late in the history of solar astronomy, perhaps because it is in contrast to the simple picture of a hot star in a cool interstellar environment. A hot corona requires energy to be pumped from low to high temperature. We shall see below that even today the problem of coronal heating is far from a satisfactory solution.

White-Light Corona. The optical coronal radiation, as observed during total eclipses, is traditionally divided into the *K corona* and the *F corona*. The F corona dominates outwards from, say, 2 or 3 solar radii. Its spectrum shows the normal dark Fraunhofer lines of the photospheric spectrum, and its name is derived from these lines. The continuum of the outer corona also resembles the photospheric spectrum, and is only weakly polarized. Hence the F corona can be explained as photospheric light scattered on dust particles. As *zodiacal light* it can be observed far into interplanetary space. In the following we shall not be concerned further with this type of coronal radiation.

The K corona derives its name from the German "Kontinuum". Its continuous spectrum also resembles the photospheric spectrum, but the Fraunhofer lines are absent (Problem 9.2 will make clear why). The light of the K corona is highly polarized, which indicates that it arises from Thomson scattering by free electrons.

The intensity of the white-light corona drops steeply with distance from the Sun, and strongly depends on the position angle. As an average over position angle Baumbach (1937), using data from 10 eclipses between 1905 and 1929, derived the formula

$$\frac{I}{I_{\odot C}} = 10^{-6} \left(\frac{0.0532}{x^{2.5}} + \frac{1.425}{x^7} + \frac{2.565}{x^{17}} \right) , \tag{9.1}$$

where $I_{\odot C}$ is the intensity at disc center, and x is the normalized radius $r/r_\odot$, projected onto the plane of the sky. Of the three terms in (9.1) the first is mainly due to the more slowly decaying F corona, while the second and third essentially describe the rapidly decaying K corona. Even during good ("coronal") eclipse conditions the sky has a radiance of $\approx 10^{-9} I_{\odot C}$; thus, according to (9.1), the mean corona becomes invisible at $x \approx 4$, the K corona even at smaller x.

From the observed intensity an estimate of the density of scattering electrons can be obtained under the assumption of a spherically symmetric corona. Let E_K be the total emission, per volume and solid angle, of scattered radiation, and let y be the coordinate along the line of sight, which is perpendicular to x. Then

$$\rho^2 \equiv (r/r_\odot)^2 - x^2 + y^2 , \tag{9.2}$$

and the observed intensity is

$$I(x) = \int_{-\infty}^{\infty} E_K(\rho) \, dy = 2 \int_0^{\infty} E_K(\rho) \, dy = 2 \int_x^{\infty} \frac{\rho \, E_K(\rho)}{\sqrt{\rho^2 - x^2}} \, d\rho . \tag{9.3}$$

Since E_K is unknown one must invert (9.3), i.e., calculate the inverse Abel transform

$$E_K(\rho) = -\frac{1}{\pi} \int_{\rho}^{\infty} \frac{dI/dx}{\sqrt{x^2 - \rho^2}} \, dx . \tag{9.4}$$

Now compare this result to the total *expected* scattered light, that is, set

$$E_K(\rho) = \sigma_s n_e \frac{1}{4\pi} \int I_\odot(\theta) \, d\Omega . \tag{9.5}$$

The integral is over the angle subtended by the solar disc as seen from the distance ρ, $I_\odot(\theta)$ is the intensity of the Sun's radiation, n_e the electron density, and σ_s the cross section of Thomson scattering, defined in (2.90). Finally, we solve (9.5) for n_e. Notice that, in addition to the assumption of spherical symmetry, it is now assumed that the scattering is isotropic. This assumption is false but may suffice for an order-of-magnitude estimate.

Allen (1947) used the method outlined here to derive the electron density of the mean K corona. His result is approximated by

$$n_e(\rho) = n_{e0} \left(\frac{1.55}{\rho^6} + \frac{2.99}{\rho^{16}} \right) , \tag{9.6}$$

Table 9.1. Degree of polarization of the white-light corona. After Shklovskii (1965) and Billings (1966)

x	Theoretical		Observed	
	at 433 nm	at 570 nm	total	K corona
1.0	0.199	0.158	0.17	0.18
1.1	0.333	0.309	0.30	0.32
1.2	0.404	0.388	0.37	0.41
1.4	0.491	0.483	0.43	0.51
2.0	0.548	0.550	0.37	0.62
3.0	0.549	0.550	0.18	0.65

where $n_{e0} = 10^{14}\,\text{m}^{-3}$. The two contributions correspond to the last two terms of (9.1).

The correct treatment of the scattering problem must of course include the anisotropy, and will predict the polarization. If the scattering angle was exactly $90°$, we would have a polarization of 100%, a consequence of the fact that both the incident and the scattered waves are *transverse*. The direction of the polarization would be tangential to the Sun (z, say). For two reasons the scattering angle is not $90°$: first, because the observer integrates over the line of sight; and second, because for any point within the corona the solar disc subtends a finite area on the sky.

If these geometrical effects are taken into account, and the observed intensity distribution is used, one obtains the theoretical results listed in Table 9.1. The degree of polarization given there is

$$P = \frac{I_z - I_x}{I_z + I_x}\,, \tag{9.7}$$

where I_z and I_x are the intensities in the two perpendicular directions of polarization. The observed polarization is listed for comparison. There is close agreement if the F corona is eliminated.

Scattering on free electrons is independent of wavelength. However, because the photospheric limb darkening is more pronounced at shorter wavelengths, the Sun looks slightly more "point-like". The polarization of the white-light corona is therefore slightly larger at smaller λ, in particular close to the Sun, at small x.

Problem 9.1. Compare the depth of coronal and photospheric Fraunhofer lines in order to separate the K corona from the F corona.

Problem 9.2. In place of the strong H and K lines rather shallow and broad dips ($\approx 20\,\text{nm}$) can be noticed in the spectrum of the K corona. Assume that the dips are in fact the broadened solar lines and derive a temperature for the scattering particles [in this way Grotrian (1931) first concluded that the corona might be hot].

Problem 9.3. Use the electron density (9.6), assume a barometric stratification, and derive the coronal electron temperature.

Like the chromosphere, the solar corona is closely related to the Sun's magnetism. A particular aspect of this relationship has been known for a long time: the variation in shape from activity minimum to activity maximum.

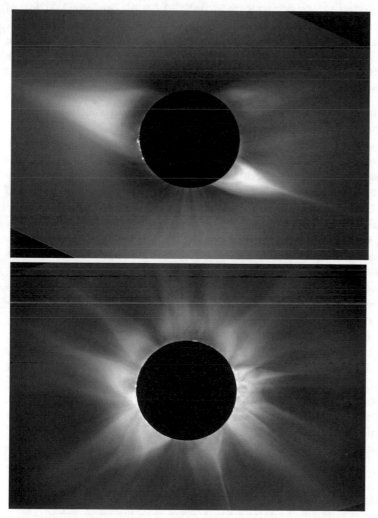

Fig. 9.8. The solar corona at activity minimum (*upper*) and maximum (*lower image*), photographed with a radially graded filter (transmission range 10^4) during the eclipses of 3 November 1994 in Putre, Chile, and 16 February 1980 in Palem, India. The vertical is heliographic north in both images. Courtesy A. Lecinski, High Altitude Observatory, NCAR, Boulder, Colorado

A typical *minimum corona* is shown in Fig. 9.8, upper panel. It is elongated along the solar equatorial direction, and around the southern pole it shows the *polar plumes*, thin hair-like spikes that resemble the lines of force of a bar magnet. In contrast, the *maximum corona* makes a more spherically symmetric appearance. However, a close look, e.g., the lower panel of Fig. 9.8, shows that the maximum corona is much more structured than the minimum corona. Only the *condensations, enhancements, helmets*, and *streamers* are distributed rather evenly over all latitudes and, projected on top of each other, produce the more circular shape in images with lesser resolution. The individual features are so well pronounced in Fig. 9.8 because a special camera, developed by G. A. Newkirk, has been used; this camera employs a radially graded filter, with a transmission range of 10^4, to compensate for the steep decline of the coronal intensity with increasing distance from the Sun.

Emission Lines in the Visible. Numerous emission lines are superposed on the continuum of the coronal spectrum. The intensity of these lines often exceeds the neighboring continuum by a factor 100, and the line width typically is of order 0.1 nm. Grotrian (1939) identified the first two of these emission lines as forbidden transitions of Fe X and Fe XI; but it was only when 19 lines were identified by Edlén (1942) as forbidden transitions of highly ionized atoms that the high temperature of the solar corona was generally recognized. Prior to this discovery exotic explanations, e.g., an element "coronium", were occasionally advanced. Today many more coronal emission lines are identified; three of the most conspicuous lines [a table of Billings (1966) contains 42] are listed in Table 9.2.

The large ionization potential of the ions responsible for the coronal emission lines directly hints to a million-degree temperature. Because of the low density and, therefore, large departures from LTE we must however not apply Saha's formula to the ionization equilibrium. Rather, the detailed processes must be put into statistical equilibrium. In the corona the relevant processes are ionization by electron collisions, radiative recombination (e.g., Elwert 1952), and *dielectronic* recombination. The latter process, which leads to an intermediate atomic state with two excited electrons, was recognized as particularly important by Burgess (1964). The result of the statistical treatment is an "ionization temperature" of the corona of 1–2×10^6 K, which agrees well with other temperature determinations (without dielectronic recombination

Table 9.2. Three coronal emission lines. From Billings (1966), with identifications of Edlén (1942); χ is the ionization potential of the preceding ion

	λ [nm]	Ion	Transition	χ [eV]
"Green line"	530.29	Fe XIV	$3s^2\,3p\;^2P_{3/2}\;-\;^2P_{1/2}$	355
"Yellow line"	569.45	Ca XV	$2s^2\,2p^2\;^3P_1\;-\;^3P_0$	820
"Red line"	637.45	Fe X	$3s^2\,3p^5\;^2P_{1/2}\;-\;^2P_{3/2}$	235

the temperature turns out too low, because then fewer electron collisions would suffice to maintain the ionization equilibrium).

Similar to the ionization equilibrium is the equilibrium that governs the line emission: excitation by collisions with electrons, and spontaneous emission of radiation. The lines are "forbidden" in the sense that, at normal densities, the probability of radiative transition would be much smaller than the probability of collisional de-excitation. In the tenuous corona collisions are rare, and the opposite is true. Formally, the selection rule $\Delta L = \pm 1$ for the total orbital angular momentum is violated for most of the transitions in question, cf. the examples of Table 9.2.

Because the coronal emission lines are so intense the combination of a coronagraph with a narrow-band filter is especially rewarding. Both tools have been developed by B. Lyot in the 1930s, so that coronal observations could be made on a routine basis independently of eclipses. Extensive observations of the emission-line corona (or E corona), have been made by Waldmeier (1951). His studies not only covered an entire cycle of solar activity, but also for the first time showed the existence of the dark *coronal holes* which are so important in modern coronal research (Waldmeier 1957, p. 275).

Problem 9.4. Estimate the coronal ion temperature from the width (≈ 1 Å) of the coronal emission lines.

The Radio Corona. We have already seen, cf. Fig. 1.6, that the quiet Sun possesses a radio spectrum of thermal nature, and that the brightness temperature changes from $\approx 10^4$ K to $\approx 10^6$ K in the wavelength range 1 cm to 1 m. Clearly, we may now attribute this transition to the temperature rise from the temperature minimum to the corona. The coronal thermal continuum mainly originates from free-free transitions.

With increasing wavelength the radio spectrum becomes dominated by the various types of *radio bursts*. We postpone this subject to Sect. 9.5.1, but note already here that the radio bursts can be used as a probe of the corona. The reason is that in a plasma such as the corona, electromagnetic waves cannot propagate if their frequency lies below the *plasma frequency*

$$\nu_P = \frac{e}{2\pi} \left(\frac{n_e}{\varepsilon_0 m_e} \right)^{1/2} . \tag{9.8}$$

Waves with lower frequency will be absorbed or reflected: on Earth, for example, we can receive distant radio stations because reflection occurs in the ionosphere. The cutoff has its complete analogue in the cutoff for acoustic wave propagation in a stratified atmosphere, which we have treated in Chap. 5. In the corona, the electron density decreases with distance from the Sun. Therefore, radio signals with longer wavelength must arise from sources which lie further outwards than the sources of short-wavelength signals.

Problem 9.5. Consider oscillations of free electrons relative to the ions (at rest) in a one-dimensional model. Use the equation of continuity, the momentum balance, and Maxwell's equations to show that such *plasma oscillations* occur with the frequency (9.8).

Problem 9.6. The radio heliograph at Nançay, France, consists of a cross-shaped array of antennas, with branches of 3200 m and 1250 m in the North-South and East-West directions, respectively. Observations are made at wavelengths between 60 cm and 2 m. Use (9.6) and (9.8) to calculate the height of the coronal sources. What is the angular resolution for the two directions?

Ultraviolet and X-rays. The spectrum shown in Fig. 9.5 demonstrates the dominance of emission lines in the extreme ultraviolet region. This continues toward shorter wavelengths. Spectral atlasses have been obtained from the SOHO instruments CDS and SUMER and cover the ranges 30–60 nm and 67–161 nm, respectively (Brekke et al. 2000, Curdt et al. 2001); an earlier table compiled by Billings (1966) contains more than 200 identified emission lines between 1.37 nm and 105.87 nm, originating from ions such as N VII, O VIII, or Fe XVII. Lines from even higher states of ionization, e.g., from Fe XVIII to Fe XXIII, are seen during flares.

Profiles of EUV emission lines far above the limb have been measured with the UV Coronagraph Spectrometer (UVCS) on board of SOHO (Kohl et al. 1997, 1998). Large line widths have been found especially for emission from coronal features having a magnetic field perpendicular to the line of sight, e.g., streamers, helmets, or coronal-hole regions above the limb. The interpretation in terms of a thermal velocity distribution yields values of the transverse kinetic temperature, $T_\perp$, of up to 2×10^8 K for the diverse ions at a distance of about $4r_\odot$. In addition to the line width, the thermal velocity distribution parallel to the magnetic field and the velocity of radial outflow from the Sun have been determined by a measurement of the *Doppler dimming* effect: the intensity of resonantly scattered light is diminished if the velocity of the scattering medium relative to the source of the incoming radiation is of the order of (or larger than) the thermal velocity of the scattering particles. Thus, with increasing speed of the solar wind in the corona, the resonance frequency becomes Doppler shifted away from the frequency of the incoming photons, and the part of the line emission that is due to resonant scattering becomes weaker. Using a given spherically symmetric density model of the corona, and applying the technique to lines of diverse ions, Cranmer et al. (1999 b) and Cranmer (2000) were able to derive parallel kinetic ion temperatures $T_\parallel$ and the outflow velocity. As a result they find that the wind speed reaches 200–300 km/s as close as one solar radius from the solar surface, and that the ion temperatures are highly anisotropic, with $T_\perp \gg T_\parallel$. The latter result will be discussed below (Sect. 9.4.2) in the context of the heating problem.

Fig. 9.9. Coronal loops at the solar limb, seen in the 17.1-nm pass band by the normal-incidence telescope on the satellite TRACE. The image (exposure 19.5 s) was taken on 6 November 1999; it shows plasma of $\approx 10^6$ K and has been evaluated by Aschwanden et al. (2000, 2001). Courtesy C. J. Schrijver

Figure 9.9 is a TRACE image that shows plasma loops in a spectral band of width ≈ 0.6 nm around 17.1 nm, containing emission lines of Fe IX and Fe X. The finest discernable threads have a width of $\approx 10^6$ m, which is close to the resolution limit; nevertheless it is remarkable that this width does not change much along the loops. Moreover, often these loops seem to be nearly isothermal, in the case of Fig. 9.9 at $\approx 10^6$ K, and for large loops the density and the pressure appear to decrease with a scale height that is several times the scale height derived from the temperature (Lenz et al. 1999, Aschwanden et al. 2000). This indicates that the loops are not in hydrostatic equilibrium.

In addition to the line spectrum, the soft x-ray region (below 10 nm, say) is characterized by a continuum of thermal form, arising from free-bound and free-free (bremsstrahlung) transitions. Figure 1.8 gives the intensity and its variation. Of course, the thermal x-ray radiation can be used to assess once more the high coronal temperature.

Since the cool photosphere has a small intensity at wavelengths below 150 nm, an x-ray telescope sees the corona with all its rich structure in front of the solar disc. Numerous coronal x-ray images were taken with the Wolter telescope flown on Skylab in 1973–1974, with the normal-incidence x-ray telescope NIXT on rocket flights (Golub et al. 1990), and with the soft x-ray telescope of the Yohkoh mission (a modified Wolter telescope). Figure 9.10 shows examples; another example was already presented in Fig. 7.7.

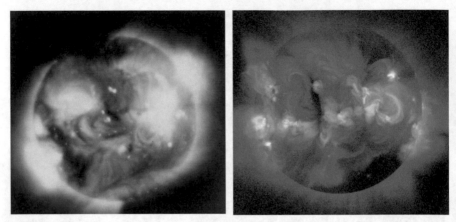

Fig. 9.10. Images of the Sun in soft x-rays (0.3–6 nm), obtained in 1973 by the Skylab mission (*left*, Vaiana et al. 1973), and in 1991 by Yohkoh (*right*)

Inspection of a coronal x-ray image immediately suggests that we must distinguish between two types of coronal regions: bright and dark. The bright regions are structured in form of *arcades, loops, helmets,* etc. We shall see below that these features outline the lines of force of the coronal magnetic field. As far as these lines close back to the solar surface, we also speak of the *closed* corona. On the other hand, the dark regions seen on the x-ray images form the most conspicuous part of the open corona. These regions essentially are the coronal holes mentioned earlier.

The brightness difference indicates a major difference in density. This is because, in the optically thin regime, the emission is essentially proportional to the square of the density. This density dependence originates from the rate of excitation of the atoms or ions by collisions with electrons, which is proportional to the particle density of either species. Moreover, apart from atomic parameters and the element abundance, the density of the particles in the excited level is proportional to the electron density (with a factor that depends on temperature), so that the total line radiation emitted from a volume V is proportional to $\int n_e^2 dV$, the *emission measure* (Pottasch 1963, 1964; Mariska 1992). Since for most lines the emission occurs in a rather limited temperature range, the T-dependent factor, which strictly should appear under the integral, usually is evaluated at the temperature of maximum emission. In a layered atmosphere the emission measure can be defined as an integral over height z:

$$EM = \int n_e^2 \, dz \; , \tag{9.9}$$

or, because a temperature range is more meaningful than a layer thickness,

$$EM = \int n_e^2 \left(\frac{dT}{dz}\right)^{-1} dT \; . \tag{9.10}$$

The integrand of expression (9.10) is the *differential emission measure*.

Typical values of T and n_e in the two kinds of coronal regions are given in Table 9.3. Just above the transition layer the difference in density is not substantial, but further out it is a factor 10 or more, which explains the difference of a factor 100 or more in emission. Also listed in the table is the mean-free-path L between two collisions of the electrons. In the inner corona L is much smaller than the distance over which T changes markedly, but only a few solar radii away from the Sun, in particular in the coronal holes, we have $L \gg r$; near $r = 5r_\odot$ the mean-free-path reaches 1 astronomical unit. We must be aware of this whenever we discuss transport effects based on collisions. Heat conduction, to be treated in Sect. 9.2.1, is an example.

Table 9.3. Electron density, n_e, and (relative) mean-free-path, L/r, in the solar corona. From Kopp (1977)

$r/r_\odot$	"Quiet" corona $T = 2.0 \times 10^6$ K		Coronal hole $T = 1.5 \times 10^6$ K	
	n_e [m^{-3}]	L/r	n_e [m^{-3}]	L/r
1.1	1.6×10^{14}	0.0029	5.4×10^{13}	0.0049
1.2	7.1×10^{13}	0.006	1.6×10^{13}	0.015
1.4	2.3×10^{13}	0.016	2.8×10^{12}	0.074
2.0	2.8×10^{12}	0.09	2.0×10^{11}	0.73
4.0	8.9×10^{10}	1.45	4.0×10^{9}	18.1
10.0	8.0×10^{9}	6.43		

As a particular feature of the x-ray corona we finally mention the *x-ray bright points*. As Figs. 9.10 and 9.12 demonstrate, these local brightenings occur both within the closed corona and within coronal holes. In the photosphere we find at their places bipolar magnetic configurations of small extent, the ephemeral active regions already mentioned in Chap. 8. Thus, enclosed by the lines of force which connect the two magnetic polarities, the x-ray bright points are in fact part of the closed corona, only at a much smaller scale than the larger loops and arcades.

9.1.4 The Wind

The first evidence for a continuous particle stream emanating from the Sun came from the ion tails of comets. These tails roughly point into the direction opposite to the Sun, but Hoffmeister (1943) found that there is a small systematic deviation: the ion tail slightly trails (in the sense of the comet's motion), so that there is a small angle, normally $< 5°$, between it and the solar radius vector. Biermann (1951 b), realizing that the radiation pressure of the sunlight could not account for the acceleration observed in the ion tails, proposed a corpuscular radiation instead, and so was able to explain Hoffmeister's discovery: the deviation angle of the tail is given by the ratio of the comet's transverse orbital velocity component to the velocity of the corpuscular radiation. From the distribution of comets it could be inferred that the postulated corpuscular radiation, later termed the *solar wind*, would blow continuously, and into all directions. What is more, the large variations in space and time of the phenomenon could already be seen by means of its variable effects upon the comets.

Problem 9.7. A comet's ion tail deviates by 4° from the radial direction. Its transverse orbital velocity is 30 km/s. What is the speed of the solar wind? Convince yourself that this wind speed is supersonic.

In order to measure the solar wind *in situ*, a spacecraft must fly above the Earth's magnetosphere, i.e., in an orbit several thousand km above ground, or in an interplanetary orbit. The data collected in 1962 by Mariner 2 on its way to Venus fully confirmed the earlier predictions: a high-speed (supersonic), continuous, but rather variable, flow of ionized matter. Electrons, protons, and α particles (3–4 percent, relative to the protons) are the predominant constituents.

As Fig. 9.11 illustrates, the solar wind velocity v often remains high (≈ 700 km/s) during a few consecutive days; the particle density n_p varies in anti-phase to v. The actual numbers show that the particle flux density $n_p v$ of the slow wind is 2–3 times that of the fast wind. It has been shown that the high-speed streams have their sources in those parts of the solar corona where the magnetic lines of force are open, i.e., essentially in the coronal holes. Figure 9.11 also shows the tendency of the high-speed streams to recur

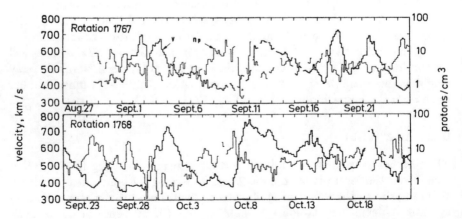

Fig. 9.11. Three-hour averages of solar wind velocity (*thick*) and proton density (*thin*), as observed by Mariner 2. Adapted from Hundhausen (1972)

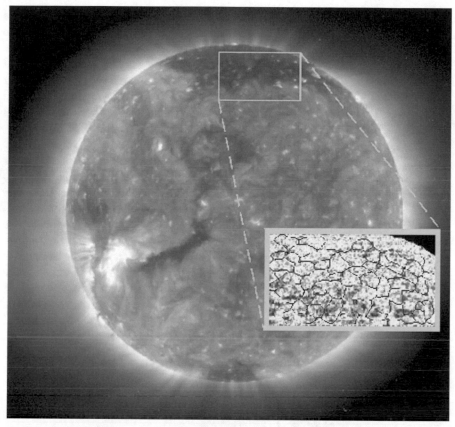

Fig. 9.12. The source region of the fast solar wind. This SOHO/EIT image in the 19.5-nm pass band shows several coronal holes. The inset is a SUMER velocity map obtained from the Doppler shift of the Ne VIII line at 77.04 nm, with outflow velocities (*blue*) of 5–20 km/s. The contours of the chromospheric network are derived from the Si II emission (Hassler et al. 1999)

after 27 days, which is the synodic equatorial period of rotation of the large-scale solar magnetic field. The sources of the solar wind coincide with the *M regions* postulated earlier as the sources on the Sun of recurrent geomagnetic perturbations.

With the spectrograph SUMER on board of the SOHO mission it has been possible to identify the source region of the solar wind in more detail (Fig. 9.12). The Ne VIII line at 77.04 nm, which is formed at $T \approx 600\,000$ K in the transition region, shows a Doppler shift corresponding to an outflow velocity of up to 20 km/s. The measurement has been made in the second spectral order, either with respect to an unshifted chromospheric line (Hassler et al. 1999) or with respect to the line position at the solar limb where no Doppler shift is expected because the outflow is transverse (Peter 1999).

As the inset of the figure illustrates, the outflow clearly originates in the coronal holes. Moreover, Hassler et al. were able to prove a relationship to the chromospheric network; the largest outflow velocities occur at the junctions of the network cell boundaries. Earlier evidence for an outflow from coronal holes, but with less spatial resolution, had already come from the Doppler shift of Si XI, O V, and Mg X lines measured during rocket flights (Cushman and Rense 1976, Rottman et al. 1982). A coronal outflow was also inferred from the Doppler shift of radar echos from the Sun (Abel et al. 1963).

The solar wind has been probed *in situ* by numerous space experiments. The distinction between the low-speed and high-speed forms of the solar wind has become clear through many space experiments, especially the two Helios missions launched in 1974 and 1976. Figure 9.13 illustrates the more recent measurements made by the Ulysses spacecraft during solar minimum conditions: at heliographic latitudes higher than ±20° the wind velocity was ≈ 750 km/s, with only small fluctuations; this characterizes the high-speed wind that originates in the coronal holes at the northern and southern poles of the Sun. In contrast to this, the wind is generally slower but much more variable in the low-latitude range. Most conspicuous is the variation caused by the rotating sector structure of the wind, with the recurrent streams of higher and lower speed. Within a fixed latitude interval, Ulysses encountered a larger number of such streams when the spacecraft was near its aphelion,

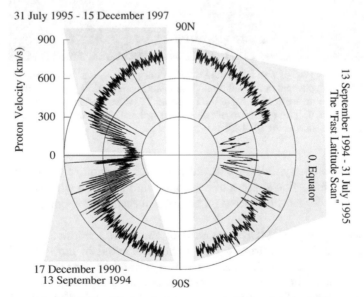

Fig. 9.13. Solar wind velocity versus heliographic latitude, measured by Ulysses. *Left*: south-bound sections of the orbit, at distances 2–5 astronomical units from the Sun; *right*: north-bound section, including the fast passage of the perihelion near 1 astronomical unit. Courtesy S. T. Suess

and therefore slow, while a smaller number was seen during the fast passage of the perihelion (Fig. 9.13, left and right, respectively).

A relatively close approach to the Sun was reached by Helios 1 and Helios 2, which had their perihelia at 0.31 and 0.29 astronomical units, respectively. The results listed in Table 9.4 are from these two spacecraft. The temperatures given there are directly derived from the velocity distributions of the diverse particles. The temperature anisotropy is defined as $T_{\parallel}/T_{\perp}$, where $T_{\parallel}$ and $T_{\perp}$ characterize the distributions parallel and perpendicular to the magnetic field. Because collisions are rare (cf. Table 9.3), and because the magnetic field guides the particles, it is quite natural that such an anisotropy occurs, and that the protons have a different temperature than the electrons (in the low-speed solar wind, where $v \approx 300\,\mathrm{km/s}$, the proton temperature is only about $4 \times 10^4\,\mathrm{K}$ at 1 astronomical unit, and has an anisotropy of ≈ 2, while the electron temperature is still $\approx 1.5 \times 10^5\,\mathrm{K}$). There is little change of the solar wind velocity between 0.3 and 1 astronomical units.

Table 9.4. Properties of high-speed streams in the solar wind, measured by the two Helios spacecraft. From Denskat (1982)

	1 astr. unit	0.3 astr. units
v [km/s]	$500 - 750$	$500 - 750$
Direction of $\boldsymbol{v}$	radial	radial
Proton density [cm^{-3}]	$3 - 4$	$20 - 40$
B [nT]	$4 - 6$	$25 - 45$
Direction of $\boldsymbol{B}$	in ecliptic plane 35° from radial	in ecliptic plane 10° from radial
Proton temp. [K]	$1.5 - 2.5 \times 10^5$	$4 - 6 \times 10^5$
Proton anisotropy	1	$0.5 - 0.8$
Electron temp. [K]	$1 - 2 \times 10^5$	$1.5 - 2.5 \times 10^5$
Electron anisotropy	1.5	1.5

The FIP Effect. A remarkable deviation from the photospheric abundance distribution of the chemical elements has been found in the solar wind, and to some degree also in the transition region and corona. Elements with a first ionization potential (FIP) lower than $\approx 10\,\mathrm{eV}$ are enriched, relative to those with higher first ionization potential, in comparison to their photospheric abundance ratio (Fig. 9.14). *In-situ* measurements of the solar wind indicate that the low-FIP elements are also enriched relative to hydrogen, cf. the two entries for H in the figure; this is less clear with respect to the spectroscopic evidence from the transition layer and corona, due to the absence of hydrogen lines.

The abundance enhancement depends on the magnetic configuration. On the average it amounts to a factor of about 4, except for the noble gases Kr and Xe. Helium, with an ionization potential of 24.6 eV, is underabundant by

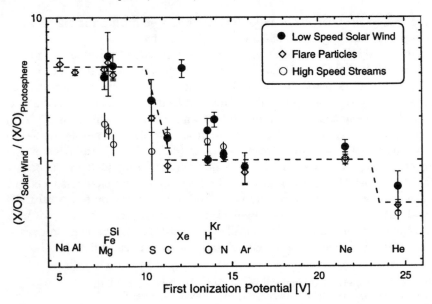

Fig. 9.14. Element abundances as a function of the first ionization potential, relative to oxygen and normalized to the photospheric ratio. The *dashed line* in an approximation to the slow-wind and flare-particle abundance ratios (except for Kr and Xe). From Geiss (1998)

a factor ≈ 1.8 relative to elements that are ionized at energies between ≈ 10 and 22 eV. In high-speed solar wind streams the effect is less pronounced, but it is clearly seen in the distribution of energetic particles originating during large flares.

The relative enhancement of the low-FIP elements was found first in the solar energetic particles (Hovestadt 1974, Meyer 1985); later it has been confirmed by many observations, e.g., by SWICS (the Solar Wind Ion Composition Spectrometer) on board of the Ulysses spacecraft (von Steiger et al. 2000), although the SWICS results yield a smaller enhancement, by a factor of somewhat less than 3.

The differentiation of the elements should take place mainly in the chromosphere, where the high-FIP elements are still neutral while those with lower FIP are partially ionized (Geiss 1982). In a partially ionized gas ions and neutral particles are subject to different diffusion velocities (which is called *ambipolar diffusion*), and to different drift velocities in a magnetic field. Alfvén (1954) has considered these effects in the context of element separation during the formation of the solar system; Lehnert (1970) first suggested their relevance to the solar wind. For the FIP effect Lyman α photons, which have an energy of 10.2 eV and abound in the chromosphere, should play a key role.

9.2 Consequences of High Temperature

9.2.1 Heat Conduction

Of the results described in the preceding section the most important one was the high temperature of the outer solar atmosphere. Due to the high temperature there are a large number of free electrons which, as we have seen in Sect. 8.1.2, give rise to a large electrical conductivity. In a similar way the electrons provide a large *thermal conductivity*.

Below we shall mainly be interested in the conduction of heat in the transition zone and the corona. Therefore, we may use the result obtained by Spitzer (1962) for *fully ionized* gases, namely

$$\kappa_\parallel = \frac{640\,\varepsilon_0^2\sqrt{2\pi}\,k(kT)^{5/2}}{\sqrt{m_e}e^4 Z \ln \Lambda}\,\varepsilon\delta_T \ . \tag{9.11}$$

This expression has been derived in analogy to the electrical conductivity (8.10). As there, the temperature dependence reflects the electron velocity distribution. The factor ε (< 1) arises because an electric field is produced by the distorting effect (on the electron velocity distribution) of a temperature gradient. The effect of this is to reduce the transport of heat by the electrons. The last factor, δ_T, of (9.11) corresponds to γ_E in expression (8.10) and corrects for the error made by the assumption of a Lorentz gas. For a charge number $Z = 1$ Spitzer gives $\varepsilon = 0.419$ and $\delta_T = 0.225$. If we use $\ln \Lambda = 20$ for the Coulomb logarithm, as it is appropriate to the corona, and substitute all the constants in (9.11), we obtain $\kappa_\parallel = 9.2 \times 10^{-12}\, T^{5/2}$ WK^{-1}m^{-1}.

In the presence of a magnetic field the electrons are guided along the field lines, and (9.11) applies only to the heat flow *parallel* to the field; this is indicated by the subscript. Transverse to the field the conductivity is greatly reduced (here heat is mainly conducted by protons, which have much larger gyration radii).

The large difference in parallel and transverse conduction of heat, together with the phenomenon of the frozen-in field, explains why we can directly see various details of the coronal structure: both matter and heat are forced to follow the magnetic field. Hence the polar plumes (Fig. 9.8), for example, indeed outline the magnetic field.

The Heat-Conduction Corona. We shall treat the problem of coronal heating below in Sect. 9.4. For the moment, let us simply assume that heat is deposited at some level r_0 in a spherically symmetric corona, and that we may model this heat input by prescribing the temperature T_0 at this level (e.g., $T_0 = 10^6$ K). We also assume that there is no other heat transport than conduction, i.e.,

$$\boldsymbol{F} = -\kappa_\parallel \nabla T \ , \tag{9.12}$$

and

$$\text{div } \boldsymbol{F} = 0 \ . \tag{9.13}$$

Substitution of (9.11) immediately yields

$$\frac{d}{dr}\left(r^2 T^{5/2}\frac{dT}{dr}\right) = 0 \ . \tag{9.14}$$

We know that in the photosphere $(r = r_\odot)$ as well as far away from the Sun $(r \to \infty)$, the temperature is much smaller than T_0; hence for the transition region and the corona the error introduced by setting $T(r_\odot) = T(\infty) = 0$ is small. The solution of (9.14) then is

$$T = T_0 \left(\frac{1 - r_\odot/r}{1 - r_\odot/r_0}\right)^{2/7} \qquad \text{for } r \leq r_0 \ , \tag{9.15}$$

and

$$T = T_0 \left(\frac{r}{r_0}\right)^{-2/7} \qquad \text{for } r \geq r_0 \ , \tag{9.16}$$

and is depicted in Fig. 9.15 for the case $r_0/r_\odot = 1.2$. This heat-conduction temperature profile correctly models both the steep gradient in the transition region and the rather flat decline towards the interplanetary space.

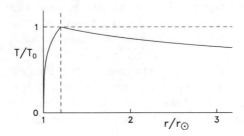

Fig. 9.15. Temperature as a function of r in a heat-conduction corona

We may now ask (Chapman 1957) whether a hydrostatic equilibrium in the corona is possible with $T(r)$ extending to infinity according to (9.16). Of course, the pressure balance is given by $dP/dr = -\rho g$, but, because we have an extended atmosphere, the decrease of $g = Gm_\odot/r^2$ must be taken into account. We eliminate $\rho = P\mu/\mathcal{R}T$ and obtain

$$\frac{1}{P}\frac{dP}{dr} = -\frac{Gm_\odot\mu}{r^2\mathcal{R}T} \ . \tag{9.17}$$

The solution to this equation,

$$P = P_0 \exp\left(-\frac{Gm_\odot\mu}{\mathcal{R}} \int\limits_{r_0}^{r} \frac{dr'}{r'^2 T(r')}\right) \ , \tag{9.18}$$

must match the interstellar pressure for $r \to \infty$. Since the interstellar pressure is many orders of magnitude smaller than P_0, this requirement means that the integral in (9.18) must diverge. Clearly, this is the case if $T(r)$ declines more rapidly than $1/r$. The heat conduction profile (9.16) does not qualify under this condition. In other words, the heat-conduction corona cannot be embedded into interstellar space under pressure equilibrium.

Problem 9.8. Calculate $P(\infty)$ for $T(r)$ given by (9.16). Take $T_0 = 10^6$ K, $P_0 = 10^{-2}$ Pa, and compare the result to the interstellar pressure at $T = 50$ K and $\rho = 10^{-20}$ kg/m^3 (cf. Problem 2.2).

9.2.2 Expansion

The difficulty encountered by the static corona model was removed when Parker (1958) replaced the hydrostatic equilibrium by a stationary *dynamic* equilibrium. In order to demonstrate the essential point let us again consider the case of spherical symmetry, with a velocity field of the form

$$\boldsymbol{v} = (v(r), 0, 0) \tag{9.19}$$

in spherical polar coordinates. Also, as the profile (9.16) is in any case too flat for hydrostatic equilibrium, we may as well seek a dynamic equilibrium for the simpler case $T = \text{const}$.

Instead of (9.17) we must now solve the equations of continuity and momentum balance, and the equation of state, viz.

$$\frac{d}{dr}(\rho r^2 v) = 0 \ , \tag{9.20}$$

$$v\frac{dv}{dr} = -\frac{1}{\rho}\frac{dP}{dr} - \frac{Gm_\odot}{r^2} \ , \tag{9.21}$$

$$P = \rho \mathcal{R} T/\mu \ . \tag{9.22}$$

Elimination of ρ and P yields the differential equation for v,

$$\frac{1}{v}\frac{dv}{dr}(v^2 - c^2) = \frac{2c^2}{r} - \frac{Gm_\odot}{r^2} \ , \tag{9.23}$$

where $c = (\mathcal{R}T/\mu)^{1/2}$ is the isothermal speed of sound. Using a coronal temperature we readily see that the right-hand side of (9.23) is negative close to the Sun, and positive at large distance from the Sun. At the distance

$$r_\mathrm{c} = \frac{Gm_\odot}{2c^2} \ , \tag{9.24}$$

the *critical radius*, there is a change of sign. Of course, the left-hand side of (9.23) must also reverse its sign at r_c; this leads to the classification of four distinct types of solution $v(r)$, as illustrated in Fig. 9.16.

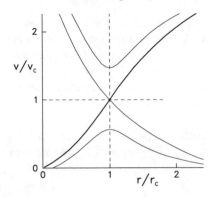

Fig. 9.16. The four solutions to (9.23). The *heavy curve* is the solar wind solution. Schematic drawing after Parker (1963 b)

Two of the solutions start in the corona with supersonic velocity. This is in contrast to observation, so these two types must be rejected. Of the other two, one always stays subsonic, and at large r even comes to rest. This also is not observed; in addition, it can be shown that this type of solution cannot match the interstellar pressure, just as the hydrostatic solution. What remains is the singular solution that turns from subsonic to supersonic as it passes through the critical radius. We must adopt this solution and, following Parker, we shall call it the *solar wind*. It is the only solution that satisfies all boundary conditions, including the requirement of matching the interstellar pressure (Problem 9.9). The adjustment to the interstellar pressure is accomplished at a distance of order 100 astronomical units from the Sun in a *termination shock*, where the velocity turns subsonic again. Still further away, at the *heliopause*, the wind merges into the interstellar medium.

The true solar wind is more complicated than the simple model treated here. In fact such a model generally fails to predict the exact values of the speed and mass flux of the wind. One must consider the effects of rotation and the magnetic field, and the variation with the angular coordinates and with time. And, instead of the assumption of isothermal expansion, one must solve a real energy equation, including heating by waves and energy loss by heat conduction back to the chromosphere. However, before we touch some of these complications let us return to the more general question of the determination and the effects of the magnetic field in the Sun's outer atmosphere.

Problem 9.9. For $T = 10^6$ K calculate the critical radius and the speed of sound. Integrate (9.23) and discuss the behavior of $v(r)$ and $P(r)$ for the various solutions at large distance from the Sun. Estimate $v(r)$ at 1 astronomical unit for the solar wind solution.

9.3 The Magnetic Field in the Outer Atmosphere

9.3.1 Magnetic Field Measurements

In principle the magnetic field could be measured by means of the Zeeman effect. However, the difficulty which already exists in the photosphere, namely that the Zeeman splitting is small in comparison to the line width, becomes almost unsurmountable in the outer atmosphere where the temperature is higher (and the lines broader) and the field is weaker (i.e., the splitting smaller). But measurements have been made in the relatively cool *prominences*. The circular polarization in *quiescent prominences* indicates a longitudinal field component of order 1–10 mT, while *active prominences* generally have a field strength above 10 mT. In the corona itself, Lin et al. (2000) were able to measure the circular polarization of emission lines. From the Stokes parameter $V(\lambda)$ of the Fe XIII near-infrared line at 1074.7 nm they could infer a field strength of 1–3 mT at a height of $\approx 10^8$ m above active regions.

A second possibility, also mainly applied to prominences, is the *Hanle effect*. The Hanle effect provides information on the transverse component of the magnetic field and thus complements the longitudinal Zeeman measurement. The field strength obtained in prominences is again of order 1 to 10 mT. It is important to notice that the Hanle effect yields magnitude and direction, but not the sign of the transverse field component. This ambiguity, which the Hanle effect has in common with the transverse Zeeman effect [cf. the dependence on γ and ϕ of the absorption matrix (3.69)], has interesting consequences for the prominence models which will be discussed below.

Some information about the coronal magnetic field is obtained from *radio* observations. Since the electrons emitting gyromagnetic radiation spiral around the lines of force, observation of moving radio sources directly yields the geometry of the guiding field. The strength and direction of the field can also be inferred by measuring the intensity and polarization of the radiation. Unfortunately there are severe limits on spatial resolution at radio frequencies. Above active regions Dulk and McLean (1978) find values of order 10 mT at the base of the corona; at $r/r_\odot = 2$ a typical field strength is 0.1 mT.

The most accurate magnetic field measurements have been made *in situ*. The Helios results have already been included in Table 9.4 above. The numbers given there apply to high-speed streams of the solar wind in or near the ecliptic plane; outside these streams the inclination is somewhat larger, so that 45° is an average value at 1 astronomical unit. Figure 9.17 illustrates the field measurements made earlier by IMP-1 ("interplanetary monitoring platform"), and their extrapolation back to the solar surface. The spiral pattern results from the fact that the field is frozen into the solar wind (Sect. 8.1.3), in combination with the effect of solar rotation. Since the wind speed does not change much over most of the distance Sun–Earth (cf. Table 9.4, or Problem 9.9), the spiral is nearly Archimedian.

The sector boundaries shown in Fig. 9.17 are the intersections of a warped *heliospheric current sheet* with the ecliptic plane. This current sheet separates the two magnetic polarities that dominate the two polar coronal holes. During phases of minimum activity such as the years 1963/64 these polar holes extend to rather low latitude, cf. Fig. 9.8. The warped current sheet, first proposed by Svalgaard and Wilcox (1976), has been confirmed by the out-of-ecliptic mission Ulysses (Fig. 9.13).

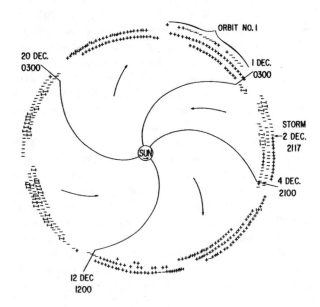

Fig. 9.17. The interplanetary magnetic field. Signs of three-hour averages of the radial field component as observed by IMP-1 in 1963/1964, and extrapolation of the sector boundaries to the solar surface. From Wilcox and Ness (1965)

9.3.2 Potential Field Extrapolation

The main outcome of the preceding section is the hardship, and often impossibility, of obtaining the magnetic field in the outer solar atmosphere. On the other hand, *photospheric* magnetograms are being made on a routine basis and, as far as the line-of-sight component of B is concerned, with good reliability. This situation has led solar astronomers to extrapolate the photospheric measurements.

The magnetic field measured in the photosphere can be ascribed to a sub-photospheric electric current. In addition to this current, there is, in general, an electric current system in the atmosphere itself which also constitutes a source for the field. But, since we know very little about the atmospheric current, we shall ignore it in a first approximation. In this case the atmospheric field has a potential,

$$B = -\nabla\Phi \,, \tag{9.25}$$

and can be calculated by the following method, first proposed by Schmidt (1964):

In complete analogy to the calculation of an electrostatic potential, the normal component B_z, which is given on the surface $z = 0$, is related to a fictitious surface charge density σ_m of magnetic "monopoles". The surface charge generates the field in the upper half-space, $z > 0$, but also, by the mirror principle, an identical field in the lower half-space, $z < 0$. The latter is of no physical interest. Thus, we have

$$\mu\sigma_m \equiv \operatorname{div} \boldsymbol{B} = 2B_z\delta(z) , \tag{9.26}$$

where $\delta(z)$ is the Dirac delta function, and, by (9.25),

$$\Delta\Phi = -2B_z\delta(z) . \tag{9.27}$$

The solution to this equation is a Poisson integral for which the z-integration can be carried out immediately; that is,

$$\Phi(\boldsymbol{r}) = \frac{1}{2\pi} \int \frac{B_z(\boldsymbol{r}')}{|\boldsymbol{r} - \boldsymbol{r}'|} df' \tag{9.28}$$

is an integral over the surface $z = 0$. We see that the normal component of $\boldsymbol{B}$, measured at the center of the solar disc via the longitudinal Zeeman effect, is just what is needed to calculate the potential field.

The potential (9.28) is appropriate if a magnetogram covering a small region of the solar surface is to be evaluated, so that the curvature can be neglected. For global field calculations the spherical geometry must be taken into account. In this case $\boldsymbol{B}$ is again given by (9.25), but the potential is now most conveniently expressed in terms of spherical harmonics. Such an expansion has been common in geomagnetism since the time of C. F. Gauss; in the solar context it was first used by Newkirk et al. (1968), and Altschuler and Newkirk (1969). Instead of (9.26) we now use div $\boldsymbol{B} = 0$ and, therefore,

$$\Delta\Phi = 0 \tag{9.29}$$

(for $r > r_\odot$), which is solved by

$$\Phi(r, \theta, \phi) = r_\odot \sum_{l=1}^{N} \sum_{m=0}^{l} f_l(r) P_l^m(\theta)(g_l^m \cos m\phi + h_l^m \sin m\phi) , \tag{9.30}$$

and

$$f_l(r) = \frac{(r_w/r)^{l+1} - (r/r_w)^l}{(r_w/r_\odot)^{l+1} - (r_\odot/r_w)^l} . \tag{9.31}$$

The radial dependence is such that $f_l(r_w) = 0$. That is, at $r = r_w$ we have a magnetic field pointing in the *radial* direction. The justification of this boundary condition is that, at some distance from the Sun, the solar wind becomes strong enough to force the frozen-in field lines into its own direction

Fig. 9.18. Photograph of the eclipse of 30 June 1973, with overlay of a potential magnetic field. From Altschuler et al. (1977)

of flow. Inspection of coronal images such as Fig. 9.8 suggests that this already happens in the outer corona. A typical choice is $r_\mathrm{w} = 2.6\,r_\odot$ (Altschuler et al. 1977); the subscript w stands for "wind". The surface $r = r_\mathrm{w}$ is also called the *source surface* because it is equivalent to an electric current in $r \geq r_\mathrm{w}$, which adds to the sources of the potential field in $r < r_\mathrm{w}$.

The functions f_l are normalized to $f_l(r_\odot) = 1$. Therefore, the amplitudes of the diverse multipoles are entirely given by the expansion coefficients g_l^m and h_l^m. These coefficients are determined by a least-square fit to the magnetic field observed in the photosphere, at $r = r_\odot$. Multipole coefficients up to $N = 90$ have been calculated in this manner. The larger the truncation index N, the more details of the magnetic structure can be represented. An example with $N = 25$ is shown in Fig. 9.18. The agreement between the magnetic field and the coronal structure is fairly good. In particular the distinction between the closed and open field regions, made in Sect. 9.1.3 above, now becomes justified. Even better agreement between field and coronal structure can be obtained by the introduction of a non-spherical source surface (Levine et al. 1982).

Problem 9.10. Confirm that (9.30), with (9.31), is a solution to the Laplace equation, (9.29). Calculate the energy distribution over the multipoles.

Problem 9.11. Show that, for given boundary conditions, the potential field represents the field of minimum magnetic energy.

9.3.3 The Force-Free Field

Let us now drop the assumption of a potential field. Thus we allow an electric current, of density j, to flow; with the current goes, at least in principle, a magnetic volume force $j \times B$. Now in the outer solar atmosphere the pressure P is everywhere small in comparison to the magnetic pressure, $B^2/2\mu$ (the plasma parameter $\beta = 2\mu P/B^2$ is small). Moreover, the dynamic pressure exerted by the solar wind is small compared to $B^2/2\mu$ unless we go beyond the Alfvén radius already introduced in Sect. 7.3.2. Hence there are no forces available to compensate the magnetic force. In an equilibrium the magnetic force must therefore essentially balance itself, i.e.,

$$j \times B = 0 . \tag{9.32}$$

Condition (9.32) characterizes the *force-free magnetic field* (Lüst and Schlüter 1954). Alternatively we may write

$$\text{curl}\, B = \alpha B , \tag{9.33}$$

where α in general is a function of the spatial coordinates. However, application of the divergence operator immediately yields

$$B \cdot \nabla \alpha = 0 , \tag{9.34}$$

i.e., α is constant along the field lines.

In the solar atmosphere a force-free magnetic field, i.e., a solution to (9.33), can be calculated if, in addition to the normal component, B_z, the value of α is known for each "foot point" of a field line (Sakurai 1981). Because α does not change along the field line, an independent value can be prescribed only at *one* foot point of any closed line.

The vertical component of (9.33) yields

$$\alpha = (\partial B_y/\partial x - \partial B_x/\partial y)/B_z . \tag{9.35}$$

Thus, the desired value of α could be obtained by observing B_z and the horizontal components, B_x and B_y, as functions of position on the solar surface, and by subsequent differentiation. The difficulty of this approach lies not only in the problem of measuring transverse field components (Sect. 3.5.4), but also in the high spatial resolution and accuracy required: without such accuracy, the process of differentiation may just produce noise. The attempts to calculate a force-free field therefore are mostly based on assumed rather than measured α distributions. Alternatively, one may adjust α in such a way that the coincidence of the visible structure with the magnetic field geometry in the corona is optimized.

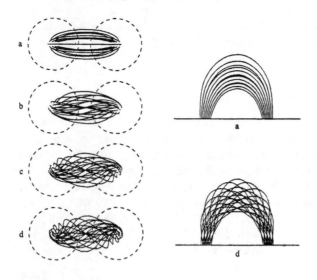

Fig. 9.19. Transition from a potential field (**a**) to a force-free field of increasing twist (**b–d**), as calculated by T. Sakurai, Tokyo Astronomical Observatory (Sakurai 1979)

The principal effect of the current in a force-free field is to introduce twist, or *current helicity*, into the field structure. The curl of a helical field has a component in the direction of the field itself. The helical structure of the force-free field is apparent in example calculations as the one by Sakurai (1979), shown in Fig. 9.19.

The origin of the helical structure must be sought in a rotation of the field-line foot points. It would be nice to observe such rotation directly in connection with the magnetic field evolution. The existence of vortex-like

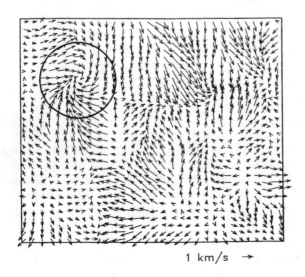

1 km/s →

Fig. 9.20. Horizontal velocity field in an area of $14.2'' \times 12.2''$ of the solar photosphere, derived from the proper motion of granules. The *circle* (diameter ≈ 3000 km) marks a prominent local vortex. From Brandt et al. (1988)

flows in the solar atmosphere has been demonstrated by the work of Brandt et al. (1988), cf. Fig. 9.20.

An interesting aspect of the force-free magnetic field is that its energy exceeds the minimum energy of the potential field belonging to the same photospheric boundary conditions (Problem 9.11). Increasing twist therefore means increasing storage of field energy. Seehafer (1994) has proposed that this energy build-up can be described in the framework of mean-field theory, with the force-free field as the mean field whose twist is increased by fluctuations originating in the photosphere. If, at some stage, the twisted field becomes unstable, the stored excess energy might be released in an explosive event. We shall return to this in Sect. 9.5.

9.3.4 Prominences

We now come to a special case where the Lorentz force $j \times B$ must *not* be balanced by itself, but rather is needed to maintain a static equilibrium. We mean the magneto-static equilibrium in a *quiescent prominence*.

According to their state of ionization, and hence their temperature, prominences belong to the chromosphere. At the limb, they offer a most beautiful view when observed in chromospheric lines (in emission), notably the lines of neutral hydrogen (Fig. 9.7). On the disc they are seen in filtergrams or spectrohcliograms taken in the same lines; here they appear as dark, thin, and rather long *filaments* (Fig. 6.22). Typical values for thickness, height, and length are 5000 km, 50 000 km, and 200 000 km, respectively.

Since the prominences protrude far above the average solar chromosphere, they are surrounded by coronal material and may thus be considered as part of the corona. Because the temperature in the prominence is $\approx 10^4$ K, about one hundred times smaller than in the surrounding corona, the lateral pressure equilibrium demands that the density is about one hundred times the coronal density.

Often prominences remain rather invariable for weeks or even months. In the following, we shall therefore consider a static equilibrium. With $T \approx 10^4$ K a pure hydrostatic equilibrium must be excluded: the scale height would be only ≈ 300 km, which is in conflict with the large vertical extent of the prominence. The problem is exactly the same which the solar astronomer had when he thought the corona was cool. In the corona, the solution came with the recognition of the high temperature. In the prominence the solution must lie in the support by an electromagnetic force.

Photospheric magnetograms show that prominences predominantly follow the lines $B_r = 0$. In fact this coincidence is so well documented that it is possible to outline the polarity of the large-scale solar magnetic field simply on the basis of Hα observations (McIntosh 1979). The assumption that the field lines connect the two polarities somewhere in the chromosphere/corona then yields the model of Kippenhahn and Schlüter (1957), Fig. 9.21 a. Due to the weight of the prominence material, the crest of the arcade-like magnetic

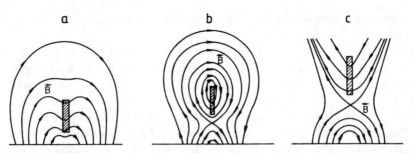

Fig. 9.21. Models for the magnetic support of a prominence (*hatched rectangle*). Adapted from Anzer (1987)

field becomes slightly depressed. The ensuing magnetic tension provides the balancing force.

More formally, let us consider an idealized prominence as a thin sheet of infinite length. In the magneto-hydrostatic equilibrium,

$$-\frac{dP}{dz} - \rho g + (\boldsymbol{j} \times \boldsymbol{B})_z = 0 , \tag{9.36}$$

we may discard the pressure gradient because in any case it is by far insufficient, as already explained. The remaining terms are integrated across the sheet (in the x direction):

$$g \int \rho \, dx = \int (\boldsymbol{j} \times \boldsymbol{B})_z \, dx \simeq \frac{1}{\mu} \int B_x \frac{\partial B_z}{\partial x} \, dx . \tag{9.37}$$

In the last expression on the right we have replaced $\boldsymbol{j}$ according to Maxwell's equation and, because the sheet is *thin*, we have only retained the derivative with respect to x. Now the normal component, B_x, is continuous across the sheet, and may be taken out of the integral. Therefore

$$g \int \rho \, dx \simeq \frac{1}{\mu} B_x [[B_z]] , \tag{9.38}$$

where $[[B_z]]$ is the jump, or discontinuity, of the vertical field component across the sheet.

For a prominence of 5000 km thickness take $\rho = 10^{-10} \, \text{kg/m}^3$ and $B_x = 10^{-3} \, \text{T}$. Equation (9.38) then yields $[[Bz]] \approx 2 \times 10^{-4} \, \text{T}$, which is a quite moderate change of the vertical field in the prominence. Too moderate, in fact, to be checked by observation.

But we may ask whether the horizontal field, B_x, really has the direction indicated in Fig. 9.21 a. Because of the above-mentioned ambiguity of transverse field measurements this question is difficult. Most limb prominences are seen edge-on, or almost edge-on, i.e., with their long (y) axis more or less parallel to the line of sight. Figure 9.22 illustrates the two possibilities of the field vector. The two angles, α_V, and α_F, are known from field measurements;

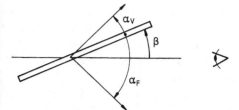

Fig. 9.22. The two possible magnetic field vectors in a prominence. After Leroy et al. (1984)

we just do not know which of the two is the true angle of the field with the prominence, and which is false.

Leroy et al. (1984) used a statistical argument to find a preferred field angle. Suppose there exists such a preferred value of α, perhaps due to an alignment effect of the solar differential rotation. Let the two possible angles be observed for a large sample of prominences, and be plotted as functions of β, the angle with the line of sight; β itself is determined by observing the prominence as a filament on the disc, with the assumption that it remains constant during the passage to the limb. Then, since

$$\alpha_F = -\alpha_V - 2\beta \tag{9.39}$$

(Fig. 9.22), the true angles, α_V, will accumulate around the (constant) preferred angle, while the false ones, α_F, will accumulate along the inclined straight line given by (9.39). For a sample of 120 relatively high prominences the result is shown in Fig. 9.23: the preferred angle is found to be $\approx -20°$. That is, the field has a large component B_y parallel to the prominence, and

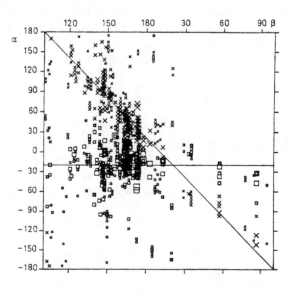

Fig. 9.23. Magnetic-field angles α as functions of β, the angle of the prominence with the line of sight. For each prominence there are two α values (*crosses* and *squares*); the size indicates the accuracy of the measurement. The data appear to cluster around two lines, as indicated. From Leroy et al. (1984)

the orientation of B_x is such that the field preferentially points from negative to positive photospheric polarity, i.e., *opposite* to the orientation of Fig. 9.21 a.

Alternative configurations of prominence fields are illustrated in Figs. 9.21 b and 9.21 c. Thanks to a magnetic neutral point, these two models (Kuperus and Raadu 1974; Malherbe and Priest 1983) predict the transverse field direction found by Leroy et al. (1984). In the open form, Fig. 9.21 c, they also account for the frequent coincidence of a prominence with an above-lying coronal streamer if the latter is interpreted by a current sheet (tangential magnetic field discontinuity). Still, the magnetostatic equilibrium in the prominence itself is the same as in the original Kippenhahn–Schlüter model, and is given by (9.38).

In contrast to the sample presented in Fig. 9.23, prominences of smaller height ($< 30\,000$ km) tend to have B_x components that conform to Fig. 9.21 a. Since the total number of low prominences is much larger than the number of high prominences, it may well be that the Kippenhahn–Schlüter configuration correctly describes the majority of cases (Leroy 1988).

Problem 9.12. Estimate the total electric current in a prominence of 50 000 km height. Which is the current direction in the models of Fig. 9.21?

We only briefly mention a number of other points that are important in prominence research. A real prominence is not a sheet, but consists of numerous fine threads. Often it is not static, but shows material motions, typically as "rain" along the threads. Real prominences form and disappear. Very probably, the formation proceeds by condensation (cooling) of coronal material. The correlation of prominences with coronal "voids", dark ray-like features (MacQueen et al. 1983), supports this model. During the eclipse of 7 March 1970 the condensation of a prominence at the base of such a void was directly cinematographed (Wagner et al. 1983).

The question of prominence *stability* will be resumed in Sect. 9.5.

9.3.5 Magnetic Braking of Solar Rotation

As long as the kinetic energy of the solar wind is small in comparison to the energy of the magnetic field, the latter determines how the wind may blow. The field keeps some parts of the corona closed altogether. In other parts, the open regions, it channels the wind along the magnetic lines of force.

Locally the channeling of the solar wind may be described by replacing the spherically symmetric equation of continuity (9.20) by a relation of form $F\rho v = $ const., where F is the cross section of a magnetic flux tube, and v is the velocity component along that tube. Because the flux, BF, is also a constant, one may determine F from B. Hence, it is possible to solve the wind equations in the case where the field is known. In a consistent treatment, however, the field B must be calculated together with the wind velocity v in a simultaneous solution of the equations of motion along $\boldsymbol{B}$ and perpendicular to $\boldsymbol{B}$. Below we shall consider only a rather special case of this general problem, which nevertheless will allow us to assess the transport of angular momentum by means of the solar wind.

The following numbers make clear that a remarkable transport of angular momentum indeed takes place. On the Sun's surface, the equatorial azimuthal velocity is $v_\phi \approx 2\,\mathrm{km/s}$. If a parcel of gas traveling with the solar wind were to conserve its angular momentum, the azimuthal velocity at 1 astronomical unit, A, would be $r_\odot/A$ times the solar surface value, or $\approx 10\,\mathrm{m/s}$. In contrast to this, *in-situ* measurements yield values between $1\,\mathrm{km/s}$ and $10\,\mathrm{km/s}$ (e.g., Hundhausen 1972; Pizzo et al. 1983). Taking the smaller of these values, and a mass flux density $\rho\,v_r \approx 3 \times 10^{-15}\,\mathrm{kg/m^2\,s}$, we find an angular momentum flux density of $\rho A\,v_\phi v_r \approx 0.5\,\mathrm{kg/s^2}$. Instead of considering the distribution of the solar wind over latitude, we simply take $2\pi A^2$ as the total permeated area. The solar loss of angular momentum is then

$$\dot{J} \simeq 2\pi A^2 \rho A\,v_\phi v_r \approx 7 \times 10^{22}\,\mathrm{kg\,m^2/s^2} \ . \tag{9.40}$$

From Problem 2.4 we know $J = 1.7 \times 10^{41}\,\mathrm{kg\,m^2/s}$. We divide this by $\dot{J}$ and obtain the (present) time scale of the Sun's rotational braking,

$$J/\dot{J} \approx 7 \times 10^{10}\,\text{years} \ . \tag{9.41}$$

Had we used the larger value of v_ϕ quoted above, the result would have been smaller by a factor of 10. In any case we may conclude that within the Sun's life solar-wind braking has been significant, in particular if we recognize that early in the Sun's history the braking was probably stronger than today (Sect. 7.3).

The simplest model that describes the solar braking is axisymmetric and stationary (Weber and Davis 1967). Moreover, the attention is restricted to the equatorial region where $\sin\theta \approx 1$.

The azimuthal balance of forces is

$$\rho(\boldsymbol{v} \cdot \nabla \boldsymbol{v})_\phi = (\boldsymbol{j} \times \boldsymbol{B})_\phi \ . \tag{9.42}$$

We eliminate $\boldsymbol{j}$ by (8.2), multiply by r^3, and obtain

$$r^2 \rho\,v_r \frac{d}{dr}(rv_\phi) = \frac{1}{\mu} r^2 B_r \frac{d}{dr}(rB_\phi) \ . \tag{9.43}$$

Now $r^2 \rho\,v_r$, essentially the mass flux, and $r^2 B_r$, the magnetic flux, are constants. Hence (9.43) may be integrated:

$$rv_\phi - \frac{rB_r B_\phi}{\rho\mu v_r} = L \ , \tag{9.44}$$

where L is a constant of integration.

Before we discuss the result (9.44) we must eliminate B_ϕ and determine L. In order to substitute B_ϕ we use the induction equation which, in the stationary and perfectly-conducting case, reduces to $\mathrm{curl}\,(\boldsymbol{v} \times \boldsymbol{B}) = 0$. We assume that $v_\theta = B_\theta = 0$ near the equatorial plane; the ϕ component of the induction equation

$$\frac{1}{r}\frac{\partial}{\partial r}\left[r(v_\phi B_r - v_r B_\phi)\right] = 0 \tag{9.45}$$

then means that $r(v_\phi B_r - v_r B_\phi)$ is a constant. We evaluate this constant at $r = r_\odot$ where $B_\phi = 0$ and $v_\phi = r_\odot\Omega$. Since $r^2 B_r$ also is a constant, we obtain $r(v_\phi B_r - v_r B_\phi) = r^2 B_r \Omega$ at *any* distance r, or

$$B_\phi = \frac{v_\phi - r\Omega}{v_r}\,B_r\ . \tag{9.46}$$

To eliminate L, we introduce the Alfvén velocity, $v_A(r) = B_r/(\rho\mu)^{1/2}$ and the Alfvén Mach number, $M_A(r) = v_r/v_A$. We thus transform (9.44) to

$$v_\phi = \frac{LM_A^2/r^2\Omega - 1}{M_A^2 - 1}\,r\Omega\ . \tag{9.47}$$

The actual values of B_r, ρ, and v_r found in the corona and at 1 astronomical unit (Sect. 9.1) tell us that M_A increases from ≈ 0.1 to ≈ 10; at some point, therefore, $M_A = 1$. This is the Alfvén point, or Alfvén radius r_A, where the energy densities of flow and field are equal. Since v_ϕ is *regular* at this point, we obtain from (9.47) a condition for L, namely

$$L = r_A^2\Omega\ . \tag{9.48}$$

Now the expression

$$\frac{M_A^2}{v_r r^2} \equiv \frac{v_r \rho\mu r^2}{B_r^2 r^4} \tag{9.49}$$

is, essentially, the mass flux divided by the square of the magnetic flux, and therefore a constant; at $r = r_A$ this constant takes the form $(v_A r_A^2)^{-1}$. Using this and (9.48), we finally find

$$v_\phi = \frac{v_r/v_A - 1}{(v_r r^2)/(v_A r_A^2) - 1}\,r\Omega\ . \tag{9.50}$$

Close to the Sun this means

$$v_\phi \simeq r\Omega\ , \tag{9.51}$$

i.e., rigid rotation with the Sun; on the other hand, at large distance from the Sun

$$v_\phi \simeq r_A^2\Omega(1 - v_A/v_r)/r\ , \tag{9.52}$$

which expresses the conservation of angular momentum (up to a magnetic correction) from the Alfvén point outwards. Thus, r_A is the effective lever arm used by the wind to brake the Sun's rotation. The present result confirms the more heuristic argument already presented in Sect. 7.3.2.

The precise value of r_A is not known. Extrapolation of Helios results yields $r_A \approx 12r_\odot$ (Pizzo et al. 1983). In this case (9.52) indeed gives an azimuthal velocity of $\approx 1\,\mathrm{km/s}$ at 1 astronomical unit.

Problem 9.13. Introduce the diverse $\sin\theta$ factors, which in the present section were omitted, and show that the solar losses of mass and angular momentum are related by

$$\dot{J} = \frac{2}{3}\,\Omega r_A^2 \dot{m}\,, \tag{9.53}$$

which is a more precise form of (7.14).

9.4 The Energy Balance

9.4.1 Needs

The outer solar atmosphere radiates and expands; this takes energy. Although the total need is only of order $3 \times 10^{22}\,\mathrm{W}$, about 10^{-4} of the solar luminosity, it is most instructive to consider the diverse processes of energy transport and energy loss in some detail.

By far the largest energy loss is *chromospheric radiation*. It occurs mainly in lines of Ca II and Mg II, in Lyman α, and the H^- continuum. A list of the main chromospheric emitters, according to the empirical models of Vernazza et al. (1981), is given in Table 9.5. The listed losses are integrals over height of the emission rates per volume. The latter are shown, as functions of height, in Fig. 9.24. The lower, broad peak is essentially due to Ca II and Mg II; the upper, narrow peak is Lyman α radiation and occurs where the temperature is high enough for the $n = 2$ level to be populated but not high enough for complete ionization for hydrogen. These results are typical for the average quiet Sun, model C of Vernazza et al. (1981). In the chromospheric network (their model F), the emission rates are higher by factors 2 to 3, and the height distribution is slightly shifted downwards.

Table 9.5. Chromospheric emission, in $\mathrm{W/m^2}$. After Avrett (1981 b)

Ca II			Mg II		
	H	490		h	430
	K	640		k	520
	866.2 nm	460	H	Lyman α	340
	849.8 nm	550	H^-	bound-free	170
	854.2 nm	680		free-free	220

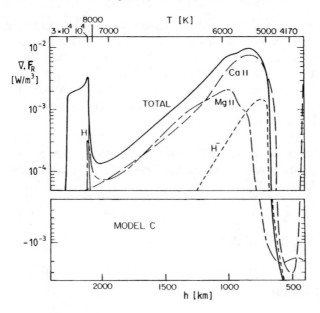

Fig. 9.24. Chromospheric energy loss. After Avrett (1981 b)

The *coronal* energy balance is more complex. In a steady state, the energy flux has zero divergence:

$$\text{div}\left[\boldsymbol{v}\left(\frac{1}{2}\rho v^2 + H - \frac{Gm_\odot\rho}{r}\right) - \kappa\nabla T + \boldsymbol{F}_\text{R} + \boldsymbol{F}_\text{H}\right] = 0 \,. \tag{9.54}$$

The contributions to this balance are, from left to right: convection of energy by the solar wind velocity $\boldsymbol{v}$, thermal conduction, radiation, and heating. The convected energy consists of kinetic energy, enthalpy, and gravitational energy. Thermal conduction is, to a smaller part, outwards from the temperature maximum, and, to the larger part, inwards toward the transition layer and upper chromosphere where it helps to cover the radiation loss in Lyman α. Thus, the chromosphere acts as a heat sink for the corona, although at the same time it provides the mass reservoir for the coronal expansion.

Table 9.6. Coronal energy losses, in W/m^2. After Withbroe and Noyes (1977)

	Coronal hole	Closed corona	
		quiet Sun	active region
Thermal conduction (both up and down)	60	200	$10^2 - 10^4$
Radiation	10	100	5000
Solar wind	700	≤ 50	≤ 100

All contributions to the coronal energy balance strongly depend on the magnetic field configuration, cf. Table 9.6. The largest losses occur in the closed corona above active regions. Here the density ρ is higher than in the open regions; because of its dependence on ρ^2, the radiative loss is also higher. Thermal conduction strongly varies even from one active region to another, because it depends on the temperature *gradient* along the various loops. Convection of energy by the solar wind is significant only in the holes.

Taken together, and in view of the fact that the regions with large losses occupy only a small fraction of the total (projected) area, the entire coronal loss is $\approx 3 \times 10^{21}$ W, about one-tenth of the whole outer solar atmosphere.

Table 9.7. Energy flux densities, in mW/m^2, at 1 astronomical unit. From Hundhausen (1972)

Convection of kinetic energy	0.22
Convection of enthalpy (electrons)	0.011
Convection of enthalpy (protons)	0.005
Convection of gravitational energy	−0.004
Heat conduction	0.007

Further out in the solar wind radiation as well as heating become unimportant. The fluxes corresponding to the other terms in (9.54) are listed for 1 astronomical unit in Table 9.7. Kinetic energy dominates. The specific enthalpy,

$$H = \frac{5}{2} nk(T_p + T_e) , \tag{9.55}$$

has contributions from protons and electrons, and contains both the internal energy density and the pressure (because work has to be done to *expand* the coronal material). The density of gravitational energy is negative; hence the corresponding flux is *inwards*. The final entry in Table 9.7 is the thermal conduction (here only outwards) by electron collisions. Conduction by protons is negligibly small.

Problem 9.14. Show that, in spite of its inward direction, the flow of gravitational energy has a positive divergence, and so constitutes a *loss*. Relate this loss to the solar mass loss, and estimate its flux density per surface area on the Sun.

9.4.2 Heating

The term div F_H in (9.54) represents the heating of the outer solar atmosphere. Besides the total magnitude of this heating, its spatial distribution

is important, in particular in the corona. The knowledge of these two ingredients would permit us to predict not only the structure and dynamics of
the solar corona, but also of the coronae of other stars (Hammer 1982a, b).
Indeed, Hammer (1984) was able to scale coronal models as functions of the
stellar mass and radius.

However, it has been impossible to assess the heat sources of the solar
outer atmosphere in a quantitative manner, even at the modest precision
with which we know the losses. Merely to identify the mechanisms of heating
is difficult. A number of processes has been proposed, but their observational
prove or disprove remains controversial, mainly because of the small space
and time scales involved.

Acoustic Waves. Heating by means of acoustic waves was first proposed
by Biermann (1946). Waves with a frequency above the atmospheric cutoff
frequency ω_A can propagate across the temperature minimum towards the
chromosphere and corona. As long as the wave amplitude u is small, the
energy flux associated with propagating waves is

$$F_H = \frac{1}{2} \rho_0 u^2 c , \tag{9.56}$$

where ρ_0 is the equilibrium density, and c is the speed of sound.

Without dissipation the wave energy is conserved. Around the temperature minimum where c changes little, this means that the amplitude increases approximately as $\rho_0^{-1/2}$. As a consequence the compression phases
of the waves steepen into shock fronts. Within the shock, that is, within a
distance of a few mean-free-paths of the particles, wave energy is dissipated
into heat by viscous friction and thermal conduction. Numerical integrations
of the non-linear wave equations have demonstrated that the radiative loss of
the lower chromosphere outside the bright network may well be compensated
by such acoustic wave heating (e.g., Ulmschneider and Muchmore 1986).

The correlation of chromospheric losses with concentrations of the magnetic field suggests that the field should play a role in the heating. In a flux
tube, the analogue to the ordinary acoustic wave is the longitudinal tube wave
(Sect. 8.2.3). The development of such waves into shocks has been studied
by Herbold et al. (1985). They found that the main effect of the magnetic
field is not the additional restoring force, but rather the geometric shape
of the tube which channels the propagating wave [the plasma parameter β,
defined in (8.31), is small, so that one may consider the field as given]. The
narrower the channeling, the stronger the upwards increase of the amplitude
and, therefore, the deeper the level where shock formation and heating occurs
(Ulmschneider et al. 1987).

Strictly, heating by dissipation of acoustic shocks must be time-dependent.
Carlsson and Stein (1995, 1997) have calculated the propagation of acoustic
perturbations in the solar atmosphere in time-dependent models, including
radiative transfer. Especially they considered the variation of the profiles of

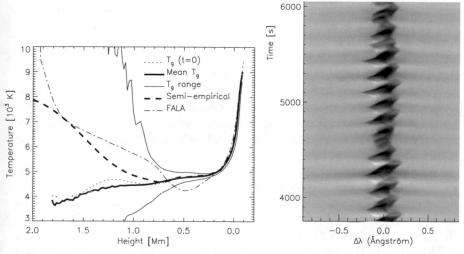

Fig. 9.25. *Left*: Temperature as a function of height: start model (*thin dashed*), time average (*thick solid*), lower and upper edges (*thin solid*) of the range attained in the calculation, the static model A (*dash-dotted*) of Fontenla et al. (1993), and a "semi-empirical" model (*thick dashed*) that fits the mean calculated emission. *Right*: Evolution of the calculated Ca II *H* line profile, with shocks occuring every 150 s on the average. From Carlsson and Stein (1995, 1997)

the Ca II lines that are characteristic for the chromosphere (Fig. 9.25, *right*). They find the remarkable result that short intervals of very high temperature are caused by the acoustic shocks. Because of the non-linear dependence of the Planck function on T this leads to a high emissivity, which can explain the bright *calcium grains* seen in spectrograms (Lites et al. 1993) and filtergrams (von Uexküll and Kneer 1995). They also find that the average chromospheric temperature continues to decrease outwards from the level where the semi-empirical models have a temperature minimum (Fig. 9.25, *left*). Nevertheless, from the calculated emission they deduce a "semi-empirical" mean temperature that resembles the models shown in Sect. 4.3. This latter result was however criticized by Kalkofen et al. (1999) who find that only an increasing mean temperature can account for all of the chromospheric emission, and that acoustic waves with periods shorter than 100 s, which were not considered by Carlsson and Stein, would yield the necessary heating.

Magnetohydrodynamic Waves. While acoustic waves seem to be adequate to heat the lower chromosphere, these waves probably cannot solve the problem of *coronal heating*. Their energy is dissipated before they reach the corona. Also, in the corona the spatial association of strong (closed) magnetic field configurations with large radiative and conductive losses is so obvious that we must expect the field to play a more central role. The heating could be due either to the dissipation of magnetohydrodynamic waves, or to direct dissipation of an electric current.

Magnetohydrodynamic waves, in particular the transverse Alfvén waves, can be generated when the foot points of flux tubes are shaken sufficiently fast by photospheric motion (Alfvén 1947). In addition, such waves could be excited (at higher frequencies than by field-line shaking) by local magnetic instabilities with impulsive field-line reconnection, like those called *nanoflares* by Parker, see below. Axford and McKenzie (1992) have proposed this mechanism and conjecture that the "explosive events", local and short-lived broadenings of transition-region emission lines (Brueckner and Bartoe 1983, Dere et al. 1989) result from many of those small-scale instabilities, although these would not be resolved individually.

Alfvén waves travel along the magnetic lines of force with the velocity $c_A = B/\sqrt{\mu\rho}$. Their restoring force is magnetic tension. These waves are transverse, with no variation of density or pressure, even in a compressible medium. The torsional Alfvén wave of Sect. 8.2.3 is an example. Alfvén waves may pass with little attenuation through the chromosphere into the corona (Osterbrock 1961). Several mechanisms of their dissipation in the corona have been proposed. One is *phase mixing* (Heyvaerts and Priest 1983), where a spectrum of randomly excited waves leads to large local gradients of the field strength and hence to enhanced ohmic dissipation. Another is coupling to longitudinal (that is, compressional) waves and subsequent shock formation (Hollweg 1982, Kudoh and Shibata 1999), and therefore again enhanced dissipation.

A third mechanism of Alfvén-wave dissipation is the *ion-cyclotron resonance*: If the frequency of an Alfvén wave is equal to the gyration frequency of a species with ion charge q and mass m,

$$\omega_G = qB/m \,, \tag{9.57}$$

a resonance occurs. At this resonance the ions are accelerated, and hence that species is preferentially heated. For any given wave frequency, and an outwards decreasing magnetic field strength B, the ions with smaller q/m will be heated first, while for those with larger q/m the wave-energy flux is reduced. The apparent decrease of the ion temperature with increasing q/m, as deduced from the width of coronal emission lines, supports this interpretation (Tu et al. 1998, Kohl et al. 1998). Marsch et al. (1982) had already discussed the ion-cyclotron resonance in the solar wind and suggested that the higher transverse temperature of the protons (Table 9.4) results from this resonance.

Electric Current Dissipation. Coronal heating by direct current dissipation may occur even for *slow* photospheric motion. This mechanism has been advanced by Parker (1972, 1983) and has been called *topological dissipation*. The name explains the principle: coronal field lines of opposite direction are brought into close contact because of the random motion of their photospheric foot points. Parker points out that the narrow sheets where the current density j is large are *unstable*. The magnetic lines of force reconnect, just as we have already seen in the case of magneto-convection, Sect. 8.2.1; the current in the sheet is dissipated. The rate of energy dissipation per volume is j^2/σ, where σ is the electrical conductivity. Since j is large, this may be significant in spite of the large value of σ, and in spite of the small volume fraction occupied by the current sheets. Presumably, the field line reconnection takes place in form of numerous unresolved events, called *nanoflares* by Parker (1988). Parker estimates the energy of a single nanoflare to 10^{17} J and less, which is only 10^{-3} (or less) of the energy released during the x-ray bursts (*microflares*) observed by Lin et al. (1984), but is perhaps manifest in the smallest individual emission spikes during those bursts.

Flow Acceleration by Waves. Acoustic and magnetohydrodynamic waves are not only able to heat the outer solar atmosphere. These waves can also directly exert a net force and so contribute to the acceleration of an outward flow. In a mean-field approach, such as already described in the context of the solar differential rotation and the solar dynamo, the fluctuating wave field u leads to a mean *wave pressure* $\rho_0 u^2/2$ whose gradient may help driving the solar wind, or the upflow seen in spicules. Belcher (1971) and Alazraki and Couturier (1971) suggested that Alfvén waves might accelerate the solar wind in this manner. The original model of Parker (1958), where the coronal expansion was based only on the thermal pressure gradient, is thus improved in that the acceleration occurs closer to the Sun, and yields a higher wind speed than was possible with increased heating alone (which essentially lead to enhanced heat conduction back to the Sun).

While wave pressure is an effect that equally occurs for damped and undamped waves, Haerendel (1992) has pointed out that damped Alfvén waves may exert a net Lorentz force. In a dense plasma the transverse field and current excursions of an Alfvén wave are 90° out of phase so that, averaged over a wave period, the Lorentz force vanishes. But when collisions become less frequent, and the gas is only partially ionized as in the solar chromosphere, the ions and neutral particles slightly slip relative to each other. This constitutes yet another damping mechanism for the Alfvén wave; moreover, it yields a non-vanishing average Lorentz force in the longitudinal direction. Haerendel has proposed that spicules might be driven by this force.

9.5 Explosive Events

The nano- and microflares mentioned in the preceeding section, and the explosive events defined by Dere et al. (1989), release energies of up to $\approx 10^{20}$ J, and the smallest of them may not be observable individually. We shall now treat more violent events. The largest have a total energy release of order 10^{26} J, and the associated energy flux may exceed the photospheric flux which, on the average, is $\approx 10^4$ times the flux originating from the outer layers of the Sun. In view of the common origin, namely an instability of a magnetic field configuration in the solar atmosphere, we shall now use the term *explosive event* with a more general meaning for small and large explosions.

9.5.1 Flares and Other Eruptions

Explosive events release energy is in various forms, e.g., accelerated particles, bulk mass motion, and radiation emitted in various spectral ranges from γ-rays to radio wavelengths. A variety of observational techniques has therefore been employed, and the phenomena seen have been given various names. Sometimes these phenomena are observed in isolation, but more often, in particular for the largest events, all or many of them are associated with each other.

Flares. The *chromospheric eruption or chromospheric flare* is easiest to observe, especially with the help of an Hα filter. Typically, the brightness increases for several minutes (sometimes called the *flash phase*), followed by a slow decay (the main or *decay phase*) which lasts 30 min to 1 hour. Often flares, in particular the larger ones, occur right after the sudden disappearance of a filament. In this case the flare generally has the form of two ribbons which lie on both sides of the location of the former filament. During the decay phase the two ribbons move apart at, typically, 10 km/s. The space between them is then usually filled with a transverse filamentary pattern seen in Hα (the *post-flare loops*, Fig. 9.26 c). This transverse pattern is also well-documented in pictures taken in transition-zone lines and coronal lines. It is quite common, however, that after some time (hours or days) a new filament is formed at the place of the previous one, and even that another (*homologous*) flare erupts.

Since the structure visible in Hα outlines the magnetic lines of force, and since filaments normally make a very small angle with the magnetic field, it appears that the field is rearranged during the flare: its horizontal component changes from nearly parallel to the original filament to nearly perpendicular. In addition to Hα, the pattern of soft x-ray emission of the corona observed with Yohkoh changes with the occurrence of flares (e.g., Tsuneta et al. 1992). A change of the magnetic field itself, derived from circular polarization measurements, has been reported first by Severny (1965). Evidence for the change of field components tangential to the solar surface comes either from vector magnetograms (e.g., Wang et al. 1994) or from measuring the longitudinal field during a flare near the solar limb (Cameron and Sammis 1999).

For other spectral ranges where flares can be seen Fig. 9.27 shows a typical scheme. Often there is a precursor, characterized by thermal radiation of up to 10^7 K. It is followed by an impulsive phase, which typically lasts 1 min and consists of bursts (1–10 s) in γ-rays, hard x-rays, EUV, and microwave radiation. The evolution of soft x-rays is more similar to the Hα flare, i.e., without an impulsive phase.

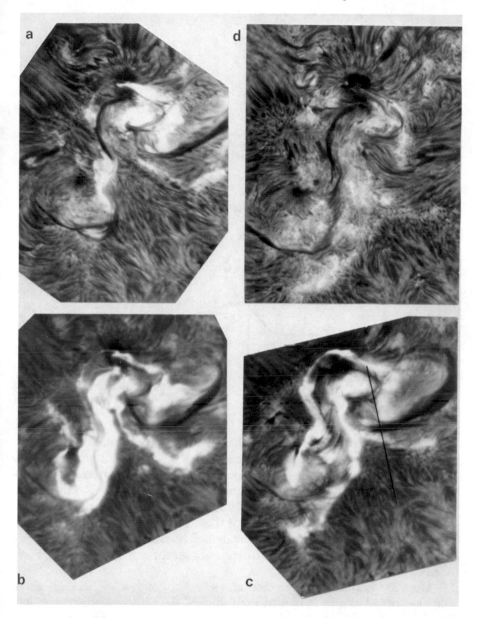

Fig. 9.26. Two-ribbon flare of 28 May 1972. (**a**): pre-flare state, 11.37 UT; (**b**): flare at maximum intensity, 13.27 UT; (**c**): expanding ribbons, 14.25 UT; (**d**): post-flare state, 29 May, 9.01 UT. Photographs in Hα, Domeless Coudé Telescope, Capri (Bruzek 1979)

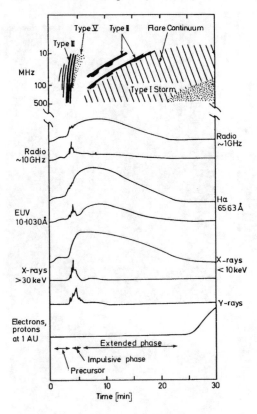

Fig. 9.27. Electromagnetic and particle radiation for a typical flare. From Dulk et al. (1985)

For events at the solar limb the height in the atmosphere can be determined by spatially resolved observations. Using this and the form of the x-ray spectra, one can explain the radiation as bremsstrahlung of electrons with energies 10–100 keV precipitating in beams from the corona down to the denser atmosphere (e.g., Haug and Elwert 1985). The microwave emission can be interpreted as synchrotron radiation of the same electron beams.

Large flares may be visible in white light. These *white-light flares* spatially coincide with the area emitting in Hα, and are synchronous with the impulsive phase.

After the impulsive phase thermal radiation dominates. Heating of the chromosphere leads to increased excitation of the atomic levels responsible for Hα and the other flaring lines. The heating may be so effective that part of the chromosphere is "evaporated", i.e., transformed into coronal material. This material could be the hot plasma, with $T \approx 3 \times 10^7$ K, responsible for the thermal radiation of the main flare phase. An alternative idea is that such a hot plasma originates directly during the primary energy release high in the atmosphere. In any case heat will be conducted downwards and so supply energy to the Hα flare.

Radio Bursts. At radio wavelengths from decimeters upwards, eruptive events have been recorded ever since the first discovery of solar radio signals around 1942 (Fig. 1.6). Many of these *radio bursts* are associated with flares, as illustrated in Fig. 9.27. The bursts have been classified by J. P. Wild and others as types I to V (Wild 1959). Types II and III are most characteristic because of their banded structure in a frequency-time diagram. The lowest (fundamental) of these bands is interpreted in terms of the local plasma frequency ν_P given by (9.8), while the higher bands are harmonics to ν_P. With this interpretation the motion of the radiating sources across the corona has been derived; when spatially resolved observations became possible, this result was confirmed. Thus, the fast frequency drift of type III bursts is translated into a velocity of 10^7 to 10^8 m/s. This is the velocity of the 10–100 keV electrons needed to explain the hard x-rays which, simultaneously with the type III radio burst, also appear during the impulsive flare phase. Type II bursts occur later and move more slowly. Their velocity, typically 10^6 m/s, is similar to the velocity measured for the coronal mass ejections described below. The common interpretation, therefore, is the emission of radio signals in a shock wave that is generated by a flare or an eruptive prominence and propagates outwards through the corona.

Energetic Particles and γ-rays. The evidence arising from the diverse radiation processes for electrons of 10–100 keV associated with solar flares is confirmed by *in-situ* measurements of these particles in the solar wind. In addition, large flares are accompanied by relativistic electrons and high-energy protons (*proton flares*); tens of MeV are common, and exceptional flares accelerate protons to energies exceeding 1 GeV.

The γ-rays, too, are characteristic for large events. Their spectrum shows very intense line radiation. One strong line, at 511 keV, originates from the annihilation of electron-positron pairs (positrons are emitted by radioactive nuclei); another, at 2.223 MeV and still stronger, results from the capture of a neutron by protons: $n + p \rightarrow d + \gamma$. The presence of radioactive nuclei and of neutrons indicates that nuclear reactions take place during flares. Further evidence for such reactions comes from the abnormal abundance of many nuclei, e.g., ^{3}He, measured in interplanetary space in the wake of flares.

Eruptive Filaments. The sudden disappearance, or *disparition brusque*, of a flare-related filament has already been mentioned. Detailed observations (e.g., Michalitsanos and Kupferman 1974) show that the disappearing filament erupts into the corona. The process usually begins with an activation, consisting of increased mass motion within the filament, expansion, or other changes. These changes indicate that the magnetic support of the filament may soon become unstable. The eruption itself typically consists of a rapidly expanding arch. Velocities of several hundred km/s may be reached. These events are most spectacular when observed at the limb as an eruptive prominence (Figs. 9.28 and 9.29).

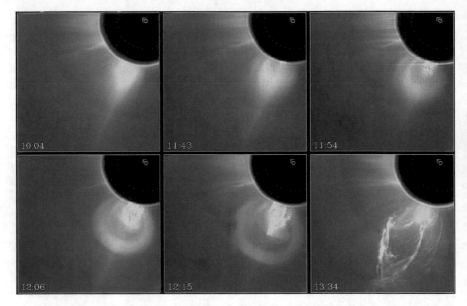

Fig. 9.28. Coronal mass ejection and prominence eruption of 18 August 1980. The originally stable helmet streamer expands with a nearly spherical front (*frames* 2–4). The eruptive prominence first appears (*frame* 2) at the edge of the occulting disc of the Solar Maximum Mission coronagraph, grows, and explodes in form of bright thin filaments (*frames* 3–6). Courtesy High Altitude Observatory, NCAR, Boulder, Colorado

Coronal Mass Ejections. *Coronal mass ejections* (CME, earlier called "coronal transients"), often bubble-shaped such as the example shown in Fig. 9.28, have been detected occasionally during a solar eclipse, but many more cases have been recorded by coronagraphs on space vehicles, beginning with the Skylab mission 1973/74. Since 1996 the SOHO instruments recorded a few per week to several per day, depending on the level of solar activity. In most cases an association with an eruptive prominence or (sometimes *and*) a flare could be established (Munro et al. 1979). The name coronal mass ejection does not necessarily imply that *coronal* mass is ejected; rather a mass ejection, e.g., prominence mass, is seen in the corona. Often the erupting prominence material shows bright filaments with a helical structure, e.g., Fig. 9.29, an event that has been studied in detail by Plunkett et al. (2000). The ejected material moves across the solar corona with a velocity of 10^5–10^6 m/s. An estimate at their mass and their frequency of occurrence suggests that CMEs contribute as much as 10% to the whole mass loss by the solar wind.

Coronal Waves and Moreton Waves. Large-scale transient coronal disturbances have been discovered by the SOHO Extreme-ultraviolet Imaging Telescope and, therefore, have been called *EIT waves*; their visibility is im-

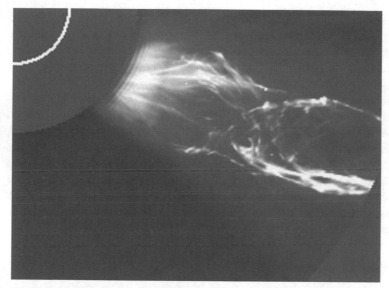

Fig. 9.29. Prominence eruption and coronal mass ejection of 2 June 1998. The front of the CME has passed the field of view, but a *faint cone* is still visible. The bright, helical *filaments* are the remainders of the prominence. White-light image, from the LASCO C2 coronagraph on SOHO. The *white circle* on the occulting disc marks the solar surface. SOHO is a cooperation between ESA and NASA

proved when the intensities observed in two exposures separated by a short interval of time are subtracted from each other (Fig. 9.30). Coronal waves often expand from the site of a CME or flare. In addition, ca. 90 % of such events are associated with type II radio bursts, i.e., with coronal shock waves, according to a catalogue of Klassen et al. (2000). The disturbance propagates with a velocity of, typically, 300 km/s, which exceeds the coronal sound speed c.

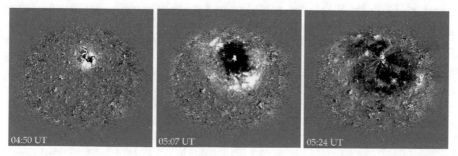

Fig. 9.30. Coronal wave, expanding from the site of a coronal mass ejection on 12 May 1997. The images show intensity differences of successive exposures (UT $4.50 - 4.35$, $5.07 - 4.50$, $5.24 - 5.07$), obtained by SOHO/EIT in the Fe XII band at 19.5 nm. The front speed is ≈ 300 km/s. SOHO is a cooperation between ESA and NASA

Thus, the phenomenon has been interpreted as a *fast magneto-acoustic* shock wave. The propagation is perpendicular to the direction of the magnetic field; the restoring force is provided by the gas pressure and the magnetic pressure together, and the propagation speed is, therefore, $\sqrt{c^2 + c_A^2}$, where c_A is the Alfvén velocity.

Thompson et al. (1999) have shown that there is a close relationship between the coronal waves observed with EIT and the *Moreton waves* discovered earlier. Moreton waves have been observed as disturbances in the wings of Hα, i.e., as a chromospheric phenomenon (Moreton and Ramsey 1960, Moreton 1965). These disturbances also originate at flare sites and travel horizontally with a speed of several hundred to over 1000 km/s. Since such a high propagation velocity cannot be explained in terms of characteristic velocities of the chromosphere, the Moreton wave must be interpreted as the chromospheric manifestation of a coronal disturbance (Meyer 1968, Uchida 1968).

9.5.2 Release of Magnetic Energy

The total energy emitted in an explosive event in the solar atmosphere may exceed 10^{25} J. As a detailed case study (Sturrock 1980) shows, most of this energy lies in the masses ejected in eruptive prominences and coronal mass ejections, and in the magnetic field carried with these masses. Only a few percent are radiated in the diverse spectral ranges.

The large majority of explosive events occurs in active regions, i.e., regions of enhanced and complex magnetic field. And, as we have seen, there is evidence that changes of the field configuration occur during flares. Finally, the energy contained in the magnetic field is of the right order: take a volume of 10^{23} m^3 and a field strength of 0.01 T, then the energy is $\approx 4 \times 10^{24}$ J, sufficient for a typical event. For these reasons one must conclude that the nature of the explosion is a sudden release of energy stored in the magnetic field. Since, for given boundary conditions, the potential field has the minimum energy (Problem 9.11), one would expect a field change towards a potential configuration. So far such is not well-documented, however. Actually cases have been studied (e.g., Wang et al. 1994) where the deviation from a potential field increases. It appears that new magnetic flux that emerges during a flare enhances the magnetic energy, as already noticed by Schmidt (1964), and so obscures the energy release during the instability proper.

The main questions, then, are the following. How is the energy stored in the magnetic field? By which instability is it released? How are the particles accelerated? And how are the masses of, e.g., coronal mass ejections set into motion? These primary questions (Sturrock 1980) must be answered by theoretical models. Synchrotron radiation in microwaves, x-ray bremsstrahlung, heating and the ensuing thermal radiation, etc., may then be considered as secondary processes whose basic physics is known (although many quantitative details remain open).

Energy storage is a *stability* problem. Sturrock distinguishes between "non-explosive" and "explosive" energy release. If *any* current-carrying magnetic field is unstable the release is non-explosive. This may heat the corona in general (Sect. 9.4.2), and also may contribute to the heating in the precursor phase of a flare. Tsuneta (1996 a) describes a case observed with Yohkoh where the anti-parallel coronal fields of two originally separated active regions in the northern and southern hemispheres reconnect across the solar equator, without much flare activity, but with significant heating. Explosive energy release, on the other hand, first requires that the field is stable so that a sufficient amount of energy can be accumulated. Or at least it should be *metastable*, which here means stable against perturbations of small amplitude.

The energy stored in the magnetic field may have a variety of sources. Shearing of field lines by photospheric motion of the foot points, shearing of field lines by the expanding (open) corona, and generation of current sheets by emergence of new magnetic flux loops in the vicinity of existing loops are specific possibilities; magnetic field configurations corresponding to these three flare models are shown in Figs. 9.19 and 9.31. As the case of homologous flares shows, the energy build-up should be achieved within a time span of the order of a day. This is no small requirement in itself: take the 4×10^{24} J from above and an area of 10^{15} m^2 to obtain $\approx 4 \times 10^4$ W/m^2 as the necessary flux; in view of the average coronal energy balance (Table 9.6) this is sizable.

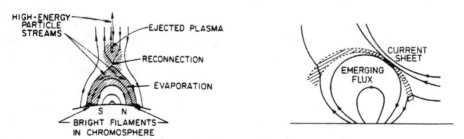

Fig. 9.31. Two magnetic field configurations which possibly lead to flares. *Left*: from Sturrock (1980); *right*: from Heyvaerts et al. (1977)

The magnetic instability itself must develop within seconds. This in the time scale of the single spikes which we see in the impulsive phase. Various magnetohydrodynamic instabilities have been proposed. A single twisted loop like the one shown in Fig. 9.19 could be *kink* unstable, i.e., develop a number of local, strongly twisted kinks that would allow rapid current dissipation. On the other hand, a current sheet could be unstable to *tearing*, i.e., the two adjacent regions of opposite field direction would tear magnetic flux across the interface and so form alternating sequences of O-type and X-type neutral points. Fig. 9.31 (*left*) shows a model that is characterized by field-line reconnection at an X-type neutral point. According to this model, which was first proposed by Kopp and Pneuman (1976), a lateral inflow towards the neutral point, and mass ejection from above the neutral point is expected. Indeed Tsuneta (1996 b) and and Yokoyama et al. (2001) report cases where the inflow can be inferred from the change of the soft x-ray and extreme UV emission, as observed by Yohkoh and SOHO/EIT. Similarly, an upward moving mass has been observed in soft x-rays in another case (Tsuneta 1997). The model is also supported by $H\alpha$ observations with a time resolution $\Delta t < 0.1$ s (Wang et al. 2000); these observations indicate that the initial energy release occurs at the top of a low-lying loop, followed by emission from loop foot points on both sides.

Quantitative models of explosive events have been two-dimensional in most cases. An unstable field configuration such as the current sheet of Fig. 9.31 (*left*) is assumed, and the instability is triggered by introducing a small region of *anomalous*, that is enhanced, electrical resistivity. Field-line reconnection then proceeds in the tearing mode (e.g., Schumacher and Kliem 1996, Magara and Shibata 1999, Roussev et al. 2001, Yokoyama and Shibata 2001). Three-dimensional models are more difficult, because of the increased computational expenditure and because of the complicated topology of magnetic reconnection (e.g., Priest and Schrijver 1999, Birn et al. 2000, Galsgaard et al. 2000, Schumacher et al. 2000, Brown and Priest 2001).

9.6 Bibliographical Notes

The textbooks of Bray and Loughhead (1974) and Athay (1976) introduce the chromosphere and corona, the books of Shklovskii (1965) and Billings (1966) are classical guides to the corona; Hundhausen (1972) is the standard text on the solar wind, while Lamers and Cassinelli (1999) treat the solar wind in the general context of stellar winds. Zirin (1988) devotes a major part of his book to the outer atmosphere of the Sun. Mariska (1992) introduces the transition region, the volumes edited by Schwenn and Marsch (1990, 1991) provide reviews of diverse topics related to the solar wind, especially with respect to the results of the two Helios probes; the lectures edited by Rozelot et al. (2000) introduce some plasma physics relevant to that subject. The monograph of Lang (2000) comprehensively covers the results of recent space missions.

Meyer (1985) and Feldman and Laming (2000) review the evidence of element abundance differentiation in dependence on the first ionization potential, Hénoux (1998) discusses the diverse models.

Chromosphere, corona, and the solar wind are the subjects for which the solar-stellar connection is most intense. The Sun, on the one hand, allows spatially resolved observations. The variety of basic parameters of the stars (surface gravity, temperature, etc.), on the other hand, and the corresponding variety of stellar outer atmospheres, may help in the understanding of those solar observations. This wide and complex field is covered by the book of

Schrijver and Zwaan (2000), and in many conference proceedings, e.g., Bonnet and Dupree (1981) or Schröter and Schüssler (1987), and the series *Cool Stars, Stellar Systems, and the Sun*, which appeared since 1980 in a two-year cycle.

Alternative techniques of calculating a potential field in the corona have been applied by Adams and Pneuman (1976), and by Elwert et al. (1982). Pneuman et al. (1978) compared the potential field to the observed coronal structure and generally found good correspondence. Hagyard and Pevtsov (1999) discuss the use of vector magnetograms for deriving the factor α that connects B and curl B in a force-free field.

Tandberg-Hanssen (1974, 1995) introduces solar prominences. Querfeld et al. (1985) and Bommier et al. (1986 a) report further attempts to measure the magnetic field vector in prominences, while Bommier et al. (1986 b) use the depolarizing effect of collisions to derive the electron density. The volumes edited by Jensen et al. (1979), by Ballester and Priest (1988), and by Priest (1989) are entirely devoted to the physics of solar prominences. Cheng and Choe (1998) construct models of current sheet and prominence formation through photospheric plasma motion. The energy balance of prominences, especially in view of their fibril structure, is discussed by Anzer and Heinzel (1999) and Heinzel and Anzer (2001).

Durney and Pneuman (1975) describe how a given magnetic field guides the solar wind. For the case of axial symmetry consistent solutions for wind *and* field have been calculated by Pneuman and Kopp (1971), Endler (1971), and Sakurai (1985). The latter of these includes the effect of solar rotation, and extends the problem posed by Weber and Davis (1967; cf. Sect. 9.3.5) to the full range of heliographic latitude. Durney and Stenflo (1972) employ the assumption that in the past the total (absolute) magnetic flux across the solar surface has varied in proportion to the Sun's angular velocity Ω; they thus calculate a rotational braking of the form $\Omega \propto t^{-1/2}$, which reproduces the $\Omega(t)$ derived from stellar clusters (Fig. 7.1).

Chromospheric and coronal heating has been reviewed by Narain and Ulmschneider (1990, 1996); Hammer (1987, 1988) discusses the physics of the whole system chromosphere, transition layer, and corona. An important part in the chromospheric energy balance could be played by molecules such as CO (Ayres 1981; Kneer 1983): molecules are strong emitters, i.e., coolants; at the same time, their formation is rather temperature-sensitive. This may lead to a two-state situation where the chromosphere proper exists only in narrow regions, perhaps confined in magnetic flux tubes, while the intermediate space is rather cool. The cooling effectivity of CO has been disputed, however, by Mauas et al. (1990).

Coronal heating by electric current dissipation is further elaborated by Browning et al. (1986) and by Browning and Priest (1986). Parker (1987) gives an illustrative account. Details of the ion-cyclotron resonance have been worked out by Cranmer et al. (1999 a) and Cranmer (2000).

Spicule models have been reviewed by Sterling (2000). Hammer and Nesis (2002) point out that spicule-driving by low-pressure regions above the chromosphere would require less fine-tuning to get the right velocity and height than acceleration from below.

Flares are introduced by Švestka (1976). The study of explosive events has been greatly advanced through the results of space missions, especially Skylab, Yohkoh, and SOHO. The volume edited by Priest (1981) concentrates on the magnetohydrodynamic aspects, that of McLean and Labrum (1985) on the physics of the radio emission. Solar radio astronomy is also introduced by Kundu (1965) and Krüger (1979). Gosling (1993) emphasizes that coronal mass ejections cause geomagnetic disturbances. Reames (1999) reviews the mechanisms of particle acceleration.

Reconnection of magnetic lines of force was first studied by Sweet (1958) and Petschek (1964). Soward and Priest (1982) presented a detailed mathematical model. The existence and stability of magnetohydrostatic equilibria as possible pre-flare states has been reviewed by Schindler et al. (1983), while Low (1984) treats special magnetohydrodynamic models of coronal mass ejections.

List of Symbols

$\boldsymbol{A}$	Vector potential
A	1. Amplitude of light wave
	2. Astronomical unit
	3. Atomic weight
	4. 10^{12} times element abundance relative to hydrogen
	5. Azimuthal component of vector potential
	6. Constant in law of differential rotation
	7. Area of sunspot
A_{UL}	Einstein coefficient for spontaneous emission
A_Z	Mean atomic weight of atoms heavier than helium
$\boldsymbol{B}$	Magnetic field vector
$\boldsymbol{B}_{\mathrm{p}}$	Poloidal magnetic field vector
$\boldsymbol{B}_{\mathrm{t}}$	Toroidal magnetic field vector
B	1. Kirchhoff–Planck function
	2. Magnitude (or component) of magnetic field
	3. Constant in law of differential rotation
B_{LU}	Einstein coefficient for radiative excitation
B_{P}	Field in equilibrium with pressure
B_{UL}	Einstein coefficient for induced emission
B_{c}	Critical magnetic field strength
B_{e}	Equipartition field strength
B_0	1. Heliographic latitude of disc center
	2. Initial magnetic field strength
C	1. Intensity contrast of granulation
	2. Constant in law of differential rotation
C_{ij}	Rate of collisional transition
D	1. Aperture (diameter) of telescope
	2. Diameter of granule
	3. Dynamo number
D_{S}	Aperture of spectrograph
$\boldsymbol{E}$	Electric field vector
E	1. Number fraction electrons/all particles
	2. Energy of nuclear reaction
$\boldsymbol{\mathcal{E}}$	$\langle \boldsymbol{u} \times \boldsymbol{b} \rangle$, vector of mean electric field

E_{ES}	Electrostatic energy of charged particles
E_{ij}	Energy of jth state of ith atom
E_j	Energy of jth atomic state
E_{max}	Energy of Gamow peak
$\boldsymbol{F}$	1. Vector of energy flux density
	2. Volume force
$\boldsymbol{F}_C$	Vector of convective energy flux density
$\boldsymbol{F}_H$	Vector of heating energy flux density
$\boldsymbol{F}_R$	Vector of radiative energy flux density
F	1. Magnitude of energy flux
	2. Cross section of flux tube
$\mathcal{F}$	Finesse of interferometer
G	Gravitational constant
H	1. Voigt function
	2. Density scale height
	3. Specific enthalpy
H_P	Pressure scale height
$\boldsymbol{I}$	Stokes vector
I	Intensity of radiation
I_C	Continuum intensity
J	1. Birefringence of Lyot filter
	2. Mean (over solid angle) of intensity
	3. Quantum number of angular momentum
	4. Total angular momentum of Sun
J_2	Solar quadrupole moment
K	Second moment of intensity
K_α	Rotational splitting kernel
L	1. Luminosity
	2. Quantum number of orbital angular momentum
	3. Linear operator of solar oscillation equations
	4. Ecliptic heliographic longitude
	5. Constant of integration for magnetic braking
L_0	Heliographic (ecliptic) longitude of disc center
L_s	Luminosity at surface of solar model
$L_\odot$	Luminosity of present Sun
L^2	$-r^2$ times angular part of Laplacian operator
LSF	Line spread function
M	Absolute magnitude of a star
M_A	Alfvén Mach number
M_J	Magnetic quantum number
MTF	Modulation transfer function
N	1. f/D, focal ratio
	2. Brunt–Väisälä frequency
	3. Truncation index of spherical harmonic expansion

N_A	Avogadro number
$O(x)$	Term of first order in x
P	1. Pressure
	2. Degree of polarization
	3. Power spectrum
	4. Position angle of northern direction on Sun
P_{ES}	Electrostatic correction to pressure
P_G	Gas pressure
P_I	Ion contribution to gas pressure
P_{PG}	Perfect gas pressure
P_R	Radiation pressure
P_e	1. Electron pressure
	2. Pressure external to flux tube
P_s	Pressure at surface of solar model
P_0	1. Equilibrium pressure
	2. Pressure on the axis of flux tube
P_1	Eulerian pressure perturbation
PSF	Point spread function
Q	1. Energy release per nuclear reaction
	2. Stokes parameter of linear polarized light
	3. $\nu/\Delta\nu$, quality factor
	4. Energy gain per volume of granular flow
Q_{ij}	Reynolds stress
Q_ν	Energy of neutrino
Q'	Energy release exclusive of neutrino energy
R	1. Reflected fraction of intensity
	2. Sunspot relative number
$\mathcal{R}$	Gas constant
R_T	Radius of flux tube
R_{ij}	Rate of transition from ith to jth state
R_m	Magnetic Reynolds number
Ro	Rossby number
S	1. Irradiance (energy flux) at mean solar distance
	2. quantum number of spin angular momentum
	3. Source function
	4. Specific entropy
$S_i(\boldsymbol{q})$	Transfer function of ith image
S_l	1. Line source function
	2. Limit frequency for acoustic oscillations
T	1. Absolute temperature
	2. Orbital period
	3. Period of solar oscillation
	4. Transmitted fraction of intensity
T_B	Brightness temperature

T_S	Sidereal period of Carrington rotation
T_e	Electron temperature
T_{eff}	Effective temperature
T_p	Proton temperature
T_s	Temperature at radius r_s
U	1. Internal energy, or specific internal energy
	2. Stokes parameter of linear polarized light
U_{ES}	Electrostatic correction to specific internal energy
V	1. Electrostatic potential
	2. Volume, or specific volume
	3. Stokes parameter of circular polarized light
	4. Acoustic potential
V_{mac}	Parameter (velocity) of macroturbulence
X	Mass fraction of hydrogen
X_i	Mass fraction of element i
Y	Mass fraction of ^{4}He
Y_0	Initial mass fraction of helium
$Y_{0\odot}$	Value of Y_0 adapted to reproduce present Sun
Y_3	Mass fraction of ^{3}He
Y_{30}	Initial mass fraction of ^{3}He
Z	1. Mass fraction of elements heavier than helium
	2. Charge number of a particle
a	1. Grating constant
	2. $4\sigma/c$, radiation constant
	3. Semi-major axis of elliptical orbit
$\boldsymbol{b}$	Fluctuating part of magnetic field vector
$\hat{\boldsymbol{b}}$	Binormal vector of thin flux tube
c	1. Velocity of light
	2. Velocity of sound
	3. Oblateness parameter
c_A	Alfvén velocity
c_P	Specific heat at constant pressure
c_T	Tube speed
c_V	Specific heat at constant volume
c_2	hc/k, radiation constant
d	1. Diameter of solar image
	2. Distance between centers of granules
	3. Width of magnetic sheet or tube
d	Deuteron
e	1. Elementary charge
	2. Thickness of first crystal in Lyot filter
e^-	Electron
e^+	Positron

f	1. Focal length
	2. Oscillator strength
	3. Filling factor for convecting parcels
$\boldsymbol{g}$	Vector of gravitational acceleration
g	1. Gravitational acceleration
	2. Landé factor of atomic state
g_L	Statistical weight of lower state
g_U	Statistical weight of upper state
$\boldsymbol{g}_\mathrm{eff}$	Vector of effective gravitational acceleration
g_{ij}	Statistical weight of jth state of ith atom
g_l^m	Coefficient of multipole expansion
g^*	g factor for atomic transition
$g_\odot$	Gravitational acceleration at present Sun's surface
h	1. Planck constant
	2. Height in solar atmosphere above the level where $\tau_{500} = 1$
h_l^m	Coefficient of multipole expansion
i	Inclination between ecliptic and solar equator
$\boldsymbol{j}$	Vector of electric current density
$\boldsymbol{k}$	Wave vector
k	1. Gaussian gravitational constant
	2. Boltzmann constant
	3. Wave number
k_Ny	Nyquist wave number
k_h	Horizontal wave number
k_r	Radial wave number
$\hat{\boldsymbol{l}}$	Tangent vector of thin flux tube
l	1. Mixing length
	2. Size of convection cell; length scale of fluctuation
	3. Degree of spherical harmonic
m	1. Mass
	2. Reduced mass
	3. Apparent magnitude of a star
	4. Order of spectrum
	5. Longitudinal order of spherical harmonic
m_H	Atomic mass of hydrogen
m_e	Electron mass
$m_\odot$	Mass of present Sun
$\hat{\boldsymbol{n}}$	Principal normal vector of thin flux tube
n	1. Number of grooves in a diffraction grating
	2. Principal quantum number
	3. Index of refraction
	4. Order (radial node number) of oscillations
	5. Coordinate in normal direction
n	Neutron

n_H Number density of hydrogen
n_L Number density of atoms in lower state
n_U Number density of atoms in upper state
n_e 1. Electron density
              2. Index of refraction for extraordinary ray
              3. Polytropic index
n_i Number density of particle species i
n_{ij} Number density of ith particle in jth ionization state
n_{ijk} Number density of ith particle in jth ionization
                  and kth excitation state
n_j^* Number density in LTE
n_o Index of refraction for ordinary ray
p Momentum of a particle
p   Proton
p_0 Impact parameter for 90° collision
$\boldsymbol{q}$ Vector of spatial frequencies
q Spatial frequency
$\boldsymbol{r}$ Radius vector
r 1. Line depression
              2. Radius
r_A Alfvén radius
r_D Debye radius
r_c 1. Reflection coefficient of retroreflector
              2. Critical radius for coronal expansion
r_e External reflection coefficient of glass plate
r_{ik} Nuclear reaction rate per mass of species i and k
r_s Surface radius of solar model
r_t Radius of reflection of an acoustic oscillation
r_v Radius where $v = 0$, i.e., base of convection zone
r_w Radius of solar wind source surface
r_0 1. Central line depression
              2. Classical electron radius
              3. Fried parameter
              4. Radius of heat input in corona
$r_{\Delta T}$ Radius where ΔT reverses sign
$r_{\Delta \nabla}$ Radius where $\Delta \nabla$ reverses sign
$r_\odot$ Radius of present Sun
s 1. Coordinate along ray
              2. $r \sin \theta$, distance from axis
s_0 Seeing parameter
$\hat{s}$ Anisotropy parameter of turbulent diffusivity
t_D Dynamical time scale
t_{KH} Kelvin–Helmholtz time
t_{ff} Time scale of free fall

$t_\odot$	Age of the Sun
$\boldsymbol{u}$	Fluctuating part of velocity vector
u	1. $h\nu/kT$
	2. Energy of particle
	3. Partition function
	4. Typical velocity in convection cell; fluctuating velocity; wave amplitude
$\boldsymbol{v}$	Velocity vector
$\boldsymbol{v}_{\mathrm{m}}$	Vector of meridional circulation
$\boldsymbol{v}_{\mathrm{p}}$	Poloidal velocity vector
$\boldsymbol{v}_{\mathrm{t}}$	Toroidal velocity vector
$\boldsymbol{v}_0$	Vector of rotational velocity
v	1. Magnitude (or component) of velocity
	2. $(\lambda - \lambda_0)/\Delta\lambda_{\mathrm{D}}$, normalized distance from line center
	3. Mean velocity of convection
v_{A}	Alfvén velocity in solar wind
v_{B}	Normalized Zeeman splitting
w_0	Angular radius of usable field
Γ_1	Adiabatic exponent
Δ	1. Focus tolerance
	2. Laplacian operator
ΔS	Total entropy change over convection zone
ΔT	1. Temperature difference between granules and intergranular space
	2. Temperature excess of displaced parcel
Δr	Difference between equatorial and polar solar radii
Δt	1. Time delay in Doppler spectroheliogram
	2. Time resolution of observed oscillation
$\Delta\lambda$	Wavelength distance to line center
$\Delta\lambda_{\mathrm{B}}$	Zeeman splitting
$\Delta\lambda_{\mathrm{D}}$	Doppler width of line (in wavelength)
$\Delta\lambda_{\mathrm{G}}$	Gravitational redshift
$\Delta\nu_{\mathrm{D}}$	Doppler width of line (in frequency)
$\Delta\rho$	Density difference of displaced parcel
$\Delta\omega_\alpha$	Rotational shift of oscillation frequency ω_α
$\Delta\nabla$	$\nabla - \nabla_{\mathrm{a}}$, excess of temperature gradient over adiabatic gradient
Λ	Ratio of Debye radius to $90°$-impact parameter
Λ_{h}	Coefficient of horizontal non-diffusive momentum flux
Λ_{r}	Coefficient of vertical non-diffusive momentum flux
Θ	1. Characteristic angle of internal gravity waves
	2. Moment of inertia
Φ	1. Gravitational potential
	2. Magnetic flux
	3. Potential of current-free magnetic field

Φ_c	Critical magnetic flux
Φ_j	Neutrino flux at Earth's orbit
Φ_1	Eulerian perturbation of gravitational potential
Ψ	1. Covariance function of oscillations at two points
	2. Effective gravitational potential
$\boldsymbol{\Omega}$	Vector of angular velocity
Ω	1. Solid angle
	2. Angular velocity of Sun
	3. Angular velocity in flux tube
α	1. Helium nucleus
	2. Ratio of mixing length to pressure scale height
	3. Angle between flux tube axis and vertical
	4. Coefficient of α effect in dynamo
	5. Coefficient of force-free field
	6. Angle of prominence with magnetic vector
α_j	Cross section of state j for photoionization
β	1. Ratio of gas pressure to total pressure
	2. $2\mu P/B^2$, Plasma parameter
	3. Contribution of turbulence to mean diffusivity
	4. Angle of prominence with line of sight
γ	1. Angle of magnetic field with line of sight
	2. Constant value of adiabatic exponent
	3. Damping constant
	4. Photon
δ	$-(\partial \ln \rho / \partial \ln T)_P$
δ_{ik}	Kronecker symbol
δP	Lagrangian pressure perturbation
δr	Radial displacement
$\delta \nu$	Asymptotic frequency separation
$\delta \rho$	Lagrangian density perturbation
ε	1. Energy release per mass
	2. Phase angle of polarized light wave
	3. Element abundance relative to hydrogen
	4. Correction factor in thermal conductivity
ε_C	Continuous emission
ε_l	Line emission
ε_0	Dielectric constant of free space
$\boldsymbol{\eta}$	Absorption matrix for Stokes vector
η	1. Efficiency of Fourier spectrometer
	2. Temperature exponent of nuclear energy release
	3. Magnetic diffusivity
	4. Ratio of line absorption to continuum absorption
	5. Degree of ionization

η_t	Mean magnetic diffusivity in turbulence
η^+	Blueshifted absorption profile
η^-	Redshifted absorption profile
θ	1. Angular distance from disc center
	2. Polar angle in spherical polar coordinate system
	3. Position angle on solar disc
	4. Angle of flux tube with horizontal direction
κ	1. Absorption coefficient
	2. Rosseland mean absorption coefficient (opacity)
	3. Curvature of thin flux tube
	4. Coefficient of thermal conduction
κ_C	Continuum absorption coefficient
κ_T	Temperature exponent of opacity
κ_l	Line absorption coefficient
κ_t	Mean heat diffusivity in turbulence
$\kappa_\parallel$	Coefficient of thermal conduction parallel to $\boldsymbol{B}$
λ	Wavelength
λ_{ik}	Nuclear reaction rate of species i and k
μ	1. Permeability of free space
	2. $\cos\theta$
	3. Mean molecular weight
	4. Muon
μ_0	Mean molecular weight of neutral gas
ν	1. Frequency of light
	2. Neutrino
	3. Frequency of solar oscillation
	4. Diffusivity
	5. ν_t/η_t, Ratio of mean diffusivities
ν_P	Plasma frequency
ν_{ij}	Tensor of turbulent diffusivity
ν_{rot}	$\Omega/2\pi$, cycle frequency of rotation
ν_t	Mean diffusivity in turbulence
ν_0	Frequency of line center
$\boldsymbol{\xi}$	Displacement vector
ξ	1. Component of displacement vector
	2. Ancillary variable in inversion problem
ξ_h	Defining scalar for horizontal components of $\boldsymbol{\xi}$
ξ_t	Parameter (velocity) of microturbulence
ρ	1. Mass density
	2. Heliocentric angular distance
	3. $r/r_\odot$ in corona
ρ_e	Density outside magnetic flux tube
ρ_0	1. Equilibrium density
	2. Density on the axis of flux tube

ρ_1	Eulerian density perturbation
ρ^*	Density of displaced parcel
σ	1. Blurring parameter
	2. Stefan–Boltzmann constant
	3. Electric conductivity
	4. Cross section
σ_s	Cross section per electron for Thomson scattering
τ	1. Optical depth
	2. Transmission coefficient of glass plate
	3. Torsion of thin flux tube
	4. Typical time scale of fluctuation
	5. Travel time of acoustic wave
τ_D	Ohmic decay time
τ_R	Radiative cooling time
τ_{500}	Optical depth at $\lambda = 500\,\mathrm{nm}$
ϕ	1. Azimuth angle of magnetic field
	2. Azimuthal coordinate in flux tube
	3. Heliographic longitude
	4. Normalized line profile; Voigt function
ϕ_C	Collisional profile
ϕ_D	Doppler profile
χ	1. Lower excitation potential
	2. Ionization potential
	3. Frequency distribution of spontaneous emission
ψ	1. Degeneracy parameter
	2. Heliographic latitude
	3. Frequency distribution of induced emission
ω	Circular frequency of oscillation
ω_A	Acoustic cutoff frequency
ω_{Ny}	Nyquist frequency
ω_α	Frequency of αth eigenoscillation
∇	1. Gradient operator
	2. $d\ln T/d\ln P$, double-logarithmic temperature gradient
∇_R	Radiative double-logarithmic temperature gradient
∇_a	Adiabatic double-logarithmic temperature gradient
∇'	Mean double-logarithmic temperature gradient of parcels
Ω	Ecliptic longitude of ascending node of Sun's equator
$\odot$	Subscript denoting present Sun
$\simeq$	approximately equal (physical relations)
$\approx$	approximately equal (numbers)
$\propto$	proportional to

References

Abdelatif, T.E., Lites, B.W., Thomas, J.H. (1984): In Keil (1984), p. 141
Abel, W.G., Chisholm, J.H., James, J.C. (1963): In *Space Research III*, ed. by W. Priester (North-Holland, Amsterdam), p. 635
Abraham, Z., Iben, I., Jr. (1971): Astrophys. J. **170**, 157
Adams, J., Pneuman, G.W. (1976): Solar Phys. **46**, 185
Adelberger, E.G., Austin, S.M., Bahcall, J.N., and 36 others (1998): Rev. Mod. Phys. **70**, 1265
Ahmad, Q.R., Allen, R.C., Andersen, T.C., and 175 others (2001): Phys. Rev. Letters **87**, 071301
Ahrens, B., Stix, M., Thorn, M. (1992): Astron. Astrophys. **264**, 673
Alazraki, G., Couturier, P. (1971): Astron. Astrophys. **13**, 380
Albregtsen, F., Maltby, P. (1978): Nature **274**, 41
Albregtsen, F., Maltby, P. (1981): In Cram and Thomas (1981), p. 127
Aldrich, L.B., Hoover, W.H. (1954): Ann. Astrophys. Obs. Smithsonian Inst. **7**
Alfvén, H. (1947): Mon. Not. R. Astron. Soc. **107**, 211
Alfvén, H. (1954): *On the Origin of the Solar System* (Clarendon Press, Oxford)
Allen, C.W. (1947): Mon. Not. R. Astron. Soc. **107**, 426
Aller, L.H. (1963): *Astrophysics, The Atmospheres of the Sun and Stars*, 2nd ed. (Ronald Press, New York)
Altschuler, M.D., Levine, R.H., Stix, M., Harvey, J. (1977): Solar Phys. **51**, 345
Altschuler, M.D., Newkirk, G., Jr. (1969): Solar Phys. **9**, 131
Anders, E., Grevesse, N. (1989): Geochimica et Cosmochimica Acta **53**, 197
Andersen, T.E., Dunn, R.B., Engvold, O. (1984): LEST Foundation, Technical Rep. No. 7
Ando, H., Osaki, Y. (1975): Publ. Astron. Soc. Japan **27**, 581
Antia, H.M. (1998): Astron. Astrophys. **330**, 336
Antia, H.M., Chitre, S.M. (1995): Astrophys. J. **442**, 434
Anzer, U. (1987): In Hillebrandt et al. (1987), p. 61
Anzer, U., Heinzel, P. (1999): Astron. Astrophys. **349**, 974
Appenzeller, I. (1982): Fundamentals of Cosmic Physics **7**, 313
Appenzeller, I., Tscharnuter, W. (1975): Astron. Astrophys. **40**, 397
Appourchaux, T., Fröhlich, C., Andersen, B., and 12 others (2000): Astrophys. J. **538**, 401
Arvesen, J.C., Griffin, R.N., Jr., Pearson, B.D., Jr. (1969): Applied Optics **8**, 2215
Aschwanden, M.J., Nightingale, R.W., Alexander, D. (2000): Astrophys. J. **541**, 1059
Aschwanden, M.J., Schrijver, C.J., Alexander, D. (2001): Astrophys. J. **550**, 1036
Asplund, M., Ludwig, H.-G., Nordlund, Å., Stein, R.F. (2000 a): Astron. Astrophys. **359**, 669
Asplund, M., Nordlund, Å., Trampedach, R., Allende Prieto, C., Stein, R.F. (2000 b): Astron. Astrophys. **359**, 729

Asplund, M., Nordlund, Å., Trampedach, R., Stein, R.F. (2000 c): Astron. Astrophys. **359**, 743
Athay, R.G. (1972): *Radiation Transport in Spectral Lines* (Reidel, Dordrecht)
Athay, R.G. (1976): *The Solar Chromosphere and Corona: Quiet Sun* (Reidel, Dordrecht)
Atkinson, R. d'E., Houtermans, F.G. (1929): Z. Phys. **54**, 656
Audouze, J., Israël, G., eds. (1985): *The Cambridge Atlas of Astronomy* (Cambridge University Press, Cambridge)
Auwers, A. (1891): Astron. Nachr. **128**, 361
Avrett, E.H. (1981a): In Cram and Thomas (1981), p. 235
Avrett, E.H. (1981b): In Bonnet and Dupree (1981), p. 173
Axford, W.I., McKenzie, J.F. (1992): In *Solar Wind Seven*, ed. by E. Marsch and R. Schwenn (Pergamon Press, Oxford), p. 1
Ayres, T.R. (1981): Astrophys. J. **244**, 1064

Babcock, H.W. (1953): Astrophys. J. **118**, 387
Babcock, H.W. (1961): Astrophys. J. **133**, 572
Backus, G., Gilbert, F. (1970): Phil. Trans. R. Soc. London A **266**, 123
Bahcall, J.N. (1979): Space Sci. Rev. **24**, 227
Bahcall, J.N. (1989): *Neutrino Astrophysics* (Cambridge University Press, Cambridge)
Bahcall, J.N. (1997): Phys. Rev. C **56**, 3391
Bahcall, J.N., Cleveland, B.T., Davis, R., Jr., Rowley, J.K. (1985): Astrophys. J. **292**, L 79
Bahcall, J.N., Davis, R., Jr. (1976): Science **191**, 264
Bahcall, J.N., Huebner, W.F., Lubow, S.H., Parker, P.D., Ulrich, R.K. (1982): Rev. Mod. Phys. **54**, 767
Bahcall, J.N., Lisi, E., Alburger, D.E., De Braeckeleer, L., Freedman, S.J., Napolitano, J. (1996): Phys. Rev. C **54**, 411
Bahcall, J.N., Loeb, A. (1990): Astrophys. J. **360**, 267
Bahcall, J.N., Pinsonneault, M.H. (1992): Rev. Mod. Phys. **64**, 885
Bahcall, J.N., Pinsonneault, M.H. (1995): Rev. Mod. Phys. **67**, 781
Bahner, K. (1967): In *Handbuch der Physik*, ed. by S. Flügge (Springer, Berlin Heidelberg), Vol. XXIX, p. 227
Bahng, J., Schwarzschild, M. (1961): Astrophys. J. **134**, 312
Baker, N., Kippenhahn, R. (1962): Z. Astrophys. **54**, 114
Baker, N., Kippenhahn, R. (1965): Astrophys. J. **142**, 868
Baker, N.H., Temesváry, S. (1966): *Tables of Convective Stellar Envelope Models*, 2nd ed., Institute for Space Studies (NASA, New York)
Baliunas, S.L., Donahue, R.A., Soon, W.H., and 24 others (1995): Astrophys. J. **438**, 269
Ballegooijen, A.A. van (1982): Astron. Astrophys. **113**, 99
Ballegooijen, A.A. van (1986): Astrophys. J. **304**, 828
Ballester, J.L., Priest, E.R., eds. (1988): *Dynamics and Structure of Solar Prominences*, Secr. Publ. Interc. Cient., Universitat de les Illes Balears
Balmforth, N.J. (1992): Mon. Not. R. Astron. Soc. **255**, 603
Balser, D.S., Bania, T.M., Rood, R.T., Wilson, T.L. (1999): Astrophys. J. **510**, 759
Balthasar, H. (1984): Solar Phys. **93**, 219
Balthasar, H. (1988): Astron. Astrophys. Suppl. **72**, 473
Balthasar, H. (1999): Solar Phys. **187**, 389
Balthasar, H., Lustig, G., Stark, D., Wöhl, H. (1986 a): Astron. Astrophys. **160**, 277
Balthasar, H., Schüssler, M., Wöhl, H. (1982): Solar Phys. **76**, 21
Balthasar, H., Stark, D., Wöhl, H. (1987): Astron. Astrophys. **174**, 359
Balthasar, H., Vázquez, M., Wöhl, H. (1986 b): Astron. Astrophys. **155**, 87

Barletti, R., Ceppatelli, G., Paternó, L., Righini, A., Speroni, N. (1977): Astron. Astrophys. **54**, 649

Bartenwerfer, D. (1973): Astron. Astrophys. **25**, 455

Basset, A.B. (1961): *A Treatise on Hydrodynamics*, republished (Dover, New York)

Basu, S., Antia, H.M. (1994): Mon. Not. R. Astron. Soc. **269**, 1137

Basu, S., Antia, H.M. (1997): Mon. Not. R. Astron. Soc. **287**, 189

Baturin, V.A., Mironova, I.V. (1998): In Korzennik and Wilson (1998), p. 717

Baumbach, S. (1937): Astron. Nachr. **263**, 121

Beck, J.G. (2000): Solar Phys. **191**, 47

Becker, U. (1954): Z. Astrophys. **34**, 129

Beckers, J.M. (1969 a): *A Table of Zeeman Multiplets*, Sacramento Peak Observatory, Rep. AFCRL-69-0115

Beckers, J.M. (1969 b): Solar Phys. **9**, 372

Beckers, J.M. (1970): Applied Optics **9**, 595

Beckers, J.M. (1971 a): In Howard (1971), p. 3

Beckers, J.M. (1971 b): Applied Optics **10**, 973

Beckers, J.M. (1976): Astrophys. J. **203**, 739

Beckers, J.M. (1977): Astrophys. J. **213**, 900

Beckers, J.M. (1981): In Jordan (1981), p. 11

Beckers, J.M. (1993): Solar Phys. **145**, 399

Beckers, J.M. (1995): In Kuhn and Penn (1995), p. 145

Beckers, J.M., Bridges, C.A., Gilliam, L.B (1976): Rep. AFGL-TR-76-0126 (II)

Beckers, J.M., Brown, T.M. (1978): Osserv. Mem. Oss. Astrofis. Arcetri **106**, 189

Beckers, J.M., Canfield, R.C. (1976): In Cayrel and Steinberg (1976), p. 207

Beckers, J.M., Dickson, L., Joyce, R.S. (1975): Sacramento Peak Observatory Rep. AFCRL-TR-75-0090

Beckers, J.M., Milkey, R.W. (1975): Solar Phys. **43**, 289

Beckers, J.M., Schröter, E.H. (1968): Solar Phys. **4**, 142

Beckers, J.M., Schröter, E.H. (1969): Solar Phys. **10**, 384

Beckers, J.M., Tallant, P.E. (1969): Solar Phys. **7**, 351

Beer, J., Tobias, S., Weiss, N. (1998): Solar Phys. **181**, 237

Belcher, J.W. (1971): Astrophys. J. **168**, 509

Bellot Rubio, L.R., Rodríguez Hidalgo, I., Collados, M., Khomenko, E., Ruiz Cobo, B. (2001): Astrophys. J. **560**, 1010

Bellot Rubio, L.R., Ruiz Cobo, B., Collados, M. (2000): Astrophys. J. **535**, 475

Belvedere, G., Godoli, G., Motta, S., Paternó, L., Zappalá, R.A. (1976): Solar Phys. **46**, 23

Belvedere, G., Kuzanyan, K.M., Sokoloff, D. (2000): Mon. Not. R. Astron. Soc. **315**, 778

Belvedere, G., Paternó, L., eds. (1978): *Workshop on Solar Rotation*, University of Catania

Belvedere, G., Paternó, L., eds. (1984): Mem. Soc. Astron. Ital. **55**, No. 1–2

Bendlin, C., Volkmer, R., Kneer, F. (1992): Astron. Astrophys. **257**, 817

Bennett, K., Roberts, B., Narain, U. (1999): Solar Phys. **185**, 41

Berger, M.A., Ruzmaikin, A. (2000): J. Geophys. Res. **105**, 10481

Berger, T.E., Löfdahl, M.G., Shine, R.A., Title, A.M. (1998): Astrophys. J. **495**, 973

Berger, T.E., Schrijver, C.J., Shine, R.A., Tarbell, T.D., Title, A.M., Scharmer, G. (1995): Astrophys. J. **454**, 531

Berrington, K.A., ed. (1997): *The Opacity Project*, Vol. 2 (Institute of Physics, Bristol and Philadelphia)

Bertello, L., Henney, C.J., Ulrich, R.K., and 8 others (2000): Astrophys. J. **535**, 1066

Berthomieu, G., Cooper, A.J., Gough, D.O., Osaki, Y., Provost, J., Rocca, A. (1980): In Hill and Dziembowski (1980), p. 307
Berthomieu, G., Provost, J., Morel, P., Lebreton, Y. (1993): Astron. Astrophys. **268**, 775
Bethe, H.A. (1939): Phys. Rev. **55**, 434
Bethe, H.A. (1986): Phys. Rev. Letters **56**, 1305
Bethe, H.A., Critchfield, C.L. (1938): Phys. Rev. **54**, 248
Bianda, M., Solanki, S.K., Stenflo, J.O. (1998): Astron. Astrophys. **331**, 760
Bianda, M., Stenflo, J.O., Solanki, S.K. (1999): Astron. Astrophys. **350**, 1060
Biermann, L. (1932): Z. Astrophys. **5**, 117
Biermann, L. (1937): Astron. Nachr. **263**, 185
Biermann, L. (1941): Vierteljahresschr. Astron. Ges. **76**, 194
Biermann, L. (1946): Naturwiss. **33**, 118
Biermann, L. (1951 a): Z. Astrophys. **28**, 304
Biermann, L. (1951 b): Z. Astrophys. **29**, 274
Biermann, L. (1977): In Spiegel and Zahn (1977), p. 4
Billings, D.E. (1966): *A Guide to the Solar Corona* (Academic Press, New York)
Birn, J., Gosling, J.T., Hesse, M., Forbes, T.G., Priest, E.R. (2000): Astrophys. J. **541**, 1078
Biskamp, D. (1993): *Nonlinear Magnetohydrodynamics* (Cambridge University Press, Cambridge)
Bodenheimer, P. (1983): Lectures in Appl. Math. **20**, 141
Bodenheimer, P. (1995): Ann. Rev. Astron. Astrophys. **33**, 199
Bodenheimer, P., Yorke, H.W., Różycka, M., Tohline, J.E. (1990): Astrophys. J. **355**, 651
Boercker, D.B. (1987): Astrophys. J. **316**, L95
Bogdan, T.J. (2000): Solar Phys. **192**, 373
Böhm-Vitense, E. (1958): Z. Astrophys. **46**, 108
Bommier, V., Leroy, J.L., Sahal-Bréchot, S. (1986 a): Astron. Astrophys. **156**, 79
Bommier, V., Leroy, J.L., Sahal-Bréchot, S. (1986 b): Astron. Astrophys. **156**, 90
Bonnet, R.M., Bruner, E.C., Jr., Acton, L.W., Brown, W.A., Decaudin, M. (1980): Astrophys. J. **237**, L47
Bonnet, R.M., Dupree, A.K., eds. (1981): *Solar Phenomena in Stars and Stellar Systems* (Reidel, Dordrecht)
Boothroyd, A.I., Sackmann, I.-J., Fowler, W.A. (1991): Astrophys. J. **377**, 318
Bord, D.J., Cowley, C.R., Mirijanian, D. (1998): Solar Phys. **178**, 221
Bouvier, J., Bertout, C., Benz, W., Mayor, M. (1986): Astron. Astrophys. **165**, 110
Bracewell, R. (1965): *The Fourier Transform and Its Applications* (McGraw-Hill, New York)
Braginsky, S.I. (1965): In *Reviews in Plasma Physics*, ed. by M.A. Leontovich, Consultants-Bureau, New York, Vol. 1, p. 205
Brajša, R., Wöhl, H., Vršnak, B., Ruždjak, V., Clette, F., Hochedez, J.-F. (2001): Astron. Astrophys. **374**, 309
Brandenburg, A., Dobler, W. (2001): Astron. Astrophys. **369**, 329
Brandenburg, A., Dobler, W., Subramanian, K. (2002): Astron. Nachr. **323**, 99
Brandenburg, A., Schmitt, D. (1998): Astron. Astrophys. **338**, L55
Brandt, P.N. (1969): Solar Phys. **7**, 187
Brandt, P.N. (1970): *Joint Org. Solar Obs.*, Ann. Rep. 1970, p. 50
Brandt, P.N., Ferguson, S., Shine, R.A., Tarbell, T.D., Scharmer, G.B. (1991): Astron. Astrophys. **241**, 219
Brandt, P.N., Mauter, H.A., Smartt, R. (1987): Astron. Astrophys. **188**, 163
Brandt, P.N., Scharmer, G.B., Ferguson, S., Shine, R.A., Tarbell, T.D., Title, A.M. (1988): Nature **335**, 238

Brandt, P.N., Schröter, E.H. (1982): Solar Phys. **79**, 3

Brault, J.W. (1985): In *High Resolution in Astronomy*, ed. by A.O. Benz, M.C.E. Huber, and M. Mayor (Geneva Observatory, Sauverny), p. 1

Braun, D.C., Duvall, T.J., Jr., LaBonte, B.J. (1987): Astrophys. J. **319**, L27

Bray, R.J., Loughhead, R.E. (1964): *Sunspots* (Chapman and Hall, London)

Bray, R.J., Loughhead, R.E. (1974): *The Solar Chromosphere* (Chapman and Hall, London)

Bray, R.J., Loughhead, R.E. (1977): Solar Phys. **54**, 319

Bray, R.J., Loughhead, R.E., Durrant, C.J. (1984): *The Solar Granulation*, 2nd ed. (Cambridge University Press, Cambridge)

Brekke, P., Thompson, W.T., Woods, T.N., Eparvier, F.G. (2000): Astrophys. J. **536**, 959

Brookes, J.R., Isaak, G.R., van der Raay, H.B. (1978): Mon. Not. R. Astron. Soc. **185**, 1

Brown, D.S., Priest, E.R. (2001): Astron. Astrophys. **367**, 339

Brown, T.M. (1981): In Dunn (1981), p. 150

Brown, T.M. (1984): Science **226**, 687

Brown, T.M. (1986): In Gough (1986), p. 199

Brown, T.M. (1991): Astrophys. J. **371**, 396

Brown, T.M., ed. (1993): *GONG 1992: Seismic Investigation of the Sun and Stars*, ASP. Conf. Ser. **42**

Brown, T.M., Christensen-Dalsgaard, J. (1998): Astrophys. J. **500**, L195

Brown, T.M., Mihalas, B.W., Rhodes, E.J., Jr. (1986): In Sturrock (1986), Vol. I, p. 177

Brown, T.M., Stebbins, R.T., Hill, H.A. (1978): Astrophys. J. **223**, 324

Browning, P.K., Priest, E.R. (1986): Astron. Astrophys. **159**, 129

Browning, P.K., Sakurai, T., Priest, E.R. (1986): Astron. Astrophys. **158**, 217

Brueckner, G.E. (1963): Z. *Astrophys*. **58**, 73

Brueckner, G.E. (1981): In *Solar Active Regions*, ed. by F. Q. Orrall, (Colorado Associated University Press, Boulder), p. 113

Brueckner, G.E., Bartoe, J.-D.F. (1983): Astrophys. J. **272**, 329

Brueckner, G.E., Howard, R.A., Koomen, M.J., and 12 others (1995): Solar Phys. **162**, 357

Bruls, J.H.M.J., von der Lühe, O. (2001): Astron. Astrophys. **366**, 281

Brune, R., Wöhl, H. (1982): Solar Phys. **75** , 75

Bruzek, A. (1967): Solar Phys. **2**, 451

Bruzek, A. (1979): Solar Phys. **61**, 35

Bruzek, A., Durrant, C.J., eds. (1977): *Illustrated Glossary for Solar and Solar-Terrestrial Physics* (Reidel, Dordrecht)

Bullard, E., Gellman, H. (1954): Phil. Trans. R. Soc. London A **247**, 213

Bumba, V. (1963): Bull. Astron. Inst. Czech. **14**, 91

Bumba, V. (1967): In *Plasma Astrophysics*, ed. by P.A. Sturrock (Academic Press, New York), p. 77

Bumba, V. (1987): Bull. Astron. Inst. Czech. **38**, 92

Bumba, V., Kleczek, J., eds. (1976): *Basic Mechanisms of Solar Activity*, IAU Symp. **71** (Reidel, Dordrecht)

Bumba, V., Suda, J. (1980): Bull. Astron. Inst. Czech. **31**, 101

Bünte, M. (1993): Astron. Astrophys. **276**, 236

Bünte, M., Steiner, O., Pizzo, V.J. (1993): Astron. Astrophys. **268**, 299

Burgers, J.M. (1969): *Flow Equations for Composite Gases* (Academic Press, New York)

Burgess, A. (1964): Astrophys. J. **139**, 776

458   References

Caligari, P., Moreno-Insertis, F., Schüssler, M. (1995): Astrophys. J. **441**, 886
Callen, H.B. (1985): *Thermodynamics and an Introduction to Thermostatistics*, 2nd ed. (Wiley, New York)
Cameron, R., Sammis, I. (1999): Astrophys. J. **525**, L61
Canfield, R.C., Beckers, J.M. (1976): In Cayrel and Steinberg (1976), p. 291
Cannon, C.J. (1985): *The Transfer of Spectral Line Radiation* (Cambridge University Press, Cambridge)
Canuto, V.M. (1993): Astrophys. J. **416**, 331
Canuto, V.M. (1996): Astrophys. J. **467**, 385
Canuto, V.M. (1999): Astrophys. J. **524**, 311
Canuto, V.M., Dubovikov, M. (1998): Astrophys. J. **493**, 834
Canuto, V.M., Goldman, I., Mazzitelli, I. (1996): Astrophys. J. **473**, 550
Canuto, V.M., Mazzitelli, I. (1991): Astrophys. J. **370**, 295
Capitani, C., Cavallini, F., Ceppatelli, G., Landi degl'Innocenti, E., Landi degl'Innocenti, M., Landolfi, M., Righini, A. (1989): Solar Phys. **120**, 173
Carlsson, M., Stein, R.F. (1995): Astrophys. J. **440**, L29
Carlsson, M., Stein, R.F. (1997): Astrophys. J. **481**, 500
Carmeron, A.G.W. (1973): In *Explosive Nucleosynthesis*, ed. by D.N. Schramm and W.D. Arnett (University of Texas Press, Austin), p. 3
Carrington, R.C. (1863): *Observations of the Spots on the Sun from November 9, 1853, to March 24, 1861, made at Redhill* (Williams and Norgate, London, Edinburgh)
Castellani, V., Degl'Innocenti, S., Fiorentini, G., Lissia, M., Ricci, B. (1997): Physics Reports **281**, 309
Cattaneo, F. (1999): Astrophys. J. **515**, L39
Cattaneo, F., Hughes, D.W. (2001): Astron. Geophys. **42**, 3.18
Caughlan, G.R., Fowler, W.A. (1988): Atomic Data and Nuclear Data Tables **40**, 283
Cavallini, F., Ceppatelli, G., Righini, A. (1992): Astron. Astrophys. **254**, 381
Cavallini, F., Ceppatelli, G., Righini, A., Meco, M., Paloschi, S., Tantulli, F. (1987): Astron. Astrophys. **184**, 386
Cayrel, R., Steinberg, M., eds. (1976): *Physique des Mouvements dans les Atmosphéres Stellaires*, CNRS, Paris
Chan, K.L., Sofia, S. (1987): Science **235**, 465
Chan, K.L., Sofia, S. (1989): Astrophys. J. **336**, 1022
Chan, K.L., Sofia, S. (1996): Astrophys. J. **466**, 372
Chandrasekhar, S. (1939): *An Introduction to the Study of Stellar Structure* (University of Chicago Press, Chicago)
Chaplin, W.J., Christensen-Dalsgaard, J., Elsworth, and 7 others (1999): Mon. Not. R. Astron. Soc. **308**, 405
Chaplin, W.J., Elsworth, Y., Howe, R., and 6 others (1996 a): Solar Phys. **168**, 1
Chaplin, W.J., Elsworth, Y., Howe, R., Isaak, G.R., McLeod, C.P., Miller, B.A., New., R. (1996 b): Mon. Not. R. Astron. Soc. **280**, 849
Chapman, G.A. (1970): Solar Phys. **14**, 315
Chapman, G.A., Cookson, A.M., Dobias, J.J. (1997): Astrophys. J. **482**, 541
Chapman, G.A., Herzog, A.D., Lawrence, J.K., Shelton, J.C. (1984): Astrophys. J. **282**, L99
Chapman, S. (1917): Mon. Not. R. Astron. Soc. **77**, 539
Chapman, S. (1957): Smithson. Contr. Astrophys. **2**, 1
Charbonneau, P., Dikpati, M., Gilman, P.A. (1999): Astrophys. J. **526**, 523
Charbonneau, P., MacGregor, K.B. (1997): Astrophys. J. **486**, 502
Charbonneau, P., Tomczyk, S., Schou, J., Thompson, M.J. (1998): Astrophys. J. **496**, 1015

Charbonnel, C., Lebreton, Y. (1993): Astron. Astrophys. **280**, 666

Charbonnel, C., Vauclair, S., Maeder, A., Meynet, G., Schaller, G. (1994): Astron. Astrophys. **283**, 155

Cheng, C.Z., Choe, G.S. (1998): Astrophys. J. **505**, 376

Childress, S., Gilbert, A.D. (1995): *Stretch, Twist, Fold: The Fast Dynamo* (Springer, Berlin, Heidelberg)

Chou, D.-Y., Chang, H.-K., Sun, M.-T., LaBonte, B., Chen, H.-R., Yeh, S.J., and the TON Team (1999): Astrophys. J. **514**, 979

Chou, D.-Y., Sun, M.-T., Huang, T.-Y., and 18 others (1995): Solar Phys. **160**, 237

Choudhuri, A.R. (1986): Astrophys. J. **302**, 809

Choudhuri, A.R. (1998): *The Physics of Fluids and Plasmas* (Cambridge University Press, Cambridge)

Choudhuri, A.R., Schüssler, M., Dikpati, M. (1995): Astron. Astrophys. **303**, L29

Christensen-Dalsgaard, J. (1982): Mon. Not. R. Astron. Soc. **199**, 735

Christensen-Dalsgaard, J. (1984): In Noels and Gabriel (1984), p. 155

Christensen-Dalsgaard, J. (1997): In Pijpers et al. (1997), p. 3

Christensen Dalsgaard, J. (1998): *Lecture notes on stellar oscillations.*, 4th ed., http://www.obs.aau.dk/~jcd/oscilnotes/

Christensen-Dalsgaard, J., Däppen, W. (1992): Astron. Astrophys. Rev. **4**, 267

Christensen-Dalsgaard, J., Däppen, W., Ajukov, S.V., and 30 others (1996): Science **272**, 1186

Christensen-Dalsgaard, J., Däppen, W., Dziembowski, W.A., Guzik, J.A. (2000): In *Variable Stars as Essential Astrophysical Tools*, ed. by C. Ibanoğlu (Kluwer, Dordrecht), p. 59

Christensen-Dalsgaard, J., Dilke, F.W.W., Gough, D.O. (1974): Mon. Not. R. Astron. Soc. **169**, 429

Christensen-Dalsgaard, J., Duvall, T.L., Jr., Gough, D.O., Harvey, J.W., Rhodes, E.J., Jr. (1985): Nature **315**, 378

Christensen-Dalsgaard, J., Frandsen, S., eds. (1988): *Advances in Helio- and Asteroseismology*, IAU Symp. **123** (Reidel, Dordrecht)

Christensen-Dalsgaard, J., Gough, D.O. (1975): Mém. Soc. R. Sci. Liège, 6e ser., **VIII**, 309

Christensen-Dalsgaard, J., Gough, D.O. (1981): Astron. Astrophys. **104**, 173

Christensen-Dalsgaard, J., Gough, D.O. (1982): Mon. Not. R. Astron. Soc. **198**, 141

Christensen-Dalsgaard, J., Gough, D.O., Morgan, J.G. (1979): Astron. Astrophys. **73**, 121

Christensen-Dalsgaard, J., Gough, D.O., Thompson, M.J. (1991): Astrophys. J. **378**, 413

Christensen-Dalsgaard, J., Monteiro, M.J.P.F.G., Thompson, M.J. (1995): Mon. Not. R. Astron. Soc. **276**, 283

Christensen-Dalsgaard, J., Proffitt, C.R., Thompson, M.J. (1993): Astrophys. J. **403**, L75

Christensen-Dalsgaard, J., Thompson, M.J. (1997): Mon. Not. R. Astron. Soc. **284**, 527

Cimino, M., Cacciani, A., Fofi, M. (1970): Solar Phys. **11**, 319

Cimino, M., Cacciani, A., Sopranzi, N. (1968): Applied Optics. **7**, 1654

Clark, A., Jr., Johnson, H.K. (1967): Solar Phys. **2**, 433

Claverie, A., Isaak, G.R., McLeod, C.P., van der Raay, H.B., Roca Cortes, T. (1979): Nature **282**, 591

Clayton, D.D. (1968): *Principles of Stellar Evolution and Nucleosynthesis* (McGraw-Hill, New York)

Clayton, D.D. (1979): Space Sci. Rev. **24**, 147

Clement, M.J. (1986): Astrophys. J. **301**, 185

Cohen, E.R., Taylor, B.N. (1987): Rev. Mod. Phys. **59**, 1121

Collados, M., Vázquez, M. (1987): Astron. Astrophys. **180**, 223

Cowling, T.G. (1934): Mon. Not. R. Astron. Soc. **94**, 39

Cowling, T.G. (1941): Mon. Not. R. Astron. Soc. **101**, 367

Cowling, T.G. (1976): *Magnetohydrodynamics* (Adam Hilger, Bristol)

Cowling, T.G. (1981): Ann. Rev. Astron. Astrophys. **19**, 115

Cox, A.N., ed. (2000): *Allen's Astrophysical Quantities*, 4th ed. (Springer, New York, Berlin, Heidelberg)

Cox, A.N., Guzik, J.A., Kidman, R.B. (1991): Astrophys. J. **342**, 1187

Cox, J.P. (1980): *Theory of Stellar Pulsation* (Princeton University Press, Princeton)

Cox, J.P., Giuli, R.T. (1968): *Principles of Stellar Structure* (Gordon and Breach, New York)

Cram, L.E., Durney, B.R., Guenther, D.B. (1983): Astrophys. J. **267**, 442

Cram, L.E., Thomas, J.H., eds. (1981): *The Physics of Sunspots* (Sacramento Peak Observatory, Sunspot, New Mexico)

Cranmer, S.R. (2000): Astrophys. J. **532**, 1197

Cranmer, S.R., Field, G.B., Kohl, J.L. (1999 a): Astrophys. J. **518**, 937

Cranmer, S.R., Kohl, J.L., Noci, G., and 27 others (1999 b): Astrophys. J. **511**, 481

Curdt, W., Brekke, P., Feldman, U., Wilhelm, K., Dwivedi, B.N., Schüle, U., Lemaire, P. (2001): Astron. Astrophys. **375**, 591

Cushman, G.W., Rense, W.A. (1976): Astrophys. J. **207**, L61

Danielson, R.E. (1964): Astrophys. J. **139**, 45

Däppen, W. (1980): Astron. Astrophys. **91**, 212

Däppen, W., Gilliland, R.L., Christensen-Dalsgaard, J. (1986): Nature **321**, 229

Däppen, W., Mihalas, D., Hummer, D.G., Mihalas, B.W. (1988): Astrophys. J. **332**, 261

Dar, A., Shaviv, G. (1996): Astrophys. J. **468**, 933

Darvann, T., Owner-Petersen, M. (1994): LEST Foundation, Technical Rep. No. 57

de Boer, C.R., Kneer, F., Nesis, A. (1992): Astron. Astrophys. **257**, L4

Degenhardt, D., Solanki, S.K., Montesinos, B., Thomas, J.H. (1993): Astron. Astrophys. **279**, L29

Degenhardt, D., Wiehr, E. (1991): Astron. Astrophys. **252**, 821

Deinzer, W. (1965): Astrophys. J. **141**, 548

Deinzer, W., Hensler, G., Schüssler, M., Weisshaar, E. (1984): Astron. Astrophys. **139**, 435

Delaboudinière, J.-P., Artzner, G.E., Brunaud, J., and 25 others (1995): Solar Phys. **162**, 291

Delache, P., Scherrer, P.H. (1983): Nature **306**, 651

Delbouille, L., Neven, L., Roland, G. (1973): *Photometric Altas of the Solar Spectrum from λ 3000 to λ 10000*, Université de Liège

Demarque, P., Guenther, D.B., van Altena, W.F. (1986): Astrophys. J. **300**, 773

Demarque, P., Mengel, J.G., Sweigart, A.V. (1973): Astrophys. J. **183**, 997

Den Hartog, E.A., Curry, J.J., Wickliffe, M.E., Lawler, J.E. (1998): Solar Phys. **178**, 239

Denker, C. (1996): Dissertation, Universität Göttingen

Denskat, K.U. (1982): Dissertation, Universität Braunschweig

Dere, K.P., Bartoe, J.-D.F., Brueckner, G.E. (1989): Solar Phys. **123**, 41

Deubner, F.-L. (1975): Astron. Astrophys. **44**, 371

Deubner, F.-L., Fleck, B. (1989): Astron. Astrophys. **213**, 423

Deubner, F.-L., Mattig, W. (1975): Astron. Astrophys. **45**, 167

Deubner, F.-L., Ulrich, R.K., Rhodes, E.J., Jr. (1979): Astron. Astrophys. **72**, 177

Dialetis, D., Macris, C., Prokakis, T., Sarris, E. (1986): Astron. Astrophys. **168**, 330

Dicke, R.H. (1974): Science **184**, 419

Dicke, R.H., Goldenberg, H.M. (1967): Phys. Rev. Letters **18**, 313

Dicke, R.H., Kuhn, J.R., Libbrecht, K.G. (1985): Nature **316**, 687

Dicke, R.H., Kuhn, J.R., Libbrecht, K.G. (1986): Astrophys. J. **311**, 1025

Diesendorf, M.O. (1970): Nature **227**, 266

Dikpati, M., Charbonneau, P. (1999): Astrophys. J. **518**, 508

Dilke, F.W.W., Gough, D.O. (1972): Nature **240**, 262

Dodd, R.T. (1981): *Meteorites* (Cambridge University Press, Cambridge)

Dorfi, E. (1989): Astron. Astrophys. **225**, 507

Dravins, D. (1982): Ann. Rev. Astron. Astrophys. **20**, 61

Dravins, D. (1987a): Astron. Astrophys. **172**, 200

Dravins, D. (1987b): Astron. Astrophys. **172**, 211

Dravins, D., Larsson, B., Nordlund, Å. (1986): Astron. Astrophys. **158**, 83

Dravins, D., Lindegren, L., Nordlund, Å. (1981): Astron. Astrophys. **96**, 345

Dulk, G.A., McLean, D.J. (1978): Solar Phys. **57**, 279

Dulk, G.A., McLean, D.J., Nelson, G.J. (1985): In McLean and Labrum (1985), p. 53

Duncombe, R.L., Fricke, W., Seidelmann, P.K., Wilkins, G.A. (1977): Trans. IAU **XVI** B, p. 49

Dunn, R.B. (1969): Sky and Telescope **38**, 368

Dunn, R.B., ed. (1981): *Solar Instrumentation: What's Next?* (Sacramento Peak Observatory, Sunspot, New Mexico)

Dunn, R.B. (1985): Solar Phys. **100**, 1

Dunn, R.B., Zirker, J.B. (1973): Solar Phys. **33**, 281

Durney, B.R. (1987): In Durney and Sofia (1987), p. 235

Durney, B.R., Cram, L.E., Guenther, D.B., Keil, S.L., Lytle, D.M. (1985): Astrophys. J. **292**, 752

Durney, B.R., Pneuman, G.W. (1975): Solar Phys. **40**, 461

Durney, B.R., Roxburgh, I.W. (1971): Solar Phys. **16**, 3

Durney, B.R., Sofia, S., eds. (1987): *The Internal Solar Angular Velocity* (Reidel, Dordrecht)

Durney, B.R., Stenflo, J.O. (1972): Astrophys. Space Sci. **15**, 307

Durrant, C.J., Mattig, W., Nesis, A., Reiss, G., Schmidt, W. (1979): Solar Phys. **61**, 251

Durrant, C.J., Mattig, W., Nesis, A., Schmidt, W. (1983): Astron. Astrophys. **123**, 319

Duvall, T.L., Jr. (1982): Nature **300**, 242

Duvall, T.L., Jr., Dziembowski, W.A., Goode, P.R., Gough, D.O., Harvey, J.W., Leibacher, J.W. (1984): Nature **310**, 22

Duvall, T.L., Jr., Gizon, L. (2000): Solar Phys. **192**, 177

Duvall, T.L., Jr., Harvey, J.W. (1983): Nature **302**, 24

Duvall, T.L., Jr., Harvey, J.W. (1984): Nature **310**, 19

Duvall, T.L., Jr., Harvey, J.W., Jefferies, S.M., Pomerantz, M.A. (1991): Astrophys. J. **373**, 308

Duvall, T.L., Jr., Harvey, J.W., Libbrecht, K.G., Popp, B.D., Pomerantz, M.A. (1988): Astrophys. J. **324**, 1158

Duvall, T.L., Jr., Harvey, J.W., Pomerantz, M.A. (1986): Nature **321**, 500

Duvall, T.L., Jr., Jefferies, S.M., Harvey, J.W., Osaki, Y., Pomerantz, M.A. (1993a): Astrophys. J. **410**, 829

Duvall, T.L., Jr., Jefferies, S.M., Harvey, J.W., Pomerantz, M.A. (1993b): Nature **362**, 430

Duvall, T.L., Jr., Kosovichev, A.G., Scherrer, P.H., and 10 others (1997): Solar Phys. **170**, 63

Dziembowski, W.A., Goode, P.R., Kosovichev, A.G., Schou, J. (2000): Astrophys. J. **537**, 1026

Dziembowski, W., Goode, P.R., Pamyatnykh, A.A., Sienkiewicz, R. (1994): Astrophys. J. **432**, 417

Dziembowski, W., Goode, P.R., Pamyatnykh, A.A., Sienkiewicz, R. (1995): Astrophys. J. **445**, 509

Dziembowski, W., Sienkiewicz, R. (1973): Acta Astron. **23**, 273

Dzitko, H., Turck-Chièze, S., Delbourgo-Salvador, P., Lagrange, C. (1995): Astrophys. J. **447**, 428

Eckart, C. (1960): *Hydrodynamics of Oceans and Atmospheres* (Pergamon Press, Oxford)

Eddy, J.A. (1976): Science **192**, 1189

Eddy, J.A. (1977): Scientific American **236**, No. 5, p. 80

Edlén, B. (1942): Z. Astrophys. **22**, 30

Eggleton, P.P. (1971): Mon. Not. R. Astron. Soc. **151**, 351

Eggleton, P.P. (1972): Mon. Not. R. Astron. Soc. **156**, 361

Elliott, J.R., Miesch, M.S., Toomre, J. (2000): Astrophys. J. **533**, 546

Elmore, D.F., Lites, B.W., Tomczyk, S., and 8 others (1992): Proc. SPIE **1746**, 22

Elwert, G. (1952): Z. Naturforsch. **7a**, 432

Elwert, G., Müller, K., Thür, L., Balz, P. (1982): Solar Phys. **75**, 205

Emonet, T., Moreno-Insertis, F. (1998): Astrophys. J. **492**, 804

Endler, F. (1971): Dissertation, Universität Göttingen

Engvold, O., Andersen, T., eds. (1990): *LEST Design* (The LEST Foundation, Stockholm)

Espagnet, O., Muller, R., Roudier, Th., Mein, N., Mein, P. (1995): Astron. Astrophys. Suppl. **109**, 79

Espagnet, O., Muller, R., Roudier, Th., Mein, P., Mein, N., Malherbe, J.M. (1996): Astron. Astrophys. **313**, 297

Evans, J.W. (1949): J. Opt. Soc. Am. **39**, 229

Evans, J.W. (1966): In *Atti del Conv. s. Campi Magn. Sol. e la Spett. ad Alta Ris.*, ed. by G. Barbéra (Firenze), p. 123

Evans, J.W. (1981): In Dunn (1981), p. 155

Evans, J.W., Michard, R. (1962): Astrophys. J. **135**, 812

Evershed, J. (1909): Mon. Not. R. Astron. Soc. **69**, 454

Faulkner, J., Gilliland, R.L. (1985): Astrophys. J. **299**, 994

Faulkner, J., Gough, D.O., Vahia, M.N. (1986): Nature **321**, 226

Faulkner, J., Swenson, F.J. (1992): Astrophys. J. **386**, L55

Feldman, U, Laming, J.M. (2000): Physica Scripta **61**, 222

Ferriz-Mas, A., Schmitt, D., Schüssler, M. (1994): Astron. Astrophys. **289**, 949

Ferriz-Mas, A., Schüssler, M. (1989): Geophys. Astrophys. Fluid Dyn. **48**, 217

Ferriz-Mas, A., Schüssler, M. (1994): Astrophys. J. **433**, 852

Ferriz-Mas, A., Schüssler, M. (1995): Geophys. Astrophys. Fluid Dyn. **81**, 233

Ferriz-Mas, A., Schüssler, M., Anton, V. (1989): Astron. Astrophys. **210**, 425

Fleck, B., Domingo, V., Poland, A., eds. (1995): *The SOHO Mission*, Reprinted from Solar Phys. **162** (Kluwer, Dordrecht)

Fligge, M., Solanki, S.K., Unruh, Y.C. (2000): Astron. Astrophys. **353**, 380

Fontenla, J.M., Avrett, E.H., Loeser, R. (1990): Astrophys. J. **355**, 700

Fontenla, J.M., Avrett, E.H., Loeser, R. (1993): Astrophys. J. **406**, 319

Forestini, M., Lumer, E., Arnould, M. (1991): Astron. Astrophys. **252**, 127

Fossat, E. (1991): Solar Phys. **133**, 1

Fossat, E., Gelly, B., Grec, G., Pomerantz, M. (1987): Astron. Astrophys. **177**, L47

Foukal, P. (1981): In Cram and Thomas (1981), p. 391
Foukal, P., Lean, J. (1986): Astrophys. J. **302**, 826
Foukal, P., Lean, J. (1988): Astrophys. J. **328**, 347
Frazier, E.N. (1970): Solar Phys. **14**, 89
Fried, D.L. (1966): J. Opt. Soc. America **56**, 1372
Friedlander, G., ed. (1978): Conf. Solar Neutrino Res., Brookhaven, N. Y., Vol. I, II
Fröhlich, C. (1988): In Christensen-Dalsgaard and Frandsen (1988), p. 83
Fröhlich, C., Huber, M.C.E., Solanki, S.K., von Steiger, R., eds. (1998): *Solar Composition and its Evolution – from Core to Corona*, Reprinted from Space Sci. Rev. **85**, (Kluwer, Dordrecht)
Fröhlich, C., Lean, J. (1998): Geophys. Res. Letters **25**, 4377
Frutiger, C., Solanki, S.K. (2001): Astron. Astrophys. **369**, 646

Gabriel, M. (1992): Astron. Astrophys. **265**, 771
Gadun, A.S., Solanki, S.K., Johannesson, A. (1999): Astron. Astrophys. **350**, 1018
Galloway, D.J., Proctor, M.R.E. (1983): Geophys. Astrophys. Fluid Dyn. **24**, 109
Galloway, D.J., Weiss, N.O. (1981): Astrophys. J. **243**, 945
Galsgaard, K., Priest, E.R., Nordlund, Å. (2000): Solar Phys. **193**, 1
Gandorfer, A. (2000): *The Second Solar Spectrum*, Vol. I (vdf Hochschulverlag, ETH Zürich)
Gandorfer, A., Povel, H.P. (1997): Astron. Astrophys. **328**, 381
Geiss, J. (1982): Space Sci. Rev. **33**, 201
Geiss, J. (1998): Space Sci. Rev. **85**, 241
Gilliland, R.L., Faulkner, J., Press, W.H., Spergel, D.N. (1986): Astrophys. J. **306**, 703
Gilman, P.A. (1974): Ann. Rev. Astron. Astrophys. **12**, 47
Gilman, P.A. (1983): Astrophys. J. Suppl. **53**, 243
Gilman, P.A., Howard, R. (1984a): Astrophys. J. **283**, 385
Gilman, P.A., Howard, R. (1984b): Solar Phys. **93**, 171
Gingerich, O., de Jager, C. (1968): Solar Phys. **3**, 5
Gingerich, O., Noyes, R.W., Kalkofen, W., Cuny, Y. (1971): Solar Phys. **18**, 347
Giovanelli, R.G. (1972): Solar Phys. **27**, 71
Glatzmaier, G.A. (1985): Astrophys. J. **291**, 300
Godier, S., Rozelot, J.-P. (2000): Astron. Astrophys. **355**, 365
Gokhale, M.H., Zwaan, C. (1972): Solar Phys. **26**, 52
Goldreich, P., Keeley, D.A. (1977a): Astrophys. J. **211**, 934
Goldreich, P., Keeley, D.A. (1977b): Astrophys. J. **212**, 243
Golub, L., Herant, M., Kalata, K., Lovas, I., Nystrom, G., Pardo, F., Spiller, E., Wilczynski, J. (1990): Nature **344**, 842
Golub, L., Rosner, R., Vaiana, G.S., Weiss, N.O. (1981): Astrophys. J. **243**, 309
Gomez, M.T., Marmolino, C., Roberti, G., Severino, G. (1987): Astron. Astrophys. **188**, 169
Gonsalves, R.A., Chidlaw, R. (1979): In *Applications of Digital Image Processing III*, ed. by A.G. Tescher, Proc. SPIE, p. 32
Goode, P.R., Strous, L.H., Rimmele, T.R., Stebbins, R.T. (1998): Astrophys. J. **495**, L27
Gosling, J.T. (1993): J. Geophys. Res. **98**, 18937
Gough, D.O. (1978): In Belvedere and Paternó (1978), p. 337
Gough, D.O. (1981): Mon. Not. R. Astron. Soc. **196**, 731
Gough, D.O., ed. (1983a): Solar Phys. **82**
Gough, D.O. (1983b): Phys. Bull. **34**, 502
Gough, D.O. (1984a): Phil. Trans. R. Soc. London A **313**, 27
Gough, D.O. (1984b): In Ulrich (1984), p. 49

Gough, D.O. (1984 c): Mem. Soc. Astron. Ital. **55**, 13

Gough, D.O., ed. (1986): *Seismology of the Sun and the Distant Stars* (Reidel, Dordrecht)

Gough, D.O., Kosovichev, A.G., Toomre, J., and 23 others (1996): Science **272**, 1296

Gough, D.O., Toomre, J. (1983): Solar Phys. **82**, 401

Gray, D.F., Linsky, J.L., eds. (1980): *Stellar Turbulence*, Lect. Notes Phys., Vol. 114 (Springer, Berlin, Heidelberg)

Grec, G., Fossat, E., Pomerantz, M.A. (1983): Solar Phys. **82**, 55

Grec, G., Fossat, E., Vernin, J. (1976): Astron. Astrophys. **50**, 221

Greenstein, J.L., Richardson, R.S. (1951): Astrophys. J. **113**, 536

Grevesse, N., Sauval, A.J. (1998): Space Sci. Rev. **85**, 161

Grossman, S.A. (1996): Mon. Not. R. Astron. Soc. **279**, 305

Grossmann-Doerth, U. (1994): Astron. Astrophys. **285**, 1012

Grossmann-Doerth, U., Kneer, F., von Uexküll, M. (1974): Solar Phys. **37**, 85

Grossmann-Doerth, U., Schüssler, M., Sigwarth, M., Steiner, O. (2000): Astron. Astrophys. **357**, 351

Grossmann-Doerth, U., Schüssler, M., Steiner, O. (1998): Astron. Astrophys. **337**, 928

Grotrian, W. (1931): Z. Astrophys. **3**, 199

Grotrian, W. (1939): Naturwiss. **27**, 214

Gruzinov, A.V., Bahcall, J.N. (1998): Astrophys. J. **504**, 996

Guenther, D.B., Demarque, P. (1997): Astrophys. J. **484**, 937

Guenther, D.B., Kim, Y.-C., Demarque, P. (1996): Astrophys. J. **463**, 382

Guzik, J.A., Cox, A.N. (1993): Astrophys. J. **411**, 394

Haber, D.A., Hindman, B.W., Toomre, J., Bogart, R.S., Thompson, M.J., Hill, F. (2000): Solar Phys. **192**, 335

Haerendel, G. (1992): Nature **360**, 241

Hagenaar, H.J., Schrijver, C.J., Title, A.M. (1997): Astrophys. J. **481**, 988

Hagyard, M.J., ed. (1985): *Measurements of Solar Vector Magnetic Fields*, NASA Conf. Publ. 2374

Hagyard, M.J., Pevtsov, A.A. (1999): Solar Phys. **189**, 25

Hale, G.E. (1924): Nature **113**, 105

Hale, G.E., Ellerman, F., Nicholson, S.B., Joy, A.H. (1919): Astrophys. J. **49**, 153

Hale, G.E., Nicholson, S.B. (1938): Publ. Carnegie Inst. No. 498 (Washington)

Hammer, R. (1982 a): Astrophys. J. **259**, 767

Hammer, R. (1982 b): Astrophys. J. **259**, 779

Hammer, R. (1984): Astrophys. J. **280**, 780

Hammer, R. (1987 a): In Schröter and Schüssler (1987), p. 77

Hammer, R. (1987 b): In Schröter et al. (1987), p. 255

Hammer, R., Nesis, A. (2003): In *Cool Stars, Stellar Systems, and the Sun: The Future of Cool-Star Astrophysics*, ed. by A. Brown, G.M. Harper, T.R. Ayres, p. 613, http://origins.colorado.edu/cs12/proceedings/poster/hammer.ps

Hammerschlag, R.A. (1981): In Dunn (1981), p. 583

Hampel, W. (1986): In *Weak and Electromagnetic Interactions in Nuclei*, ed. by H.V. Klapdor (Springer, Berlin, Heidelberg), p. 718

Handy, B.N., Acton, L.W., Kankelborg, C.C., and 45 others (1999): Solar Phys. **187**, 229

Hanle, W. (1924): Z. Physik **30**, 93

Hansen, C.J., Cox, J.P., Van Horn, H.M. (1977): Astrophys. J. **217**, 151

Hardorp, J. (1982): Astron. Astrophys. **105**, 120

Harrison, R.A., Sawyer, E.C., Carter, M.K., and 36 others (1995): Solar Phys. **162**, 233

Hart, A.B. (1956): Mon. Not. R. Astron. Soc. **116**, 38

Harvey, J.W. (1973): Solar Phys. **28**, 9

Harvey, J. (1977): Highlights of Astronomy **4**, Part II, p. 223

Harvey, J., and the GONG Instrument Development Team (1988): In *Seismology of the Sun and Sun-like Stars*, ed. by E.J. Rolfe, ESA SP-286, p. 203

Harvey, J.W., Hill, F., Hubbard, R.P., and 14 others (1996): Science **272**, 1284

Harvey, K.L., ed. (1992 a): *The Solar Cycle*, ASP Conf. Ser. **27**

Harvey, K.L. (1992 b): In Harvey (1992 a), p. 335

Hassler, D.M., Dammasch, I.E., Lemaire, P., Brekke, P., Curdt, W., Mason, H.E., Vial, J.-C., Wilhelm, K. (1999): Science **283**, 810

Hathaway, D.H. (1987): Solar Phys. **108**, 1

Haug, E., Elwert, G. (1985): Solar Phys. **99**, 219

Heasley, J.N., Milkey, R.W. (1978): Astrophys. J. **221**, 677

Heath, D.F., Thekaekara, M.P. (1977): In White (1977), p. 193

Heinzel, P., Anzer, U. (2001): Astron. Astrophys. **375**, 1082

Henning, H.M., Scherrer, P.H. (1986): In Gough (1986), p. 55

Hénoux, J.-C. (1969): Astron. Astrophys. **2**, 288

Hénoux, J.-C. (1998): Space Sci. Rev. **85**, 215

Henyey, L.G., Wilets, L., Böhm, K.H., LeLevier, R., Levee, R.D. (1959): Astrophys. J. **129**, 628

Herbold, G., Ulmschneider, P., Spruit, H.C., Rosner, R. (1985): Astron. Astrophys. **145**, 157

Hess, W.N., ed. (1964): *The Physics of Solar Flares*, NASA Sp-50 (NASA, Washington)

Heyvaerts, J., Priest, E.R. (1983): Astron. Astrophys. **117**, 220

Heyvaerts, J., Priest, E.R., Rust, D.M. (1977): Astrophys. J. **216**, 123

Hill, F. (1988): Astrophys. J. **333**, 996

Hill, F. (1989): Astrophys. J. **343**, L69

Hill, H.A., Dziembowski, W.A., eds. (1980): *Nonradial and Nonlinear Stellar Pulsation*, Lect. Notes Phys., Vol. 125 (Springer, Berlin, Heidelberg)

Hill, H.A., Stebbins, R.T. (1975): Astrophys. J. **200**, 471

Hillebrandt, W., Meyer-Hofmeister, E., Thomas, H.-C., eds. (1987): *Physical Processes in Comets, Stars, and Active Galaxies* (Springer, Berlin, Heidelberg)

Hirzberger, J., Bonet, J.A., Vázquez, M., Hanslmeier, A. (1999 a): Astrophys. J. **515**, 441

Hirzberger, J., Bonet, J.A., Vázquez, M., Hanslmeier, A. (1999 b): Astrophys. J. **527**, 405

Hirzberger, J., Vázquez, M., Bonet, J.A., Hanslmeier, A., Sobotka, M. (1997): Astrophys. J. **480**, 406

Hoeksema, J.T., Domingo, V., Fleck, B., Battrick, B., eds. (1995): *Helioseismology*, ESA SP-376, Vol. I, II

Hoekzema, N.M., Brandt, P.N. (2000): Astron. Astrophys. **353**, 389

Hoekzema, N.M., Brandt, P.N., Rutten, R.J. (1998): Astron. Astrophys. **333**, 322

Hoffmeister, C. (1943): Z. Astrophys. **22**, 265

Hollweg, J.V. (1982): Astrophys. J. **257**, 345

Holst, G.C. (1998): *CCD Arrays, Cameras, and Displays*, 2nd ed. (SPIE Optical Engineering Press, Bellingham)

Holweger, H. (1967): Z. Astrophys. **65**, 365

Holweger, H., Livingston, W., Steenbock, W. (1983): Nature **302**, 125

Horn, T., Staude, J., Landgraf, V. (1997): Solar Phys. **172**, 69

Houdek, G., Balmforth, N.J., Christensen-Dalsgaard, J., Gough, D.O. (1999): Astron. Astrophys. **351**, 582

Hovestadt, D. (1974): In *Solar Wind Three*, ed. by C.T. Russell (University of California, Los Angeles), p. 2

Howard, R., ed. (1971): *Solar Magnetic Fields*, IAU Symp. **43** (Reidel, Dordrecht)

Howard, R. (1974): Solar Phys. **38**, 283

Howard, R., Gilman, P.A. (1986): Astrophys. J. **307**, 389

Howard, R., Gilman, P.A., Gilman, P.I. (1984): Astrophys. J. **283**, 373

Howard, R., LaBonte, B.J. (1980): Astrophys. J. **239**, L33

Howard, R., LaBonte, B.J. (1981): Solar Phys. **74**, 131

Howard, R., LaBonte, B.J. (1983): In Stenflo (1983), p. 101

Howe, R., Christensen-Dalsgaard, J., Hill, F., Komm, R.W., Larsen, R.M., Schou, J., Thompson, M.J., Toomre, J. (2000 a): Science **287**, 2456

Howe, R., Christensen-Dalsgaard, J., Hill, F., Komm, R.W., Larsen, R.M., Schou, J., Thompson, M.J., Toomre, J. (2000 b): Astrophys. J. **533**, L163

Howe, R., Komm, R., Hill, F. (1999): Astrophys. J. **524**, 1084

Howell, S.B. (2000): *Handbook of CCD Astronomy* (Cambridge University Press, Cambridge)

Hoyt, D.V., Schatten, K.H. (1996): Solar Phys. **165**, 181

Huebner, W.F. (1986): In Sturrock (1986), Vol. I, p. 33

Huebner, W.F., Merts, A.L., Magee, N.H., Jr., Argo, M.F. (1977): *Astrophysical Opacity Library*, Los Alamos Scientific Laboratory Rep. LA-6760-M

Hummer, D.G., Mihalas, D. (1988): Astrophys. J. **331**, 794

Hundhausen, A.J. (1972): *Coronal Expansion and Solar Wind* (Springer, Berlin, Heidelberg)

Hurlburt, N.E., Toomre, J., Massaguer, J.M. (1984): Astrophys. J. **282**, 557

Iglesias, C.A., Rogers, F.J. (1996): Astrophys. J. **464**, 943

Illing, R.M.E., Landman, D.A., Mickey, D.L. (1975): Astron. Astrophys. **41**, 183

Jahn, K. (1989): Astron. Astrophys. **222**, 264

Jahn, K., Schmidt, H.U. (1994): Astron. Astrophys. **290**, 295

Janssens, T.J. (1970): Solar Phys. **11**, 222

Jefferies, J.T. (1968): *Spectral Line Formation* (Blaisdell, Waltham)

Jefferies, J., Lites, B.W., Skumanich, A. (1989): Astrophys. J. **343**, 920

Jensen, E. (1955): Ann. d'Astrophys. **18**, 127

Jensen, E., Maltby, P., Orrall, F.Q., eds. (1979): *Physics of Solar Prominences*, IAU Coll. **44** (Institute of Theoretical Astrophysics, Oslo)

Jiménez, A., Pallé, P.L., Pérez, J.C., Régulo, C., Roca Cortés, T., Isaak, G.R., McLeod, C.P., van der Raay, H.B. (1988): In Christensen-Dalsgaard and Frandsen (1988), p. 205

Jones, H.P., Duvall, T.L., Jr., Harvey, J.W., Mahaffey, C.T., Schwitters, J.D., Simmons, J.E. (1992): Solar Phys. **139**, 211

Jordan, S., ed. (1981): *The Sun as a Star*, NASA SP-450 (NASA, Washington)

Junker, M., D'Alessandro, A., Zavatarelli, S., and 16 others (1998): Phys. Rev. C **57**, 2700

Kaisig, M., Durrant, C.J. (1982): Astron. Astrophys. **116**, 332

Kaisig, M., Knölker, M., Stix, M. (1984): In Noels and Gabriel (1984), p. 239

Kalkofen, W., ed. (1984): *Methods in Radiative Transfer* (Cambridge University Press, Cambridge)

Kalkofen, W., ed. (1987): *Numerical Radiative Transfer* (Cambridge University Press, Cambridge)

Kalkofen, W., Ulmschneider, P., Avrett, E.H. (1999): Astrophys. J. **521**, L141

Kayser, B. (1981): Phys. Rev. D **24**, 110

Keil, S.L. (1980): Astrophys. J. **237**, 1035

Keil, S.L., ed. (1984): *Small-Scale Dynamical Processes in Quiet Stellar Atmospheres* (Sacramento Peak Observatory, Sunspot, New Mexico)

Keil, S.L., Canfield, R.C. (1978): Astron. Astrophys. **70**, 169

Keil, S.L., Rimmele, T.R., Keller, C.U., and the ATST Team (2001): In Sigwarth (2001 a), p. 597

Keller, C.U. (1995): Rev. Mod. Astron. **8**, 27

Keller, C.U., and the SOLIS Team (2001): In Sigwarth (2001 a), p. 16

Keller, C.U., von der Lühe, O. (1992): Astron. Astrophys. **261**, 321

Kentischer, T.J., Schmidt, W., Sigwarth, M., von Uexküll, M. (1998): Astron. Astrophys. **340**, 569

Keppens, R., Martínez Pillet, V. (1996): Astron. Astrophys. **316**, 229

Khomenko, E.V., Kostik, R.I., Shchukina, N.G. (2001): Astron. Astrophys. **369**, 660

Kiefer, M., Grabowski, U., Mattig, W., Stix, M. (2000 a): Astron. Astrophys. **355**, 381

Kiefer, M., Stix, M., Balthasar, H. (2000 b): Astron. Astrophys. **359**, 1175

Kiepenheuer, K.O. (1953): Astrophys. J. **117**, 447

Kiepenheuer, K.O., ed. (1968): *Structure and Development of Solar Active Regions*, IAU Symp. **35** (Reidel, Dordrecht)

Kiepenheuer, K.O., Schellenberger, W. (1973): *Second Aircraft Campaign (1973)*, Rep. JOSO SIT 17 (Fraunhofer-Institut, Freiburg)

Kim, Y.-C., Demarque, P., Guenther, D.B. (1991): Astrophys. J. **378**, 407

Kim, Y.-C., Fox, P.A., Demarque, P., Sofia, S. (1996): Astrophys. J. **461**, 499

King, H.C. (1979): *The History of the Telescope* (Dover, New York)

Kippenhahn, R. (1963): Astrophys. J. **137**, 664

Kippenhahn, R., Schlüter, A. (1957): Z. Astrophys. **43**, 36

Kippenhahn, R., Thomas, H.-C. (1970): In *Stellar Rotation*, ed. by A. Slettebak (Reidel, Dordrecht), p. 20

Kippenhahn, R., Weigert, A. (1990): *Stellar Structure and Evolution* (Springer, Berlin, Heidelberg)

Kippenhahn, R., Weigert, A., Hofmeister, E. (1967): Meth. Comp. Phys. **7**, 129

Kitchatinov, L.L., Rüdiger, G. (1993): Astron. Astrophys. **276**, 96

Kitchatinov, L.L., Rüdiger, G. (1995): Astron. Astrophys. **299**, 446

Kitchatinov, L.L., Rüdiger, G. (1999): Astron. Astrophys. **344**, 911

Kitchin, C.R. (1984): *Astrophysical Techniques* (Adam Hilger, Bristol)

Kitchin, C.R. (1995): *Optical Astronomical Spectroscopy* (Institute of Physics, Bristol)

Klapdor-Kleingrothaus, H.V., Zuber, K. (1997): *Teilchenastrophysik* (Teubner, Stuttgart)

Klassen, A., Aurass, H., Mann, G., Thompson, B.J. (2000): Astrophys. J. Suppl. **141**, 357

Kneer, F. (1978): Osserv. Mem. Oss. Astrofis. Arcetri, Fasc. **106**, 204

Kneer, F. (1983): Astron. Astrophys. **128**, 311

Kneer, F., Heasley, J.N. (1979): Astron. Astrophys. **79**, 14

Kneer, F., von Uexküll, M. (1993): Astron. Astrophys. **274**, 584

Knobloch, E. (1977): J. Fluid Mech. **83**, 129

Knobloch, E. (1978): Astrophys. J. **225**, 1050

Knobloch, E., Tobias, S.M., Weiss, N.O. (1998): Mon. Not. R. Astron. Soc. **297**, 1123

Knölker, M., Stix, M. (1984): Mem. Soc. Astron. Ital. **55**, 305

Knox, K.T., Thompson, B.J. (1974): Astrophys. J. **193**, L45

Kohl, J.L., Esser, R., Gardner, L.D., and 37 others (1995): Solar Phys. **162**, 313

Kohl, J.L., Noci, G., Antonucci, E., and 23 others (1997): Solar Phys. **175**, 613
Kohl, J.L., Noci, G., Antonucci, E., and 27 others (1998): Astrophys. J. **501**, L127
Komm, R.W., Howard, R.F., Harvey, J.W. (1993 a): Solar Phys. **143**, 19
Komm, R.W., Howard, R.F., Harvey, J.W. (1993 b): Solar Phys. **147**, 207
Komm, R., Mattig, W., Nesis, A. (1991 a): Astron. Astrophys. **243**, 251
Komm, R., Mattig, W., Nesis, A. (1991 b): Astron. Astrophys. **252**, 827
Kopecký, M. (1957): Bull. Astron. Inst. Czech. **8**, 71
Kopecký, M. (1966): Bull. Astron. Inst. Czech. **17**, 270
Kopecký, M., Soytürk, E. (1971): Bull. Astron. Inst. Czech. **22**, 154
Kopp, R.A. (1977): In Zirker (1977), p. 179
Kopp, R.A., Pneuman, G.W. (1976): Solar Phys. **50**, 85
Korzennik, S., Wilson, A., eds. (1998): *Structure and Dynamics of the Interior of the Sun and Sun-like Stars*, ESA SP-418, Vol. I, II
Kosovichev, A.G. (1996): Astrophys. J. **461**, L55
Kosovichev, A.G., Duvall, T.L., Jr., Scherrer, P.H. (2000): Solar Phys. **192**, 159
Kosovichev, A.G., Schou, J., Scherrer, P.H., and 31 others (1997): Solar Phys. **170**, 43
Kosovichev, A.G., Zharkova, V.V. (1998): Nature **393**, 317
Kotov, V.A. (1985): Solar Phys. **100**, 101
Kotov, V.A., Haneychuk, V.I., Tsap, T.T., Hoeksema, J.T. (1997): Solar Phys. **176**, 45
Koutchmy, S., Lebecq, C. (1986): Astron. Astrophys. **169**, 323
Kovitya, P., Cram, L. (1983): Solar Phys. **84**, 45
Krafft, M. (1968): Solar Phys. **5**, 462
Krause, F. (1967): Habilitationsschrift, Universität Jena
Krause, F., Rädler, K.-H. (1980): *Mean-Field Magnetohydrodynamics and Dynamo Theory* (Akademie-Verlag, Berlin)
Krause, F., Rädler, K.-H., Rüdiger, G., eds. (1993): *The Cosmic Dynamo*, IAU Symp. **157** (Kluwer, Dordrecht)
Krause, F., Rüdiger, G. (1975): Solar Phys. **42**, 107
Krieg, J., Wunnenberg, M., Kneer, F., Koschinsky, M., Ritter, C. (1999): Astron. Astrophys. **343**, 983
Krüger, A. (1979): *Introduction to Solar Radio Astronomy and Radio Physics* (Reidel, Dordrecht)
Kubičela, A. (1973): In *Solar Activity and Related Interplanetary and Terrestrial Phenomena*, ed. by J. Xanthakis (Springer, Berlin, Heidelberg), p. 123
Kubičela, A. (1976): Solar Phys. **47**, 551
Kudoh, T., Shibata, K. (1999): Astrophys. J. **514**, 493
Kuhn, J.R., Bush, R.I., Scheick, X., Scherrer, P. (1998): Nature **392**, 155
Kuhn, J.R., Penn, M.J., eds. (1995): *Infrared Tools for Solar Astrophysics: What's Next?* (World Scientific, Singapore)
Kumar, P. (1994): Astrophys. J. **428**, 827
Kumar, P., Lu, E. (1991): Astrophys. J. **375**, L35
Kundu, M.R. (1965): *Solar Radio Astronomy* (Interscience, New York)
Kuperus, M., Raadu, M.A. (1974): Astron. Astrophys. **31**, 189
Kupke, R., LaBonte, B.J., Mickey, D.L. (2000): Solar Phys. **191**, 97
Kurucz, R.L. (1979): Astrophys. J. Suppl. **40**, 1
Küveler, G. (1983): Solar Phys. **88**, 13

Labeyrie, A. (1970): Astron. Astrophys. **6**, 85
LaBonte, B.J., Howard, R., Gilman, P.A. (1981): Astrophys. J. **250**, 796
LaBonte, B.J., Mickey, D.L., Leka, K.D. (1999): Solar Phys. **189**, 1
Labs, D., Neckel, H. (1962): Z. Astrophys. **55**, 269

Laclare, F., Delmas, C., Coin, J.P., Irbah, A. (1996): Solar Phys. **166**, 211
Lamers, H.J.G.L.M., Cassinelli, J.P. (1999): *Introduction to Stellar Winds* (Cambridge University Press, Cambridge)
Landau, L.D., Lifschitz, E.M. (1966 a): *Statistische Physik* (Akademie-Verlag, Berlin)
Landau, L.D., Lifschitz, E.M. (1966 b): *Hydrodynamik* (Akademie-Verlag, Berlin)
Landau, L.D., Lifschitz, E.M. (1967): *Elektrodynamik der Kontinua* (Akademie-Verlag, Berlin)
Landi degl'Innocenti, E., Landi degl'Innocenti, M. (1972): Solar Phys. **27**, 319
Landolfi, M., Landi degl'Innocenti, E. (1982): Solar Phys. **78**, 355
Landolfi, M., Landi degl'Innocenti, E. (1986): Astron. Astrophys. **167**, 200
Landolfi, M., Landi degl'Innocenti, E., Arena, P. (1984): Solar Phys. **93**, 269
Landolt-Börnstein (1981): New Series, Group VI, **2a** (Springer, Berlin, Heidelberg)
Lang, K.R. (1999): *Astrophysical Formulae*, 3rd ed., Vols. I, II (Springer, Berlin, Heidelberg)
Lang, K.R. (2000): *The Sun from Space* (Springer, Berlin, Heidelberg)
Langhans, K., Schmidt, W., Rimmele, T., Sigwarth, M. (2001): In Sigwarth (2001 a), p. 439
Lantz, S.R., Fan, Y. (1999): Astrophys. J. Suppl. **121**, 247
Lapwood, E.R., Usami, T. (1981): *Free Oscillations of the Earth* (Cambridge University Press, Cambridge)
Lavely, E.M., Ritzwoller, M.H. (1992): Phil. Trans. R. Soc. London A **339**, 431
Lavely, E.M., Ritzwoller, M.H. (1993): Astrophys. J. **403**, 810
Lazrek, M., Pantel, A., Fossat, E., and 10 others (1996): Solar Phys. **166**, 1
Lean, J. (1997): Ann. Rev. Astron. Astrophys. **35**, 33
Lebovitz, N.R. (1966): Astrophys. J. **146**, 946
Lebreton, Y., Maeder, A. (1987): Astron. Astrophys. **175**, 99
Ledoux, P. (1962): Bull. Acad. Roy. Belg., Cl. Sc., 5me série, **48**, 240
Ledoux, P., Walraven, T. (1958): In *Handbuch der Physik*, ed. by S. Flügge (Springer, Berlin, Heidelberg), Vol. LI, p. 353
Lehnert, B., ed. (1958): *Electromagnetic Phenomena in Cosmical Physics*, IAU Symp. **6** (Cambridge University Press, Cambridge)
Lehnert, B. (1970): Cosmic Electrodynamics **1**, 397
Leibacher, J.W., Noyes, R.W., Toomre, J., Ulrich, R.K. (1985): Scientific American **253**, No. 3, p. 34
Leighton, R.B. (1969): Astrophys. J. **156**, 1
Leighton, R.B., Noyes, R.W., Simon, G.W. (1962): Astrophys. J. **135**, 474
Leka, K.D., Steiner, O. (2001): Astrophys. J. **552**, 354
Lenz, D.D., DeLuca, E.E., Golub, L., Rosner, R., Bookbinder, J.A. (1999): Astrophys. J. **517**, L155
Leroy, J.L. (1988): In *Solar and Stellar Coronal Structure and Dynamics*, ed. by R.C. Altrock (Sacramento Peak Observatory, Sunspot, New Mexico), p. 422
Leroy, J.L., Bommier, V., Sahal-Bréchot, S. (1984): Astron. Astrophys. **131**, 33
LeVeque, R.J., Mihalas, D., Dorfi, E.A., Müller, E. (1998): *Computational Methods for Astrophysical Fluid Flow*, ed. by O. Steiner and A. Gautschy (Springer, Berlin, Heidelberg)
Levine, R.H., Schulz, M., Frazier, E.N. (1982): Solar Phys. **77**, 363
Libbrecht, K.G. (1988): Astrophys. J. **334**, 510
Libbrecht, K.G., Kaufman, J.M. (1988): Astrophys. J. **324**, 1172
Libbrecht, K.G., Woodard, M.F., Kaufman, J.M. (1990): Astrophys. J. Suppl. **74**, 1129
Libbrecht, K.G., Zirin, H. (1986): Astrophys. J. **308**, 413
Liggett, M., Zirin, H. (1985): Solar Phys. **97**, 51
Lighthill, M.J. (1952): Proc. R. Soc. London A **211**, 564

Lin, H., Penn, M.J., Tomczyk, S. (2000): Astrophys. J. **541**, L83
Lin, H., Rimmele, T. (1999): Astrophys. J. **514**, 488
Lin, R.P., Schwartz, R.A., Kane, S.R., Pelling, R.M., Hurley, K.C. (1984): Astrophys. J. **283**, 421
Lites, B.W., Rutten, R.J., Kalkofen, W. (1993): Astrophys. J. **414**, 345
Lites, B.W., Thomas, J.H. (1985): Astrophys. J. **294**, 682
Lites, B.W., Thomas, J.H., Bogdan, T.J., Cally, P.S. (1998): Astrophys. J. **497**, 464
Lites, B.W., White, O.R., Packman, D. (1982): Astrophys. J. **253**, 386
Livingston, W.C. (1982): Nature **297**, 208
Livingston, W.C., Harvey, J., Pierce, A.K., Schrage, D., Gillespie, B., Simmons, J., Slaughter, C. (1976 a): Applied Optics **15**, 33
Livingston, W.C., Harvey, J., Slaughter, C., Trumbo, D. (1976 b): Applied Optics **15**, 40
Löfdahl, M.G., Scharmer, G.B. (1994): Astron. Astrophys. Suppl. **107**, 243
Longcope, D.W., Fisher, G.H., Pevtsov, A.A. (1998): Astrophys. J. **507**, 417
Lorenz, E.N. (1963): J. Atmos. Sci. **20**, 130
Low, B.C. (1984): Astrophys. J. **281**, 392
Lubow, S.H., Rhodes, E.J., Jr., Ulrich, R.K. (1980): In Hill and Dziembowski (1980), p. 300
Lühe, O. von der (1983): Astron. Astrophys. **119**, 85
Lühe, O. von der (1984): J. Opt. Soc. America A **1**, 510
Lühe, O. von der (1985): Astron. Astrophys. **150**, 229
Lühe, O. von der (1987): In Schröter et al. (1987), p. 156
Lühe, O. von der (1988): J. Opt. Soc. America A **5**, 721
Lühe, O. von der, ed. (1989): *High Spatial Resolution Solar Observations* (Sacramento Peak Observatory, Sunspot, New Mexico)
Lühe, O. von der, Schmidt, W., Soltau, D., Berkefeld, T., Kneer, F., Staude, J. (2001): Astron. Nachr. **322**, 353
Lühe, O. von der, Widener, A.L., Rimmele, Th., Spence, G., Dunn, R.B., Wiborg, P. (1989): Astron. Astrophys. **224**, 351
Lumer, E., Forestini, M., Arnould, M. (1990): Astron. Astrophys. **240**, 515
Lüst, R., ed. (1965): *Stellar and Solar Magnetic Fields*, IAU Symp. **22** (North-Holland, Amsterdam)
Lüst, R., Schlüter, A. (1954): Z. Astrophys. **34**, 263
Lydon, T.J., Fox, P.A., Sofia, S. (1992): Astrophys. J. **397**, 701
Lydon, T.J., Fox, P.A., Sofia, S. (1993): Astrophys. J. **403**, L79
Lydon, T.J., Sofia, S. (1996): Phys. Rev. Letters **76**, 177
Lynden-Bell, D., Ostriker, J.P. (1967): Mon. Not. R. Astron. Soc. **136**, 293
Lyot, B. (1932): Z. Astrophys. **5**, 73
Lyot, B. (1933): Comptes Rendus **197**, 1593
Lyot, B. (1944): Ann. d'Astrophys. **7**, 31

MacQueen, R.M., Sime, D.G., Picat, J.-P. (1983): Solar Phys. **83**, 103
Macris, C.J., Rösch, J. (1983): Comptes Rendus **296**, 265
Magain, P. (1986): Astron. Astrophys. **163**, 135
Magara, T., Shibata, K. (1999): Astrophys. J. **514**, 456
Maier, E., Twigg, L.W., Sofia, S. (1992): Astrophys. J. **389**, 447
Malherbe, J.M., Priest, E.R. (1983): Astron. Astrophys. **123**, 80
Maltby, P., Avrett, E.H., Carlsson, M., Kjeldseth-Moe, O., Kurucz, R.L., Loeser, R. (1986): Astrophys. J. **306**, 284
Mankin, W.G. (1977): In White (1977), p. 151
Manson, J.E. (1977): In White (1977), p. 261

Mariska, J.T. (1992): *The Solar Transition Region* (Cambridge University Press, Cambridge)

Márquez, I., Bonet, J.A., Vázquez, M., Wöhl, H. (1996): Astron. Astrophys. **305**, 316

Martin, S.F., Harvey, K.L. (1979): Solar Phys. **64**, 93

Martínez Pillet, V. (2000): Astron. Astrophys. **361**, 734

Martínez Pillet, V., Collados, M., Sánchez Almeida, J., and 8 others (1999): In Rimmele et al. (1999), p. 264

Martínez Pillet, V., Moreno-Insertis, F., Vázquez, M. (1993): Astron. Astrophys. **274**, 521

Mathys, G., Stenflo, J.O. (1987): Astron. Astrophys. Suppl. **67**, 557

Mattig, W. (1966): In *Atti del Conv. s. Macchie Sol.*, ed. by G. Barbéra (Firenze), p. 194

Mattig, W. (1980): Astron. Astrophys. **83**, 129

Mattig, W. (1983): Solar Phys. **87**, 187

Mattig, W., Mehltretter, J.P., Nesis, A. (1981): Astron. Astrophys. **96**, 96

Mauas, P.J., Avrett, E.H., Loeser, R. (1990): Astrophys. J. **357**, 279

Maunder, W. (1922): Mon. Not. R. Astron. Soc. **82**, 534

McIntosh, P.S. (1979): *Annotated Atlas of Hα Synoptic Charts*, Rep. UAG-70 (World Data Center A, Boulder)

McIntosh, P.S. (1981): In Cram and Thomas (1981), p. 7

McIntosh, P.S. (1990): Solar Phys. **125**, 251

McLean, D.J., Labrum, N.R., eds. (1985): *Solar Radiophysics* (Cambridge University Press, Cambridge)

Mehltretter, J.P. (1973): Solar Phys. **30**, 19

Mehltretter, J.P. (1974): Solar Phys. **38**, 43

Mehltretter, J.P. (1978): Astron. Astrophys. **62**, 311

Mein, P. (1971): Solar Phys. **20**, 3

Meister, C.-V., Staude, J., Pregla, A.V. (1999): Astron. Nachr. **320**, 43

Meyer, F. (1968): In Kiepenheuer (1968), p. 485

Meyer, F., Schmidt, H.U. (1968): Z. Angew. Math. Mech. **48**, T218

Meyer, F., Schmidt, H.U., Simon, G.W., Weiss, N.O. (1979): Astron. Astrophys. **76**, 35

Meyer, F., Schmidt, H.U., Weiss, N.O. (1977): Mon. Not. R. Astron. Soc. **179**, 741

Meyer, F., Schmidt, H.U., Weiss, N.O., Wilson, P.R. (1974): Mon. Not. R. Astron. Soc. **169**, 35

Meyer, J.-P. (1985): Astrophys. J. Suppl. **57**, 151

Michalitsanos, A.G., Kupferman, P. (1974): Solar Phys. **36**, 403

Michaud, G., Charbonneau, P. (1991): Space Sci. Rev. **57**, 1

Mickey, D.L., Canfield, R.C., LaBonte, B.J., Leka, K.D., Waterson, M.F., Weber, H.M. (1996): Solar Phys. **168**, 229

Miesch, M.S. (2000): Solar Phys. **192**, 59

Miesch, M.S., Elliott, J.R., Toomre, J., Clune, T.L., Glatzmaier, G.A., Gilman, P.A. (2000): Astrophys. J. **532**, 593

Mihalas, B.W., Toomre, J. (1981): Astrophys. J. **249**, 349

Mihalas, B.W., Toomre, J. (1982): Astrophys. J. **263**, 386

Mihalas, D. (1978): *Stellar Atmospheres*, 2nd ed. (Freeman, San Francisco)

Mihalas, D., Auer, L.H., Mihalas, B.R. (1978): Astrophys. J. **220**, 1001

Mihalas, D., Däppen, W., Hummer, D.G. (1988): Astrophys. J. **331**, 815

Mihalas, D., Hummer, D.G., Mihalas, B.W., Däppen, W. (1990): Astrophys. J. **350**, 300

Mihalas, D., Mihalas, B.W. (1984): *Foundations of Radiation Hydrodynamics* (Oxford University Press, New York)

472   References

Moffatt, H.K. (1978): *Magnetic Field Generation in Electrically Conducting Fluids* (Cambridge University Press, Cambridge)
Monteiro, M.J.P.F.G., Christensen-Dalsgaard, J., Thompson, M.J. (1994): Astron. Astrophys. **283**, 247
Montesinos, B., Thomas, J.H. (1997): Nature **390**, 485
Moore, R.L. (1981): In Cram and Thomas (1981), p. 259
Morel, P., Provost, J., Berthomieu, G. (1997): Astron. Astrophys. **333**, 444
Moreno-Insertis, F., Emonet, T. (1996): Astrophys. J. **472**, L53
Moreno-Insertis, F., Spruit, H.C. (1989): Astrophys. J. **342**, 1158
Moreno-Insertis, F., Vázquez, M. (1988): Astron. Astrophys. **205**, 289
Moreton, G.E. (1965): In Lüst (1965), p. 371
Moreton, G.E., Ramsey, H.E. (1960): Publ. Astron. Soc. Pacific **72**, 357
Müller, D.A.N., Steiner, O., Schlichenmaier, R., Brandt, P.N. (2001): Solar Phys. **203**, 211
Muller, R. (1973): Solar Phys. **32**, 409
Munro, R.H., Gosling, J.T., Hildner, E., MacQueen, R.M., Poland, A.I., Ross, C.L. (1979): Solar Phys. **61**, 201
Münzer, H., Hanslmeier, A., Schröter, E.H., Wöhl, H. (1989): Astron. Astrophys. **213**, 431

Nagasawa, S. (1955): Publ. Astron. Soc. Japan **7**, 9
Nagatani, K., Dwarakanath, M.R., Ashery, D. (1969): Nucl. Phys. A **128**, 325
Namba, O., Diemel, W.E. (1969): Solar Phys. **7**, 167
Narain, U., Ulmschneider, P. (1990): Space Sci. Rev. **54**, 377
Narain, U., Ulmschneider, P. (1996): Space Sci. Rev. **75**, 453
Neckel, H. (1986): Astron. Astrophys. **159**, 175
Neckel, H. (1995): Solar Phys. **156**, 7
Neckel, H. (1996): Solar Phys. **167**, 9
Neckel, H. (1997): Solar Phys. **171**, 257
Neckel, H., Labs, D. (1984): Solar Phys. **90**, 205
Neckel, H., Labs, D. (1994): Solar Phys. **153**, 91
Nesis, A. (1985): Dissertation, Universität Berlin
Nesis, A., Bogdan, T.J., Cattaneo, F., Hanslmeier, A., Knölker, M., Malagoli, A. (1992): Astrophys. J. **399**, L99
Nesis, A., Hammer, R., Hanslmeier, A., Schleicher, H., Sigwarth, M., Staiger, J. (1996): Astron. Astrophys. **310**, 973
Nesis, A., Hammer, R., Roth, M., Schleicher, H. (2001): Astron. Astrophys. **373**, 307
Nesis, A., Hanslmeier, A., Hammer, R., Komm, R., Mattig, W., Staiger, J. (1993): Astron. Astrophys. **279**, 599
Nesis, A., Mattig, W. (1989): Astron. Astrophys. **221**, 130
Nesme-Ribes, E., Ferreira, E.N., Vince, I. (1993): Astron. Astrophys. **276**, 211
Newkirk, G., Altschuler, M.D., Harvey, J. (1968): In Kiepenheuer (1968), p. 379
Newton, H.W., Nunn, M.L. (1951): Mon. Not. R. Astron. Soc. **111**, 413
Nigam, R., Kosovichev, A.G. (1998): Astrophys. J. **505**, L51
Nigam, R., Kosovichev, A.G. (1999): Astrophys. J. **514**, L53
Noels, A., Gabriel, M., eds. (1984): *Theoretical Problems in Stellar Stability and Oscillations*, 25th Liège Coll., Université de Liège
Noels, A., Scuflaire, R., Gabriel, M. (1984): Astron. Astrophys. **130**, 389
Nordlund, Å. (1982): Astron. Astrophys. **107**, 1
Nordlund, Å. (1984 a): In Keil (1984), p. 174
Nordlund, Å. (1984 b): In Keil (1984), p. 181
Nordlund, Å. (1985): Solar Phys. **100**, 209

Norton, A.A., Ulrich, R.K., Bush, R.L., Tarbell, T.D. (1999): Astrophys. J. **518**, L123
November, L.J. (1986): Applied Optics **25**, 392
November, L.J., ed. (1991): *Solar Polarimetry* (Sacramento Peak Observatory, Sunspot, New Mexico)
November, L.J., Stauffer, F.R. (1984): Applied Optics **23**, 2333
November, L.J., Toomre, J., Gebbie, K.B. (1979): Astrophys. J. **227**, 600
November, L.J., Toomre, J., Gebbie, K.B. (1982): Astrophys. J. **258**, 846
November, L.J., Toomre, J., Gebbie, K.B., Simon, G.W. (1981): Astrophys. J. **245**, L123
Noyes, R.W., Rhodes, E.J., Jr., eds. (1984): *Probing the Depths of a Star: The Study of Solar Oscillations from Space*, NASA/JPL 400-237
Núñez, M., Ferriz-Mas, A., eds. (1999): *Stellar Dynamos: Nonlinearity and Chaotic Flows*, ASP Conf. Ser. **178**
Nye, A.H., Thomas, J.H. (1974): Solar Phys. **38**, 399

Oda, N. (1984): Solar Phys. **93**, 243
Ogura, Y., Phillips, N.A. (1962): J. Atmos. Sci. **19**, 173
Öhman, Y. (1938): Nature **141**, 157
Öhman, Y. (1956): Stockholms Obs. Ann. **19**, 3
Ossendrijver, M.A.J.H. (2000): Astron. Astrophys. **359**, 364
Ossendrijver, M., Stix, M., Brandenburg, A. (2001): Astron. Astrophys. **376**, 713
Osterbrook, D.E. (1961): Astrophys. J. **134**, 347
Owocki, S.P., Auer, L.H. (1980): Astrophys. J. **241**, 448

Pap, J. (1985): Solar Phys. **97**, 21
Pap, J.M. (1997): In *Past and Present Variability of the Solar-Terrestrial System: Measurement, Data Analysis and Theoretical Models*, ed. by G. Cini Castagnoli, A. Provenzale, IOS Press Amsterdam, p. 3
Pap, J.M., Kuhn, J.R., Fröhlich, C., Ulrich, R., Jones, A., Rozelot, J.P. (1998): In *A Crossroads for European Solar and Heliospheric Physics*, ed. by E.R. Priest, F. Moreno-Insertis, R.A. Harris, ESA SP-417, p. 267
Parker, E.N. (1955 a): Astrophys. J. **121**, 491
Parker, E.N. (1955 b): Astrophys. J. **122**, 293
Parker, E.N. (1958): Astrophys. J. **128**, 664
Parker, E.N. (1963 a): Astrophys. J. **138**, 552
Parker, E.N. (1963 b): *Interplanetary Dynamical Processes* (Interscience, New York)
Parker, E.N. (1966): Astrophys. J. **145**, 811
Parker, E.N. (1972): Astrophys. J. **174**, 499
Parker, E.N. (1978): Astrophys. J. **221**, 368
Parker, E.N. (1979 a): *Cosmical Magnetic Fields* (Clarendon Press, Oxford)
Parker, E.N. (1979 b): Astrophys. J. **230**, 905
Parker, E.N. (1979 c): Astrophys. J. **234**, 333
Parker, E.N. (1983): Astrophys. J. **264**, 642
Parker, E.N. (1987): Physics Today **40**, No. 7, p. 36
Parker, E.N. (1988): Astrophys. J. **330**, 474
Parker, E.N. (1993): Astrophys. J. **408**, 707
Parker, P.D., Kavanagh, R.W. (1963): Phys. Rev. **131**, 2578
Paxman, R.G., Seldin, J.H. (1993): In Radick (1993), p. 112
Paxman, R.G., Seldin, J.H., Löfdahl, M.G., Scharmer, G.B., Keller, C.U. (1996): Astrophys. J. **466**, 1087
Pérez Garde, M., Vázquez, M., Schwan, H., Wöhl, H. (1981): Astron. Astrophys. **93**, 67
Perot, A., Fabry, C. (1899): Astrophys. J. **9**, 87

Peter, H. (1996): Mon. Not. R. Astron. Soc. **278**, 821
Peter, H. (1999): Astrophys. J. **516**, 490
Petrovay, K.G. (1990): Astrophys. J. **362**, 722
Petrovay, K., van Driel-Gesztelyi, L. (1997): Solar Phys. **176**, 249
Petrovay, K., Moreno-Insertis, F. (1997): Astrophys. J. **485**, 398
Petschek, H.E. (1964): In Hess (1964), p. 425
Pidatella, R.M., Stix, M. (1986): Astron. Astrophys. **157**, 338
Pidatella, R.M., Stix, M., Belvedere, G., Paternó, L. (1986): Astron. Astrophys. **156**, 22
Pierce, A.K., Allen, R.G. (1977): In White (1977), p. 169
Pierce, A.K., Slaugther, C.D. (1977): Solar Phys. **51**, 25
Pijpers, F.P. (1998): Mon. Not. R. Astron. Soc. **297**, L76
Pijpers, F.P., Christensen-Dalsgaard, J., Rosenthal, C.S., eds. (1997): *SCORe'96: Solar Convection and Oscillations and their Relationship* (Kluwer, Dordrecht)
Pinsonneault, M. (1997): Ann. Rev. Astron. Astrophys. **35**, 557
Pinsonneault, M., Kawaler, S.D., Sofia, S., Demarque, P. (1989): Astrophys. J. **338**, 424
Pizzo, V.J. (1986): Astrophys. J. **302**, 785
Pizzo, V.J. (1990): Astrophys. J. **365**, 764
Pizzo, V.J., MacGregor, K.B., Kunasz, P.B. (1993 a): Astrophys. J. **404**, 788
Pizzo, V.J., MacGregor, K.B., Kunasz, P.B. (1993 b): Astrophys. J. **413**, 764
Pizzo, V.J., Schwenn, R., Marsch, E., Rosenbauer, H., Mühlhäuser, K.H., Neubauer, F.M. (1983): Astrophys. J. **271**, 335
Plunkett, S.P., Vourlidas, A., Šimberová, S., Karlický, M., Kotrč, P., Heinzel, P., Kupryakov, Yu.A., Guo, W.P., Wu, S.T. (2000): Solar Phys. **194**, 371
Pneuman, G.W., Hansen, S.F., Hansen, R.T. (1978): Solar Phys. **59**, 313
Pneuman, G.W., Kopp, R.A. (1971): Solar Phys. **18**, 258
Pneuman, G.W., Solanki, S.K., Stenflo, J.O. (1986): Astron. Astrophys. **154**, 231
Pottasch, S.R. (1963): Astrophys. J. **137**, 945
Pottasch, S.R. (1964): Space Sci. Rev. **3**, 816
Povel, H.P., Keller, C.U., Yadigaroglu, I.-A. (1994): Applied Optics **33**, 4254
Priest, E.R., ed. (1981): *Solar Flare Magnetohydrodynamics* (Gordon and Breach, New York)
Priest, E.R. (1982): *Solar Magnetohydrodynamics* (Reidel, Dordrecht)
Priest, E.R., ed. (1989): *Dynamics and Structure of Quiescent Solar Prominences* (Kluwer, Dordrecht)
Priest, E.R., Schrijver, C.J. (1999): Solar Phys. **190**, 1
Proctor, M.R.E., Matthews, P.C., Rucklidge, A.M., eds. (1993): *Solar and Planetary Dynamos* (Cambridge University Press, Cambridge)
Proctor, M.R.E., Weiss, N.O. (1982): Rep. Prog. Phys. **45**, 1317
Proffitt, C.R., Michaud, G. (1991): Astrophys. J. **380**, 238
Pulkkinen, P., Tuominen, I. (1998): Astron. Astrophys. **332**, 755

Querfeld, C.W., Smartt, R.N., Bommier, V., Landi degl'-Innocenti, E., House, L.L. (1985): Solar Phys. **96**, 277

Rachkovsky, D.N. (1962 a): Izv. Krym. Astrofiz. Obs. **27**, 148
Rachkovsky, D.N. (1962 b): Izv. Krym. Astrofiz. Obs. **28**, 259
Radick, R.R., ed. (1993): *Real Time and Post Facto Solar Image Correction* (Sacramento Peak Observatory, Sunspot, New Mexico)
Rädler, K.-H. (1990): In *Inside the Sun*, ed. by G. Berthomieu, M. Cribier (Kluwer, Dordrecht), p. 385
Ramsay, J.V., Kobler, H., Mugridge, E.G.V. (1970): Solar Phys. **12**, 492

Rast, M.P. (1995): Astrophys. J. **443**, 863

Rast, M.P., Nordlund, Å., Stein, R.F., Toomre, J. (1993): Astrophys. J. **408**, L53

Reames, D.V. (1999): Space Sci. Rev. **90**, 413

Rees, D.E., López Ariste, A., Thatcher, J., Semel, M. (2000): Astron. Astrophys. **355**, 759

Rees, D.E., Murphy, G.A., Durrant, C.J. (1989): Astrophys. J. **339**, 1093

Reeves, E.M., Huber, M.C.E., Timothy, J.G. (1977): Applied Optics **16**, 837

Reiter, J., Walsh, L., Weiss, A. (1995): Mon. Not. R. Astron. Soc. **274**, 899

Rempel, M., Schüssler, M., Tóth, G. (2000): Astron. Astrophys. **363**, 789

Renzini, A. (1987): Astron. Astrophys. **188**, 49

Reynolds, O. (1895): Phil. Trans. R. Soc. London A **186**, 123

Rhodes, E.J., Jr., Kosovichev, A.G., Schou, J., Scherrer, P.H., Reiter, J. (1997): Solar Phys. **175**, 287

Ribes, E., Mein, P., Mangeney, A. (1985): Nature **318**, 170

Ribes, J.C., Nesme-Ribes, E. (1993): Astron. Astrophys. **276**, 549

Rieutord, M., Roudier, T., Malherbe, J.M., Rincon, F. (2000): Astron. Astrophys. **357**, 1063

Rieutord, M., Zahn, J.-P. (1995): Astron. Astrophys. **296**, 127

Rimmele, T.R. (1995): Astron. Astrophys. **298**, 260

Rimmele, T.R., Balasubramaniam, K.S., Radick, R.R., eds. (1999): *High Resolution Solar Physics: Theory, Observation, and Techniques*, ASP Conf. Ser. **183**

Rimmele, T.R., Goode, P.R., Harold, E., Stebbins, R.T. (1995): Astrophys. J. **444**, L119

Rimmele, T., Schröter, E.H. (1989): Astron. Astrophys. **221**, 137

Ritzwoller, M.H., Lavely, E.M. (1991): Astrophys. J. **369**, 557

Robe, H. (1968): Ann. d'Astrophys. **31**, 475

Roberts, B., Webb, A.R. (1978): Solar Phys. **56**, 5

Roberts, P.H. (1967): *An Introduction to Magnetohydrodynamics* (Longmans, London)

Roberts, P.H., Soward, A.M. (1975a): J. Math. Phys. **16**, 609

Roberts, P.H., Soward, A.M. (1975b): Astron. Nachr. **296**, 49

Roddier, F. (1975): Comptes Rendus, Ser. B **281**, 93

Roddier, F. (1981): In *Progress in Optics*, ed. by E. Wolf (North-Holland, Amsterdam), Vol. XIX, p. 281

Rogers, E.H. (1970): Solar Phys. **13**, 57

Rogers, F.J. (1984): In Ulrich (1984), p. 357

Roggemann, M.C., Welsh, B. (1996): *Imaging through Turbulence* (CRC Press, Boca Raton)

Rohlfs, K., Wilson, T.L. (2000): *Tools of Radio Astronomy* (Springer, Berlin, Heidelberg)

Rood, R.T., Bania, T.M., Wilson, T.L. (1984): Astrophys. J. **280**, 629

Rosenthal, C.S (1998): Astrophys. J. **508**, 864

Rosenthal, C.S., Christensen-Dalsgaard, J., Nordlund, Å., Stein, R.F., Trampedach, R. (1999): Astron. Astrophys. **351**, 689

Ross, J.E., Aller, L.H. (1976): Science **191**, 1223

Roth, M., Stix, M. (1999): Astron. Astrophys. **351**, 1133

Rottman, G.J., Orrall, F.Q., Klimchuk, J.A. (1982): Astrophys. J. **260**, 326

Roudier, Th., Malherbe, J.M., November, L., Vigneau, J., Coupinot, G., Lafon, M., Muller, R. (1997): Astron. Astrophys. **320**, 605

Roudier, Th., Muller, R. (1986): Solar Phys. **107**, 11

Roussev, I., Galsgaard, K., Erdélyi, R., Doyle, J.G. (2001): Astron. Astrophys. **370**, 298

Rozelot, J.-P., Klein, L., Vial, J.-C., eds. (2000): *Transport and Energy Conversion in the Heliosphere* (Springer, Berlin, Heidelberg)
Rozelot, J.-P., Rösch, J. (1997): Solar Phys. **172**, 11
Rucklidge, A.M., Weiss, N.O., Brownjohn, D.P., Matthews, P.C., Proctor, M.R.E. (2000): J. Fluid Mech. **419**, 283
Rüdiger, G. (1980): Geophys. Astrophys. Fluid Dyn. **16**, 239
Rüdiger, G. (1983): Geophys. Astrophys. Fluid Dyn. **25**, 213
Rüdiger, G., Hasler, K.-H., Kitchatinov, L.L. (1997): Astron. Nachr. **318**, 173
Rüdiger, G., Kitchatinov, L.L. (1996): Astrophys. J. **466**, 1078
Rüdiger, G., von Rekowski, B., Donahue, R.A., Baliunas, S.L. (1998): Astrophys. J. **494**, 691
Rüdiger, G., Tuominen, I. (1987): In Durney and Sofia (1987), p. 361
Rüedi, I., Solanki, S.K., Stenflo, J.O., Tarbell, T., Scherrer, P.H. (1998): Astron. Astrophys. **335**, L97
Ruiz Cobo, B., del Toro Iniesta, J.C. (1992): Astrophys. J. **398**, 375
Ruiz Cobo, B., del Toro Iniesta, J.C., Rodríguez Hidalgo, I., Collados, M., Sánchez Almeida, J. (1996): In *Cool Stars, Stellar Systems, and the Sun*, ed. by R. Pallavicini, A.K. Dupree, ASP Conf. Ser. **109**, p. 155
Rutten, R.J., Milkey, R.W. (1979): Astrophys. J. **231**, 277
Rutten, R.J., Schrijver, C.J., eds. (1994): *Solar Surface Magnetism*, NATO ASI Ser. C, **433** (Kluwer, Dordrecht)
Ruzmaikin, A.A., Sokoloff, D.D., Starchenko, S.V. (1988): Solar Phys. **115**, 5

Sackmann, I.-J., Boothroyd, A.I., Fowler, W.A. (1990): Astrophys. J. **360**, 727
Sackmann, I.-J., Boothroyd, A.I., Kraemer, K.E. (1993): Astrophys. J. **418**, 457
Sakao, T., Tsuneta, S., Hara, H., Shimizu, T., Kano, R., Kumagai, K., Yoshida, T., Nagata, S., Kobayashi, K. (1999): Solar Phys. **187**, 303
Sakurai, T. (1979): Publ. Astron. Soc. Japan **31**, 209
Sakurai, T. (1981): Solar Phys. **69**, 343
Sakurai, T. (1985): Astron. Astrophys. **152**, 121
Salpeter, E.E. (1954): Australian J. Phys. **7**, 373
Sánchez Almeida, J., Landi degl'Innocenti, E., Martínez Pillet, V., Lites, B.W. (1996): Astrophys. J. **466**, 437
Schatzman, E. (1959): In *The Hertzsprung–Russell Diagram*, ed. by J.L. Greenstein, IAU Symp. **10** (Mount Wilson and Palomar Observatories, Pasadena), p. 129
Schatzman, E., Maeder, A., Angrand, F., Glowinski, R. (1981): Astron. Astrophys. **96**, 1
Scherrer, P.H., Bogart, R.S., Bush, R.I., 9 others, and the MDI Engineering Team (1995): Solar Phys. **162**, 129
Scherrer, P.H., Bogart, R., Hoeksema, J.T., Yoshimura, H. (1986): In Gough (1986), p. 93
Scherrer, P.H., Wilcox, J.M., Christensen-Dalsgaard, J., Gough, D.O. (1983): Solar Phys. **82**, 75
Schindler, K., Birn, J., Janicke, L. (1983): Solar Phys. **87**, 103
Schlattl, H., Weiss, A. (1999): Astron. Astrophys. **347**, 272
Schlattl, H., Weiss, A., Ludwig, H.G. (1997): Astron. Astrophys. **322**, 646
Schleicher, H. (1976): Dissertation, University of Göttingen
Schlichenmaier, R., Collados, M. (2002): Astron. Astrophys. **381**, 668
Schlichenmaier, R., Jahn, K., Schmidt, H.U. (1998 a): Astrophys. J. **493**, L121
Schlichenmaier, R., Jahn, K., Schmidt, H.U. (1998 b): Astron. Astrophys. **337**, 897
Schlichenmaier, R., Schmidt, W. (1999): Astron. Astrophys. **349**, L37
Schlichenmaier, R., Schmidt, W. (2000): Astron. Astrophys. **358**, 1122
Schlichenmaier, R., Stix, M. (1995): Astron. Astrophys. **302**, 264

Schlüter, A., Temesváry, S. (1958): In Lehnert (1958), p. 263

Schmidt, H.U. (1964): In Hess (1964), p. 107

Schmidt, H.U. (1991): Geophys. Astrophys. Fluid Dyn. **62**, 249

Schmidt, H.U., Spruit, H.C., Weiss, N.O. (1986): Astron. Astrophys. **158**, 351

Schmidt, W., Hofmann, A., Balthasar, H., Tarbell, T.D., Frank, Z.A. (1992): Astron. Astrophys. **264**, L27

Schmidt, W., Kentischer, T. (1995): Astron. Astrophys. Suppl. **113**, 363

Schmidt, W., Stix, M. (1983): Astron. Astrophys. **118**, 1

Schmidt, W., Stix, M., Wöhl, H. (1999): Astron. Astrophys. **346**, 663

Schmieder, B. (1976): Solar Phys. **47**, 435

Schmieder, B. (1977): Solar Phys. **54**, 269

Schmieder, B., Hofmann, A., Staude, J., eds. (1999): *Third Advances in Solar Physics Euroconference: Magnetic Fields and Oscillations*, ASP Conf. Ser. **184**

Schmieder, B., del Toro Iniesta, J.C., Vázquez, M., eds. (1997): *First Advances in Solar Physics Euroconference: Advances in the Physics of Sunspots*, ASP Conf. Ser. **118**

Schmitt, D. (1998): Geophys. Astrophys. Fluid Dyn. **89**, 75

Schmitt, J.H.M.M., Rosner, R., Bohn, H.U. (1984): Astrophys. J. **282**, 316

Schmitz, N. (1997): *Neutrinophysik* (Teubner, Stuttgart)

Schou, J., Antia, H.M., Basu, S., and 21 others (1998): Astrophys. J. **505**, 390

Schou, J., Kosovichev, A.G., Goode, P.R., Dziembowski, W.A. (1997): Astrophys. J. **489**, L197

Schrijver, C.J., Hagenaar, H.J., Title, A.M. (1997): Astrophys. J. **475**, 328

Schrijver, C.J., Zwaan, C. (2000): *Solar and Stellar Magnetic Activity* (Cambridge University Press, Cambridge)

Schröder, W. (1988): Meteorol. Atmos. Phys. **38**, 246

Schröter, E.H. (1957): Z. Astrophys. **41**, 141

Schröter, E.H. (1973): Mitt. Astron. Ges. **32**, 55

Schröter, E.H. (1985): Solar Phys. **100**, 141

Schröter, E.H., Schüssler, M., eds. (1987): *Solar and Stellar Physics*, Lect. Notes Phys., Vol. 292 (Springer, Berlin, Heidelberg)

Schröter, E.H., Vázquez, M., Wyller, A., eds. (1987): *The Role of Fine-Scale Magnetic Fields on the Structure of the Solar Atmosphere* (Cambridge University Press, Cambridge)

Schröter, E.H., Wöhl, H. (1976): Solar Phys. **49**, 19

Schröter, E.H., Wöhl, H., Soltau, D., Vázquez, M. (1978): Solar Phys. **60**, 181

Schumacher, J., Kliem, B. (1996): Phys. Plasmas **3**, 4703

Schumacher, J., Kliem, B., Seehafer, N. (2000): Phys. Plasmas **7**, 108

Schüssler, M. (1979): Astron. Astrophys. **71**, 79

Schüssler, M. (1983): In Stenflo (1983), p. 213

Schüssler, M. (1987a): In Durney and Sofia (1987), p. 303

Schüssler, M. (1987b): In Schröter et al. (1987), p. 223

Schüssler, M., Schmidt, W., eds. (1994): *Solar Magnetic Fields* (Cambridge University Press, Cambridge)

Schwabe, H. (1844): Astron. Nachr. **21**, 233

Schwan, H., Wöhl, H. (1978): Astron. Astrophys. **70**, 297

Schwarzschild, K. (1906): Nachr. Kgl. Ges. Wiss. Göttingen, Math.-Phys. Klasse, No. 1, p. 41

Schwarzschild, M. (1958): *Structure and Evolution of the Stars* (Princeton University Press, Princeton)

Schwenn, R., Marsch, E., eds. (1990, 1991): *Physics of the Inner Heliosphere*, Vol. I, II (Springer, Berlin, Heidelberg)

Sears, F.H. (1913): Astrophys. J. **38**, 99

478   References

Seaton, M.J., ed. (1995): *The Opacity Project*, Vol. 1 (Institute of Physics, Bristol and Philadelphia)
Seehafer, N. (1994): Astron. Astrophys. **284**, 593
Seehafer, N. (1996): Phys. Rev. E **53**, 1283
Severino, G., Roberti, G., Marmolino, C., Gomez, M.T. (1986): Solar Phys. **104**, 259
Severny, A.B. (1965): In Lüst (1965), p. 358
Seykora, E.J. (1993): Solar Phys. **145**, 389
Shack, R.V., Platt, B.C. (1971): J. Opt. Soc. America **61**, 656
Shaviv, G., Salpeter, E.E. (1973): Astrophys. J. **184**, 191
Sheeley, N.R., Jr. (1964): Astrophys. J. **140**, 731
Shelke, R.N., Pande, M.C. (1985): Solar Phys. **95**, 193
Shibahashi, H., Noels, A., Gabriel, M. (1983): Astron. Astrophys. **123**, 283
Shibahashi, H., Osaki, Y. (1981): Publ. Astron. Soc. Japan **33**, 713
Shibahashi, H., Osaki, Y., Unno. W. (1975): Publ. Astron. Soc. Japan **27**, 401
Shimabukuro, F.I. (1977): In White (1977), p. 133
Shine, R.A., Simon, G.W., Hurlburt, N.E. (2000): Solar Phys. **193**, 313
Shklovskii, I.S. (1965): *Physics of the Solar Corona*, English Translation of 2nd ed. (Pergamon Press, Oxford)
Shurcliff, W.A. (1962): *Polarized Light* (Harvard University Press, Cambridge, Ma.)
Sigwarth, M. (1999): Dissertation, Universität Freiburg
Sigwarth, M., ed. (2001 a): *Advanced Solar Polarimetry: Theory, Observation, and Instrumentation*, ASP Conf. Ser. **236**
Sigwarth, M. (2001 b): Astrophys. J. **563**, 1031
Sigwarth, M., Balasubramaniam, K.S., Knölker, M., Schmidt, W. (1999): Astron. Astrophys. **349**, 941
Simon, G.W., Brandt, P.N., November, L.J., Scharmer, G.B., Shine, R.A. (1994): In Rutten and Schrijver (1994), p. 261
Simon, G.W., Leighton, R.B. (1964): Astrophys. J. **140**, 1120
Simon, G.W., Weiss, N.O. (1968): Z. Astrophys. **69**, 435
Singh, H.P., Roxburgh, I.W., Chan, K.W. (1994): Astron. Astrophys. **281**, L73
Singh, H.P., Roxburgh, I.W., Chan, K.W. (1995): Astron. Astrophys. **295**, 703
Singh, H.P., Roxburgh, I.W., Chan, K.W. (1998): Astron. Astrophys. **340**, 178
Skaley, D., Stix, M. (1991): Astron. Astrophys. **241**, 227
Skumanich, A. (1972): Astrophys. J. **171**, 565
Skumanich, A., Lean, J.L., White, O.R., Livingston, W.C. (1984): Astrophys. J. **282**, 776
Skumanich, A., Lites, B.W. (1987): Astrophys. J. **322**, 473
Skumanich, A., Smythe, C., Frazier, E.N. (1975): Astrophys. J. **200**, 747
Slettebak, A. (1970): In *Stellar Rotation*, ed. by A. Slettebak (Reidel, Dordrecht), p. 3
Snodgrass, H.B. (1984): Solar Phys. **94**, 13
Snodgrass, H.B. (1985): Astrophys. J. **291**, 339
Snodgrass, H.B., Kress, J.M., Wilson, P.R. (2000): Solar Phys. **191**, 1
Snodgrass, H.B., Ulrich, R.K. (1990): Astrophys. J. **351**, 309
Sobotka, M. (1999): In *Motions in the Solar Atmosphere*, ed. by A. Hanslmeier and M. Messerotti (Kluwer, Dordrecht), p. 71
Sobotka, M., Brandt, P.N., Simon, G.W. (1997 a): Astron. Astrophys. **328**, 682
Sobotka, M., Brandt, P.N., Simon, G.W. (1997 b): Astron. Astrophys. **328**, 689
Sobouti, Y. (1980): Astron. Astrophys. **89**, 314
Sofia, S., Heaps, W., Twigg, L.W. (1994): Astrophys. J. **427**, 1048
Solanki, S.K. (1986): Astron. Astrophys. **168**, 311
Solanki, S.K. (1993): Space Sci. Rev. **63**, 1

Soltau, D., Schröter, E.H., Wöhl, H. (1976): Astron. Astrophys. **50**, 367
Souffrin, P. (1966): Ann. d' Astrophys. **29**, 55
Soward, A.M., Priest, E.R. (1982): J. Plasma Phys. **28**, 335
Spiegel, E.A. (1957): Astrophys. J. **126**, 202
Spiegel, E.A., Zahn, J.P., eds. (1977): *Problems of Stellar Convection*, Lect. Notes Phys., Vol. 71 (Springer, Berlin, Heidelberg)
Spiegel, E.A., Zahn, J.-P. (1992): Astron. Astrophys. **265**, 106
Spitzer, L., Jr. (1962): *Physics of Fully Ionized Gases*, 2nd ed. (Interscience, New York)
Spruit, H.C. (1976): Solar Phys. **50**, 269
Spruit, H.C. (1981 a): Astron. Astrophys. **98**, 155
Spruit, H.C. (1981 b): In Jordan (1981), p. 385
Spruit, H.C. (1981 c): In Cram and Thomas (1981), p. 480
Spruit, H.C. (1987): In Durncy and Sofia (1987), p. 185
Spruit, H.C. (1997): Mem. Soc. Astron. Ital. **68**, 397
Spruit, H.C., van Ballegooijen, A.A. (1982): Astron. Astrophys. **106**, 58
Spruit, H.C., Nordlund, Å., Title, A.M. (1990): Ann. Rev. Astron. Astrophys. **28**, 263
Spruit, H.C., Zweibel, E.G. (1979): Solar Phys. **62**, 15
Staiger, J. (1987): Astron. Astrophys. **175**, 263
Stanchfield II, D.C.H., Thomas, J.H., Lites, B.W. (1997): Astrophys. J. **477**, 485
Stark, D., Wöhl, H. (1981): Astron. Astrophys. **93**, 241
Staude, J. (1972): Solar Phys. **24**, 255
Staude, J. (1999): In Schmieder et al. (1999), p. 113
Steenbeck, M., Krause, F. (1969): Astron. Nachr. **291**, 49
Steenbeck, M., Krause, F., Rädler, K.-H. (1966): Z. Naturforsch. **21a**, 361
Steffen, M., Ludwig, H.-G., Krüß, A. (1989): Astron. Astrophys. **213**, 371
Steiger, R. von, Schwadron, N.A., Fisk, L.A., Geiss, J., Gloeckler, G., Hefti, S., Wilken, B., Wimmer-Schweingruber, R.F., Zurbuchen, T.H. (2000): J. Geophys. Res. **105** 27217
Stein, R.F., Nordlund, Å. (1998 a): Astrophys. J. **499**, 914
Stein, R.F., Nordlund, Å. (1998 b): In Korzennik and Wilson (1998), p. 693
Steiner, O. (2000): Solar Phys. **196**, 245
Steiner, O., Hauschildt, P.H., Bruls, J. (2001): Astron. Astrophys. **372**, L13
Stenflo, J.O. (1971): In Howard (1971), p. 101
Stenflo, J.O. (1976): In Bumba and Kleczek (1976), p. 69
Stenflo, J.O. (1982): Solar Phys. **80**, 209
Stenflo, J.O., ed. (1983): *Solar and Stellar Magnetic Fields: Origins and Coronal Effects*, IAU Symp. **102** (Reidel, Dordrecht)
Stenflo, J.O. (1986): Mitt. Astron. Ges. **65**, 25
Stenflo, J.O. (1994): *Solar Magnetic Fields* (Kluwer, Dordrecht)
Stenflo, J.O. (2001): In Sigwarth (2001 a), p. 97
Stenflo, J.O., Gandorfer, A., Keller, C.U. (2000 a): Astron, Astrophys. **355**, 781
Stenflo, J.O., Harvey, J.W., Brault, J.W., Solanki, S. (1984): Astron. Astrophys. **131**, 333
Stenflo, J.O., Keller, C.U. (1997): Astron, Astrophys. **321**, 927
Stenflo, J.O., Keller, C.U., Gandorfer, A. (2000 b): Astron, Astrophys. **355**, 789
Stenflo, J.O., Keller, C.U., Povel, H.P. (1992): LEST Foundation, Technical Rep. No. 54
Stenflo, J.O., Nagendra, K.N., eds. (1996): Solar Phys. **164**
Stepanov, V.E. (1958): Izv. Krym. Astrofiz. Obs. **18**, 136
Sterling, A.C. (2000): Solar Phys. **196**, 79
Stewart, J.C., Pyatt, K.D., Jr. (1966): Astrophys. J. **144**, 1203

Stix, M. (1970): Astron. Astrophys. **4**, 189
Stix, M. (1972): Astron. Astrophys. **20**, 9
Stix, M. (1974): Astron. Astrophys. **37**, 121
Stix, M. (1976 a): Astron. Astrophys. **47**, 243
Stix, M. (1976 b): In Bumba and Kleczek (1976), p. 367
Stix, M. (1981 a): Astron. Astrophys. **93**, 339
Stix, M. (1981 b): Solar Phys. **74**, 79
Stix, M. (1984): Astron. Nachr. **305**, 215
Stix, M. (1987 a): In Durney and Sofia (1987), p. 329
Stix, M. (1987 b): In Schröter and Schüssler (1987), p. 15
Stix, M. (1989): Rev. Mod. Astron. **2**, 248
Stix, M. (1991): Geophys. Astrophys. Fluid Dyn. **62**, 211
Stix, M. (2000 a): Solar Phys. **196**, 19
Stix, M. (2000 b): Naturwissenschaften **87**, 524
Stix, M. (2001): Astron. Astrophys. Trans. **20**, 417
Stix, M., Knölker, M. (1987): In Hillebrandt et al. (1987), p. 67
Stix, M., Rüdiger, G., Knölker, M., Grabowski, U. (1993): Astron. Astrophys. **272**, 340
Stix, M., Zhugzhda, Y.D. (1998): Astron. Astrophys. **335**, 685
Straus, Th., Bonaccini, D. (1997): Astron. Astrophys. **324**, 704
Straus, Th., Deubner, F.-L., Fleck, B. (1992): Astron. Astrophys. **256**, 652
Stroke, G.W. (1967): In *Handbuch der Physik*, ed. by S. Flügge (Springer, Berlin, Heidelberg), Vol. XXIX, p. 426
Strous, L.H., Goode, P.R., Rimmele, T.R. (2000): Astrophys. J. **535**, 1000
Sturrock, P.A., ed. (1980): *Solar Flares* (Colorado Associated University Press, Boulder)
Sturrock, P.A., ed. (1986): *Physics of the Sun*, Vol. I–III (Reidel, Dordrecht)
Svalgaard, L., Wilcox, J.M. (1976): Nature **262**, 766
Švestka, Z. (1976): *Solar Flares* (Reidel, Dordrecht)
Sweet, P.A. (1958): In Lehnert (1958), p. 123

Takata, M., Shibahashi, H. (1998): Astrophys. J. **504**, 1035
Tandberg-Hanssen, E. (1974): *Solar Prominences* (Reidel, Dordrecht)
Tandberg-Hanssen, E. (1995): *The Nature of Solar Prominences* (Kluwer, Dordrecht)
Tarbell, T.D., Smithson, R.C. (1981): In Dunn (1981), p. 491
Tassoul, J.-L. (1978): *Theory of Rotating Stars* (Princeton University Press, Princeton)
Tassoul, M. (1980): Astrophys. J. Suppl. **43**, 469
Tatarski, V.I. (1961): *Wave Propagation in a Turbulent Medium* (Dover, New York)
Tatarski, V.I. (1971): *The Effects of the Turbulent Atmosphere on Wave Propagation* (Israel Prog. Sci. Trans., Jerusalem)
Thomas, J.H. (1981): In Cram and Thomas (1981), p. 345
Thomas, J.H. (1985): Aust. J. Phys. **38**, 811
Thomas, J.H. (1988): Astrophys. J. **333**, 407
Thomas, J.H., Cram, L.E., Nye, A.H. (1984): Astrophys. J. **285**, 368
Thomas, J.H., Weiss, N.O., eds. (1992): *Sunspots: Theory and Observations*, NATO ASI Ser. C, **375** (Kluwer, Dordrecht)
Thompson, B.J., Gurman, J.B., Neupert, W.M., and 7 others (1999): Astrophys. J. **517**, L151
Thoul, A.A., Bahcall, J.B., Loeb, A. (1994): Astrophys. J. **421**, 828
Thuillier, G., Hersé, M., Simon, P.C., Labs, D., Mandel, H., Gillotay, D. (1997): Solar Phys. **171**, 283

Timothy, A.F., Krieger, A.S., Vaiana, G.S. (1975): Solar Phys. **42**, 135
Title, A.M., Frank, Z.A., Shine, R.A., Tarbell, T.D., Topka, K.P., Scharmer, G., Schmidt, W. (1993): Astrophys. J. **403**, 780
Title, A.M., Ramsey, H.E. (1980): Applied Optics **19**, 2046
Title, A.M., Tarbell, T.D., Simon, G.W., and the SOUP Team (1986): Adv. Space Res. **6**, No. 8, p. 253
Title, A.M., Tarbell, T.D., Topka, K.P., Shine, R.A., Simon, G.W., Zirin, H., and the SOUP Team (1987): In Schröter and Schüssler (1987), p. 173
Tobias, S.M. (1997): Astron. Astrophys. **322**, 1007
Tolansky, S. (1948): *Multiple-Beam Interferometry of Surfaces and Films* (Clarendon Press, Oxford)
Tomczyk, S., Schou, J., Thompson, M.J. (1995 b): Astrophys. J. **448**, L57
Tomczyk, S., Streander, K., Card, G., Elmore, D., Hull, H., Cacciani, A. (1995 a): Solar Phys. **159**, 1
Tripathy, S.C., Christensen-Dalsgaard, J. (1998): Astron. Astrophys. **337**, 579
Tritschler, A., Schmidt, W. (2002): Astron. Astrophys. **382**, 1093
Trujillo-Bueno, J., Moreno-Insertis, F., Sánchez, F., eds. (2002): *Astrophysical Spectropolarimetry* (Cambridge University Press, Cambridge)
Tsinganos, K.C., ed. (1996): *Solar and Astrophysical Magnetohydrodynamic Flows*, NATO ASI Ser. C, **481** (Kluwer, Dordrecht)
Tsuneta, S. (1996 a): Astrophys. J. **456**, L63
Tsuneta, S. (1996 b): Astrophys. J. **456**, 840
Tsuneta, S. (1997): Astrophys. J. **483**, 507
Tsuneta, S., Acton, L., Bruner, M., and 10 others (1991): Solar Phys. **136**, 37
Tsuneta, S., Hara, H., Shimizu, T., Acton, L.W., Strong, K.T., Hudson, H.S., Ogawara, Y. (1992): Publ. Astron. Soc. Japan **44**, L63
Tu, C.-Y., Marsch, E., Wilhelm, K., Curdt, W. (1998): Astrophys. J. **503**, 475
Tuominen, J. (1941): Z. Astrophys. **21**, 96
Turck-Chièze, S., Lopes, I. (1993): Astrophys. J. **408**, 347

Uchida, Y. (1968): Solar Phys. **4**, 30
Uexküll, M. von, Kneer, F. (1995): Astron. Astrophys. **294**, 252
Ulmschneider, P., Muchmore, D. (1986): In: *Small Scale Magnetic Flux Concentrations in the Solar Photosphere*, ed. by W. Deinzer, M. Knölker and H.H. Voigt (Vandenhoeck & Ruprecht, Göttingen), p. 191
Ulrich, R.K. (1970): Astrophys. J. **162**, 993
Ulrich, R.K. (1982): Astrophys. J. **258**, 404
Ulrich, R.K., ed. (1984): *Solar Seismology from Space*, NASA/JPL 84-84
Ulrich, R.K. (1986): In Sturrock (1986), Vol. I, p. 161
Ulrich, R.K., Rhodes, E.J., Jr. (1977): Astrophys. J. **218**, 521
Ulrich, R.K., Rhodes, E.J., Jr. (1983): Astrophys. J. **265**, 551
Ulrich, R.K., Rhodes, E.J., Jr., Däppen, W., eds. (1995): *GONG'94: Helio- and Asteroseismology from the Earth and Space*, ASP Conf. Ser. **76**
Unno, W. (1956): Publ. Astron. Soc. Japan **8**, 108
Unno, W., Osaki, Y., Ando, H., Saio, H., Shibahashi, H. (1989): *Non-radial Oscillations of Stars*, 2nd ed. (University of Tokyo Press, Tokyo)
Unsöld, A. (1930): Z. Astrophys. **1**, 138
Unsöld, A. (1955): *Physik der Sternatmosphären*, 2nd ed. (Springer, Berlin, Heidelberg)
Unsöld, A., Baschek, B. (1999): *Der neue Kosmos*, 6th ed. (Springer, Berlin, Heidelberg)

Vaiana, G.S., Davis, J.M., Giacconi, R., Krieger, A.S., Silk, J.K., Timothy, A.F., Zombeck, M. (1973): Astrophys. J. **185**, L47

Vandakurov, Yu.V. (1967): Astron. Zh. **44**, 786, English Translation: Sov. Astron. **11**, 630

Vaughan, A.H., Baliunas, S.L., Middelkoop, F., Hartmann, L.W., Mihalas, D., Noyes, R.W., Preston, G.W. (1981): Astrophys. J. **250**, 276

Vaughan, J.M. (1989): *The Fabry–Perot Interferometer* (Adam Hilger, Bristol)

Vernazza, J.E., Avrett, E.H., Loeser, R. (1973): Astrophys. J. **184**, 605

Vernazza, J.E., Avrett, E.H., Loeser, R. (1976): Astrophys. J. Suppl. **30**, 1

Vernazza, J.E., Avrett, E.H., Loeser, R. (1981): Astrophys. J. Suppl. **45**, 635

Vitense, E. (1953): Z. Astrophys. **32**, 135

Voigt, H.-H. (1956): Z. Astrophys. **40**, 157

Vrabec, D. (1971): In Howard (1971), p. 329

Wagner, W.J., Gilliam, L.B. (1976): Solar Phys. **50**, 265

Wagner, W.J., Newkirk, G., Jr., Schmidt, H.U. (1983): Solar Phys. **83**, 115

Waldmeier, M. (1937): Z. Astrophys. **14**, 91

Waldmeier, M. (1951): *Die Sonnenkorona I* (Birkhäuser, Basel)

Waldmeier, M. (1955): *Ergebnisse und Probleme der Sonnenforschung*, 2nd ed. (Akademische Verlagsgesellschaft, Leipzig)

Waldmeier, M. (1957): *Die Sonnenkorona II* (Birkhäuser, Basel)

Wambsganss, J. (1988): Astron. Astrophys. **205**, 125

Wang, H. (1989): Solar Phys. **123**, 21

Wang, H., Ewell, M.W., Jr., Zirin, H. (1994): Astrophys. J. **424**, 436

Wang, H., Qiu, J., Denker, C., Spirok, T., Chen, H., Goode, P.R. (2000): Astrophys. J. **542**, 1080

Ward, F. (1965). Astrophys. J. **141**, 534

Wasserburg, G.J. (1995): Rev. Mod. Phys. **67**, 781 (Appendix)

Weber, E.J., Davis, L., Jr. (1967): Astrophys. J. **148**, 217

Weigelt, G. (1977): Optics Comm. **21**, 55

Weiss, A., Keady, J.J., Magee, Jr., N.H. (1990): Atomic Data and Nuclear Data Tables **45**, 209

Weiss, N.O. (1964): Phil. Trans. R. Soc. London A **256**, 99

Weiss, N.O. (1965): Observatory **85**, 37

Weiss, N.O. (2001): Astron. Geophys. **42**, 3.11

Weiss, N.O., Brownjohn, D.P., Matthews, P.C., Proctor, M.R.E. (1996): Mon. Not. R. Astron. Soc. **283**, 1153

Weiss, N.O., Cattaneo, F., Jones, C.A. (1984): Geophys. Astrophys. Fluid Dyn. **30**, 305

Weizsäcker, C.F. von (1938): Phys. Z. **39**, 633

Westendorp Plaza, C., del Toro Iniesta, J.C., Ruiz Cobo, B., Martínez Pillet, V., Lites, B.W., Skumanich, A. (1997): Nature **389**, 47

White, O.R., ed. (1977): *The Solar Output and Its Variation* (Colorado Associated University Press, Boulder)

Wiehr, E. (1999): In Schmieder et al. (1999), p. 86

Wielen, R., Fuchs, B., Dettbarn, C. (1996): Astron. Astrophys. **314**, 438

Wiese, W.L., Smith, M.W., Glennon, B.M. (1966): *Atomic Transition Probabilities*, Vol. I, NSRDS-NBS 4 (Washington)

Wiese, W.L., Smith, M.W., Miles, B.M. (1969): *Atomic Transition Probabilities*, Vol. II, NSRDS-NBS 22 (Washington)

Wigley, T.M.L. (1981): Solar Phys. **74**, 435

Wilcox, J.M., Ness, N.F. (1965): J. Geophys. Res. **70**, 5793

Wild, J.P. (1959): Rend. Scuola Int. Fisica, Corso **XII**, 296

Wilhelm, K., Curdt, W., Marsch, E., and 13 others (1995): Solar Phys. **162**, 189
Willson, R.C., Gulkis, S., Janssen, M., Hudson, H.S., Chapman, G. A. (1981): Science **211**, 700
Willson, R.C., Hudson, H.S. (1981): Astrophys. J. **244**, L 185
Willson, R.C., Hudson, H.S., Fröhlich, C., Brusa, R.W. (1986): Science **234**, 1114
Wilson, A., ed. (2001): *Helio- and Asteroseismology at the Dawn of the Millennium*, ESA SP-464
Wilson, O.C. (1978): Astrophys. J. **226**, 379
Winkler, K.-H., Newman, M.J. (1980): Astrophys. J. **236**, 201
Wissink, J.G., Hughes, D.W., Matthews, P.C., Proctor, M.R.E. (2000 a): Mon. Not. R. Astron. Soc. **318**, 501
Wissink, J.G., Matthews, P.C., Hughes, D.W., Proctor, M.R.E. (2000 b): Astrophys. J. **536**, 982
Withbroe, G.L., Noyes, R.W. (1977): Ann. Rev. Astron. Astrophys. **15**, 363
Wittmann, A. (1974 a): Solar Phys. **35**, 11
Wittmann, A. (1974 b): Solar Phys. **36**, 65
Wittmann, A.D. (1997): Solar Phys. **171**, 231
Wittmann, A., Schröter, E.H. (1969): Solar Phys. **10**, 357
Wöhl, H. (1971): Solar Phys. **16**, 362
Wöhl, H. (1978): Astron. Astrophys. **62**, 165
Wöhl, H., Brajša, R. (2001): Solar Phys. **198**, 57
Woodard, M., Hudson, H.S. (1983): Nature **305**, 589
Worden, S.P. (1975): Solar Phys. **45**, 521

Xiong, D.R. (1985): Astron. Astrophys. **150**, 133
Xiong, D.R., Chen, Q.L. (1992): Astron. Astrophys. **254**, 362

Yerle, R. (1986): Astron. Astrophys. **161**, L5
Yokoyama, T., Akita, K., Morimoto, T., Inoue, K., Newmark, J. (2001): Astrophys. J. **546**, L69
Yokoyama, T., Shibata, K. (2001): Astrophys. J. **549**, 1160
Yorke, H.W., Bodenheimer, P., Laughlin, G. (1993): Astrophys. J. **411**, 274
Yoshimura, H. (1975 a): Astrophys. J. Suppl. **29**, 467
Yoshimura, H. (1975 b): Astrophys. J. **201**, 740
Yoshimura, H. (1976): Solar Phys. **50**, 3

Zahn, J.-P. (1991): Astron. Astrophys. **252**, 179
Zahn, J.-P. (1992): Astron. Astrophys. **265**, 115
Zeldovich, Ya.B., Ruzmaikin, A.A., Sokoloff, D.D. (1983): *Magnetic Fields in Astrophysics* (Gordon and Breach, New York)
Zhang, J., Wang, J., Wang, H., Zirin, H. (1998): Astron. Astrophys. **335**, 341
Zhevakin, S.A. (1953): Astron. Zh. **30**, 161
Zhugzhda, Y.D. (1996): Phys. Plasmas **3**, 10
Zhugzhda, Y.D., Stix, M. (1994): Astron. Astrophys. **291**, 310
Zirin, H. (1971): In Howard (1971), p. 237
Zirin, H. (1988): *Astrophysics of the Sun* (Cambridge University Press, Cambridge)
Zirin, H., Stein, A. (1972): Astrophys. J. **178**, L85
Zirker, J.B., ed. (1977): *Coronal Holes and High Speed Wind Streams* (Colorado Associated University Press, Boulder)
Zombeck, M.V., Vaiana, G.S., Haggerty, R., Krieger, A.S., Silk, J.K., Timothy, A. (1978): Astrophys. J. Suppl. **38**, 69
Zwaan, C. (1987): Ann. Rev. Astron. Astrophys. **25**, 83

Index

ASTRONOMY AND ASTROPHYSICS LIBRARY

Series Editors: I. Appenzeller · G. Börner · A. Burkert · M. A. Dopita
T. Encrenaz · M. Harwit · R. Kippenhahn · J. Lequeux
A. Maeder · V. Trimble

ASTRONOMY AND ASTROPHYSICS LIBRARY

Series Editors: I. Appenzeller · G. Börner · A. Burkert · M. A. Dopita
T. Encrenaz · M. Harwit · R. Kippenhahn · J. Lequeux
A. Maeder · V. Trimble

The Early Universe Facts and Fiction
4th Edition By G. Börner

**The Design and Construction of Large
Optical Telescopes** By P. Y. Bely

The Solar System 4th Edition
By T. Encrenaz, J.-P. Bibring, M. Blanc,
M. A. Barucci, F. Roques, Ph. Zarka

**General Relativity, Astrophysics,
and Cosmology** By A. K. Raychaudhuri,
S. Banerji, and A. Banerjee

Stellar Interiors Physical Principles,
Structure, and Evolution 2nd Edition
By C. J. Hansen, S. D. Kawaler, and V. Trimble

Asymptotic Giant Branch Stars
By H. J. Habing and H. Olofsson